Synthesis
and
Characterization
of
OLIGOMERS

Constantin V. Uglea, Ph.D.
Head, Genetic Laboratory
Center of Biological Research
Iasi, Romania

Ioan I. Negulescu, Dr.Eng.
Head, Research Laboratories on Polymers
Institute of Macromoloecular Chemistry
Iasi, Romania

CRC Press
Boca Raton Ann Arbor Boston London

Library of Congress Cataloging-in-Publication Data

Uglea, Constantin V.
    Synthesis and characterization of oligomers / authors, Constantin
V. Uglea, Ioan I. Negulescu.
      p. cm.
    Includes bibliographical references and index.
    ISBN 0-8493-4954-0
    1. Oligomers. 2. Polymerization.   I. Negulescu, Ioan I., 1942-
    II. Title.
QD382.043U34   1991
547.7—dc20
                        90-23320
                        CIP

   Direct all inquiries to CRC Press, Inc., 2000 Corporate Blvd., N.W., Boca Raton, Florida 33431.

© 1991 by CRC Press, Inc.

International Standard Book Number 0-8493-4954-0

Library of Congress Card Number 90-23320
Printed in the United States

To our doctoral and
postdoctoral professors

FOREWORD

The objective of this book is to provide a reference work on the synthesis and characterization of oligomers. Since the chemistry of oligomers encompasses all of polymer synthesis, with the added criteria of precise molecular weight control and end group functionalization, the authors faced a formidable challenge. Further, each chapter assumes very little background knowledge in the field so the book could serve as an introduction to the field for an industrial chemist or a student with no prior training in polymer chemistry and is as well an excellent reference work for the practicing polymer chemist. Coming from Eastern Europe and being conversant in Russian, the authors were in a unique position to summarize the Russian and Eastern European literature as well as the Western literature, including patents. Thus, material not readily accessible to Western readers comprises a major part of the book. The reader will be impressed with the truly global perspective of the review. The net result is a very thorough coverage of oligomer syntheses.

The book is organized along traditional lines, i.e., condensation and radical and ionic oligomerization processes. The introductory discussion of the historical development along with a review of the proper oligomer nomenclature is quite enlightening. The chapter on radical processes contains a good summary of the fundamentals of free radical reactions including kinetics; the chemistry of initiation and chain transfer are well treated, because these reactions are critical in controlling telomerization. Perusing the chapter on oligocondensation, one is reminded of the wealth of organic chemistry associated with condensation processes. The section on oligomers with unusual architectures, which includes crown ethers, cryptands, rotaxanes, cascade molecules, and calixarenes, is particularly fascinating.

Ionic oligomerizations are the most important approaches to the preparation of functionalized molecules with narrow molecular weight distributions. Discussion of all aspects of this chemistry comprises the largest chapter in the book; the survey is based upon over 600 references. Although this subject can be quite complex, the practical approach taken by the authors renders the material comprehensible and relatively expansive. After absorbing the material in this chpater, the reader would be ready to make a contribution to the oligomerization field.

The book concludes with a section on characterization. Although not as comprehensive as the discussions on synthesis, the fundamental procedures are reviewed.

Perhaps one's first reaction to a book on oligomers is to query the need for a tome on the subject. After reading the book one gains an appreciation for the complexity of the field and the beauty of the chemistry required. Controlled preparation of an oligomer is indeed a challenge. However, modern RIM processing techniques and environmental considerations, which are emphasizing all types of solvent-free processes, have placed oligomers in the forefront of polymer research. I applaud the foresight of Constantin V. Uglea and Ioan I. Negulescu in electing to write such a comprehensive volume on a very timely subject.

William H. Daly
Professor of Chemistry
Macromolecular Studies Group
Louisiana State University

THE AUTHORS

Constantin V. Uglea, Ph.D., is Head of the Bioregulator Synthesis Department of the Institute of Biological Research of "Alexandru Ioan Cuza" University in Iasi, Romania.

Dr. Uglea received the M.S. degree in chemistry in 1958 and a Ph.D. in polymer chemistry in 1971 from the University of Iasi. From 1958 until 1972, he was associated with the "Petru Poni" Institute of Macromolecular Chemistry in Iasi as a research scientist and later as head of the polymer chemistry division. In 1972 Dr. Uglea joined the Chemical Research Center of Rîmnicu Vîlcea as Head of the Laboratory for Fine Organic Chemical Synthesis. During this period he served also as an Adjunct Professor of Chemistry at the Polytechnic Institute of Bucharest. In 1979 he assumed his present position in Iassy.

Dr. Uglea is a member of the Association of Romanian Scientists. In 1976 he received the "Gheorghe Spacu" Award for Organic Technology from the Romanian Academy of Sciences.

Dr. Uglea has published over 80 technical papers and holds 25 patents. He is the author of two books, entitled *Gel Permeation Chromatography—New Domains of Application* and *Polymer Characterization,* and is co-author of a book entitled *Fractionation of Macromolecular Components.*

His current major research interests are in the areas of bioactive oligomers and polymers and the characterization of polymers.

Ioan I. Negulescu, Dr.Eng., Head of the Laboratory of Organic Polymers of the "Petru Poni" Institute of Macromolecular Chemistry, Iasi, Romania, and Research Professor of Polymer Science and Engineering at Louisiana State University in Baton Rouge, received an M.S. degree in polymer technology in 1965 and a Dr.Eng. degree in polymer science and engineering in 1973, both from the Polytechnic Institute of Iasi.

In 1973 to 1974 he was associated with the Department of Polymer Science and Engineering of the University of Massachusetts in Amherst as a Fulbright postdoctoral fellow. After a short association with PETROCHIM Institute in Ploiesti, in 1968 Dr. Negulescu came to the Institute of Macromolecular Chemistry in Iasi where he assumed his present position in 1975. Since 1990 he has also been associated with the Macromolecular Studies Group at Louisiana State University.

His research efforts focus on electroactive polymers and composites, polymers of medical interest, and, more recently, thermotropic polypeptides. Dr. Negulescu has 75 publications and 12 patents in the polymer field to his credit. He has also co-authored a book on addition polymerizations. Dr. Negulescu serves on the advisory boards of the Scientific Memoirs of the Romanian Academy, *Bioactive and Biocompatible Polymers,* and the *Journal of Macromolecular Science-Chemistry.*

He is a member of the Association of Romanian Scientists, the Romanian Society of Medical Engineering, the American Chemical Society (Polymer and Polymeric Materials Science and Engineering Divisions), and the New York Academy of Sciences.

TABLE OF CONTENTS

GENERAL CONSIDERATIONS

Learning originally from nature and following up on the established principles, scientists and engineers succeeded in producing a wide spectrum of polymeric materials which outdo their original native examples in many ways and, in most cases, are more accessible and less expensive. All this gave a tremendous lift to the important industries of man-made fibers, films, plastics, rubbers, coatings, and adhesives and has made everybody's life richer, safer, and more comfortable. As a consequence, synthetic organic polymers have become a significant factor in the economy of all industrialized countries in the world.[1]

I. HISTORICAL BACKGROUND

Semantically, a *polymer* is composed of many (*poly-*, from the Greek *polys*) parts (*mer*, from the Greek *meros*). The term polymer was introduced by Berzelius in 1833,[2] and today it designates a chemical compound containing an indefinite number (more than one) of a specified substance which is the monomer.[3] When the compound contains few monomer units, it is an *oligomer*. As pointed out by Hendry,[4] the first use of the prefix *oligo* (from the Greek *oligos,* meaning *few*) with reference to low molecular mass polymers apparently dates back to 1930 when Burckardt et al.[5] proposed that the term *oligosaccharides* be used to designate those carbohydrates having molecular weights between those of monosaccharides and those of polysaccharides. In tracking down the first reference to *oligopeptides* in *Chemical Abstracts,* Hendry was surprised to learn that Burckardt also was the originator of it.[6] It seems, however, that the term oligomer was coined in the 1940s by D'Alelio at the suggestion of Larsen.[7] D'Alelio used this term in print in his book on polymer chemistry published in 1952,[8] giving credit to Staudinger for the term already in existence, oligosaccharide. However, Heidemann[9] considers that the term oligomer was introduced by van de Want and Stavermann in 1952,[10] also by analogy to the nomenclature generally used in the chemistry of natural products and adopted by Kern in the same year.[11]

At present, *oligomer* is understood to mean any of the terms of a homologous series having a molecular mass higher than that of the dimer and lower than that of the corresponding statistical segment ($\overline{M}_n \sim 10^4$). Although there is not yet a unified point of view regarding the upper limit of the molecular mass, which is set by different authors at 20 monomer units,[12] or 1000,[13] 6000,[14] or 10,000 Da.[15,16] The last limit is a convenient number for designating a compound as a polymer, although it should be clear that this cutoff is arbitrary. This domain of molecular masses was later subdivided into true oligomers and *pleinomers* (from the Greek *pleios,* meaning *large*).[17] The former are the low members of an oligomer series that may be easily obtained or separated in the form of individual compounds which show sufficiently clear differences in their physical properties to permit distinguishing among them. Pleinomers are members with higher molecular masses, also separable as molecularly homogeneous entities, whose scalar properties differ only slightly from the properties of polymers. A natural barrier separates true oligomers from pleinomers, i.e., a degree of polymerization (DP) of 10 to 13 marks the modification of the dependence between intrinsic viscosity and the number of statistical segments from a given species.[18] Molecular masses of about 10^4 are high enough to determine the appearance of vectorial properties which can be considered as characteristics of high polymers, such as the ability to form fiber or films. Therefore, the oligomers represent not only quantitative intermediary steps between simple chemical derivatives and macromolecular compounds, but also distinct forms specific to these domains of molecular mass.

At the other end of the scale of polymers, constructed according to their molecular mass (dimensions), are pleistomers (Figure 1). The term *pleistomer* (from the Greek *pleistos,* meaning *very many*) was introduced recently by Simionescu et al.[19] and refers to very high

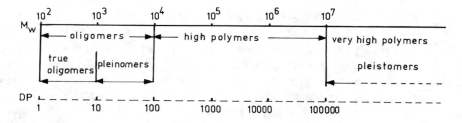

FIGURE 1.   Nomenclature of polymers according to their molecular mass.

molecular mass polymers ($M_w \geq 10^7$), the size of which is much larger than the wavelength of the light used in the light-scattering technique.

II. APPLICATIONS

Coating technology is a prime example of applied polymer science. Dating from pre-history and Cro-Magnon man, it developed slowly from artistry to technology. Some organic pigments were developed before 6000 B.C., and vehicles came from a variety of sources: gum arabic, egg white, gelatin, and beeswax. The introduction of polymers as a main component in coatings led to the development of lacquer, which was used in China at least as early as the Chou dynasty (1122 to 221 B.C.) for carriages and weapons.[20] The Egyptians employed pitches and balsams to seal ships, a practice followed by the rest of the world. Much later, in the Middle Ages, the naturally drying vegetable oils became the backbone of what has come to be known as the oleoresinous type of coating. As early as the eleventh century, the monk Theophylus discussed in his *Diversarium artium schedula* the use of oil in paint. And events since the Industrial Revolution have been responsible for the emergence of coatings as a high-technology industry. Synthetic resins are now the heart of the coating industry (the practical side of polymer coatings requires that they be of high molecular mass and cross-linked, at least for many applications). There are two basic ways to produce a rigid film on a substrate: (1) continual growth of an oligomer until it crosses the threshold between liquid and solid and (2) aggregation of finely dispersed polymer that has already terminated its growth as a molecule. One could start with an oligomer, as in the case of natural drying oils, and expect it to polymerize after the coating has been established. Therefore, the history of protective coatings is, indeed, a history linked to natural oligomers.

Used as adhesives, the oligomers are the diplomats of the polymer world.[21] They exist for the purpose of bringing other materials together. In nature, organic thermoplastic adhesives applied as hot melts are represented by rosin and other resins, waxes, and bitumens. The grandfather of synthetic resin adhesives is the phenol formaldehyde condensate, obtained either inadvertently or intentionally by von Baeyer in 1872 and directly established as a molding material by Baekeland in 1907.[22] Five years later, it occurred to Baekeland that novolacs (methylol-terminated soluble oligomers) might be useful for bonding wood veneers. Today, the bonding of plywood by phenol formaldehyde remains the largest single end use of this adhesive material. In 1930, Drew patented a development of major significance: pressure-sensitive tapes whose masscoats are blends of rubber with low molecular mass resins such as the rosin esters. In a half century, pressure-sensitive tapes and labels have become ubiquitous for holding, bonding, masking, sealing, protecting, reinforcing, splicing, stenciling, identifying, packaging, and insulating.

In 1933, the first liquid elastomer, DPR, obtained by depolymerization of natural rubber, was marketed, although only for a short period, due to its poor characteristics. Twenty years later, the liquid polysulfides, made by Thiocol Corporation, took a firm market, developing into a rather cheap and accessible material in the following years.

In the 1940s and 1950s, three classes of thermosetting resins were welcomed because they could be cross-linked without formulation of volatile byproducts: unsaturated polyesters, epoxides, and polyurethanes. The general-purpose polyesters are usually liquid oligomers of phthalic anhydride, maleic anhydride, and a molar excess of propylene glycol, cross-linked by copolymerization with styrene. Unsaturated polyesters are the main binders for the reinforced plastic industries, of importance in automobile bodies, boat hulls, building panels, chemical piping, etc. Epoxy resins were first patented in 1930 by Castan,[23] but the development of epoxy coatings from oligomers containing hydroxyl as well as epoxide groups was the accomplishment of Greenlee and associates in the late 1940s and early 1950s.[24] The main applications of these oligomers are in construction (including binders and grouts), and electrical/electronic, furniture, aircraft, aerospace, automotive, and other product assembly.

In the 1960s, the liquid oligomers developed into a strong industry, so that by 1972 at least eight structural families and more than 20 marks could be inventoried.[24-27]

In the last 2 decades, interest in the chemistry of oligomers increased to the extent that today they are a distinct part of polymer science. This is mainly due to the fact that most oligomers are reactive (i.e., it is possible to realize the transition from liquid or fusible oligomers to high polymers with linear, branched, or cross-linked structures) and thus open new possibilities for the technology of polymers and composite materials. Moreover, oligomers having dimensions close to the value of statistical segments allow the production of polymers with properties characteristic of high molecular compounds having the same structure as the initial oligomer. This feature of oligomers solves the problem of obtaining materials with a given complex of properties. Thus, the applications of oligomers opened a new stage in the technology of polymers based on the direct transition of liquid compounds (i.e., oligomers) into articles (finite products) without synthesis, separation, and processing of high molecular compounds. In other words, in this case, polymer formation and article production are combined in the same process, e.g., *reaction injection molding, RIM,* for the production of urethane foams and elastomers. (RIM is also referred to in many cases as high-pressure impingement mixing, HPIM, or liquid reaction molding, LRM. The RIM process can be defined as the precise metering of two highly reactive liquid oligomers, generally using direct impingement with intimate mixing of the two components at relatively high pressure to a mixing head, followed by injection into a closed mold and completion of the reaction within the mold). Many types of oligomers are capable not only of combination with each other, but also with high polymers. In the latter case, they play the part of "reactive" plasticizers which may increase the ability of high molecular substances to flow into new composition materials in the processing stage. Moreover, the high-ordered regions existing in oligomer liquids may be included as a whole in high polymers, thus determining their supermolecular structures and properties to a considerable extent. By changing the chemical nature and the dimensions of the oligomer blocks, as well as the order of location and the types of reactive groups, one can purposely influence the structure and properties of oligomer liquids, the kinetics of transition from oligomer into a high polymer, and the complex of its properties. The transition from oligomers to cross-linked polymers has particular importance because it requires simple technological methods for the production of large-size articles with high thermostability, strength, chemical resistance, and other properties required in various branches of engineering and particularly in aircraft and rocket construction, the electromechanical engineering industry, building, radio location, and automobile and ship construction, etc.

III. ORIGINS AND NOMENCLATURE

Oligomers may be purposely synthesized or they may be separated from the reaction medium where they accompany the polymer as "natural" components. By analogy to *polymerization* (semantically, as a process of encatenation, regardless of the mechanism of reaction), the formation of oligomers is called *oligomerization*.

The equilibrium between the addition to a growing oligomer molecule (oligomerization)

$$\sim M_i^* + M \rightleftharpoons \sim M_{i+1}^* \tag{1}$$

is, like every equilibrium, independent of the reaction path and therefore of the mechanism (and the chemical nature of the growing end, denoted by the asterisk, which can be an anion, a cation, or a radical).

The molar Gibbs energy of encatenation for the transition m → o (monomer — oligomer) can be calculated from the equilibrium constant K_e as:

$$\Delta G_{mo}^m = -RT \ln K_e \tag{2}$$

where

$$K_e = [\sim M_{i+1}^*]_e / [M]_e [\sim M_i^*]_e \tag{3}$$

From the second law of thermodynamics, one can write

$$\Delta G_{mo}^m = \Delta H_{mo}^m - T \Delta S_{mo}^m \tag{4}$$

and if both ΔH_{mo}^m and ΔS_{mo}^m are negative, the term $- T \Delta S_{mo}^m$ becomes more positive as the temperature increases. The molar Gibbs energy will equal zero at a specific thermodynamic ceiling temperature, $T_c = \Delta H_{mo}^m / \Delta S_{mo}^m$, above which no encatenation (polymerization) is possible. According to Equation 2, at T_c, the equilibrium constant K_e is unitary (ln K_e = 0). When $T > T_c$, in order to get $\Delta G_{mo}^m = 0$ from Equation 2, K_e should be smaller than unity (ln $K_e < 0$). For $1 > K_e > 0$, however, the chain concentration $[\sim M_{i+1}^*]_e$ is, according to Equation 3, not equal to zero. "No encatenation" means, therefore, no polymerization to high molecular mass compounds — production of dimers, trimers, etc. (true oligomers) can occur.[28] To produce oligomers at $T < T_c$, the rate of propagation must always be smaller than the sum of the rates of all the reactions that terminate the individual chains. The degree of polymerization is lowered drastically if the transfer occurs to a low molecular mass species. In general, only oligomers are obtained with very efficient transfer reactions.

Polymerization by both step-growth reactions and chain-growth reactions may generate an equilibrium of cyclic oligomers and acyclic polymer (ring-chain equilibrium).[29] The starting materials are bifunctional compounds for which, under the chosen dilution conditions (cyclic molecules generally are prepared by the high dilution principle technique — *vide infra*), the probability of the intramolecular reaction is significantly larger than that of the intermolecular reaction.

In polyreactions, the distribution of cyclic oligomers in the thermodynamic equilibrium is controlled by the probability that two "ends" approach each other to make a reaction possible, and this for each step of the reaction. However, not only may chain ends react with each other (end-biting reaction), but the reaction may also occur between an active end and a reactive group within the same chain (back-biting reaction). The distribution of cycles under kinetic control generally differs from that under thermodynamic control, i.e., the formation of cyclics may be enhanced or depressed in the early stages of the polymerization (polycondensation).

Typical examples for step reactions are the polycondensation reaction of diacids and diols (or hydroxyacids) and of diacid dichlorides and diamines. As pointed out above, not only the ends of the molecule may react with each other. In a similar way, the end group of a molecule may react with an internal ester (or amide) group or two internal groups may react with each other following a transesterification (or transamidation) reaction. In all cases, intramolecular reactions result in cyclic molecules.

On the other hand, ring-opening polymerization reactions can establish ring-chain equilibria as well. Two prominent examples are the cationic ring-opening polymerization of heterocycles and the metathesis reaction of cycloolefins. Here, the active end (i.e., cationic species or transition metal carbene) may react with each "functional group" (i.e., heteroatom in the case of cationic polymerization or C=C double bond in the case of the metathesis reaction) to form cyclic molecules. In the polymerization of ring-shaped molecules, a whole series of cyclic oligomers is formed as well as the actual polymers. Thus, the 14-membered cyclic dimer ring, the 21-membered cyclic trimer ring, etc. are produced from the 7-membered cyclic monomer, ε-caprolactam ring. The 10-membered cyclic dimer ring, the 15-membered cyclic trimer ring, etc. are produced from the 5-membered 1,3-dioxolane ring. In the polymerization of octamethyl cyclotetrasiloxane, not only the simple multiples (16-, 24-membered rings etc.,), but also the 6-, 10-, 12-, and 14-membered rings are formed by equilibria exchange reactions, because the smallest basic element is the siloxane group, $-O-Si(CH_3)_2-$, which has two members.[28]

The mass spectra of the oligomers of cyclooctene and cyclododecene show high-intensity molecular ion peaks, the most prominent fragments having the general formula $C_nH_{2(n-p)-1}$ with $0 \leq p \leq x$, where x is the degree of polymerization of the respective oligomer (or the number of double bonds in it) and p represents the number of double bonds in the fragment.[29,30]

It might be asked,[29] "How many skeletal atoms (on average) must there be in the ring molecule before we have a cyclic polymer"? There are indications that cyclic poly(dimethyl siloxane) with about 100 skeletal atoms shows the properties expected of a polymer, whereas ring fractions containing substantially fewer skeletal atoms do not. The term *macrocyclic* (from the Greek *makros,* meaning *long*) is being used in the literature to describe rings with relatively few skeletal atoms, such as 15 or 20. These "macrocyclics" do not show macromolecular behavior and the term is a misnomer. Semlyen proposed for these medium rings the term *mesocyclic* (from the Greek *mesos,* meaning *middle* or intermediate).[31] The term *macrocyclic* could then be reserved for cyclic oligomers (pleinomers) and ring macromolecules.

The distribution of cyclics under kinetic control — chain formation is generally favored in kinetically controlled reactions — differs from that under thermodynamic control, i.e., the formation of cyclics may be enhanced or depressed in the early stages of polymerization (polycondensation). In the melt, for example, at the beginning of encatenation (low yield), the probability that a chain end will meet another molecule is much greater than the probability that it will meet its other end or internal functional group.

The ratio of ring to chain formation is given by the rates of formation of rings, v_r, and chains, v_c.[28] Chain formation is bimolecular, since, for example, for an irreversible polycondensation of ω-amino acids $H_2N-(CH_2)_x-COOH$

$$v_c = k_c[\sim M^*][M] \tag{5}$$

where $[\sim M^*]$ and $[M]$ represent the concentration of the growing chain and amino acid, respectively. Equation 5 is not restricted to polycondensation. For example, for the free-radical reaction of a 1,6 diene with polymer radical M· and monomer M, Equation 6 is valid for chain formation, although the proportionality constant is now $2k_c$, not k_c, because the reaction relates to two double bonds.

$$v_c = 2k_c[\sim M\cdot][M] \tag{6}$$

On the other hand, the ring formation is a monomolecular process, since it is an intramolecular reaction, and, consequently, only the concentration of active ends is important:

$$v_r = k_r[\sim M^*] \tag{7}$$

The number fraction of rings f_r is given by the ratio of the rate of ring formation to the total rate:

$$f_r = \frac{v_r}{v_r + v_c} \tag{8}$$

If Equations 6 and 7 are inserted into Equation 8, one obtains on rearrangement:

$$\frac{1}{f_r} = 1 + \frac{k_c}{k_r}[M] = 1 + \frac{1}{v_c}[M] \tag{9}$$

The higher the monomer concentration, the smaller will be f_r. This is the quantitative definition of the Ruggli-Ziegler principle.[28]

The systematization of oligomers must take into account that their chemical nature is determined by the structure of the main chain and the type of functional groups, the character of the distribution of functional groups in macromolecules, the ability of oligomers for transformation into high polymers, and the type of polyreaction responsible for this transformation.[32] *Polyreactive* oligomers and *inert* or unpolyreactive oligomers can be distinguished, therefore, by taking into consideration the presence or absence of reactive centers in their molecules.[33] In accordance with the distribution of reactive centers in macromolecules, the polyreactive oligomers may be further divided into oligomers with terminal, regular-alternating, and statistically distributed functional groups. The fragment of the chain between reactive centers has been named the oligomer block by Alfred Berlin.[34]

Linear macromolecules carrying at their chain end (ends) some polymerizable function are generally referred to as macromolecular monomers, *macromonomers,* or *Macromers®*.[35] The last two terms, originating from the patents of Ralph Milkovich,[36] were not used until 1978.[37] Macromer® is a trademark of CPC, Inc., for a family of macromolecular monomers.[38] In most cases, this function is an unsaturation or a heterocycle, e.g., $H-M_n-CH=CH_2$ or

$$H-M_n-CH \overset{\displaystyle \diagdown \diagup}{\underset{\displaystyle O}{\rule{0pt}{0pt}\hspace{2em}}} CH_2$$

that can undergo polymerization. Polymer chains bearing at one end two functions, $H-M_n-CHX_2$, which are able to participate in a stepwise growth process (polycondensation reaction) can also be referred to as macromonomers.[35] A variety of functional groups can be selected to tailor the relative reactivity of the macromer desired with a particular comonomer. Macromonomers described in the literature generally exhibit rather low molecular masses. This ensures adequate characterization of the species and provides for a sufficient reaction probability of the terminal groups in an encatenation process. However, a low molecular mass is by no means a necessary limitation.

The molecular mass of the macromonomer retards homopolymerization, but does not alter the reactivity of its terminal functional groups with conventional comonomers. Homopolymerization and copolymerization of macromonomers produce comb- and graft-like structures, respectively, where the pendant chain is the macromonomer.

Oligomers and polymers bearing two functions (either identical or antagonistic) at their ends are called telechelics. The term *telechelic* (from the Greek *tele,* meaning *far off* or distant, and *chelos,* meaning *claw*) has been proposed by Uranek et al.[39] to describe polymers carrying reactive groups at both chain ends. Telechelic species can participate in further events (such as step polymerization) to yield much larger yet linear macromolecules. Chain extension does not involve branching, and that is why such species should be referred to as telechelic oligomers or α-ω-functionalized oligomers (thus keeping in mind that the whole chain can be incorporated into a linear polymer), but not as macromers.[35]

Kennedy broadened the scope of the term telechelic to include not only terminally functional species, $X-M_n-Y$, as implied by Uranek et al.,[39] but also branched oligomers and pleinomers, $X-M_n-X$, and recommended distinguishing between *homotelechelic* (or

$$\overset{\displaystyle |}{X}$$

simply telechelic) species, the terminal functions of which are identical, and *heterotelechelic* or *asymmetric telechelic* species, $X-M_n-Y$, which carry dissimilar terminal functions. This terminology only concerns end groups and does not indicate the nature of the polymer that carries the functional groups. The last may be any kind of reactive function, such as $-OH$, $-NH_2$, $-NCO$, $-SH$, $-CHO$, $-COOH$, $-SiH$, etc., suitable for further derivatization or end linking, either directly or by means of linking agents. Kennedy also introduced the symbol $\overline{F}_n$ to express the average number of end groups of telechelics.[40] A linear telechelic may have $\overline{F}_n = 1.0$ (if only one terminus is functional) or $\overline{F}_n = 2$, and if it is branched, $\overline{F}_n = 2, 3, 4, \ldots$. This symbol should not be confused with the functionality-type distribution (FDT) introduced by Evreinov et al.[41] and Entelis et al.,[42] which makes it possible to take into account the fraction of molecules of polyreactive oligomers with the number of functional groups differing from the expected number. The introduction of this parameter is necessary because the deviation of the number of functional groups from that expected has a great influence on the process in which the oligomers are used.

More specifically, telechelic refers to those macromolecular species in which the reactive end groups have been deliberately introduced. By convention, intermediates that arise during step-growth polymerizations, and therefore have end groups of the same type as their monomers, fall outside the scope of telechelic systems. Typical telechelics are, for example, liquid diols used for the synthesis of polyurethanes, which are prepared by derivatization of oligodienes.

When the polymer formed, say $X-M_n-Y$ (n is a small number), is of very low molecular mass and is derived by utilizing a chain-transfer agent $X-Y$, the reaction is called *telomerization* (from the Greek *telos,* meaning *end*) and the polymer is a *telomer*.[43] The best known example is the formation of a compound of the type $Cl_3C(CH_2CH_2)_nCl$ (n = 12) from CCl_4 and ethylene;[44] CCl_4 is the *telogen*. Telomerization is a special case of free radical transfer reactions which has grown in recent years, especially for halogen-containing monomers such as vinyl chloride, vinylidene chloride, and fluorine-related monomers. The process can be initiated by the usual free radical initiators or by redox systems. Telogens include CCl_4, CCl_3Br, or CCl_3R (R = alchyl).[45] The molecular mass may be controlled by proper choice of the molar ratio of monomer to telogen and of the reaction temperature. Telomerization is one way to produce macromonomers by a two-step process, i.e., telomerization with functional telogens and subsequent reaction of the terminal functions with an unsaturated reagent.[45]

As with organic nomenclature, the nomenclature of oligomers describes chemical structure rather than substances. It is realized that, as in the case of polymers,[46] oligomeric substances ordinarily include many structures, and that a complete description of even a single oligomer molecule would include an itemization of terminal groups, branching, random impurities, degree of steric regularity, chain imperfection, etc. Therefore, the nomenclature of oligomers is generally based upon their origin, structure, functional groups which they might carry, or their destination. Based on the repeating functional groups of the chain, they can be oligoamides, oligoacrylics, or, more specifically, oligostyrenes. In the last case, the monomers which have been or might have been used in synthesis are indicated. A more accurate method would indicate whether the oligomer is cyclic or linear and, if linear, the groups at the ends of the chain molecules. For example, *cyclic poly(dimethyl siloxane)* is a precise name for the ring oligomers, but the name *poly(dimethyl siloxane),* meaning linear species, makes no reference to HO–, $(CH_3)_3Si$–, or other final groups which it usually carries. Well-defined species (i.e., true oligomers such as trimers, tetramers, pentamers, etc.) are named, wherever possible, according to the IUPAC nomenclature of organic chemistry.[47] For example, 3-methyl-l-phenylindane as well as 1,4-diphenyl-l-butene and 1,2-diphenylcyclobutane represent dimers of styrene, while α,ω-bis(acryloyloxyethyleneoxy)adipoyl and α,ω-bis(methacryloyloxyethylenephenyliminocarbonyloxy)ethylene define the following oligoethers, respectively:[48]

$$CH_2{=}CH{-}COOCH_2CH_2OCOCH_2CH_2CH_2COOCH_2CH_2OOC{-}CH{=}CH_2$$

and

$$CH_2{=}C{-}COOCH_2CH_2{-}N{-}COOCH_2CH_2OOC{-}N{-}CH_2CH_2OOC{-}C{=}CH_2$$
$$\overset{|}{CH_3}\qquad\overset{|}{C_6H_5}\qquad\overset{|}{C_6H_5}\qquad\overset{|}{CH_3}$$

The nomenclature system of oligomers which do not represent homogeneous species rests upon the selection of a preferred constitutional repeating unit of which the oligomer is a multiple; the name of the oligomer is simply the name of this repeating unit, prefixed by *oligo*. The unit itself is named according to the nomenclature of organic chemistry.[47] End groups may be specified by prefixes placed ahead of the name of the oligomer. The end group designated by α is that to the left side of the constitutional repeating unit; the other end group is designated by ω, such as in the monofunctional methacrylic oligoisobutylene macromonomer α-phenyl-ω-methacryloyloxyoligo(1,1-dimethylethylene) and the bifunctional reactive oligoester α,ω-bis-(methacryloyloxyethyleneoxy)oligo(carbonyloxyethyleneoxy), respectively:

$$C_6H_5{-}({-}\overset{\overset{\textstyle CH_3}{|}}{\underset{\underset{\textstyle CH_3}{|}}{C}}{-}CH_2{-})_n{-}OCOC{=}CH_2 \quad \text{rather than} \quad C_6H_5{-}({-}CH_2{-}\overset{\overset{\textstyle CH_3}{|}}{\underset{\underset{\textstyle CH_3}{|}}{C}}{-})_n{-}OCOC{=}CH_2$$
$$\qquad\qquad\qquad\qquad\overset{|}{CH_3}$$

and

$$CH_2{=}C{-}COOCH_2CH_2O{-}({-}COOCH_2CH_2O{-})_n{-}COOCH_2CH_2OCO{-}C{=}CH_2$$
$$\overset{|}{CH_3}\qquad\qquad\qquad\qquad\qquad\qquad\qquad\qquad\qquad\overset{|}{CH_3}$$

A number of common polymers have semisystematic or trivial names well established by usage, which are recognized by the Commission on Macromolecular Nomenclature of IUPAC.[46] Therefore, systematic and source names may be used for their oligomers, too. For example, the telomer $Cl_3C{-}(CH{-}CH_2{-})_n{-}Cl$, α-(trichloromethyl)-ω-chlorooligo(1-phen-
$$\overset{|}{C_6H_5}$$

ylethylene) may also be called α-(trichloromethyl)-ω-chlorooligo(styrene), and α-methacryloyl-ω-methacryloyl-bis(oxytehylene)oxyoligo[carbonylphenylenecarbonyl-bis(oxyethylene)-oxy], with the formula,

$$CH_2=C–COOCH_2CH_2OCH_2CH_2O(COC_6H_4COOCH_2CH_2OCH_2CH_2O)_nCOC=CH_2$$

with CH_3 groups on each end,

may alternatively be named α,ω-bis(methacryloyl)oligo(diethyleneglycolphthalate).[48]

The main trends for synthesis of oligomers parallel the methods of obtaining high polymers. Therefore, according to the type of oligomerization, this process may be subdivided into oligoaddition and oligocondensation. According to the mechanism of reactions, each of these processes may involve a homolytic or a heterolytic scission of bonds, i.e., may be radicalic or ionic (coordinative). Depolymerization of high polymers and chemical modification of oligomeric species are additional routes for the synthesis of oligomers. All these processes are the topics of the following sections.

REFERENCES

1. **Mark, H.,** Coming to an age of polymers in science and technology, *J. Macromol. Sci. Chem.,* A15, 1065, 1981.
2. **Berzelius, J.,** *Jahresberichte,* 12, 63, 1833; as cited in **Simionescu, C. and Vasiliu-Oprea, C.,** *Treatise of Chemistry of Macromolecular Compounds,* Vol. 1, EDP, Bucharest, 1973, 74 (in Romanian).
3. *Webster's Seventh New Collegiate Dictionary,* G & C Merriam, Springfield, MA, 1972, 657.
4. **Hendry, R. A.,** More on the term "oligomer", *Chem. Eng. News,* 12, 85, 1984.
5. **Burckard, H., Bohn, E., and Winkler, S.,** Ungesättige Derivate von Gentiobiose und Cellobiose, *Berichte,* 63B, 989, 1930.
6. **Burckard, H. and Grünert, H.,** N-methan-sulfonyl Derivate von amino-Säuren und Oligopeptiden, *Ann.,* 545, 178, 1940.
7. **Larsen, L. V.,** Origin of the term "oligomer", *Chem. Eng. News,* 23, 58, 1984.
8. **D'Alelio, G. F.,** *Fundamental Principles of Polymer Chemistry,* John Wiley & Sons, New York, 1952.
9. **Heidemann, G.,** Oligomers, in *Encyclopedia of Polymer Science and Technology,* Vol. 9, Kirk and Othmer, Eds., John Wiley & Sons, New York, 1968.
10. **Van der Vant, G. M. and Staverman, A. J.,** Synthesis of N(ε-aminocaproyl-)ε-aminocaproic acid, *Rec. Trav. Chim. Pays-Bas,* 71, 379, 1952.
11. **Kern, W.,** Oligomere und Pleinomere von synthetischen faserbildenden Polymeren, *Chem. Ztg.,* 76, 667, 1952.
12. **Elias, H. G.,** *Macromolecules. Structure and Properties,* Plenum Press, New York, 1977, 6.
13. **Hiemenez, P. C.,** *Polymer Chemistry. The Basic Concepts,* Marcel Dekker, New York, 1984, 8.
14. **Krakovitsch, G. A. and Bezkorovajnyi, K. G.,** *Pulverization of Oligomeric and Polymeric Powdery Materials,* Khimija, Leningrad, 1980, 28 (in Russian).
15. **Shur, A. M.,** *High-Molecular Compounds,* Vysshaya Shkola, Moscow, 1981, 266 (in Russian).
16. **Entelis, S. G., Evreinov, V. V., and Kuzaev, A. I.,** *Fundamental Reactive Oligomers,* Khimija, Moscow, 1985, 9 (in Russian).
17. **Zahn, H. and Gleitsmann, G. B.,** Oligomere und Pleinomere von synthetischen faserbildenden Polymeren, *Angew. Chem.,* 75, 772, 1963.
18. **Vilenchic, L. Z., Zhmakina, T. P., Belenkii, B. G., and Frenkel, S. Ya.,** On the natural physical boundary between oligomers and proper polymers in the hydrodynamic behavior of macromolecules, *Acta Polym.,* 36, 125, 1985.
19. **Simionescu, C. I., Simionescu, B. C., and Ioan, S.,** Plasma induced living radical copolymerization, *J. Macromol. Sci. Chem.,* A22, 765, 1985.
20. **Myers, P. L.,** History of coating science and technology, *J. Macromol. Sci. Chem.,* A15, 1133, 1981.
21. **Skeist, I. and Miron, J.,** History of adhesives, *J. Macromol. Sci. Chem.,* A15, 1151, 1981.
22. **Seymur, R. B.,** History of development and growth of thermosetting polymers, *J. Macromol. Sci. Chem.,* A15, 1165, 1981.

23. **Milewski, J. V.,** History of reinforced plastics, *J. Macromol. Sci. Chem.,* A15, 1303, 1981.
24. **Greenlee, S. O.,** U.S. Patent 2,456,408, 1946.
25. **Patrick, J. C. and Ferguson, H. R.,** U.S. Patent 2,466,963, 1946.
26. **Rochow, E. G.,** U.S. Patent 2,380,995, 1945.
27. **Daury, J. D.,** Synthesis and study of polyisoprene hydroxytelechelic liquid, *Cautchucs Plast.,* No. 571, 85, 1977.
28. **Elias, H. G.,** *Macromolecules. Synthesis and Materials,* Vol. 2, Plenum Press, New York, 1977, chap. 16.
29. **Semlyen, J. A., Ed.,** *Cyclic Polymers,* Elsevier, London, 1986.
30. **Höcker, H. and Riebel, K.,** Mass spectrometrical behavior of hydrocarbons. II. Oligododecenylenes and corresponding alkanes, *Makromol. Chem.,* 179, 1765, 1978.
31. **Semlyen, J. A.,** Linear polymers and cyclic polymers, in *Cyclic Polymers,* Elsevier, London, 1986, 1.
32. **Berlin, A. A. and Matveyeva, N. G.,** The progress in the chemistry of polyreactive oligomers and some trends of its developments. I. Synthesis and physico-chemical properties, *Macromol. Rev.,* 12, 1, 1977.
33. **Berlin, A. A.,** Supermolecular aggregates in liquid oligomers and their effect on kinetics of polymerization and properties of high polymers, *Vysokomol. Soedin. Ser. A,* 12, 2313, 1970.
34. **Berlin, A. A.,** The synthesis of mixed polyesters of the acryl series (polyesteracrylates), *Dokl. Akad. Nauk. S.S.S.R.,* 123, 282, 1958.
35. **Rempp, P. F. and Franta, E.,** Macromonomers. Synthesis, characterization and applications, *Adv. Polym. Sci.,* 58, 1, 1984.
36. **Milkovich, R. and Chiang, M. T.,** U.S. Patent 3,786,116, 1974.
37. **Vogl, O.,** Development in radical polymerization, *J. Polym. Sci. Polym. Symp.,* 64, 1, 1978.
38. **Milkovich, R.,** Synthesis of controlled polymer structures, *Polym. Prepr.,* 21, 40, 1980.
39. **Uraneck, C. A., Hsieh, H. L., and Buck, O. G.,** Telechelic polymers, *J. Polym. Sci.,* 46, 535, 1960.
40. **Kennedy, J. P.,** Synthesis of telechelic polymers by cationic techniques and applications of the products, *J. Macromol. Sci. Chem.,* A21, 929, 1984.
41. **Evreinov, V. V., Gerbich, V. I., Sarynina, L. I., and Entelis, S. G.,** Molecular weight distribution and functionality of oligomeric polyethyleneglycoladipates, *Vysokomol. Soedin. Ser. A,* 12, 829, 1970.
42. **Entelis, S. G. and Kazanskii, K. S.,** *Progress in Chemistry and Physics of Polymers,* Khimija, Moscow, 1970, 324.
43. **Peterson, M. D. and Weber, A. G.,** U.S. Patent 2,395,592, 1946.
44. **Joyce, R. M., Hanford, W., and Harnen, J.,** Free radical initiated reaction of ethylene with carbon tetrachloride, *J. Am. Chem. Soc.,* 70, 2529, 1948.
45. **Boutevin, B., Hugon, J. P., Pietrasanta, Y., and Sideris, A.,** Télomérisation par catalyse redox. XIII Synthése d'alcools fluorés a partir des téloméres des acétates de vinyle et d'allyle avec des télogenes fluorés, *Eur. Polym. J.,* 14, 353, 1978.
46. International Union of Pure and Applied Chemistry, Macromolecular Division, Commission on Macro- molecular Nomenclature, Nomenclature of regular single-strand organic polymers, *Pure Appl. Chem.,* 48, 373, 1976.
47. **Rigandi, J. and Klesney, S. P.,** *International Union of Pure and Applied Chemistry Nomenclature of Organic Chemistry. Sections A, B, C, D, E, F and H.,* Pergamon Press, Oxford, 1979.
48. **Berlin, A. A., Korolev, G. V., Kefely, T. Ya., and Siverin, Yu. M. M.,** *Acrylic Oligomers and Derived Materials,* Khimija, Moscow, 1983.

PART I

General Methods for the Synthesis of Oligomers

INTRODUCTION

Generally, taking into account that oligomers are low polymers, they may be divided classically, paralleling the polymers,[1] into two main groups on the basis of a comparison of the structure of the repeating unit of the oligomer with the structure of the monomer from which the oligomer was derived: *addition oligomers and condensation oligomers.*

The original classification into addition polymers (i.e., polymers in which the molecular formula of the repeating unit of the polymer is a simple summation of the molecular masses of all combined monomer units in the chain) and condensation polymers (i.e., polymers in which the repeating unit contains fewer atoms than that of the monomer or monomers, and, necessarily, the molecular mass of the polymer is less than the sum of the molecular masses of all the original monomer units which combined to form the polymer chain) was suggested by Carothers in 1929, and these definitions are still widely used.[2-4]

However, in a classification system based on the mechanism of growth reaction, the two major types of encatenation are *chain-growth polymerization* and *step-growth polymerization.*

Chain-growth polymerization occurs by introducing an active growth center of a free-radical or ionic nature into a monomer, followed by the addition of monomers in that center by a chain-type kinetic mechanism. Therefore, chain-growth polymerization is a reaction in which each polymer chain, once initiated, grows rapidly and, once terminated, is incapable of further growth. This type of encatenation occurs through a series of reactions (initiation, propagation, and termination) in which the rates and mechanisms are generally different. Other reactions — notably, chain transfer and chain inhibition — also need to be considered to give a more fully developed picture of chain-growth polymerization. Termination reactions always predominate if oligomeric species are to be obtained.

As the name implies, step-growth polymerization occurs one step at a time through a series of simple organic reactions, such as carbonyl addition-elimination reactions used for the preparation of polyesters and polyamides, or addition to multiple bonds, used for the preparation of polyurethanes and poly(alkylene sulfides).[1] In step-growth polymerization reactions, the functional group on the end of a growing polymer chain is normally assumed to have the same reactivity as the functional group in the monomer unit. As a result, monomer units can react with other monomer units or with polymer chains with the same ease and, consequently, the initiation, propagation, and termination reactions are usually considered to be identical in both rate and mechanism. The consequences of this are that both high yield and high molecular mass require extensive reaction to occur.

Another kind of classification of polymerizations takes into account the mechanism of the bonding of the chain with another monomer or chain in the propagation reactions, i.e., a monomer or other chain can be *added onto* or *inserted into* the growing chain. According to this criterion, a distinction can be made among additional polymerization (free radical and some ionic), polyinsertion, and polycondensation. This classification is, however, more useful, when discussing how to obtain high molecular polymers, than for the general methods for the synthesis of oligomers. Therefore, the oligomerization reactions will be classified in two major classes only: chain-growth oligomerization and step-growth oligomerization.

REFERENCES

1. **Lenz, R. W.,** *Organic Chemistry of Synthetic High Polymers,* Interscience, New York, 1967.
2. **Carothers, W. H.,** *J. Am. Chem. Soc.,* 51, 2548, 1929; Untersuchungen über Polymerization und Ringbildung. I. Eine Einführung zur allgemeinen Theorie von Kondensationpolymeren, *Chem. Ztg.,* 2, 1641, 1929.
3. **Shur, A. M.,** *High Molecular Compounds,* Vysshaya Shkola, Moscow, 1981 (in Russian).
4. **Hiemenez, P. C.,** *Polymer Chemistry—The Basic Concepts,* Marcel Dekker, New York, 1984.

Chapter 1.1

RADICAL OLIGOMERIZATION

Most radical oligomerization processes are started by free radicals that have been produced by homolytic dissociation of covalent bonds. Free radicals are unstable species containing an unpaired electron generally believed to reside in a pure *p*-orbital. The short lifetimes of free radicals are only a reflection of the facility with which they react with one another, and their instability is, therefore, a kinetic rather than a thermodynamic characteristic. Free radicals can undergo several types of reactions, including transfer or abstraction, elimination, addition, combination or coupling, and substitution, which account for essentially all of the products obtained in solution or gas-phase free radical reactions in general and for the mechanism and products of free radicals, especially chain-growth polymerization reactions.

Virtually all free radical chain reactions require a separate initiation step in which radical species are generated in the reaction mixture. Even if the chain reactions are initiated by adding directly to the reactants a stable free radical that shows little or no tendency for self-combination, a separate initiation step is still involved. Therefore, according to the monomer in which the first radical species is formed, one can divide radical initiation into (1) homolytic decomposition of covalent bonds by energy absorption and (2) electron transfer from ions or atoms containing unpaired electrons, followed by bond dissociation in the acceptor molecule.

I. RADICAL OLIGOMERIZATION INITIATED BY CHEMICAL AGENTS

Organic compounds may be dissociated into two or more free radical fragments by absorption of energy in almost any form, including thermal, chemical (redox systems), electromagnetic, particulate, electrical, sonic, and mechanical, and all of these forms of energy have been used for the initiation of radical chain-growth polymerization reactions. For free-radical oligomerization, the most important of these is thermal energy.

The *initiation* of free radical processes involves the homolytic dissociation of radical-forming agents (initiators), I_2

$$I_2 \xrightarrow{k_i} 2I^{\bullet} \tag{1}$$

and the start reaction (actual initiation) in which derived free radicals are formed through the addition of primary free radicals $I^{\bullet}$ to the monomer:

$$I^{\bullet} + M \xrightarrow{k_i} IM^{\bullet} \tag{2}$$

In the *propagation* or chain-growth reaction, the monomer is repeatedly attacked by the growing free radicals:

$$IM^{\bullet} + M \xrightarrow{k_p} P_2^{\bullet}$$

$$P_2^{\bullet} + M \xrightarrow{k_p} P_3^{\bullet}$$

$$\cdots\cdots\cdots\cdots\cdots\cdots$$

$$P_{n-1}^{\bullet} + M \xrightarrow{k_p} P_n^{\bullet} \tag{3}$$

Termination means the interruption of the kinetic chain of reactions and might imply an array of possibilities, e.g., recombination of radicals, the results of which are always the formation of inactive (dead) polymeric species:

$$P_m^{\cdot} + P_n^{\cdot} \xrightarrow{\ k_{t(c)}\ } P_{m+n} \tag{4}$$

Some of the most widely used initiation systems are compounds which have bond dissociation energies in the range of 125 to 175 kJ/mol. This narrow range of dissociation energies limits the types of useful compounds to those containing fairly specific labile bonding systems, such as the peroxides and the azo compounds, e.g., benzoyl peroxide, *t*-butyl peroxide, *t*-butyl hydroperoxide, cumen hydroperoxide, *t*-butyl perbenzoate, 2,2′-azobisisobutyronitrile, etc., alone or activated with organic salts or transitional metals such as naphthenates of Co, Mn, Mo, or V.[1,2] Homolytic decomposition of peroxides and hydroperoxides gives rise — according to their structure — to a series of radicals, such as $RCOO^{\cdot}$, $RO^{\cdot}$, $R^{\cdot}$, and $^{\cdot}OH$. In the case of azo compounds, the homolysis is derived by the liberation of the very stable N_2 molecule, despite the relatively high dissociation energy of the $C-N$ bond.

Redox systems are another source of free radicals and have the advantage that they can operate at rather low positive temperatures.

The initiation process is a two-stage sequence: (1) the dissociation of the initiator, I_2, to generate two radical fragments, $I^{\cdot}$, and (2) the addition of one of these fragments to a monomer molecule, M, to start the growth of a polymer chain. However, not every free radical formed by dissociation of the initiator lives long enough to add to a monomer molecule, and the mole fraction of initiator radicals formed that successfully add to the monomer molecule to initiate polymer chains is termed the *initiator efficiency*, f.

$$f = \frac{[IM^{\cdot}] \text{ from Equation 2}}{[I] \text{ from Equation 1}} \tag{5}$$

The initiator efficiency is generally between 0.3 and 0.8,[3] sometimes higher,[4] and depends on the conditions of the oligomerization, including the solvent. The second stage may have a much higher rate constant than the first, but since $f \leq 1$, it is assumed that $k_d = k_i$. Therefore, the general approximation made driving kinetic expressions for chain-type encatenation is that the dissociation reaction (Equation 1) and the initial addition reactions (Equation 2) have equal rates:

$$R_i = 2fk_d[I] = k_i[I^{\cdot}][M] \tag{6}$$

Equation 6 is written for initiation by peroxides or azo compounds. It transforms in Equation 7 if the initiation is made using redox systems:

$$kf[Ox][Red] = k_i[I^{\cdot}][M] \tag{7}$$

where [Ox] and [Red] describe the concentration of oxidizing and reducing agents, respectively, and k is the rate constant for the particular reactants.

The concentration of radicals in the oligomerization system is not determined solely by the initiation stages. Simultaneous destruction of two radicals, regardless of their degree of polymerization, as illustrated by Equation 4, will terminate the kinetic chain of reactions. The propagation reactions (Equation 3) do not change the radical concentration.

Termination by combination (Equation 4) results in an inactive oligomer with a molecular mass higher than that of the average molecular mass of the growing radicals. At high initiator concentration, the growing radicals combine with the initiator free radicals. The oligomeric product molecule contains two initiator fragments per molecule by this mode of termination.

Termination by disproportionation, which implies a transfer of an atom, usually hydrogen, from one growing chain to another, as illustrated by Equation 8, introduces an unsaturation in one molecule:

$$M_{m-1}\text{--}CH_2\text{--}\overset{\cdot}{C}HX + {\cdot}CHX\text{--}CH_2\text{--}M_{n-1} \xrightarrow{\;k_{t(d)}\;}$$

$$M_{m-1}\text{--}CH_2\text{--}CH_2X + HXC\text{=}CH\text{--}M_{n-1} \tag{8}$$

The effect of this transfer on the molecular mass of the reacting radicals is practically nil, and the resulting oligomer molecules contain one initiator fragment.

The relative amount of combination and disproportionation is an important consideration in both polymer and kinetic chain termination. For example, one method for obtaining telechelic oligomers and polymers is to use a functionalized initiator which will supply the function at one or both ends of the polymeric species. Two important structural factors which affect the ratio of these two termination reactions are the number of atoms available for disproportionation (for example, α-methyl vinyl monomers, such as methyl methacrylate or α-methyl styrene, have five hydrogen atoms, versus two for corresponding vinylic monomers, i.e., methyl acrylate and styrene) and the number and bulkiness of the substituents on the radical atom.[4] For example, the poly(methyl methacrylate) radical carries two bulky groups (methyl and ester groups) and the combination of two growing radicals would create a strained bond. Therefore, there is an activation energy difference between the two possible reactions of about 20 kJ/mol in favor of disproportionation.

Polar effects are also important in radical-radical termination reactions, e.g., termination in the polymerization of p-chlorostyrene is by combination, while that of p-methoxystyrene is partly by disproportionation.

Temperature is another factor which favors one termination process or another in radicalic encatenation. Since bond breaking is required only in disproportionation, the activation energy of this process, $E_{t(d)}$, is expected to be greater than $E_{t(c)}$.[3] Indeed, styrene polymerization terminates almost entirely by combination at low temperatures and by disproportionation at high temperature.[3,5]

The kinetic analysis of termination, which is a bimolecular reaction, gives the following equation for the rate of disappearance of radicals:

$$R_t = -\frac{d[M^{\cdot}]}{dt} = 2k_t[M^{\cdot}]^2 \tag{9}$$

where $M^{\cdot}$ represents the total concentration of growing radicals, regardless of their degree of polymerization, and k_t is the general rate constant for termination:

$$k_t = k_{t(c)} + k_{t(d)} \tag{10}$$

When the rate of appearance of radicals in the systems in the initiation stage is balanced by the rate of disappearance of these species in the termination stage, a steady state is reached and therefore $R_i = R_t$, or more explicitly:

$$2fk_d[I] = 2k_t[M^{\cdot}]^2 \tag{11}$$

which, on rearrangement, gives a first view of the dependence of the molecular mass of the polymer upon initiator concentration

$$[M^{\cdot}] = \left(\frac{fk_d}{k_t}\right)^{0.5} [I]^{0.5} \tag{12}$$

since each radical is a growth center.

Considering the multiple reaction (Equation 3) under steady-state conditions, assuming that k_p is a constant independent of the size of the growing chain and the extent of monomer conversion, the rate at which monomer is polymerized, R_p, can be expressed as:

$$R_p = \frac{-d[M]}{dt} = k_p[M][M^{\cdot}] \tag{13}$$

or

$$R_p = k_p[M]\left(\frac{fk_d}{k_t}\right)^{0.5} [I]^{0.5} \tag{14}$$

Defining the kinetic chain length $\bar{\nu}$ as the ratio of the number of propagation reactions (Equation 3) to the rate of initiation, regardless of the mode of termination:

$$\bar{\nu} = \frac{R_p}{R_i^{\cdot}} \tag{15}$$

or using the steady-state condition $R_i = R_t$:

$$\bar{\nu} = \frac{R_p}{R_t} \tag{16}$$

Thus, the kinetic chain length gives the average number of monomer molecules $\bar{n}$ that can be added to a growing free radical before the free radical is destroyed by a termination process. It is obvious that for termination by disproportionation $\bar{\nu} = \bar{n}$, while for termination by combination, $\bar{\nu} = 0.5\,\bar{n}$. Therefore, the average degree of polymerization, $\overline{DP}$, is equal to the kinetic chain length for a free radical polymerization started by the thermal decomposition of an initiator and terminated by mutual deactivation of two growing radicals, and it is twice the kinetic chain length for termination by combination.

Insertion of the expressions for R_p (Equation 13), R_t (Equation 9), and the concentration of growing free radicals (Equation 12) in Equation 16 leads to:

$$\bar{\nu} = \frac{k_pM}{2(fk_tk_d)^{0.5}I^{0.5}} \tag{17}$$

The kinetic chain length and, consequently, the degree of polymerization depends not only on the nature and concentration of the monomer, but also on the nature and concentration of the initiator.

The higher the initiator concentration, the lower will be the number-average molecular mass.

Since the kinetic constants are highly sensitive to temperature, as shown by the Arrhenius equation

$$dlnk = -d\left(\frac{E^*}{R_t}\right) \tag{18}$$

the kinetic chain length can be expressed in terms of temperature (T) and activation energy (E*), taking first the logarithms of Equation 17:

$$ln\bar{v} = lnk_p(k_tk_d)^{0.5} + ln\left(\frac{[M]}{2(f[I])^{0.5}}\right) \tag{19}$$

Differentiating the obtained expression with respect to T (assuming that the temperature dependence of [M] and [I] is negligible compared to that of the rate constants)

$$\frac{\partial\bar{v}}{\partial t} = dlnk_p - 0.5\,dlnk_t - 0.5\,dlnk_d \tag{20}$$

and inserting the Arrhenius dependence (Equation 18)

$$\frac{\partial\bar{v}}{\partial t} = -d\left(\frac{E_p^*}{RT}\right) + 0.5d\left(\frac{E_t^*}{KT}\right) + 0.5d\left(\frac{E_d^*}{RT}\right) \tag{21}$$

or explicating

$$\frac{\partial\bar{v}}{\partial t} = \frac{E_p^* - 0.5(E_t^* + E_d^*)}{RT^2}\,dT \tag{22}$$

The activation energies for the initiator decomposition reactions E_d^* vary between 120 and 180 kJ/mol (40 to 60 kJ/mol for redox initiation); for propagation, E_p^*, they have values between 15 and 35 kJ/mol and for termination reactions, between 0 and 25 kJ/mol.[6] Using, at T = 50°C, the average values of the activation energies implied in Equation 22 for a series of initiators and vinyl monomers, it has been calculated that $\bar{v}$ *decreases* about 6.5% per degree Celsius for thermal initiation and *increases* about 2% per degree for photoinitiation.[3]

When a polymer chain is physically terminated without the destruction of the kinetic chain, an effective *chain transfer* occurs in the polymerization system. Chain transfer can take place with any molecule in the system and arises when a hydrogen or some other atom X is transferred from some molecule RX in the system (e.g., initiator, monomer, solvent, polymer, or any other species) to the polymer radical:[7]

$$M_n^{\cdot} + RX \xrightarrow{k_{tr}} M_nX + R^{\cdot} \tag{23}$$

One of the most important consequences of chain transfer reactions is the decrease of the kinetic chain length, i.e., the average molecular mass, with no noticeable decrease in the overall rate of polymerization if the rate constant of the reinitiation, k_{ri}, is comparable to k_p:

$$R^{\cdot} + M \xrightarrow{k_{ri}} RM^{\cdot} \tag{24}$$

If $k_p \ll k_{tr}$ and $k_{ri} \simeq k_p$ or $k_p \ll k_{tr}$ and $k_p > k_{ri}$, the polymerization products are mainly oligomers.

The transfer reactions follow second-order kinetics, and the general rate is

$$R_{tr} = k_{tr}[M_n^{\cdot}][Rx] \qquad (25)$$

The kinetic chain length in the presence of chain transfer is redefined, taking into account all reactions that compete with propagation:

$$\bar{\nu}_{tr} = \frac{R_p}{R_t + R_{tr}} \qquad (26)$$

Since several different molecules from the reaction mixture are involved in the chain-transfer process, Equation 26 becomes, after explication of the specific reaction rates:

$$\bar{\nu}_{tr} = \frac{k_p[M^{\cdot}][M]}{2k_t[M^{\cdot}]^2 + \sum\limits_{\text{all RX}} k_{tr,RX}[RX][M^{\cdot}]} \qquad (27)$$

where $\sum\limits_{\text{all RX}} k_{tr,\,RX} [RX] [M^{\cdot}]$ represents the summation over all pertinent RX species:

$$\sum\limits_{\text{all RX}} k_{tr,RX}[RX][M^{\cdot}] = k_{tr,I}[I][M^{\cdot}] + k_{tr,M}[M][M^{\cdot}]$$

$$+ k_{tr,S}[S][M^{\cdot}] + k_{tr,P}[P][M^{\cdot}] + k_{tr,Y}[Y][M^{\cdot}] \qquad (28)$$

where $k_{tr,\,I}$, $k_{tr,\,M}$, $k_{tr,\,S}$, $k_{tr,\,P}$, and $k_{tr,\,Y}$ represent the rate constants for chain transfer to initiator, monomer, solvent, polymer, and any other species present in the reaction mixture, respectively, and [I], [M], [S], [P], and [Y] are the corresponding concentrations. For example, the kinetic chain length when an effective transfer to the solvent (S) occurs — all other transfer processes being insignificant — is given by the expression:

$$\bar{\nu}_{tr,S} = \frac{k_p[M]}{2k_t[M^{\cdot}] + k_{tr,S}[S]} \qquad (29)$$

The reciprocal of Equation 27 is generally used for ascertaining the chain-transfer activity of the various components present in a polymerization system

$$\frac{1}{\bar{\nu}_{tr}} = \frac{2k_t[M^{\cdot}]}{k_p[M]} + \frac{\sum\limits_{\text{all RX}} k_{tr,RX}RX}{k_p[M]} \qquad (30)$$

since the first term on the right-hand side represents the kinetic chain in the complete absence of transfer reactions

$$\frac{1}{\bar{\nu}_{tr}} = \frac{1}{\bar{\nu}} + \sum\limits_{\text{all RX}} C_{RX} \frac{[RX]}{[M]} \qquad (31)$$

where C_{RX} is the chain-transfer constant for the monomer-to-molecular species RX:

$$C_{RX} = \frac{k_{tr,RX}}{k_p} \qquad (32)$$

The more readily transferable atoms there are per molecule, the weaker the bond, and the more resonance stabilized the resulting radical, the higher is the transfer constant. Consequently, the transfer constants may vary by several order of magnitude, depending also on the pair transfer agent/monomer (Table 1). In both poly(styrene) and poly(methyl methacrylate) at 60°C, $C_p > C_M$ in homopolymerization situations.

Transfer constants to polymer, C_p, are also independent of the degree of polymerization in poly(styrene), but not in poly(methyl methacrylate), where C_p depends on the end groups and decreases as the degree of polymerization falls. The large transfer constants of many compounds are useful in the oligomerization process.

Since the number-average degree of polymerization is equal to or twice the kinetic chain length, depending on the type of termination, it is apparent from Equation 31 that the larger the sum of chain-transfer terms becomes, the lower will be the molecular mass of the polymer.

Transfer, which sometimes can be very marked, is used industrially to regulate the degree of polymerization. According to Equation 17, higher initiator concentrations would, indeed, lower the degree of polymerization, but at the same time, as shown by Equation 14, it would increase the rate of polymerization, and too high a rate of polymerization may lead to a loss of control over the entire process. Chain-transfer agents which form radicals of low reactivity are therefore used as regulators. A compound which shows inhibitor-type activity but is not entirely efficient in trapping and stopping the kinetic chain will cause a significant decrease in the rate of polymerization. However, the reaction rate is only retarded, since once the added transfer agent is consumed, polymerization proceeds at the same rate as that registered in the absence of retardants.[20] Since regulators have very high transfer constants, they only need to be added in small concentrations. Mercaptans in particular are suitable regulators. Moreover, since retardation and inhibition depend on the relative reactivities of the growing radical, monomer, and added material, a substance that acts as a transfer agent can act as a retarding agent in another reaction and as an inhibitor in a third.

A reaction of particular interest, regarding the transfer to the monomer, is the abstraction of an allylic hydrogen atom:[4]

$$\text{\textbraceleft\textbraceright}CH_2-\overset{\cdot}{C}H-CH_2X + CH_2{=}CH-CH_2X \xrightarrow{k_1} \text{\textbraceleft\textbraceright}CH_2-CH_2-CH_2X + CH_2{=}CH-\overset{\cdot}{C}HX \quad (33)$$

which is generally quite facile because of stabilization imparted to the allylic radical formed through the resonance interaction with the adjacent double bond:

$$CH_2{=}CH-\overset{\cdot}{C}HX \rightleftharpoons CH_2\cdots CH\cdots CHX \quad (34)$$

Therefore, the interaction of free radicals with allylic compounds is a competition between addition (Equation 35) and abstraction (Equation 36) reactions:

$$R^{\cdot} + CH_2{=}CH-CH_2X \rightarrow R-CH_2-\overset{\cdot}{C}H-CH_2X \quad (35)$$

$$R^{\cdot} + CH_2{=}CH-CH_2X \rightarrow RH + CH_2{=}\overset{\cdot}{C}-CH_2X \quad (36)$$

The alkyl radical formed in the addition reaction (Equation 35) is very reactive and can continue the kinetic chain by another abstraction (Equation 33) or by addition:

$$R-CH_2-\overset{\cdot}{C}H-CH_2X + CH_2{=}CH-CH_2X \rightarrow R-CH_2-CH-CH_2-\overset{\cdot}{C}H \quad (37)$$
$$\underset{CH_2X}{|} \quad \underset{CH_2X}{|}$$

TABLE 1
Transfer Constants C_{RX} for Polymerization of Styrene and Methyl Methacrylate at 60°C

Transfer agent	Styrene					Methyl methacrylate				
	C_I	C_M	C_S	C_P	Ref.	C_I	C_M	C_S	C_P	Ref.
Benzoyl peroxide	0.048				8	0.00				8
1,1-Dimethylbenzylhydroperoxide	0.063				8	0.33				8
t-Butyl hydroperoxide	0.035				8	1.27				8
Azobisisobutyronitrile	0.000				8	0.00				8
Styrene		0.00006			9					
Methyl methacrylate							0.00001			8, 10, 11
Benzene			0.000018		12			0.000083		13
Carbon tetrachloride			0.0092		12			0.0005		14
Carbon tetrabromide			1.36		12			0.27		15
n-Butyl mercaptan			22.0		12			0.66		16
Poly(styrene)				0.00019	17				0.00022	18
Poly(methyl methacrylate):										
Backbone				0.00003[a]	19				0.00015	19
End groups				0.11[a]	19				0.036	19

[a] At 50°C.

The allylic radical formed in the abstraction reactions is too unreactive to attack another allylic compound and, instead, will undergo combination reactions with radicals from the system:

$$CH_2=\overset{\cdot}{C}H-CH_2X + R^{\cdot} \rightarrow CH_2=\underset{\underset{R}{|}}{C}-CH_2X \tag{38}$$

Such an abstraction reaction followed by radical termination (Equation 38) is a degradative chain transfer since it can be classified kinetically as termination but chemically as transfer, and the growing radical commits "suicide".[16] The abstraction reaction, which is followed by addition of the allylic radical to a monomer molecule, is considered to be an effective chain transfer.

Considering the effective and degradation chain transfers as distinct reactions, the allylic oligomerization can be described by the following equations:

$$\text{Initiator } (I_2) \xrightarrow{k_1} 2R^{\cdot} \quad \text{(initiator decomposition)} \tag{39}$$

$$R^{\cdot} + M \xrightarrow{k_i} RM_i^{\cdot} \quad \text{(initiation)} \tag{40}$$

$$R^{\cdot} + M \xrightarrow{k_d} R + M' \quad \text{(degradative transfer)} \tag{41}$$

$$R^{\cdot} + M \xrightarrow{k_t} R + M^{\cdot} \quad \text{(effective transfer)} \tag{42}$$

$$RM_i^{\cdot} + M \xrightarrow{k_2} RM_{i+1}^{\cdot} \quad \text{(propagation)} \tag{43}$$

$$RM_i^{\cdot} + M \xrightarrow{k_d} RM_i + M' \quad \text{(degradative transfer)} \tag{44}$$

$$RM_i^{\cdot} + M \xrightarrow{k_T} RM_i + M_i^{\cdot} \quad \text{(effective transfer)} \tag{45}$$

$$M_i^{\cdot} + M \xrightarrow{k_2} M_{i+1} \quad \text{(propagation)} \tag{46}$$

$$M_i^{\cdot} + M \xrightarrow{k_d} M_i + M' \quad \text{(degradative transfer)} \tag{47}$$

$$M_i^{\cdot} + M \xrightarrow{k_T} M_i + M_i^{\cdot} \quad \text{(effective transfer)} \tag{48}$$

$$R^{\cdot} + R^{\cdot} \xrightarrow{k_3} Y \quad \text{(termination)} \tag{49}$$

where $R^{\cdot}$ is a radical formed by decomposition of initiator, M is an allylic monomer, M' is an inactive radical produced by a degradative chain transfer, $RM_i^{\cdot}$ and RM_i are a growing radical and a stable polymer molecule, respectively, on which an initiation fragment is attached, and $M_i^{\cdot}$ and M_i are a growing radical and a stable polymer molecule, respectively, without an attached initiator fragment.[21]

By means of an appropriate steady-state assumption,

$$\left(\frac{d[M]}{d[I]}\right)_{[I]\to 0} = \frac{2(k_i + k_t)(k_2 + k_T + k_D)}{(k_i + k_d + k_t)k_D} = \text{constant} \tag{50}$$

and assuming $k_i/k_2 = k_d/k_D = k_t/k_T$, one obtains

$$\left(\frac{d[M]}{d[I]}\right)_0 = \frac{2(k_2 + k_T)}{k_D} \tag{51}$$

The degree of polymerization can be reduced to

$$\overline{DP} - 1 = \frac{k_2}{k_D + k_T} = \text{constant} \tag{52}$$

The combination of the last two equations leads to

$$\frac{k_D}{k_D + k_T} = \frac{2\overline{DP}}{(d[M]/d[I])_0 + 2} \tag{53}$$

The left-hand side of Equation 53 gives the fraction of degradative chain transfer in the total transfer reaction, since both $\overline{DP}$ and $(d[M]/d[I])_0$ are measurable quantities. The experimentally observed constancy of $d[M]/d[I]$ in any given polymerization of allylic monomers was one of the reasons for the postulation of degradative chain transfer as a characteristic of allyl polymerization.[22] The reason for the observed constancy of $d[M]/d[I]$ is that $[I]/[M]^2$ is nearly constant in the course of polymerization. The value of $d[M]/d[I]$ decreases with increasing initiator concentration. A constant $d[M]/d[I]$ does not require the occurrence of degradative chain transfer as the exclusive or even the predominant chain termination reaction.

Whereas in the polymerization of allyl acetate 75% of the chain transfer is degradative, in other allylic monomers degradative chain transfer varies from 3 to 86% of the total transfer reactions. Both the ratio of addition to abstraction reactions and the ratio of degradative to effective chain transfers are strongly dependent upon the nature of the allylic monomer (i.e., substituent X), and this dependence is reflected by the degree of polymerization, which is about 14 for allyl acetate or carbonate and only 4 for allyl ethyl ether.[22] The low reactivity of allyl monomers and their radicals is an additional reason for the low $\overline{DP}$ of polymers.[23]

Propylene (X = H), unlike ethylene, cannot be polymerized to a solid product of high molecular mass by the radical mechanism, even at high pressures, because of its readiness to undergo chain transfer, with formation of the allyl radical.

The dependence of the rate constant k of a chemical reaction on the pressure P at the absolute temperature T is approximated by the equation:[24-26]

$$\left(\frac{\partial \ln k}{\partial P}\right)_T = \frac{-\Delta V^\ddagger}{RT} \tag{54}$$

where $\Delta V^\ddagger$ represents the volume change which accompanies the formation of the transition state from the reactants. The isothermal variation of k with P is determined by the sign of magnitude of $\Delta V^\ddagger$. A value of -20 cm^3/mol for a reaction at 333 K would, from Equation 54, correspond to an increase in the rate constant by a factor of 9 at 3 kb, and a factor of 1500 at 10 kb.[26]

The shrinkage for addition polymerization varies from 11.8% for diallyl phthalate and 14.5% for styrene up to 66% for ethylene.[27]

Brown and Wall[28] found that the rate of the γ-ray-initiated polymerization of propylene increased about 100 times between 5 and 16 kb ($\Delta V^{\ddagger}$ is about -10 cm³/mol), but the molecular mass of the polymer was increased over the same pressure range by a factor of only 2 (from 1500 to 3000). The degree of polymerization given by the ratio of propagation rate to transfer rate decreased — as expected from Equation 22 — while T increased (from 294 to 356 K).

Zharov and co-workers,[29] who used α,α′-azobis(isobutyronitrile) (AIBN) initiator, found the exponent of [I] in the rate equation (Equation 14) to be nearly 1 below 3 kb, but to be 0.55 at 6.5 kb. The rate increased with pressure between 3 and 8 kb, but the increase in the molecular mass of the polymer did not exceed 30%. The near constancy of the molecular mass, in comparison with the large pressure dependence of the rate, shows that the transfer reaction with the monomer is accelerated nearly as much as chain propagation by an increase of pressure. Another example of the effect is the polymerization of allyl acetate.[30] The average molecular mass remains at about 2000 for reaction pressures up to 8.5 kb.

The application of all equations describing the kinetics of free radical polymerization initiated by chemical agents requires that generation of radicals from labile initiator species be the only original source of free radicals for the polymerization reaction. For certain monomers, however, there is an appreciable, spontaneous initiation reaction brought about by a co-reaction of the monomer molecule alone, especially when the reaction temperature is raised. This spontaneous initiation is termed thermal initiation (see Section II.B) and must be taken into account in the term for the rate of propagation in the chain-transfer equation (Equation 26) which gives the kinetic chain length. The corrected value for ν_{tr} is given by

$$\nu_{tr} = \frac{(R_{p,ob}^2 - R_{p,th}^2)^{0.5}}{R_t + R_{tr}} \tag{55}$$

in which $R_{p,ob}$ is the observed rate of polymerization and $R_{p,th}$ is the rate of polymerization caused by thermal initiation alone.[31] The use of squared terms in the correction is necessary because the rates of initiation, R_i, are additive and R_i is proportional to R_p^2.[4]

Ito[32] proposed the following expression for the rate of polymerization of styrene in bulk initiated by both the primary radicals formed by decomposition of the initiator and those formed thermally:

$$R_p = (2fk_d[I] + k_{th}[M]^\sigma)^{1/2}(k_p/k_t^{1/2})[M] \tag{56}$$

where k_{th} represents the thermal initiation rate constant and σ is the order of thermal initiation in the monomer, i.e., σ = 2 or 3, depending mostly on $[M]_o$. The conversion-time curves observed at 100°C in both the presence of 2-phenyl-azo-2,4-dimethyl-4-methoxyvaleronitrile (PADMV) as an initiator ($[I]_o$ = 15 mmol/l) and its absence ($[I]_o$ = 0) are shown in Figure 1.

The negative entropies of addition reactions predict that these reactions should become increasingly reversible with increasing temperature. Consequently, at higher temperatures, the elimination of a monomer from the active ends of growing radicals, i.e., the reverse of propagation (Equation 4), which is termed *depropagation* or depolymerization, becomes increasingly important:

$$P_n^\bullet \underset{k_p}{\overset{k_{dp}}{\rightleftharpoons}} P_{n-1}^\bullet + M \tag{57}$$

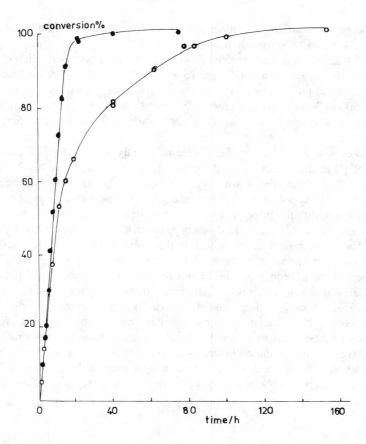

FIGURE 1.   Conversion time curves obtained at 100°C at the polymerization of styrene. (●) chemically initiated, I_o = 15 mM/l, and (○) thermally initiated, I_o = 0. (Adapted from Ito, K., *Polym. J.*, 11, 877, 1986.)

At the equilibrium

$$k_{dp}[P_n^{\cdot}] \ = \ k_p[P_{n-1}^{\cdot}][M]_e \tag{58}$$

taking first the logarithms of Equation 58 and expressing k_{dp} and k_p in terms of activation energies and temperature using the Arrhenius equation (Equation 18), and taking into account the constancy radical concentration in the system, one obtains

$$-d\left(\frac{E_{dp}^*}{RT_c}\right) \ = \ -d\left(\frac{E_p^*}{RT_c}\right) \ + \ \ln[M]_e \tag{59}$$

The temperature at which equilibrium (Equation 57) holds is the ceiling temperature. Combining Equation 59 with T_c obtained from the second law of thermodynamics, T_c = $\Delta H_p/\Delta S_p$, gives

$$T_c \ = \ \frac{\Delta H_p}{\Delta S_p^{\circ} \ + \ R\ln[M]_e} \tag{60}$$

where ΔS_p^o is the entropy of polymerization at the standard state. Rearrangement of Equation 60 leads to

$$\ln[M]_e = \Delta HS_p/RT_c - \Delta S_p^o/R \tag{61}$$

which shows that no chain-growth polymerization can ever be driven quantitatively to completion because at each polymerization temperature $T_p \leq T_c$, there is a specific monomer concentration, $[M]_e$, which will exist in equilibrium with the active polymeric species.

Oligomerization of unsaturated monomers (olefins, dienes, and vinyl derivatives) initiated by free radicals is limited by the chain character of the reactions, which determines the achievement of high molecular masses in the early stages of the polymerization.

Extensive chain-transfer processes and/or high initiator concentration may, however, drastically shorten the kinetic chain, allowing the formation of oligomers as main reaction products.

Radical polymerization of olefins, vinyl monomers, and dienes in alcohols initiated by hydrogen peroxide at 90 to 120°C were found by Pinazzi et al.[33-40] to yield oligomers ($\overline{M}_n \leq 6000$) with hydroxyl end groups. The value of the constant of the chain transfer to initiator determined for the polymerization of vinyl acetate, isoprene, and butadiene indicated that the reactions are essentially initiated by hydrogen peroxide.[38]

The polymerization of butadiene in a mixture of alcohol and benzene[34] gives oligomers and liquid polymers ($\overline{M}_n$ = 500 to 11,000) in which the monomeric units have mainly 1,4 structures *(trans)*. The polymers contain two or three, and in certain cases up to six, hydroxy groups per molecule. Both degree of polymerization and the conversion in oligomers are dictated by the alcohol/benzene ratio which, in turn, determines the actual concentration of the initiator, i.e., the presence of an alcohol in a heterogeneous mixture composed of an aqueous phase (hydrogen peroxide) and an organic phase (benzene and butadiene) assures the miscibility of the monomer and initiator. An example is given in Figure 2.

By homopolymerization of 1-butene, 2-butene, and isobutene in the same conditions,[36] only low molecular mass hydroxylated oligomers were obtained in low polymerization yields. The copolymerization of these olefins with a diene monomer such as butadiene was easily carried out, but again only oligomers were formed due to transfer reactions. The polymerization and copolymerization rates of 2-butene were higher than those of isobutene, but lower than those of 1-butene. The authors proposed for these copolymerizations the mechanisms shown in Figure 3.

Reactive oligomers were obtained by polymerization and copolymerization of dienes and olefins and polar unsaturated monomers (alcohols and acids) initiated by organic peroxides at 80°C.[41] Mixed aliphatic-aromatic resins ($\overline{M}_n$ = 800 to 3000) were formed by free radical oligomerization initiated by peroxides of cracked oil fractions of nonhomogeneous compositions (e.g., 6 to 7% dienes, 2 to 4% cycloolefins, 2 to 6% monoolefins, and 16 to 20% arylalkenes).[42]

Piperylene (1,3-pentadiene), a monomer much less reactive than isoprene or butadiene, can be oligomerized in bulk with cumene hydroperoxide to give unsaturated liquid polymers which can be subsequently epoxidized by peracids.[43]

The possibility of the formation of telechelic oligoethylenes ($\overline{M}_n$ = 750) was investigated by Guth and Heitz by reacting ethylene with peroxides or azo derivatives.[44,45] In the dihexanoyl peroxide/ethylene system,[45] mainly linear paraffins having an even number of C atoms (combination products) were formed at high temperatures (>200°C). A reaction temperature of 95°C, however, gave mainly the uneven-membered paraffins, due to transfer reactions to initiator, i.e., induced decomposition of dihexanoyl peroxide caused by growing alkyl radicals abstracting H at the β-position to the carbonyl group. The radical polymerization of ethylene using high concentrations of AIBN[44] did not predominantly result in the expected oligomers with two nitrile end groups, but, rather, those with one. The proposed

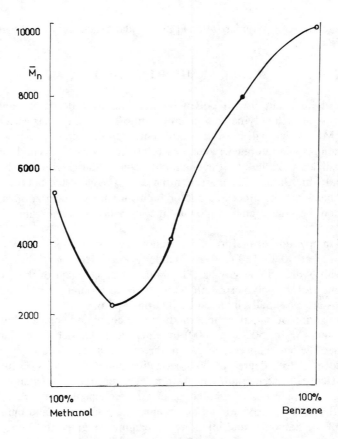

FIGURE 2.   Dependence of the average molecular mass of liquid polybutadiene on the methanol/benzene ratio. Reaction conditions: initiator, H_2O_2, 10 g; monomer, 60 g; temperature, 120°C; time, 5 h. (From Pinazzi, C., Legeay, G. and Brosse, J. C., *Makromol. Chem.*, 176, 2509, 1974. With permission.)

mechanism is presented in Figure 4. Caused by a dominating H-abstraction of the macro-radical from the isobutyronitrile radical, the nitrile functionality average was $\bar{f}_{CN} = 1.7$. The use of dialkylperoxydicarbonates gave, for similar reasons, oligoethylenes with $\bar{f}_{alkoxycarbonyloxy} \simeq 1.1$.

II. RADICAL OLIGOMERIZATION INITIATED BY PHYSICAL MEANS

These radical polymerizations include primarily encatenation reactions started or propagated by electromagnetic radiations or thermally.

A. RADIATION-ACTIVATED POLYMERIZATIONS

These processes are classified as radiation-initiated polymerizations and radiation polymerizations. The former reactions occur when the radiation starts a polyreaction, but each individual propagation reaction, i.e., each new addition of a monomer unit to a growing molecule, proceeds without the direct action of radiation. In radiation polymerization, each individual propagation reaction is affected by the radiation. The radiation used to initiate polymerization can be of high energy (α-, β-, and γ-radiations, slow neutrons) or of low energy (visible or ultraviolet light). Polymerizations started by low-energy radiation are

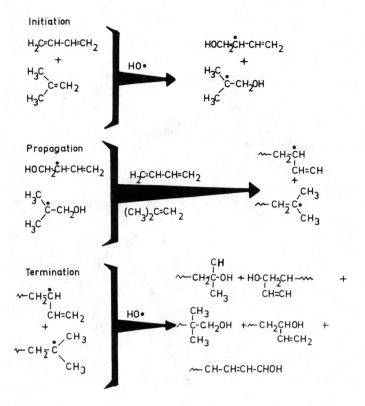

FIGURE 3. The mechanism of copolymerization of 1,3-butadiene and isobutene.

termed photoactivated polymerizations, i.e., photoinitiated polymerizations and photo-polymerizations.

High-energy radiations are nonselective and produce a wide array of initiating species. Even though the high-energy radiations give rise to both ionic or free radical species, the polymerizations that are so initiated follow the free radical mechanism almost exclusively, except at very low temperatures, where ionic intermediates become more stable. According to ESR measurements, the start reaction appears to involve a disproportionation:[16]

$$2CH_2=CHR \rightarrow CH_3-\dot{C}HR + CH_2=\dot{C}R \tag{62}$$

These high-energy sources of initiating radicals will not be treated further here because they are of little use in the synthesis of oligomers.

When the photoradiation is absorbed by a monomer, its energy is immediately assimilated and the molecule is converted into the excited state. The excited monomer either dissociates into free radicals or dissipates the energy by fluorescence, phosphorescence, or nonradiative processes such as collision deactivation. All these processes constitute the initial photo-chemical act.

Photochemical initiation of polymerization may occur in different ways: through energy transfer of an excited sensitizer molecule (Sm) to either a monomer or foreign molecule (A):

$$Sm \xrightarrow{h\nu} Sm^* \xrightarrow{+A} Sm + A^* \tag{63}$$

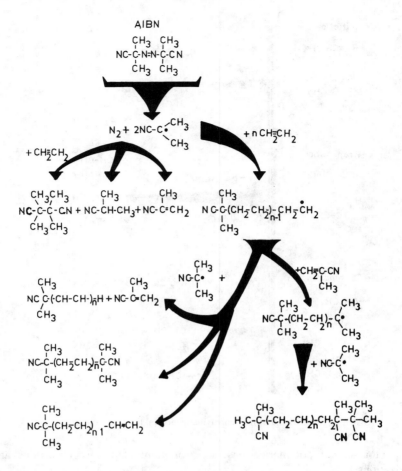

FIGURE 4.   The mechanism of ethylene oligomerization in the presence of AIBN.

or through direct excitation

$$A \xrightarrow{h\nu} A^* \tag{64}$$

followed by a reaction creating an initiating species:

$$A^* \longrightarrow A^\circ \tag{65}$$

$$A^\circ + M \xrightarrow{k_i} AM^\circ \tag{66}$$

$$AM^\circ + nM \xrightarrow{k_p} A(M)_n M^\circ \tag{67}$$

If the sensitizer (Sm) only absorbs light, the rate of formation of the initiating species (A°) can, in general, be written as

$$R_{A^\circ} = \alpha I_o (1 - e^{-\epsilon_s l}[Sm]) \tag{68}$$

where I_o is the incident light intensity, [Sm] and ϵ_s are the sensitizer concentration and its molar extinction coefficient, respectively, and 1 is the path length, while α is the overall quantum yield of formation of the initiating species as measured by actinometry; this takes into account the quantum yield of formation of the excited state, the energy-transfer yield, and the efficiency of forming initiating species. If the product $\epsilon_s l$ [Sm] is small, this equation becomes

$$R_{A^o} = \alpha I_o \epsilon l [Sm] \qquad (69)$$

In addition to permitting the use of a wider variety of initiators with stronger covalent bonds, photoinitiation of chain-growth polymerization reactions can be carried out at any desired temperature and can be started and stopped instantly. Its main feature is the rate of formation of the initiating species and the molecular weight distribution.

If direct irradiation results only in formation of the singlet or triplet excited state of the monomer, oligomerization (dimerization) into cyclobutane derivatives could be a competitive process.

In actual photopolymerization, each individual propagation reaction is photochemically activated, i.e., the chain lengthening occurs through a repetitive photochemical act. Singlet states are involved in the photopolymerization of anthracene derivatives, which represents a $(4\pi + 4\pi)$ cycloaddition, while triplet states occur in the four-center polymerization of 2,5-distyrylpyrazines, which is a $(2\pi + 2\pi)$ cycloaddition:

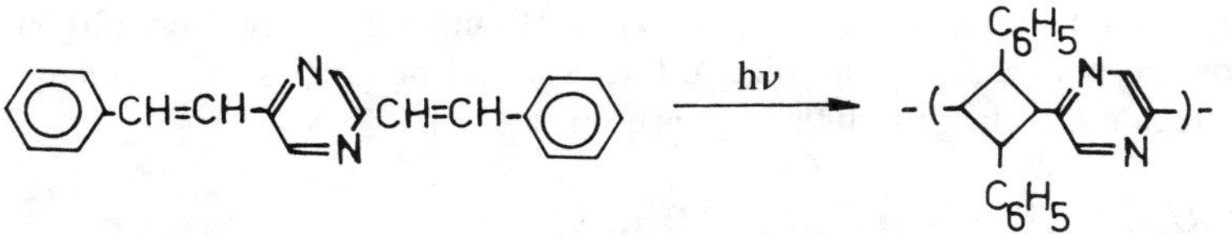

(70)

Kinetic studies[46] showed that photopolymerization (Equation 70) is, however, a true stepwise polyaddition reaction. This was proved by the influence of irradiation time on the molecular mass of the polymer, and by its Gaussian distribution, which corresponds to a former polyaddition and not to a photoinitiated polymerization.

Formation of high molecular mass compounds is strongly dependent on the monomer concentration and on the solubility of monomer and polymer. For example, reductive photopolymerization of bis-benzophenones in isopropanol, as shown below,

$$C_6H_5-\underset{\underset{O}{\|}}{C}-C_6H_4-R-C_6H_4-\underset{\underset{O}{\|}}{C}-C_6H_5 \xrightarrow[i\text{-}C_3H_7OH]{h\nu}$$

$$\sim\left[-C_6H_4-\overset{\overset{OH}{|}}{\underset{\underset{C_6H_5}{|}}{C}}-\overset{\overset{OH}{|}}{\underset{\underset{C_6H_5}{|}}{C}}-C_6H_4-R-\right]_n-C_6H_4-\overset{\overset{OH}{|}}{\underset{\underset{C_6H_5}{|}}{C}}-\overset{\overset{OH}{|}}{\underset{\underset{C_6H_5}{|}}{C}}-C_6H_4-R\sim \qquad (71)$$

resulted in oligomeric polyalcohols ($\overline{M}_n$ = 3000 to 3500), if the monomer was a phthalophenone, or higher polymers ($\overline{M}_n$ = 11,000 to 30,000) when a *p,p'*-dibenzoylphenylalkene was photoreacted.[47]

By UV irradiation of maleic anhydride in dioxan at 35°C without initiators or sensitizers, only low oligomers ($n \leq 4$) were formed.[48] Oligomers are also the only reaction products when the polymerization of this monomer is chemically initiated with benzoylperoxide (AIBN is inefficient), but the average degree of encatenation is much higher ($\overline{M}_n = 625$ to 710).[49]

Hydrogen peroxide cleaves by UV irradiation (253.7 nm) into two hydroxyl radicals which can be used as initiators of polymerization in order to give hydroxytelechelic oligomers. Yields and average molecular mass are influenced by the presence of an alcohol in the reaction medium introduced in order to homogenize the system.[37] If the hydrogen peroxide is cleaved thermally (120°C), the thermal initiation *(vide infra)* may compete with the chemical initiation brought about by hydroxyl radicals.[39]

Both thermally and photochemically initiated oligomerizations may also be carried out in aqueous systems, but the process is conditioned by the amphiphilic character of monomers and their concentration.

Haimiel and Sherrington[50] investigated the polymerization of a series of quaternary ammonium (meth)acrylates in water at concentrations both above and below their critical micelle values, and also in isotropic alcoholic solutions. Oligomeric species with DP $\leq$ 25 were formed readily from micellized solutions, using both photochemical and thermal initiation systems. Below the critical micelle concentration, however, polymerizations seemed very inefficient and conversions were very low. Even in isotropic alcoholic solutions, polymerization remained sluggish and the DP value was typically very low, about 214. The rate of exchange of monomer surfactant between micelles was much faster than the rate of oligomerization propagation and so a topochemical polymerization of a micelle, i.e., the occurrence of polymerized micelles, was extremely unlikely. However, oligomeric species could be formed which displayed micelle-like physical properties and showed a great tendency to aggregate into large units and phase up.

B. THERMAL-INITIATED OLIGOMERIZATION

A true thermal polymerization is a reaction in darkness with total exclusion of oxygen and other reagents (impurities, e.g., those coming from the container walls) that might lead to the formation of active polymerizing species. Very few monomers, however, exhibit pure thermal initiation. Spontaneous, thermally induced free radical polymerization of vinyl monomers in bulk has received considerable interest, not only in industrial applications, but also with reference to theoretical aspects in the field of closed-shell π-orbital interactions.[51]

Styrene and methyl methacrylate are the primary examples, and the former is by far the most active one known. At 127°C, styrene showed a thermal self-initiation rate of 14%/h, while methyl methacrylate has an activity only about 1% of that of styrene. In contrast, vinyl acetate, methyl acrylate and vinyl chloride show no measurable polymerization rate at 100°C, and other monomers undergo only a Diels-Alder reaction at elevated temperatures.[52]

In the thermal polymerization of styrene at high temperatures (120 to 200°C), a significant level of small molecular products (dimers and trimers) is formed, but depending on the experimental conditions, oligomers with a molecular mass up to 4500 may be formed.[53,54] Studies to determine the structure and composition of the mixture of oligomers accompanying the polystyrene in the thermal polymerization of styrene have been carried out.[55] The most essential aspect of these investigations concerned the molecular processes of free radical formation. Based on kinetic measurements and product analysis, Mayo[54] was first to postulate the intermediate existence of a somewhat unconventional 4 + 2 Diels-Alder adduct between two styrene molecules, Z, which, by reaction with other monomer molecules, was expected to produce free radicals (Reaction 74) capable of propagating to form polystyrene chains:

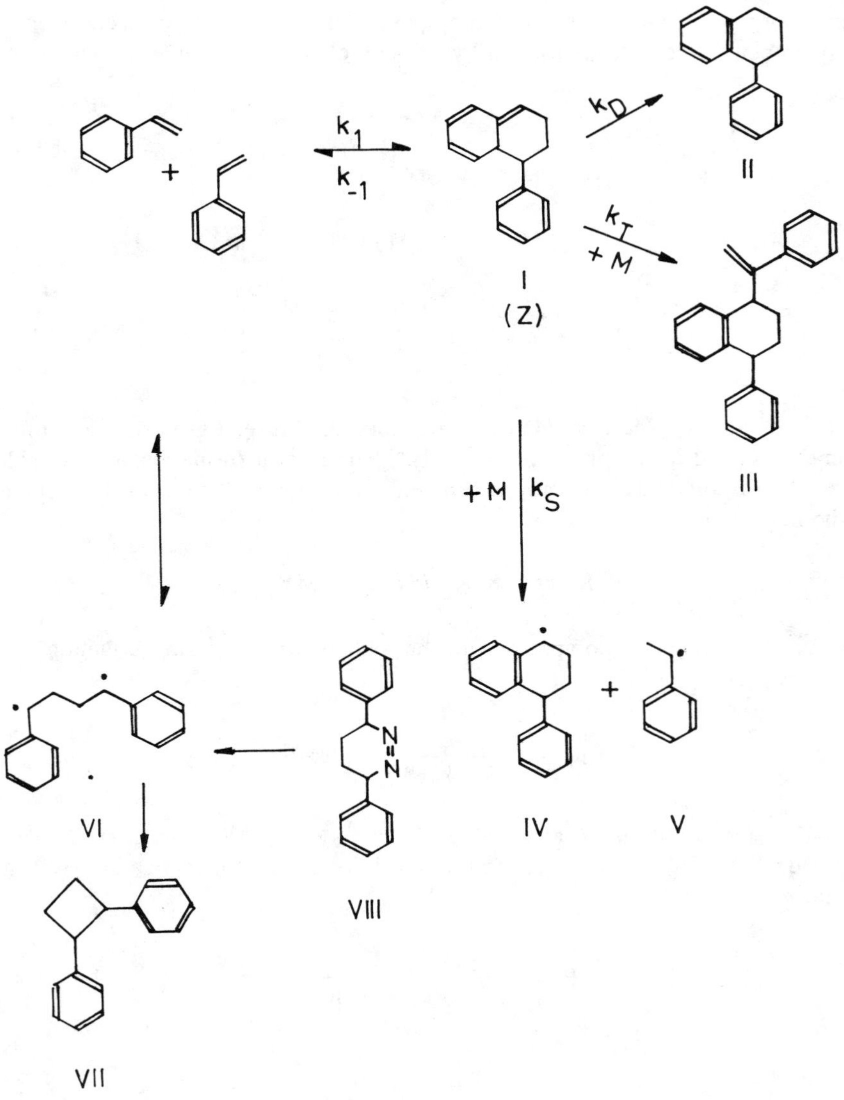

(72)

(73)

(74)

(75)

The reaction of Z with styrene (Reaction 74) can explain the observation that the start of polymerization is second or third order with respect to the monomer, depending on the monomer concentration.[56-60] The following equation has been proposed by Mayo[56] for the rate of initiation, R_i (mol/l/s), which correlates rather well with experimental data for [M] < 2 mol/l:

$$R_i = 1.32 \times 10^6 \, e^{-121/RT} [M]^3 \qquad (76)$$

For [M] > 2 mol/l, the initiation is second order, i.e., $R_i \sim [M]^2$.

The vinyl double bonds of two molecules can react β, β or α, β. The intermediary Flory-type diradical VI from Reaction 74 definitely does not start any polymerization, since the same diradical produced by the decomposition of the corresponding azo compound VIII does not.[61]

Since benzyl and phenylpropenyl fragments were identified as end groups of thermally synthesized polystyrene chains, it has been proposed[62] that the initiating species were benzyl and phenylpropenyl free radicals formed by scission of the linear trimer biradical:

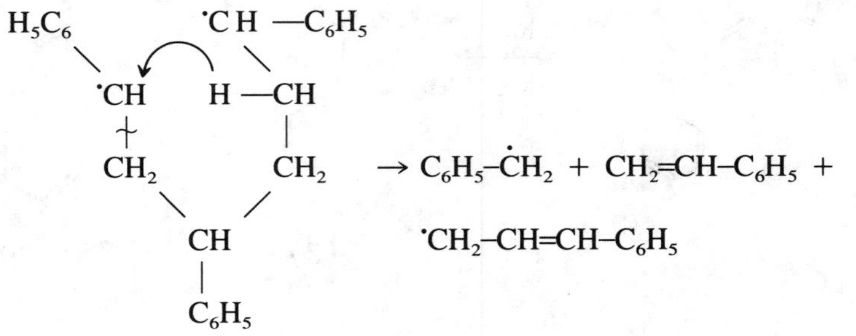

$$\rightarrow C_6H_5\text{-}\overset{\cdot}{C}H_2 + CH_2\text{=}CH\text{-}C_6H_5 + \qquad (77)$$

$$\overset{\cdot}{C}H_2\text{-}CH\text{=}CH\text{-}C_6H_5$$

Radicals IV and V (Reaction 74) have been identified as end groups of low molecular polystyrenes obtained in the presence of $FeCl_3$.[62] The rate of formation of low molecular derivatives, R_L (l/mol/s) has been proposed by the same author[53,56] to be of the second order against the monomer:

$$R_L = 1.8 \times 10^5 e^{-95/RT}[M]^2 \qquad (78)$$

This equation should be compared with that obtained by rewriting Equation 56 when the initiator is absent:

$$R_{pth} = \frac{k_p}{(k_t \cdot k_{th})^{1/2}} [M]^{(1+\sigma/2)} \qquad (79)$$

where σ is 2 at $[M] > 0.7$ mol/l and 3 otherwise.[63]

From Equations 76, 78, and 79, the rates of initiation and trimerization (R_T) may be expressed as:

$$R_i = \frac{k_1 k_s[M]^3}{k_D + (k_s + k_T)[M]} \qquad (80)$$

and

$$R_T = \frac{k_1 k_T[M]^3}{k_D + (k_s + k_T)[M]} \qquad (81)$$

Reaction 75 for the formation of 1,2 diphenylcyclobutane (*cis, trans*) is second order with respect to the monomer and has an activation energy of 104 kJ/mol.[64]

A more complex kinetic treatment based on Mayo's scheme takes into consideration both formation of initiating radicals from dimer Z and polymerization (the rate constants were determined at 60°C):[65]

Initiation

$$M + M \underset{k_{-1}}{\overset{k_1}{\rightleftharpoons}} Z \quad (k_1 = 10^{-9} \text{ l/mol/s}) \qquad (82)$$

$$(k_{-1} = 1.3 \times 10^{-4} \text{ l/mol/s})$$

$$Z + M \xrightarrow{k_s} IV + V \quad (k_s = 10^{-8} \text{ l/mol/s}) \tag{83}$$

$$Z + M \xrightarrow{k_T} III \tag{84}$$

$$IV + M \xrightarrow{k_A} R_1 \tag{85}$$

$$V + M \xrightarrow{k_B} R_1 \tag{86}$$

Propagation

$$R_1 + M \xrightarrow{k_p} R_2 \quad (k_p = 1.45 \times 10^2 \text{ l/mol/s}) \tag{87}$$

$$\cdots\cdots\cdots\cdots\cdots\cdots$$

$$R_n + M \xrightarrow{k_p} R_{n+1} \tag{88}$$

Termination

$$R_X + R_Z \xrightarrow{k_t} R_{X+Z} \quad (k_t = 1.7 \times 10^7 \text{ l/mol/s}) \tag{89}$$

Chain transfer

$$R_X + M \xrightarrow{k_{tr,M}} P_X + R_1 \quad (k_{tr,M} = 10^{-3} \text{ l/mol/s}) \tag{90}$$

$$R_X + Z \xrightarrow{k_{tr,Z}} P_X + R_1 \quad (k_{tr,Z} = 1.5 \times 10^2 \text{ l/mol/s}) \tag{91}$$

The rate of thermal initiation, deduced from Equations 82 to 91, is

$$R_i = 2k_S[Z][M] \tag{92}$$

which, expressed only in terms of monomer concentration, becomes:

$$R_i = \frac{2k_S k_1[M]^3}{k_{-1} + (k_S + k_T)[M] + k_{tr,Z}(v_i/k_t)^{1/2}} \tag{93}$$

where

$$v_i = (k_A[IV] + k_B[V])[M] = R_i \tag{94}$$

It follows that if

$$(k_S + k_T)[M] \gg k_{-1} + k_{tr,Z}(v_i/k_t)^{1/2} \tag{95}$$

then

$$R_i = \left(\frac{2k_S k_1}{k_S + k_t}\right)[M]^2 \tag{96}$$

i.e., the initiation is second order with respect to the monomer, and if

$$k_{-1} \gg (k_S + k_T)[M] + k_{tr,z}(v_i/k_t)^{1/2} \tag{97}$$

then

$$R_i = \left(\frac{2k_S k_i}{k_{-1}}\right)[M]^3 \tag{98}$$

which means that the initiation is third order in the monomer. Experimentally, however, it is difficult to make a distinction between these two limit cases.

Kinetic reaction spectroscopy of polymerizing styrene systems in the near UV has revealed the intermediary nature of the Diels-Alder dimer.[66] Moreover, the spectroscopic data analysis[67] has led to a more refined concept of two diastereoisomeric $4\pi + 2\pi$ intermediates, ZA and ZB:

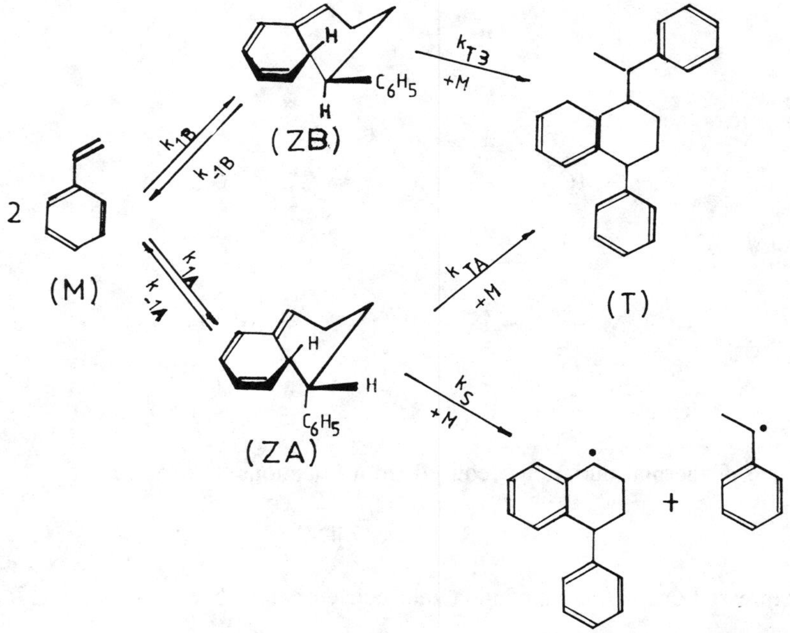

$$(99)$$
$$(100)$$
$$(101)$$

The modified Mayo mechanism of radical initiation in the thermal polymerization of styrene, proposed by Olaj et al.,[68] supposes that these two intermediates are of different reactivity. Both intermediates (ZA and ZB) can form cyclic trimers (Equations 99 and 100) such as 1-phenyl-4-phenylethyl-(1,2,3,4)-tetrahydronaphthalene. The more reactive adduct ZA rapidly attains its stationary-state level and is responsible for radical generation (Equation 101) and the chain-transfer reaction:[69]

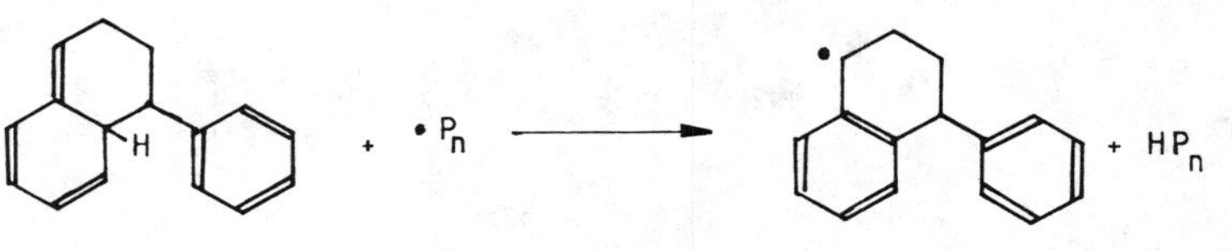

$$(102)$$

Kirchner and Riederle[70] followed quantitatively, by gas chromatography, the formation of oligomeric styrenes during the thermal polymerization of styrene at high conversions. The total concentration of the two intermediates (ZA and ZB) was measured by UV spectrometric analysis at 325 nm. Under the polystyrene synthesis conditions (137 and 180°C and 0 to 97% conversion of monomer), five dimers and, in substantially higher quantities, five trimers were formed in addition to polymers. The main constituents of the dimeric fraction (80 to 90%) were *cis-* and *trans-*1,2-diphenylcyclobutanes (VII, Reaction 75). Also formed were 1,3-diphenylbutene-3, a dimer not clearly identified, and a dimer assumed to be 1-phenyl tetralin. Dimers with a total of up to 0.1 wt% were formed at 137°C and a monomer conversion of 97%. At 180°C, the total dimer level increased to 0.3 wt% at the same monomer conversion. There was clearly a strong temperature dependence for dimer formation. About 96% of the trimers formed were stereoisomers of 1-phenyl-4-phenyl tetralin. The trimers reached 0.65 wt% at 137°C and 57% conversion. At 180°C and 97% conversion, the trimer content increased very slightly to 0.7 wt%. The appropriate rate expression (Equation 103), which is second order in the monomer, fit the data very well at both temperature levels:

$$\frac{d[DCB]}{dt} = k_{DCB}[M]^2 \tag{103}$$

where [DCB] represents the concentration of diphenyl cyclobutanes.

The kinetic model for trimer formation is

$$\frac{d[T]}{dt} = \left[\left(1 - \frac{k_{TB}}{k_{TA}} \right) k_{1A}[M] + k_{TB}[Z] \right][M] \tag{104}$$

where [T] represents the trimer concentration and [Z] = [ZA] + [ZB] was derived by applying the stationary-state hypothesis to the reactive intermediate ZA.

$$[ZA] = \frac{k_{1A}[M]}{k_{TA}} \tag{105}$$

The fit of experimental points was considerably improved over the simpler second-order model, $d[T]/dt = k_T[M]^2$, but still somewhat lacking.

Recently, an improved S-type approximation curve was claimed by Hang et al.[71] to agree better with experimental data, compared to the representation of Equation 100.

An analytical solution for the time dependence of [Z], containing the rate constants $k'_{1A} = (1 - k_{TB}/k_{TA})k_{1A}$, k_{TA}, and k_{TB}, was derived. The rate constants k_{1A} and k_{TB} were known from the measurements of [T] vs. time and, hence, measurements of [Z] vs. time permit one to estimate k_{1B} and k_{TA}. The fit was excellent (Figure 5). The slow change of ZB with time is clearly evident and the rapid rise to a stationary-state level for the reactive intermediate [ZA] is as expected.[70] It has been calculated that the reactions responsible for dimer and trimer formation as well as for radical generation are not diffusion controlled, a fact of considerable practical importance to polystyrene reactor engineering.[72]

α-Methylstyrene in bulk undergoes at ordinary pressures a variety of thermally induced processes, but not polymerization because of the low ceiling temperature (61°C at p = 10^5 N m^{-2}). Efficient depropagation at T > T_c prohibits the formation of long chains. To obtain a total oligomer concentration of approximately 10^{-3} mol/l, reaction periods of 150 to 1500 h for temperatures of 70 to 90°C are necessary. Kauffmann et al.[73] identified, by GC-MS coupling techniques and independent synthesis, nine dimers and eight trimers. No higher oligomers were detected. Two isomeric cyclobutanes, i.e., *cis-* and *trans-*1,2,-

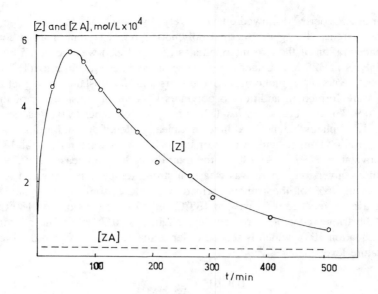

FIGURE 5. Concentration of intermediates Z and ZA vs. polymerization time at 137°C. (○) experimental points; ---- estimated concentration. (From Kirchner, K. and Riederle, K., *Angew. Makromol.Chem.*, 111, 1, 1983. With permission.)

dimethyl-1,2-diphenylcyclobutane, and one of the open-chain unsaturated dimers, 2,5-di-phenyl-hexene-1, were shown to be formed by a conventional two-step cycloaddition in which a Flory-type diradical (DR) was the common intermediate:

(106)

In contrast to these $2\pi + 2\pi$ products, the majority of the remaining oligomers could not be interpreted on the basis of π-electron interactions between closed-shell molecules, and their structures were compatible with the free radical process in which cumyl (MH$\cdot$) and 1,4-dimethyl-1-phenyl-tetrahydronaphthaleneyl (THN$\cdot$) — consecutive products of $4\pi + 2\pi$ and $2 + 2$ interactions, respectively — were involved in addition and transfer reactions.

Two types of stabilization steps, hydrogen transfer to monomer

$$R_nH\cdot + M \rightarrow R_n + MH\cdot \tag{107}$$

and hydrogen abstraction from a hypothetical $4\pi + 2\pi$ intermediate I

$$R_nH\cdot + I \rightarrow R_nH_2 + THN\cdot \tag{108}$$

were thus attributed to the high rates of formation of the following unsaturated and saturated oligomers: 2-methyl-2,4-diphenyl-pentane; 4-methyl-2,4-diphenyl-pentene-1; 1,4-dimethyl-1-phenyl-(1,2,3,4)-tetrahydronaphthalene (two diastereomers); 1-methyl-4-methylene-1-phenyl-(1,2,3)-trihydronaphthalene and 1,4-dimethyl-1-phenyl-(1,2)-dihydronaphthalene — all being dimers; and 2,4-dimethyl-2,4,6-triphenylheptane (two diastereomers); 4,6-dimethyl-2,4,6-triphenyl heptene; 1,4-dimethyl-4-(2-phenyl-propyl)-(1,2,3,4)-tetrahydronaphthalene (four diastereomers) and 1,4-dimethyl-4-(2-phenyl-2-propenyl)-(1,2,3,4)-tetrahydronaphthalene (two diastereomers) — all being trimers.

Since ethylene cannot undergo a Diels-Alder reaction, its thermal polymerization in the gaseous phase has been proposed to involve the formation of methylene and methyl radicals, the former by decomposition of excited ethylene molecules, which act as initiating species in oligomerization and polymerization processes.[74] Polymerization occurs at temperatures lower than 400°C. The first reaction products are butenes ($\sim$32%) and cyclobutane ($\sim$13%) as well as propene and higher olefins. The activation energy for the formation of lower molecular hydrocarbons is $E_a = 475$ kJ/mol and the reaction order in the early stages of the process lies in the range of 2.3 to 2.5.[75]

Thermal polymerization of ethylene in the liquid phase (solution in naphthalene) is characterized by a lower activation energy, i.e., $E_a = 170$ kJ/mol, and a reaction order of 2.6. The initiating species seems to be the tetramethylene biradical:

$$2\,CH_2{=}CH_2 \rightarrow \dot{C}H_2CH_2CH_2\dot{C}H_2 \begin{array}{l} \longrightarrow n{-}C_4H_{10} + cyclo{-}C_4H_8 \\[1em] \longrightarrow 2R\cdot \rightarrow polymer \end{array} \tag{109}$$

and the initiation reaction is third order with respect to the monomer:[76]

$$R_i = 10^{15}\,e^{-272/RT}[M]^3 \tag{110}$$

The only oligomer which accompanies thermal polymerization of chlorotrifluoroethylene and tetrafluoroethylene is the corresponding cyclobutane dimer.

Thermal polymerization of methyl methacrylate is accompanied by formation of appreciable amounts of low oligomers (dimers and trimers), the actual initiating species being a dimeric biradical:[77-79]

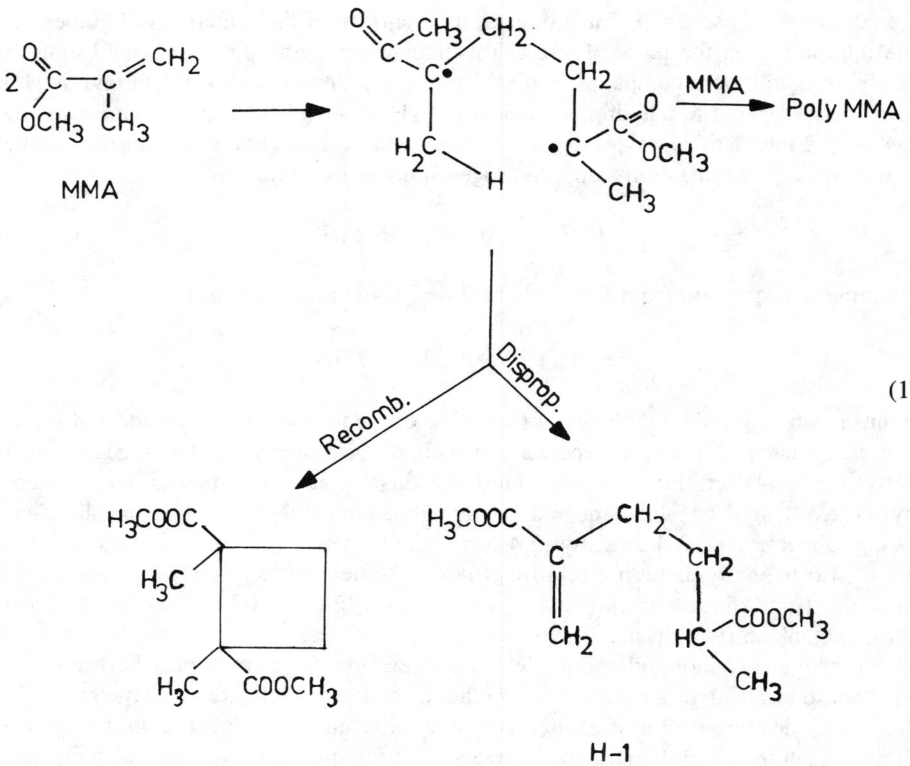

(111)

H-1

The main oligomer product is the linear dimer dimethyl-1-hexene-2,5-dicarboxylate, H-1, the formation of which is characterized by the following rate constant:[77]

$$k_{H-1} = 1.3 \times 10^{13} e^{-107.2/RT} \text{ (l/mol/s)} \tag{112}$$

Heterocyclic dimers are also formed, supposedly through intramolecular recombination of another dimeric radical:

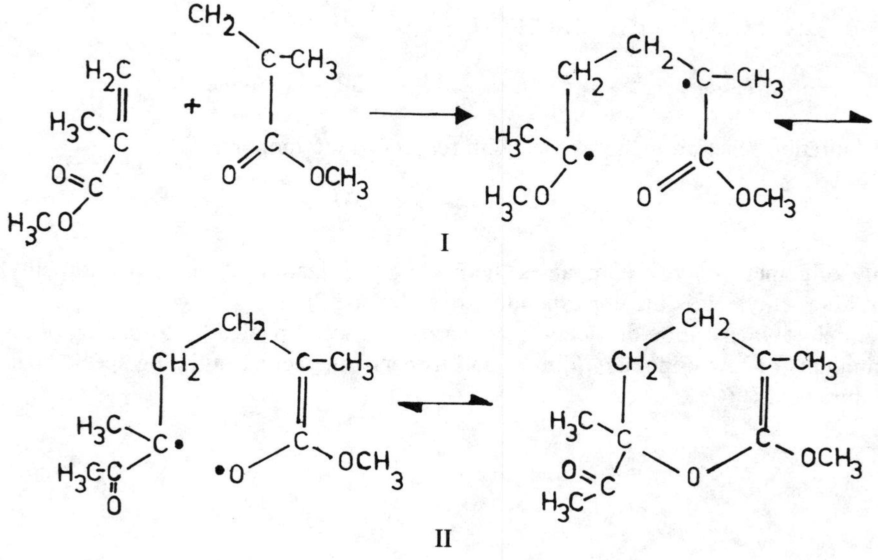

I

(113)

II

The formation of high polymers, polyMMA, is initiated by dimer biradicals. The linear dimer may further react with a monomer molecule to give three types of linear trimers:

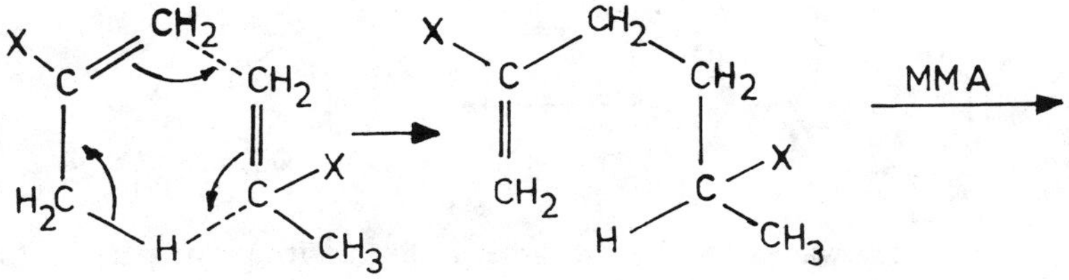

$$\tag{114}$$

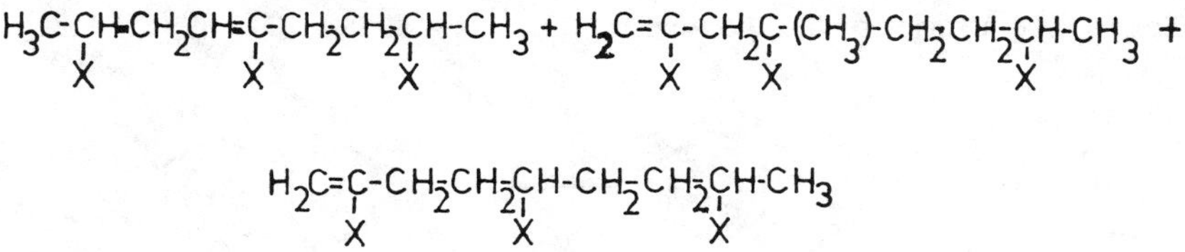

$$H_3C\text{-}CH\text{-}CH_2CH\text{=}C\text{-}CH_2CH_2CH\text{-}CH_3 + \underset{2}{}H_2C\text{=}C\text{-}CH_2\overset{}{C}\text{-}(CH_3)\text{-}CH_2CH_2CH\text{-}CH_3 +$$
$$\qquad\ \, X\qquad\quad X\qquad\quad X\qquad\qquad X\qquad X\qquad\qquad\qquad X$$

$$H_2C\text{=}C\text{-}CH_2CH_2CH\text{-}CH_2CH_2CH\text{-}CH_3$$
$$\qquad X\qquad\quad X\qquad\quad X$$

where X = $-C(O)-CH_3$.

The rate of homopolymerization of H-1 (Equation 111) is very low even in the presence of initiating radicals. The transfer constant to monomer H-1 is about $C_{H-1} = 0.003$ at 80°C and the termination is essentially by disproportionation. It plays no role in initiating polymerization of MMA.

Thermal oligomerization of metallic salts of methacrylic acid was investigated in the temperature range of 150 to 240°C in the solid state.[80] Sodium methacrylate gave a large amount of polymers and a small amount of oligomer (about 3.5%), while calcium, barium, or strontium methacrylates gave a large amount of dimers (about 33%) and trimers (about 23%). The identified dimers were (1) α-methylene-γ,γ-dimethylglutarate and (2) α-methylene-δ-methyladipate, and the trimers were (3) 1-nonene-2,5,8-tricarboxylate and (4) 4-methyl-1-octene-2,4,7-tricarboxylate, all of linear form.

The occurrence of individual species was dependent on the monomer and the reaction temperature. For example, calcium methacrylate gave dimers 1 and 2 below 190°C (but 1 was not formed above 210°C) and trimer 3 above 170°C, while barium methacrylate gave both dimers above 170°C and trimers 3 and 4 above 190°C.

Thermopolymerization of acrylic acid also supposes an initiating biradical species. Since the acroylhydroacrylic acid, the unsaturated dimer, was identified as the main low oligomer species:[62]

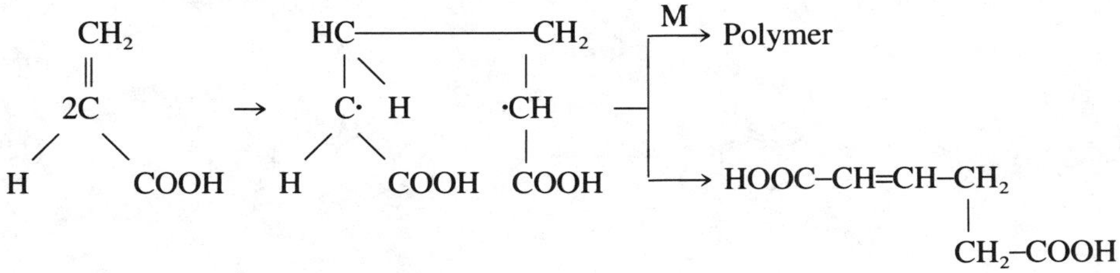

$$\tag{115}$$

On the other hand, the thermal polymerization of vinyl acetate is started by an activated monomer molecule and the rate of initiation was found to be third order with respect to the vinyl acetate concentration (between 80 and 110°C):[62]

$$R_i = 4.8 \times 10^{18} e^{-167.2/RT} M^3 \tag{116}$$

α,β-Unsaturated aldehydes and ketones undergo a Diels-Alder type of reaction to produce cyclic dimers. Methacrylaldehyde easily forms 2-formyl-2,5-dimethyl-3,4-dihydropyran, even at moderate temperature:[81]

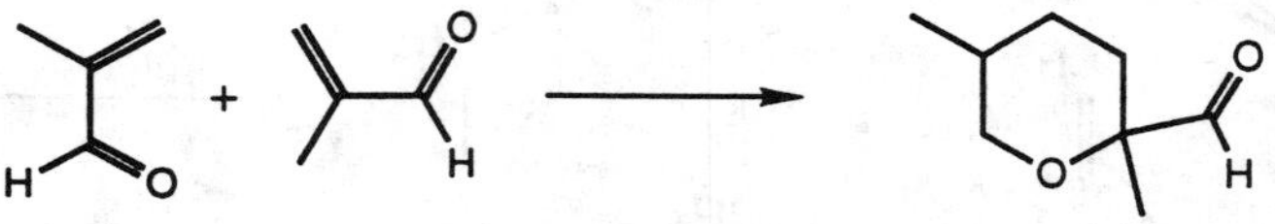

(117)

Practically no tendency was shown by the dimer to form a trimer by participating as a "dienophile", and it retards the formation of higher oligomers because of allylically activated hydrogen atoms in the 4-position and in the 5-methyl group.

The acrolein dimer, 2-formyl-2,3-dihydropyran, however, readily reacts with monomer molecules to give a series of low polymers:[82]

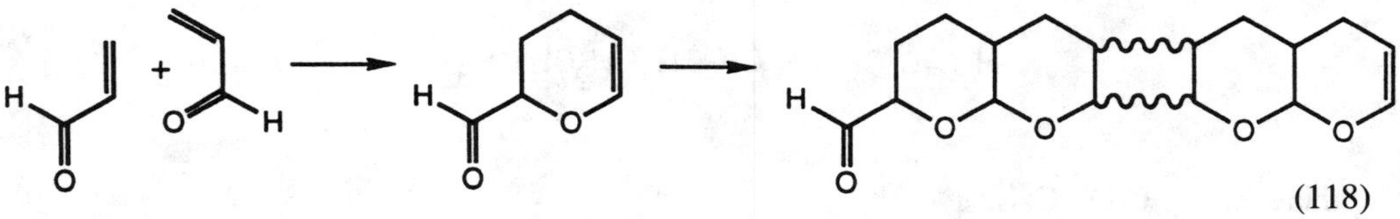

(118)

In the presence of inhibiting hydroquinone (perhaps large amounts), the dimer was the sole reaction product of acrolein. The activation energy for thermal polymerization of this monomer jumped from 65.4 kJ/mol in the absence of inhibitor to 118.8 kJ/mol when nitrogen-containing inhibitors (unspecified) were present, indicating the decisive role of the dimer in the process of polymerization.[83]

Vinyl ketones polymerize readily in bulk even at room temperature.[84] The mechanism is not entirely radicalic, since high amounts of hydroquinone did not prevent polymerization of acrylophenone on standing.[85] Chandhuri[86] has reported a detailed study of the thermally initiated polymerization of methyl isopropenyl ketone, in bulk and in solutions at 80°C. The author observed an acceleration in rate at the onset of polymer precipitation due to the preferential absorption of monomer at the active radical end. This prevented chain transfer and consequently increased the DP and the overall reaction rate. The order of reaction with respect to monomer was less than 2 in homogeneous systems, and greater than 2 in heterogeneous systems.

The role of oligomeric radicals has been evidenced in thermal polymerization of α,β-unsaturated ketones. Only Diels-Alder dimers were isolated in the case of radically homopolymerizable vinyl ketones, e.g., α-methylacrylophenone,[87] which was shown to have the expected head-to-head structure by nuclear magnetic resonance (NMR) analysis:

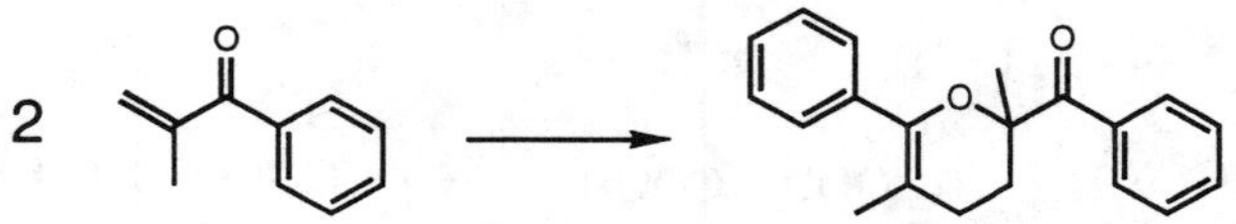

(119)

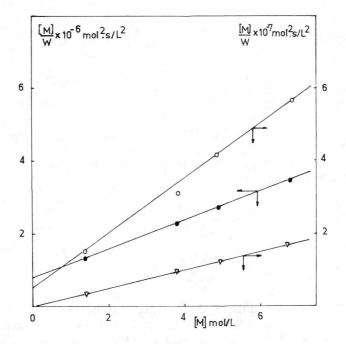

FIGURE 6. The observed dependence upon the monomer concentration of the inverse rate constants for oligomerization of phenyl acetylene (●), formation of 1,2,4-triphenylbenzene (○), and formation of 1,3,5-triphenylbenzene at 140°C. (Adapted from Bantsyrev, G. I., Shcherbakova, I. M., Cherkoshyn, M. I., Kalykhman, I. D., Tschigir, A. N., and Berlin, A. A., *Izv. Akad. Nauk S.S.S.R. Ser. Khim.*, 8, 1762, 1970 [in Russian].)

Thermal polymerization of acetylenic monomers results practically only in low molecular ﹢lymers ($\overline{M}_n = 10^3$).[88-90] The highest molecular mass of fractionated thermal poly(phenyl ﹢etylene) was 1200.[90] Substituted aromatic trimers account for up to 10% of reaction ﹢oducts. Phenyl acetylene gives both 1,3,5- and 1,2,4-triphenyl benzenes (TPhB), with a ﹢tio of isomers varying from 85/15 (at 135°C) to 75/25 (at 175°C), and the $\overline{M}_n$ of unfrac-﹢nated polymers does not exceed 700 Da.[88-90]

In the case of 3-ethynyl phenyl ether, it was apparent that $\overline{M}_n$ and $\overline{M}_w$ were insensitive both conversion and temperature.[88,89] The reactions are initiated by a diradical from two ﹢onomers which requires the lowest energy of activation. Thermal oligomerization of phe-﹢lacetylene and 2-methyl-5-ethynyl pyridine is second order and dependent on monomer ﹢ncentration. The following empirical equation proposed by Bantsyrev et al.[88] correlates ﹢ry well with the experimental findings.

$$\frac{[M]^3}{w} = a + bM \tag{120}$$

﹢here w is the rate (velocity) of the process and a and b are constants (Figure 6).

The authors supposed that the mechanism of encatenation implies an intermediary product (unidentified) which gives the initiating radicals R through a macromolecular reaction, or ﹢meric TPhB, through a bimolecular process:

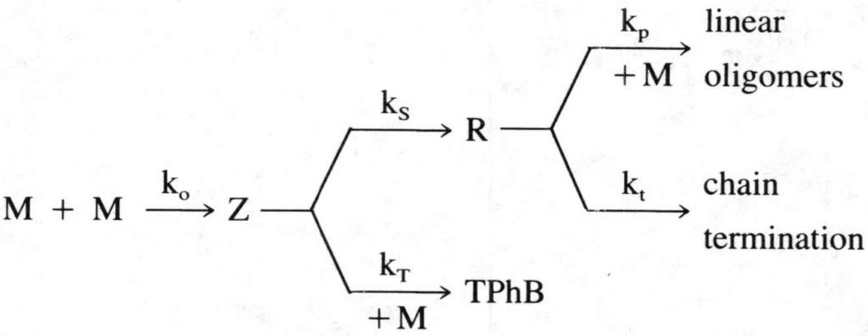

$$M + M \xrightarrow{k_o} Z \left\{ \begin{array}{l} \xrightarrow{k_S} R \left\{ \begin{array}{l} \xrightarrow[+M]{k_p} \text{linear oligomers} \\ \xrightarrow{k_t} \text{chain termination} \end{array} \right. \\ \xrightarrow[+M]{k_T} \text{TPhB} \end{array} \right. \tag{121}$$

With the usual steady state assumption, the rate of oligomerization, R_o, and the rate of trimerization, R_T, may be expressed as

$$R_o = \frac{k_o k_p k_S [M]^3}{k_t(k_S + k_T)[M]} \tag{122}$$

$$R_T = \frac{k_o k_T [M]^3}{k_S + k_T [M]} \tag{123}$$

Equation 122 coincides practically with empirical relation 120 proposed by the same authors.[88] The rate constant k_o, determined from the linear dependence shown in Figure 6, was 1.5×10^{-7} l/mol/s for the formation of 1,2,4-TPhB and 5×10^{-7} l/mol/s for the formation of 1,3,5-TPhB.

The biradical nature of initiating species was proved by ESR measurements.[90]

Thermal polymerization of vinyl acetylene gave only the dimer corresponding to the biradical formed by vinylic encatenation, i.e., diethynyl cyclobutane, while the higher oligomers have a linear polyenic structure.[62,91]

The oligomers formed during thermal polymerization of conjugated dienes are the result of 2 + 4 cyclization processes. Both their nature and concentration depend on the monomer structure.

Thermal polymerization of butadiene is accompanied by formation of 1,2-divinyl cyclobutane, 1-vinyl cyclohexene-3, and 1,5-cyclooctadiene,[92] while that of isoprene is accompanied by formation of the following cyclic dimers of four, six, or eight members (R is –CH$_3$):[93,94]

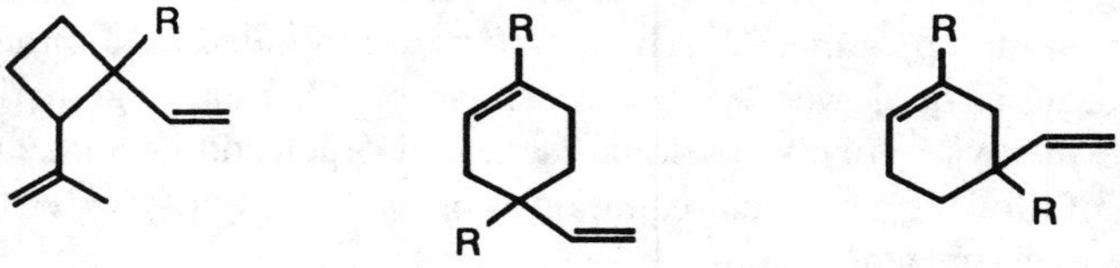

I   II   III

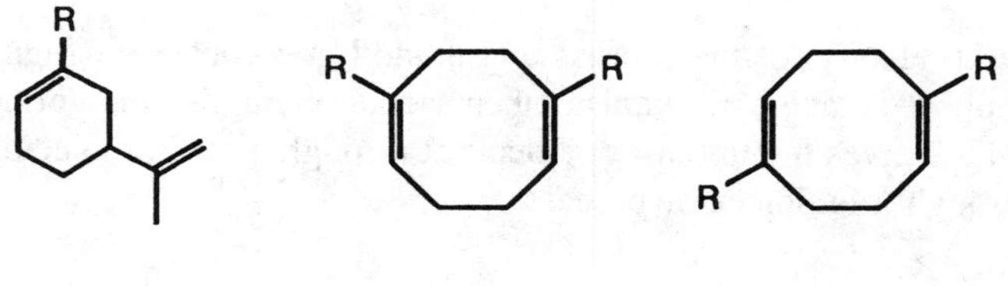

IV   V   VI   (124)

cis-Piperylene (1,3 pentadiene) forms, by thermal oligomerization, only six- and eight-membered cyclic dimers, while 80% of the dimers formed by thermal polymerization of the *trans* isomer are substituted cyclobutanes.[94]

Chloroprene polymerizes thermally at relatively low temperatures (35°C), giving a series of cyclic dimers of types I to VI (Equation 124), where R is Cl and linear *cis*-polychloroprene.[95-97] At high temperatures (95°C), the four- and eight-membered dimers isomerize in vinyl cyclohexene derivatives of types II and III (Equation 124).[97]

Diisopropenyl gives, by thermal oligomerization, only the six-membered dimethyl-4,6-dimenthen.[62]

Cyclopentadiene does not polymerize thermally. It forms only dimers and some trimers by the reaction of dimers with cyclopentadiene. Cyclohexadienes give both dimers and thermal polymers.[98]

The formal encatenation mechanism of conjugated dienes in which thermal dimers, D, or higher oligomers and polymers, P, are formed simultaneously can be written as:[62]

$$M + M \xrightarrow{k_o} Z \begin{cases} \xrightarrow{k_S} R^{\cdot} \\ \xrightarrow{k_D} D \end{cases} \tag{125}$$

The rate of thermal initiation is accordingly:

$$R_i = k_o k_S [M]^2/(k_S + k_D) \tag{126}$$

Then, taking $k_S \ll K_D$ and adopting the mean experimental values for the activation energy of dimerization $E_D \simeq 84$ to 92 kJ/mol and frequency factor $A_D \simeq 10^6$ to 10^7 l/mol/s, it follows that $E_s - E_D \simeq 0$ and $A_S/A_D \simeq 10^{-6}$ to 10^{-4}. The order of the initiation reaction is 2.5, and the rate constant k_S actually characterizes a bimolecular process.[62]

III. TELOMERIZATION

Telomerization is a special case of free radical reactions.[99] These reactions involve the formation of short-chain polymers *(telomers)* by an addition polymerization reaction which is prematurely terminated by a chain-transfer reactions.[100] In most cases of telomerization reactions of a monomer M, termed *taxogen,* the chain transfer occurs on an active transfer agent molecule (usually the solvent), XY, termed *telogen,* and both the telogen and taxogen radicals are reactive enough to reinitiate another addition reaction and maintain the life of the kinetic chain:

$$XY + nM \xrightarrow{\text{initiator } (I_2)} X-(M)_n-Y \tag{127}$$

The terms taxogen, telogen, and telomer were introduced by Petersen and Webster in 1946.[100] The telomers produced invariably contain a distribution of low oligomers (usually n = 1 to 10) to polymer molecules of different sizes.

Telomerization can be initiated either by free radicals, produced chemically by thermal decomposition of initiators (peroxides, peracids, or azo derivatives)[101] or physically by irradiation[102] or by redox processes first investigated by Asher and Vofsi,[103-105] and developed by Boutevain and Piétrasanta.[106-108] Redox systems generally consist of a metal salt (Fe^{3+} or Cu^{2+}) and a reducing agent such as amines or benzoin. Telogens include CCl_4 or CCl_3Br

and trichloromethyl derivatives such as CCl_3-R, some of which are themselves monoadducts obtained by telomerization between CCl_4 and an unsaturated compound.[109]

The degree of polymerization of the telomers largely depends on the system chosen and the experimental conditions. Free radical and redox systems differ very much in this respect.[110] Using redox systems, it is generally easier to obtain adducts or well-defined oligomers which can be separated by fractional distillation. Using free radical initiation, macromolecules with much higher molecular mass are formed. The molecular mass may sometime be controlled by proper choice of the molar ratio of taxogen (monomer) to telogen (solvent) and of reaction temperature. Structural and polar effects in taxogen molecules may play a decisive role in the competition, *simple addition of telogen/higher telomer formation,* as well as in the rate of the overall addition reaction. Thus, substituents stabilizing the radicals derived from taxogen also increase its reactivity, so the tendency to add another taxogen molecule does not change greatly with structure. On the other hand, resonance stabilization of radicals decreases the rate of the displacement reaction, so telomer formation is favored. That is why ethylene, vinyl acetate, and other monomers which polymerize well alone but react via highly reactive, unstabilized radicals give 1:1 products in many systems. In contrast, styrene and methyl acrylate usually give only telomers of large $\overline{DP}_n$. Nonterminal olefins, in which polymerization is restricted by steric hindrance, show little tendency to form telomeric products, although the 1:1 addition may not occur in high yield.

The mechanism of free radical telomerization may be represented by the following equations:[110]

Initiation

$$I_2 \xrightarrow{k_d} 2I^{\cdot} \tag{128}$$

$$I^{\cdot} + XY \longrightarrow IY + X^{\cdot} \tag{129}$$

$$X^{\cdot} + M \longrightarrow XM^{\cdot} \tag{130}$$

Propagation

$$XM_n^{\cdot} + M \rightarrow XM_{n+1}^{\cdot} \tag{131}$$

Chain transfer

$$XM_n^{\cdot} + XY \xrightarrow{k_{tr}} X-M_n-Y + X^{\cdot} \tag{132}$$

Termination

$$X^{\cdot} + X^{\cdot} \xrightarrow{k_t} X-X \tag{133}$$

$$XM_n^{\cdot} + R^{\cdot} \longrightarrow XM_nR \tag{134}$$

where $R^{\cdot}$ denotes any free radical species present in the system.

Redox telomerization catalyzed by transition metal (Me) salts, such as $FeCl_3$, and reducing amines or benzoin may be described as follows:[103,110]

Initiation

$$XY + Me^{m+} \xrightarrow{k_i} [Me(Y)]^{(m+1)+} \tag{135}$$

where M^{m+} is the metallic ion, but not in its highest oxidation state, e.g., Fe^{2+}.

$$X^{\cdot} + M \rightarrow XM^{\cdot} \tag{136}$$

Propagation

$$XM^{\cdot} + M \xrightarrow{k_p} XM_2^{\cdot}$$

$$\ldots\ldots\ldots\ldots\ldots\ldots\ldots$$

$$XM_n^{\cdot} + M \longrightarrow XM_{n+1}^{\cdot} \tag{137}$$

Chain transfer

$$XM_n^{\cdot} + XY \xrightarrow{k_{tr}} X-M_n-Y + X^{\cdot} \tag{138}$$

Termination

$$XM_n^{\cdot} + Me(Y)^{(m+1)+} \rightarrow X-M_n-Y + Me^{m+} \tag{139}$$

A typical free radical telomerization reaction is that between styrene and bromotrichloromethane, in the latter as a solvent, initiated by a peroxide:[111]

Initiation

$$ROOR \xrightarrow{k_d} 2RO^{\cdot} \tag{140}$$

$$RO^{\cdot} + BrCCl_3 \longrightarrow ROBr + Cl_3C^{\cdot} \tag{141}$$

$$\tag{142}$$

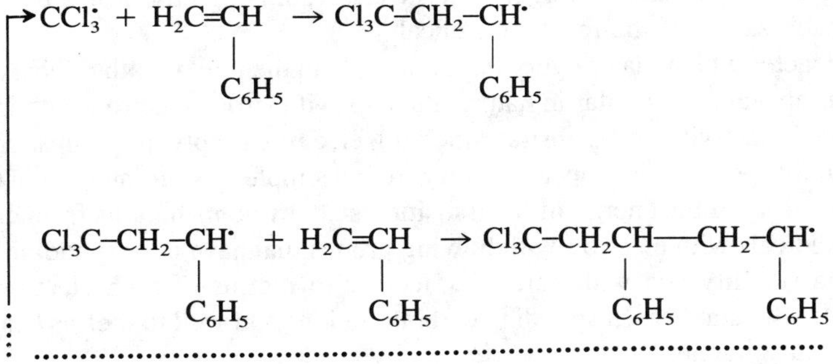

Propagation

$$Cl_3C-(CH_2-CH-)_{n-1}-CH_2CH^{\bullet} \underset{C_6H_5}{\overset{}{|}} \underset{C_6H_5}{\overset{}{|}} + CH_2=CH \underset{C_6H_5}{\overset{}{|}} \xrightarrow{k_p} \tag{143}$$

$$Cl_3C-(-CH_2-CH-)_n-CH_2CH^{\bullet} \underset{C_6H_5}{\overset{}{|}} \underset{C_6H_5}{\overset{}{|}}$$

Chain transfer

$$Cl_3C-(-CH_2-CH-)_n-CH_2-CH^{\bullet} \underset{C_6H_5}{\overset{}{|}} \underset{C_6H_5}{\overset{}{|}} + BrCCl_3 \xrightarrow{k_{tr}}$$

$$Cl_3C-(-CH_2-CH-)_{n+1}Br + {}^{\bullet}CCl_3 \underset{C_6H_5}{\overset{}{|}} \tag{144}$$

Termination

$$Cl_3C^{\bullet} + Cl_3C^{\bullet} \rightarrow Cl_3C-CCl_3 \tag{145}$$

Actually, the propagation step consists of two types of reactions, addition (Equations 137 and 143) and abstraction or chain transfer (Equations 138 and 144).

In a series of free radical telomerizations in which the carbon radical X⋅ was held constant and Y was changed, reactivity generally increased in the sequence H < Cl < Br < I.[112] Thus, Cl_3C-Br undergoes addition more readily than Cl_3C-Cl, and F_3C-I more readily than F_3C-Br. The differentiation between Cl− and H− is less clear-cut and depends to some extent on the particular system involved. Thus, chloroform is usually less reactive than carbon tetrachloride and adds as Cl_3C-H. Bromoform is more reactive and adds as Br_2HC-Br. Steric hindrance affects free radical telomerizations, particularly in the addition step, and seems to be less important in the displacement step (chain transfer). In fact, the usual direction of addition probably is largely a steric effect, although it is also aided by the greater resonance stabilization of the resulting radical.

Another factor which plays a very important role in the rate of both radical displacements and additions appears to be polar in nature: radicals with strong electron-withdrawing groups show enhanced reactivity with substrates bearing electron-supplying groups, and vice versa. The nature of this polar effect appears to vary from a simple dipolar interaction to, in extreme cases, the lowering of the energy of the transition state by contributions from charge-transfer structures. Radicals with electron-withdrawing groups that have corresponding negative ions of reasonable stability add with particular facility to olefins of high electron availability, while radicals with stable corresponding carbonium ions add well to olefins bearing electron-withdrawing substituents.[112]

The average value of the degree of polymerization, $\overline{DP}_n$, of the telomers produced will depend primarily upon two factors: the molar ratio of monomer to telogen and the ratio of the rate constants for addition and subtraction, k_p/k_{tr}.

TABLE 2
Variation of Chain Transfer Constants for Different
Pairs of Monomer/CCl₄ with Chain Length[112]

Monomer	Temperature (°C)	C_T		
		n = 1	n = 2	n
Ethylene	70	0.08	1.9	3.2[a]
Propylene	100	1.3	—	5.1
Isobutylene	100	1.4	—	17 ± 3
Allyl chloride	100	0.01	0.1	0.5
Allyl acetate	100	0.01	0.5	2.0
Styrene	76	0.0006	0.0025	0.012

[a] This is the value for n = 3.

The reciprocal of Equation 29, applied for telomerization, will give, taking into account Equation 17 and the fact that the kinetic chain length is equal to the degree of polymerization:

$$\frac{1}{DP_n} = \frac{2(f \cdot k_d \cdot k_t)^{0.5}}{k_p} \cdot \frac{[I_2]^{0.5}}{[M]} + C_T \frac{[T]}{[M]} \qquad (146)$$

where T represents the concentration of the telogen XY, DP_n denotes the degree of polymerization n, and

$$C_T = \frac{k_{tr}}{k_p} \qquad (147)$$

In general, Equation 146 does not apply for n below a value of 5 because C_T in most cases is not a constant, but varies greatly with the degree of encatenation in going from the 1:1 adduct to the 5:1 adduct (Table 2).[112] This is because the group located at the end of the chain has a certain influence on the behavior of the growing radical.

Some telomerization reactions have shown a maximum or minimum in the variation of C_T with n.[111] For styrene/BrCCl₃ at 35°C, for example, C_T varies from 7.7 to 280 to 30 in going from the dimeric (2:1 adduct) to the trimeric (3:1 adduct) to the pentameric (5:1 adduct) radicals.[112] With $C_T < 1$, high yields of inferior telomers can be obtained with a small excess of XY. In principle, good yields can also be obtained in systems where $C_T < 1$ by working at very high ratios of XY/M, or by adding the monomer slowly during the reaction, although kinetic chains may be short, thus requiring a relatively large amount of initiator. Since small transfer constants generally increase with temperature, better yields of higher adducts may also be obtained at lower temperatures.

Transfer constants vary with structure in the manner suggested by the previous discussion. In the telogen XY, they increase with changes in Y from H < Cl < Br < I and with substituents in X that increase its resonance stabilization, since all these changes increase the rate of the abstraction. Thus, Cl₃CBr gives good yields of 1:1 products in reactions with equimolecular quantities of most taxogens, while CCl₄ must be used at high telogen/taxogen ratios.

Polyhalomethanes with less than three halogens in the molecule generally give low yields of 1:1 products unless the methanes are further activated by nitrile, carbonyl, or similar groups.[112]

In order to calculate the real degree of polymerization or cumulative degrees of poly-

TABLE 3
The Change of the Degree of Polymerization during Telomerization of Vinyl Acetate in Chloroform at 60°C[115]

	Monomer conversion (%)					
	0	20	40	60	80	100
Intervals (i)	1	2	3	4	5	6
$(DP_n)_j$	65	52	40	27	13	1
$(\overline{DP}_n)_{cum}$	65	57.8	50.3	41.4	28.8	5.1
$(\overline{DP}_w)_{cum}$	65	58.3	55.8	51.8	46.9	40.8
Polymolecularity[a]	1	1.01	1.11	1.26	1.63	8

[a] $(\overline{DP}_w)_{cum}/(\overline{DP}_n)_{cum}$

merization, $(DP_n)_{cum}$, one has to know the laws of variation in terms of the concentration of the initiator,[113]

$$I_2 = [I_2]_o \cdot e^{-k_d \cdot t} \tag{148}$$

the monomer[114]

$$\log \frac{[M]_o}{[M]} = 2k_p \left(\frac{f \cdot [I_2]_o}{k_d k_t} \right)^{0.5} (1 - e^{-(k_d \cdot t)/2}) \tag{149}$$

and the telogen:[115]

$$\log \frac{[T]_o}{[T]} = C_T \cdot \log \frac{[M]_o}{[M]} \tag{150}$$

The cumulative number-average degree of polymerization can be written as:

$$(\overline{DP}_n)_{cum} = \frac{i}{\sum\limits_{j=0}^{j=i-1} \frac{1}{(DP_n)_j}} = \frac{[M]_o - [M]}{[T]_o - [T]} \tag{151}$$

where $(DP_n)_j$ are instantaneous degrees of polymerization calculated for a certain time according to Equation 146. The cumulative weight-average degree of polymerization is accordingly:[115]

$$(DP_w)_{cum} = \sum\limits_{j=0}^{j=i-1} \frac{(DP_n)_j}{i} \tag{152}$$

where i represents the number of intervals chosen for monomer conversion. Application of the last two equations is shown in Table 3 for the telomerization of vinyl acetate in chloroform at 60°C initiated by benzoyl peroxide. The instantaneous degree of polymerization for this particular telomerization is given by Equation 153, which was obtained by inserting in Equation 146 the values of the corresponding constants:

$$\frac{10^3}{(DP_n)_j} = 5.8 \frac{[I_2]^{0.5}}{[M]} + \frac{[CHCl_3]}{[M]} \tag{153}$$

TABLE 4
Transfer Constants in Redox Telomerization of
Certain Taxogen/Telogen/Catalyst Systems[110]

Telomerization system	C_{Me}	C_T
$CH_2=CFCl/CCl_4/FeCl_3$	$75(Fe^{2+})$	1.3×10^{-2}
$CH_2=CH-COOH/CCl_4/FeCl_3$	$1.5(Fe^{2+})$	10^{-4}
$CH_2=CH-COOC_2H_5/CCl_4/FeCl_3$	$8(Fe^{2+})$	10^{-4}
$CH_2=CH-COOC_2H_5/CCl_4/CuCl_2$	$800(Cu^{2+})$	10^{-4}

In the redox telomerization, the following general equation can be derived by an analogous treatment:[110]

$$\frac{1}{(DP_n)_j} = C_{Me} \frac{[Me^{(m+1)}]}{[M]} + C_T \frac{[XY]}{[M]} \tag{154}$$

where $C_{Me} = k_t/k_p$.

The variation with time of telogen (XY or T) and taxogen (M) concentrations can be written as:

$$[T] = [T]_o \cdot e^{(\phi-1) \cdot f \cdot k_i [Me^{(m+1)+}]_o t} \tag{155}$$

and

$$\log \frac{[M]_o}{[M]} = \frac{[T]_o \cdot k_p}{\phi \cdot k_t [Me^{(m+1)+}]_o} \, 1 \, - \, e^{(\phi-1) \cdot f \cdot k_i [Me^{(m+1)+}]_o t} \tag{156}$$

where $\phi = [Me^{(m+1)+}]/[Me^{(m+1)+}]_o$

Boutevin et al.[116] have established that in the reaction which bring redox catalysts into play, the influence of the metallic ion is much more important than that of the telogen. This fact can actually be seen in the transfer constant values C_{Me} and C_T of different telomerization systems (Table 4).

If F_n is defined as the molar fraction of the telomer having a $\overline{DP}_n$ equal to n, compared with telomers, one obtains:

$$F_n = \frac{v_{T_n}}{\sum_i v_{T_i}} \tag{157}$$

where V_{T_i} is the rate of formation of the telomer of $DP_n = i$. For $n = 1$, one has

$$v_{T_1} = k_{tr}^1 [T_1^\cdot][XY] \tag{158}$$

and

$$\Sigma v_{T_i} = k_{tr}^1 [T_1^\cdot][XY] + k_p^1 [T_1^\cdot][M] \tag{159}$$

where $[T_1^\cdot] = [YM^\cdot]$, while $k_p^1 [T_1^\cdot][M]$ is the formation rate of all telomers for which n is higher than 1. By saying $C_T^1 = k_{tr}^1/k_p^1$ and combining Equations 157 and 158, one obtains

$$F_1 = \frac{C_T^1 \dfrac{[XY]}{[M]}}{C_T^1 \dfrac{[XY]}{[M]} + 1} \cdot \tag{160}$$

TABLE 5

**Transfer Constants of CTFE Radicals to Telogen (C_T^n) and to
Metallic Salt(C_{Me}^n) at the Telomerization of
Chlorotrifluoroethylene with CCl_4 and $FeCl_3$/Benzoin System
in Acetonitrile at 110°C[107]**

$CCl_3(CF_2CFCl)_nCl$	Degree of polymerization (n)							
	1	2	3	4	5	6	7	8
$10^3 \cdot C_T^n$	12	20	20	20	22	22	50	100
C_{Me}^n	50	78	90	94	96	108	108	147

which leads to the general formula

$$F_n = \frac{C_T^n \dfrac{[XY]}{[M]}}{\displaystyle\int_{i=1}^{i=n} \left(C_T^i \dfrac{[XY]}{[M]} + 1 \right)} \tag{161}$$

first proposed by Walling.[117]

A similar expression was derived by Boutevin and Piétrasanta[110] for redox telomerization:

$$F_n = \frac{\phi \cdot C_{Me}^n \cdot C_o}{\displaystyle\int_{i=0}^{i=n} (1 + \phi \cdot C_{Me}^i \cdot C_o)} \tag{162}$$

where $C_o = [Me^{(m+1)+}]_o/[M]_o$, and C_{Me}^i is the transfer constant to the metallic ion and refers to the formation of the telomer $DP_n = i$, i.e., $C_{Me}^i = k_t^i/k_p^i$.

The transfer constants are related to F_n according to Equation 163, and both C_T^n and C_{Me}^n increase with n, but since $C_t^n < C_{Me}^n$, the termination is caused essentially by the metallic ion intervention:[107]

$$\frac{F_n}{\displaystyle\int_{i=n+1}^{i=\infty} F_i} = C_{Me}^n \frac{[XMe^{(m+1)+}]}{[M]} + C_T^n \frac{[XY]}{[M]} \tag{163}$$

Table 5 presents the transfer constants of chlorotrifluoroethylene (CTFE) radicals to telogen (CCl_4) and to $FeCl_3$ for different degree of polymerization. The ferric chloride/benzoin mixture in solution in acetonitrile provided the ferrous ions (Fe^{2+}) necessary for the redox telomerization of this particular system.[116]

Since $C_{Me} > C_T$, the molecular mass of telomers obtained by redox telomerization is much lower than that obtained when the same telomerization is initiated by free radicals.

Practically any monomer which can be polymerized radically can be telomerized. Monoadducts are, however, the only products when the taxogen is of the allylic type, irrespective of the kind of telomerization (free radical or redox), and the reactions are quantitative.

A large variety of derivatives can be used as telogens. A certain taxogen/telogen couple may, however, give different telomers since the telogen fragments ending the telomer molecule may vary, depending on the initiator or catalyst used in telomerization, e.g., Equation 164 vs. Equation 165, and Equation 166 vs. Equation 170. The following telomerizations of chlorotrifluoroethylene[110,118-122] exemplify the type of bond broken in the telogen molecule together with the large variety of oligomers (n = 1 to 20) obtainable starting from only one monomer.

Breaking of C–F bond

$$Cl_3C-F + nCF_2=CFCl \xrightarrow{AlCl_3} Cl_3C(CF_2-CFCl)_nF \qquad (164)$$

Breaking of C–Cl bond

$$FCl_2C-Cl + nCF_2=CFCl \xrightarrow{CuCl} FCl_2C(CF_2-CFCl)_nCl \qquad (165)$$

$$HCl_2C-Cl + nCF_2=CFCl \xrightarrow{FeCl_3} HCl_2C(CF_2-CFCl)_nCl \qquad (166)$$

$$Cl_3C-Cl + nCF_2=CFCl \xrightarrow{RhCl_2, P(C_6H_5)_3} Cl_3C(CF_2-CFCl)_nCl \qquad (167)$$

Breaking of C–Br bond

$$Cl_3C-Br + nCF_2=CFCl \xrightarrow{I_2} Cl_3C(CF_2-CFCl)_nBr \qquad (168)$$

Breaking of C–I bond

$$C_6F_{13}-I + nCF_2=CFCl \xrightarrow{I_2} C_6F_{13}(CF_2-CFCl)_nI \qquad (169)$$

Breaking of C–H bond

$$Cl_3C-H + nCF_2=CFCl \xrightarrow{I_2} Cl_3C(CF_2-CFCl)_nH \qquad (170)$$

Breaking of S–H bond

$$HO(CH_2)_2S-H + nCF_2=CFCl \xrightarrow{I_2} HOCH_2CH_2S(CF_2-CFCl)_nH \qquad (171)$$

Breaking of P–H bond

$$(H_5C_2O)_2P(O)-H + nCF_2=CFCl \rightarrow (H_5C_2O)_2P(O)(CF_2-CFCl)_nH \qquad (172)$$

Functional polymers have been prepared by telomerization with functional telogens and subsequent reactions of the terminal functions with an unsaturated reagent.[101] For example, CTFE was reacted in acetonitrile with carbon tetrachloride in the presence of the Fe^{3+}/ benzoin redox system (Equation 166), whereby oligomers with degrees polymerization ranging from 1 to 20 were obtained. These oligomers were thereafter used as telogens with allyl alcohol or allyl acetate in order to form monoadducts:

$$Cl-CCl_2(CF_2-CFCl)_n + HOCH_2CH=CH_2 \rightarrow$$

$$\underset{\underset{Cl}{|}}{HOCH_2CH}-CH_2-CCl_2(CF_2-CFCl)_nCl \qquad (173)$$

To fit these ω-hydroxy telomers with terminal double bonds, they can be esterified with acryloyl chloride:

$$CH_2=CH-C-Cl \ + \ HOCH_2CHClCH_2CCl_2(CF_2-CFCl)_nCl \ \rightarrow$$
$$\| $$
$$O$$

$$CH_2=CH-COCH_2CHClCH_2CCl_2(CF_2-CFCl)_nCl \qquad\qquad (174)$$
$$\|$$
$$O$$

The oligomers (macromers, $\overline{M}_n < 10^3$) formed in Equation 174 copolymerize readily with the usual vinyl monomers.

REFERENCES

1. **Barton, J. and Borsig, E.,** *Complexes in Free-Radical Polymerization,* Elsevier, New York, 1988, chap. 1.
2. **Liakumovich, A. V.,** Russian Patent 952.865, 1982.
3. **Hiemenez, P. C.,** *Polymer Chemistry — The Basic Concepts,* Marcel Dekker, New York, 1984, 7.
4. **Lenz, R. W.,** *Organic Chemistry of Synthetic High Polymers,* Interscience, New York, 1967, chap. 10.
5. **Bevington, J. C.,** *Radical Polymerization,* Academic Press, New York, 1961, 148.
6. **Brandup, J., Immergut, E. H., and McDowell, W.,** Eds., *Polymer Handbook,* 2nd ed., John Wiley & Sons, New York, 1975, chap. 2.
7. **Mayo, R. F.,** Chain transfer in the polymerization of styrene: the reaction of solvents with free radicals, *J. Am. Chem. Soc.,* 65, 2324, 1943; **Flory, P. J.,** The mechanism of vinyl polymerizations, *J. Am. Chem. Soc.,* 59, 241, 1937.
8. **Baysal, B. and Tobolski, A. V.,** Rate of initiation in vinyl polymerization, *J. Polym. Sci.,* 8, 529, 1952.
9. **Tobolski, A. V. and Offenbach, J.,** Kinetic constants for styrene polymerization, *J. Polym. Sci.,* 16, 311, 1955.
10. **Saha, N. G., Nandi, V. S., and Palit, S. R.,** Peroxides as initiators of polymerization of methyl methacrylate, *J. Chem. Soc.,* p. 427, 1956.
11. **Palit, S. R., Nandi, V. S., and Saha, W. G.,** Studies in chain transfer. III. Determination of chain transfer coefficients from catalyzed polymerization data, *J. Polym. Sci.,* 14, 295, 1954.
12. **Simionescu, C. I. and Vasiliu, O. C.,** *Treatise of Chemistry of Macromolecular Compounds,* Vol. 1, EDP, Bucharest, 1973, 450 (in Romanian).
13. **Lipatov, Yu. S., Nesterov, A. E., Gritzenko, T. M., and Veselovskyi, P. A.,** *Handbook of Polymer Chemistry,* Naukova Domka, Kiev, 1971, chap. I/2 (in Russian).
14. **Clarke, J. T., Howard, R. O., and Stockmayer, W. T.,** Chain transfer in vinyl acetate polymerization, *Makromol. Chem.,* 44/46, 427, 1961.
15. **Fuhrman, N. and Mesobian, R. B.,** Chain transfer of vinyl monomers with carbon tetrabromide, *J. Am. Chem. Soc.,* 76, 3281, 1954.
16. **Elias, H. G.,** *Macromolecules. II. Synthesis and Characterization,* John Wiley & Sons, London, 1977, sect. 20.3.
17. **Cantow, M., Meyerhoff, G., and Schulz, G. V.,** Verzweigunggrad und Viskositüatszahl bei Polystyrolen, *Makromol. Chem.,* 49, 1, 1961.
18. **Morton, M. and Purma, I.,** The branching reaction. II. Styrene and methyl methacrylate, *J. Am. Chem. Soc.,* 50, 5596, 1958.
19. **Schultz, G., Henrici, G., and Olivé, S.,** Chain transfer constants of poly(methyl methacrylate) in the polymerization of methyl methacrylate and styrene, *J. Polym. Sci.,* 17, 45, 1955.
20. **Bevington, J. C., Ghanem, N. A., and Melville, H. W.,** The mechanism of retardation and inhibition in radical polymerizations. II. The effect of *p*-benzoquinone upon the sensitized polymerization of styrene, *J. Chem. Soc.,* p. 2822, 1955.
21. **Gaylord, N. G. and Eirich, F. R.,** Allyl polymerization. III. Kinetics of polymerization of allyl esters, *J. Am. Chem. Soc.,* 74, 337, 1952.

22. **Gaylord, N. G.,** Allyl polymerization. IV. Effective chain transfer in polymerization of allylic monomers, *J. Polym. Sci.,* 22, 71, 1956.
23. **Rånby, B.,** Free radical reactions of allyl and methallyc monomers, *Appl. Polym. Symp.,* 26, 327, 1975.
24. **Haman, S. D.,** in *High Pressure Physics and Chemistry,* Vol. 2, Bradley, R. S., Ed., Academic Press, New York, 162.
25. **Allen, P. E. M. and Patrick, C. R.,** *Kinetics and Mechanisms of Polymerization Reactions. Application of Physicochemical Principles,* Ellis Harwood, Chichester, 1974, 27.
26. **Weale, K. E.,** The influence of pressure on polymerization reactions, in *Reactivity Mechanism and Structure in Polymer Chemistry,* Jenkins, A. D. and Ledwith, A., Eds., John Wiley & Sons, London, 1974, 160.
27. **Bailey, W. J.,** Cationic polymerization with expansion in volume, *J. Macromol. Sci. Chem.,* A9, 849, 1975.
28. **Brown, D. W. and Wall, L. A.,** Radiation induced polymerization of propylene at high pressure, *J. Phys. Chem.,* 67, 1016, 1963.
29. **Zharov, A. A., Berlin, A. A., and Enikolopyan, N. I.,** The investigation of polymerization at high pressures, *J. Polym. Sci. Part C,* 16, 2313, 1967.
30. **Walling, C. and Pellon, J.,** Organic reactions under high pressures. II. The polymerization of allyl acetate, *J. Am. Chem. Soc.,* 79, 4782, 1957.
31. **Pryor, W. A. and Pultinas, E. P., Jr.,** Reactions of radicals. II. The rates of spontaneous and induced decomposition of propyl peroxides, *J. Am. Chem. Soc.,* 85, 133, 1963.
32. **Ito, K.,** Evidence for third order initiation in monomers in thermal polymerization of styrene, *Polym. J.,* 11, 877, 1986.
33. **Pinazzi, C., Legeay, G., and Brosse, J. C.,** Synthése par voie radicalaire des polmères à extrémitś hydroxylées. I. Synthése de polyispréne, *Makromol. Chem.,* 176, 1307, 1975.
34. **Pinazzi, C., Legeay, G., and Brosse, J. C.,** Synthése par voie radicalaire des polymères à extremités hydroxylées. II. Synthése de polybutadiene, *Makromol. Chem.,* 176, 2509, 1974.
35. **Pinazzi, C., Legeay, G., and Brosse, J. C.,** Synthése par voie radicalaire des polymères à extrémités hydroxylées. III. Synthése de poly(acetate de vinyle), *Makromol. Chem.,* 177, 2661, 1976.
36. **Pinazzi, C., Legeay, G., and Brosse, J. C.,** Synthése par voie radicalaire de polymères à extrémitées hydroxylées. IV. Polymèrisation et copolymèrisation de butène, *Makromol. Chem.,* 177, 2877, 1976.
37. **Pinazzi, C., Legeay, G., and Brosse, J. C.,** Synthése par voie radicalaire de polymères à extrémitées hydroxylées. V. Polymérisation du méthacrylate de méthyl sous rayonnement ultraviolet, *Makromol. Chem.,* 177, 3139, 1976.
38. **Pinazzi, C., Legeay, G., and Brosse, J. C.,** Synthése par voie radicalaire de polymères à extrémités hydroxylées. VI. Etude des reactions de transfert, *Makromol. Chem.,* 179, 79, 1978.
39. **Pinazzi, C., Legeay, G., and Brosse, J. C.,** Synthése par voie radicalaire de polymères à extrémités hydroxylées VII. Etude des parametres kinetiques, *Makromol. Chem.,* 181, 1737, 1980.
40. **Pinazzi, C., Legeay, G., and Brosse, J. C.,** Synthése par voie radicalaire de polyméres à extrémites hydroxylées VIII. Synthése de polybutadiènes: étude de différents paramètres influençant la polymolécularité, *Makromol. Chem.,* 182, 3457, 1981.
41. **Junghaus, W. and Topisch, H.,** German Patent 61,102, 1968.
42. **Aliev, S. M. and Mamedaliev, V. C.,** British Patent 2,068,005, 1981.
43. **Ceauşescu, E., Corciovei, M., and Donescu, D.,** The chemistry of hydrocarbon resin synthesis, in *Hydrocarbon Resins,* Vol. 1, Academic Editorial House, Bucharest, 1988, 89 (in Romanian).
44. **Guth, W. and Heitz, W.,** Telechelic oligomere, II. Oligoäthylene durch radicalische polymerisation mit AIBN oder Dialkylperoxidicarbonat als Initiator, *Makromol. Chem.,* 177, 1835, 1976.
45. **Guth, W. and Heitz, W.,** Telechele oligomere, III. Oligoäthylene durch radicalische Polymerization mit Diacylperoxiden als Initiator, *Makromol. Chem.,* 177, 3159, 1976.
46. **Bawn, C. E. H. and Ledwith, A.,** Stereoregular addition polymerization, *Q. Rev.,* 16, 361, 1962.
47. **De Schryver, F. C. and Smets, G.,** Interaction of light with monomers and polymers, in *Reactivity, Mechanism and Structure in Polymer Chemistry,* Jenkins, A. D. and Ledwith, A., S., John Wiley & Sons, London, 1974, 446.
48. **Nagahiro, I., Nishihara, K., and Sakota, N.,** Photoinduced polymerization of maleic anhydride in dioxane, *J. Polym. Sci. Polym. Chem. Ed.,* 12, 785, 1974.
49. **Sharabash, M. M. and Guile, R. L.,** Homopolymerization of maleic anhydride, *J. Macromol. Sci. Chem.,* A10, 1017, 1976.
50. **Haimiel, S. M. and Sherrington, D. C.,** Novel quaternary ammonium amphiphilic (meth)acrylates. II. Thermally and photochemically initiated polymerizations, *Polymer,* 28, 332, 1987.
51. **Pryor, W. A. and Lasswell, L. D.,** *Advances in Free Radical Chemistry,* Williams, G. H., Ed., Academic Press, New York, 1975, 27 (and references cited therein).
52. **Willing, C.,** *Free Radicals in Solution,* John Wiley & Sons, New York, 1957, 180.
53. **Mayo, F. R.,** Thermal oligomerization of styrene, *J. Am. Chem. Soc.,* 76, 55, 1955.
54. **Mayo, F. R.,** The dimerization of styrene, *J. Am. Chem. Soc.,* 90, 1289, 1968.

55. **Uglea, C. V.**, Gel permeation chromatography of styrene oligomers, *Makromol. Chem.*, 166, 275, 1973 (and references 5 to 12 cited therein).

56. **Mayo, F. R.**, Chain transfer in the polymerization of styrene. VIII. Chain transfer of brombenzene and mechanism of thermal initiation, *J. Am. Chem. Soc.*, 75, 6133, 1953.

57. **Flory, P. J.**, The mechanism of vinyl polymerization, *J. Am. Chem. Soc.*, 59, 241, 1937.

58. **Kirchner, K.**, Kinetics of the thermal polymerization of styrene, *Makromol. Chem.*, 128, S150, 1969.

59. **Hiatt, R. R. and Bartlett, P. D.**, The thermal reaction of styrene with ethyl thioglycolate; evidence for the termolecular thermal initiation of styrene polymerization, *J. Am. Chem. Soc.*, 81, 1149, 1959.

60. **Ito, K.**, Evidence for third order initiation in monomers in thermal polymerization of styrene, *Polym. J.*, 11, 877, 1986.

61. **Ebdon, J. R.**, Thermal polymerization of styrene—a critical review, *Br. Polym. J.*, 3, 9, 1971.

62. **Kurbatov, V. A.**, Thermal polymerization and oligomerization of monomers, *Usp. Khim.*, 56, 865, 1987 (and references cited therein) (in Russian).

63. **Bengough, W. J. and Park, G. B.**, Thermal polymerization of styrene, *Eur. Polym. J.*, 14, 431, 1976.

64. **Kirchner, K. and Bucholtz, K.**, Zur Oligomerbildung bei der thermischen Styrolpolymerisation, *Angew. Makromol. Chem.*, 13, 127, 1970.

65. **Mihail, R.**, *Kinetic Model of Polymerizations*, Editura Stiinţifică şi Enciclopedică, Bucharest, 1986, chap. 6 (in Romanian).

66. **Bucholtz, K. and Kirchner, K.**, Evidence for an unstable intermediate in thermal styrene polymerization, *Makromol. Chem.*, 177, 935, 1976.

67. **Kauffmann, H. F. and Olaj, O. F.**, Spectroscopic measurements on spontaneously polymerizing styrene. Evidence for the formation of two Diels-Alder isomers of different stability, *Makromol. Chem.*, 177, 939, 1976.

68. **Olaj, O. F., Kauffmann, H. F., and Breitenbach, J. W.**, Thermally initiated oligomerization of styrene, *Makromol. Chem.*, 178, 2707, 1977.

69. **Kauffmann, H. F.**, Untersuchungen zur Thermolyse und zur Sauerstoffstabilität der 4-2 Diels-Alder Addukte des Styrols, *Makromol. Chem.*, 178, 3007, 1977.

70. **Kirchner, K. and Riederle, K.**, Thermal polymerization of styrene. The formation of oligomers and intermediates. I. Discontinuous polymerization up to high conversions, *Angew. Makromol. Chem.*, 111, 1, 1983.

71. **Hang, M., Zhang, X., and Tu, Z.**, Simulation kinetics of oligomer reactions during thermal polymerization of styrene, *Gaoding Xue Xiao Huaxue Xuebao*, 7, 251, 1986 (in Chinese); *Chem. Abstr.*, 106. 50710, 1987.

72. **Hamielec, A. E.**, Recent developments in polymer reactor engineering, in Plenary and Invited Lectures, Part 2, IUPAC Macro '83, Bucharest, September, 1983, 335.

73. **Kauffmann, H. F., Harms, H., and Olaj, O. F.**, Ground-state dynamics of α-methylstyrene. I. Thermally induced oligomerization in bulk, *J. Polym. Sci. Polym. Chem. Ed.*, 20, 2943, 1982.

74. **Yang, K. and Manno, P. J.**, A mechanism for the thermal polymerization of ethylene, *J. Polym. Sci.*, 35, 548, 1959.

75. **Walling, C.**, *Free Radicals in Solution*, John Wiley & Sons, New York, 1957, 80.

76. **Walling, C. and McElhill, E. A.**, The over-all rate of copolymerization of styrene-diethylfumarate, *J. Am. Chem. Soc.*, 73, 2819, 1951.

77. **Stickler, M. and Meyerhoff, G.**, Die thermische Polymerisation von Methylmethacrylate. Bildung des ungesättigen dimeren, *Makromol. Chem.*, 181, 131, 1980.

78. **Brand, E., Stickler, M., and Meyerhoff, G.**, Die thermiche Polymerisation von Methylmethacrylate. Werhalten des ungesättigen Dimeren, bei der Polymerisation, *Makromol. Chem.*, 181, 913, 1980.

79. **Stickler, M. and Meyerhoff, G.**, The spontaneous thermal polymerization of methyl methacrylate. Experimental study and computer simulation of the high conversion reaction at 130°C, *Polymer*, 22, 928, 1981.

80. **Naruchi, K. and Miura, M.**, Thermally initiated oligomerization of metallic salts of methacrylic acid in solid state, *Polymer*, 22, 1716, 1981.

81. **Stoner, G. G. and McNulty, J. C.**, Methacrylaldehyde dimer—derivatives obtained through Canizzaro reaction, *J. Am. Chem. Soc.*, 72, 1531, 1950.

82. **Schulz, R. C.**, Polymere Acroleine. Mitt Untersuchungen Über die Polymerisation des Acroleins, *Makromol. Chem.*, 17, 62, 1955.

83. **Hardin, A. P., Vol'dman, D. I., Derbischer, V. E., Panfilov, B. I., Bukalov, V. I., and Bal'dman, A. I.**, Determination of activation energy of process of thermopolymerization of acrolein from thermokinetic interpretation, *J. Prikladn. Khim.*, 49, 2562, 1976 (in Russian).

84. **Lyons, A. R.**, Polymerization of vinyl ketones, *J. Polym. Sci. Part D*, 6, 251, 1972.

85. **Mulvaney, J. E. and Dillon, J. G.**, Anionic polymerization of acrylophenone, *J. Polym. Sci. Part A*, 6, 1849, 1968.

86. **Chandhuri, A. K. and Basu, S.,** Kinetic of the polymerization of methyl isopropenyl ketone. I. Uncatalyzed and catalyzed, *Makromol. Chem.,* 29, 48, 1959.
87. **Mulvaney, J. E., Dillon, J. G., and Laverty, J. L.,** Polymerization and copolymerization of methacrylophenone, *J. Polym. Sci. Part A,* 6, 2841, 1968.
88. **Bantsyrev, G. I., Shcherbakova, I. M., Cherkashyn, M. I., Kalykhman, I. D., Tschigir, A. N., and Berlin, A. A.,** On thermal polymerization of phenylacetylene and 2-methyl-5-ethynyl pyridine, *Izv. Akad. Nauk S.S.S.R., Ser. Khim.,* 8, 1762, 1970 (in Russian).
89. **Pickard, J. M. and Grant, J.,** Kinetics of the bulk thermal polymerization of 3-ethynil phenyl ether, *Polym. Prepr. Am. Chem. Soc. Div. Polym. Chem.,* 19(2), 591, 1978.
90. **Oancea, D., Schuster, R. H., Caragheorgheopol, A., Nicolau, A., Ionescu, M., and Popescu, D.,** Einige Aspecte der termischen Polymerisation von Phenylazetylen, *Plaste Kautsch.,* 26, 213, 1979.
91. **Matnishyan, A. A., Grigoryan, S. G., Arzumyan, A. M., and Matsoyan, S. G.,** On the mechanism of formation of vinylacetylene polymers, *Dokl. Akad. Nauk S.S.S.R.,* 257, 1384, 1981.
92. **Roberts, J. D. and Sharts, C. M.,** Cyclobutane derivatives from thermal cycloaddition reactions, in *Organic Reactions,* Vol. 12, Cope, A. C., Ed., John Wiley & Sons, New York, 1962, 53.
93. **Nazarov, I. N., Kuznetsova, A. I., and Kuznetsov, N. V.,** Dimerization of isoprene, (Russ.), *Zh. Obshch. Khim.,* 25, 307, 1955 (in Russian).
94. **Dushek, Ch., Höbold, W., Pritzkow, W., Rothenhausser, H., Schmidt, H., Engler, W., Estel, D., Hanthal, H. G., Korn, J., and Zimmerman, G.,** Über die Homodimerisierung der conjugieren C_5-Diene, *Z. Prakt. Chem.,* 312, 15, 1970.
95. **Billingham, N. C., Leeming, P. A., Lehrle, R. S., and Robb, J. C.,** A cyclobutane derivative from chloroprene dimerization, *Nature,* 213, 494, 1967.
96. **Billingham, N. C., Leeming, P. A., Lehrle, R. S., and Robb, J. C.,** The polymerization of chloroprene. III. Characteristics of the initiation in thermal polymerization, *J. Polym. Sci. Part C,* 16, 3424, 1968.
97. **Billingham, N. C., Ebdon, J. R., Lehrle, R. S., and Robb, J. C.,** Polymerization of chloroprene. IV. Influence of oxygen on the thermal reactions, *Trans. Faraday Soc. II,* 66, 421, 1970.
98. **Shantarovich, P. S. and Shlyapnikova, I. A.,** Polymerization kinetics of cyclohexadiene, *Vysokomol. Soedin.,* 3, 1364, 1961 (in Russian).
99. **Starks, C. M.,** *Free Radical Telomerization,* Academic Press, New York, 1974, chap. 1.
100. **Peterson, M. D. and Weber, A. G.,** U.S. Patent 2,395,292, 1946.
101. **Rempp, R. F. and Franta, E.,** Macromonomers: synthesis, characterization and application, *Adv. Polym. Sci.,* 58, 1, 1984 (and references cited therein).
102. **David, C. and Gosselain, P. A.,** Etude de la reaction de télomérisation de l'éthylène et du tétrachlorure de carbone initiée par rayonement gamma, *Tetrahedron,* 18, 639, 1962.
103. **Asscher, M. and Vofsi, D.,** Chlorine activation by redox-transfer. I. The reaction between aliphatic amines and carbon tetrachloride, *J. Chem. Soc.,* p. 2261, 1961.
104. **Asscher, M. and Vofsi, D.,** Chlorine-activation by redox-transfer. II. The addition of carbon tetrachloride to olefins, *J. Chem. Soc.,* p. 1887, 1963.
105. **Asscher, M. and Vofsi, D.,** Chlorine-activation by redox-transfer. III. The "abnormal" addition of chloroform to olefins, *J. Chem. Soc.,* p. 3929, 1963.
106. **Piétrasanta, Y. and Rigal, G.,** Télomerisation par catalyse redox. I. Télomerisation du styrene et des halogen-methanes, *Eur. Polym. J.,* 10, 933, 1974.
107. **Boutevin, B., Maubert, C., Piétrasanta, Y., and Sierra, P.,** Telomerisation kinetics by redox catalysis: a study of reactions leadings to telomers with low degrees of polymerization, *J. Polym. Sci. Polym. Chem. Ed.,* 19, 511, 1981.
108. **Boutevin, B. and Pieétrasanta, Y.,** Cinetique de télomerisation. III. Détermination des constants des transfert en catalyse redox, *Makromol. Chem.,* 186, 831, 1985.
109. **Boutevin, B., Hugon, J. B., Piétrasanta, Y., and Sideris, A.,** Télomérisation per catalyse redox. XIII. Synthése dçools fluorés à partir des télomérs des acétates de vinyle et d'allyle avec des télomérs fluorés, *Eur. Polym. J.,* 14, 353, 1978.
110. **Boutevin, B. and Piétrasanta, Y.,** Préparation de nouveaux oligomèrs réactifs au moyen des réactions de télomerisation, *Bull. Soc. Chim. Fr.,* p. 734, 1987.
111. **Lenz, R. W.,** *Organic Chemistry of Synthetic High Polymers,* Wiley-Interscience, New York, 1967, 295.
112. **Walling, C. and Huyser, E. S.,** Free radical additions to olefins to form carbon-carbon bonds, in *Organic Reactions,* Vol. 13, Cope, A. C., Ed., John Wiley & Sons, New York, 1963, 91.
113. **Robb, J. C. and Vofsi, D.,** The photochemical telomerisation of styrene with bromotrichloromethane, *Trans. Faraday Soc.,* 55, 558, 1959.
114. **Kirkham, W. J. and Robb, J. C.,** Telomerization of styrene, *Trans. Faraday Soc.,* 57, 1757, 1961.
115. **Boutevin, B., Piétrasanta, Y., and Bauduin, G.,** Cinétique de télomérisation. I. Détermination des lois applicable à la synthèese des télomérs à $\overline{DP}a$ moyens en télomérisation radicalaire, *Makromol. Chem.,* 186, 283, 1985.

116. **Boutevin, B., Maubert, C., Mebkhout, A., and Piétrasanta, Y.,** Telomerization kinetics by redox catalysis, *J. Polym. Sci. Polym. Chem. Ed.,* 19, 499, 1981.
117. **Walling, C.,** *Free Radicals in Solution,* John Wiley & Sons, New York, 1957, 245.
118. **Haszeldine, R. N. and Steele, B. R.,** The addition of free radicals to unsaturated systems. III. Chloro-fluoroethylene, *J. Chem. Soc.,* p. 1592, 1953.
119. **Ehrenfeld, R. L.,** U.S. Patent, 1,778,375, 1957.
120. **Botevin, B. and Piétrasanta, Y.,** Cinétique de télomérisation radicalaire, *Eur. Polym. J.,* 12, 219, 1975.
121. **Paleta, O. and Posta, P.,** Über die Addition von Tetrachloromethan mit Trifluorchloräthylen, *Coll. Czech. Chem. Commun.,* 31, 2389, 1966.
122. **Bittles, J. A. and Joyce, A. M.,** U.S. Patent, 2,559,754, 1951.

Chapter 1.2

OLIGOCONDENSATION

I. INTRODUCTION

Condensation polymerization is a method for the synthesis of polymers which implies the interaction of bi- or multifunctional compounds. The reactions follow a stepwise, or step-growth kinetics, i.e., the molecular weight of the polymers increases in a slow, steplike manner — one step at a time — as reaction time increases. High molecular weight materials result from a large number of steps, and two reaction centers are consumed in each step-growth process according to the following general equation:

$$na–A–a + nb–B–b \rightarrow a–(A–B)_n–b + (2n - 1)ab \qquad (1)$$

where a and b are functional groups and A and B are the rest of the monomer structure.

These reactions are in contrast to the propagation step of an *addition* polymerization reaction, when the reactive site (e.g., radical or ion) is regenerated by the addition of one monomer unit.

The development of the step-growth processes can be represented as shown below:[1]

$$monomer \ + monomer \leftrightarrow dimer$$

$$dimer + monomer \leftrightarrow trimer$$

$$dimer + dimer \leftrightarrow tetramer$$

$$trimer + dimer \leftrightarrow pentamer$$

$$trimer + monomer \leftrightarrow tetramer$$

$$x\text{-mer} + monomer \leftrightarrow (x + 1)\text{-mer}$$

$$\overline{n\text{-mer} + m\text{-mer} \leftrightarrow (n + m)\text{-mer}} \qquad (2)$$

Most condensation polymers are formed from systems exhibiting step-growth kinetics. Each step is independent and does not influence significantly the step to follow.

A basic assumption in approaching the kinetics of polycondensation is that the rate of reaction of a functional group is independent of the size of the molecule to which it is attached. Therefore, in a polycondensation system, all existing species, i.e., monomers, oligomers, and n-mers, are equally reactive. This may not seem reasonable at first glance, since the frequency of collision between molecules at a given temperature is normally inversely proportional to the square root of the mass. The rates of polycondensation in the melt, in solution, and in emulsion are controlled primarily by the kinetics of the reactions between the functional groups involved. Interfacial and solid state polycondensation are largely dependent on the rate of diffusion of reacting species to the site of reaction. While increasing molecular weight may reduce the ease with which reactive ends come together, the larger molecules also serve as a "cage", reducing the ease with which reactive ends diffuse away from each other. Except for the very high molecular weights in the melt polycondensation — which is beyond the scope of this book — the two effects seem essentially to cancel, so that the assumption of equal reactivity agrees reasonably well with experimental results.[2]

Equation(s) 2 give by summation:

$$P_i + P_j = P_{i+j} + \text{condensation products} \tag{3}$$

where $i = 1, 2, \ldots$, $j = 1, 2, \ldots$ and P_i, P_j, and P_{i+j} represent polymeric molecules with i, j, and i + j monomer units, respectively.

The monomers are practically consumed in the early stages of condensation when *oligomers* are formed. Therefore, formation of oligomers represents a necessary step of the polycondensation processes which are the main source of oligomers.

Because of the random interaction of all oligomer molecules (P_i, P_j, P_{i+j}) that generally occurs in many step-growth polymerizations, most polymer samples contain a broad distribution of molecular weight species. The ratio of the weight-average degree of polymerization, $\overline{M}_w$, to the number-average degree of polymerization, $\overline{M}_n$, is a measure of the heterogeneity of the polydisperse system. The distribution of molecular weights in polycondensations, where all functional groups are assumed to have an equal probability of reaction, is termed the *most probable distribution*. In the case of condensation polymers, $\overline{M}_w/\overline{M}_n = 2$ at high degrees of polymerization.

There is a large overlap between the terms "addition polymers" and "chain-growth kinetics" (characteristic of radical and ionic polymerizations) and the terms "condensation polymers" and "stepwise kinetics."[3] For example, the formation of polyurethanes typically occurs in the bulk solution through kinetics that are clearly stepwise, and the polymer backbound is heteroatomed, yet no byproduct is released through the condensation of the isocyanate with the diol or diamine because condensation occurs through internal rearrangement and shift of the hydrogen — neither necessitating expulsion of a byproduct.

$$n\,O{=}C{=}N{-}(CH_2)_6{-}N{=}C{=}O \;+\; n\,HO{-}(CH_2)_4{-}OH \rightarrow$$

$$-\left[\begin{array}{c} \underset{\displaystyle \underset{O}{\|}}{-C}{-}NH{-}(CH_2)_6{-}NH{-}\underset{\displaystyle \underset{O}{\|}}{C}{-}O{-}(CH_2)_4{-}O{-} \end{array} \right]_n - \tag{4}$$

Nylon-6, clearly a condensation polymer, is readily formed from either the internal amide (lactam) through a stepwise addition (hydrolytic polymerization of ε-caprolactam),

$$n\left[\begin{array}{c} (CH_2)_5 \\ \underset{\displaystyle \underset{O}{\|}}{C}{-}NH \end{array} \right] \xrightarrow{H_2O} HO{-}\left[\underset{\displaystyle \underset{O}{\|}}{C}{-}(CH_2)_5{-}NH{-} \right]_n {-}H \tag{5}$$

or from the stepwise reaction of the α-amino acid (bimolecular condensation):

$$n\,HOOC{-}(CH_2)_5{-}NH_2 \xrightarrow{\Delta} HO{-}\left[\underset{\displaystyle \underset{O}{\|}}{C}{-}(CH_2)_5{-}NH{-} \right]_n {-}H \;+\; (n-1)H_2O \tag{6}$$

Similarly, internal esters (lactones) are readily polymerized by acid-catalyzed ring openings without expulsion of a by-product (chainwise polyaddition), yet the resulting polyester is clearly a condensation polymer exhibiting a heteroatomed backbone

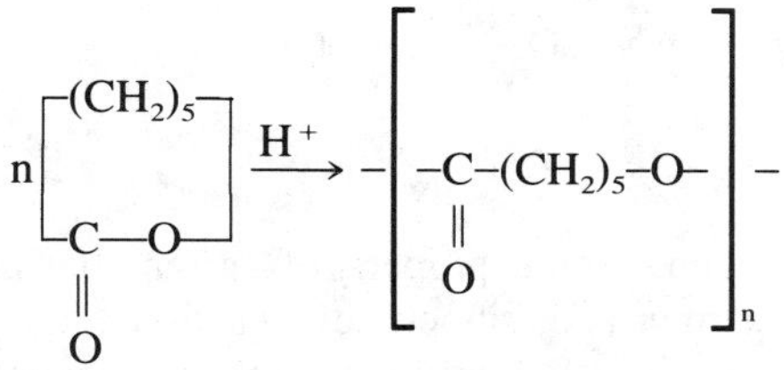

$$n \begin{bmatrix} (CH_2)_5 \\ | \\ C\!-\!O \\ \| \\ O \end{bmatrix} \xrightarrow{H^+} - \begin{bmatrix} -C-(CH_2)_5-O- \\ \| \\ O \end{bmatrix}_n - \qquad (7)$$

while the same 6-carbon polyester is also formed using typical stepwise polycondensation of ω-hydroxycarboxylic acid:

$$nHOOC\!-\!(CH_2)_5\!-\!OH \xrightarrow{\Delta} HO\!-\!\begin{bmatrix} -C-(CH_2)_5-O- \\ \| \\ O \end{bmatrix}_n\!-\!H + (n-1)H_2O \qquad (8)$$

Therefore, the step-growth polymerizations are the polymerizations of functional groups, and these reactions fall into two classes:[4,5] polycondensation and polyaddition.

II. FUNCTIONALITY

Since functionality defines the number of positions per molecule capable of reacting under specific conditions, it has the same meaning in both chain-addition polymerization and step-growth polymerization. Functionality can assume all values from zero upward, including fractions, but chain molecules are formed only if the value is at least two.

The requirements for formation of condensation polymers are twofold:[6] (1) the monomers must possess functional groups capable of reacting to form the linkage and (2) they ordinarily require more than one reactive group to generate a chain structure. The functional groups can be distributed such that two difunctional monomers with different functional groups react or a single monomer reacts which is difunctional with one group of each kind. In the latter case especially, but also with condensation polymerization in general, the tendency to form cyclic products from intramolecular reactions may compete with the formation of polymer. The presence of monofunctional reagents introduces the possibility of a reaction product forming which would not be capable of further growth. Step-growth polymerizations are therefore especially sensitive to impurities. If the functionality is greater than two, on the other hand, branching becomes possible. When reagents of functionality less than or greater than two are added in carefully measured and controlled amounts, the size and geometry of polymer molecules can be monitored.

A lone isocyanate group, $-NCO$, in a monomer is, for example, monofunctional with respect to a lone $-OH$ group if both groups are present at about the same concentration. Therefore, to form polyurethanes with the urethane group $-NH-CO-O-$, diisocyanates must react with diols (stepwise kinetics, Equation 4). With an excess of isocyanate groups, however, the urethane groups can convert to allophanates:

$$-NCO + -NH-CO-O- \rightarrow -NH-CO \qquad (9)$$
$$| $$
$$-N-CO-O-$$

Since two isocyanate groups react with one hydroxyl group to form allophanates, the −NCO group is semifunctional. With polymerization initiators (chain-growth kinetics), however, the isocyanate group is always bifunctional:

$$n \, \underset{\underset{R}{\big|}}{N{=}CO} \xrightarrow{\text{bases}} -\underset{\underset{R}{\big|}}{(N{-}CO{-})_n}- \tag{10}$$

Thus, functionality is not an absolute property of a group, but always has to be considered in relation to the reaction partner and, sometimes, reaction conditions *(vide infra)*.

The chemical structure of the resulting macromolecule, moreover, will be decided not only by the functionality of the groups capable of encatenation, but also by the functionality of the molecule. The strained heterocycle of lactams is bifunctional with respect to cationic initiators. However, neutral and protonated polymer amide groups can also take part in the disproportionation, viz.:

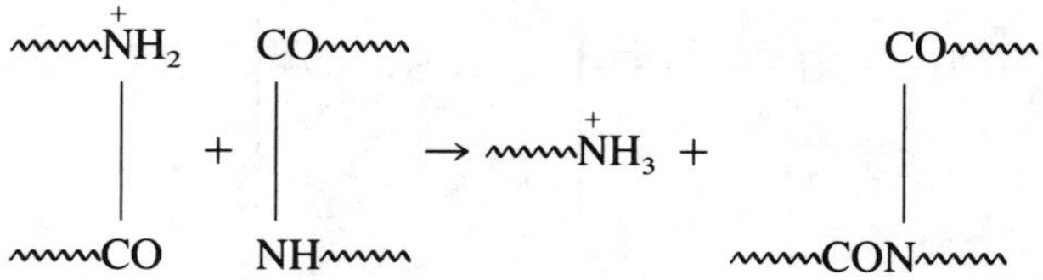

$$(11)$$

and

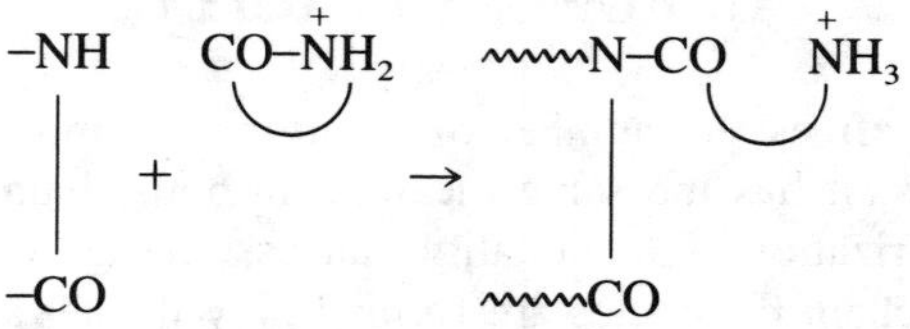

$$(12)$$

The new polymer cation can, in turn, initiate lactam polymerization again. These side reactions increase the mean functionality of the base units to more than two.

The mean functionality of the monomer, however, can also be smaller than the sum of the functionalities of the groups contained in its molecule. For example, in the Diels-Alder condensation of a bisdiene with benzoquinone (where no byproduct is formed, yet the polymer is formed through a stepwise kinetic process), the total functionality of the bisdiene in chain formation is two per unbranched macromolecule:

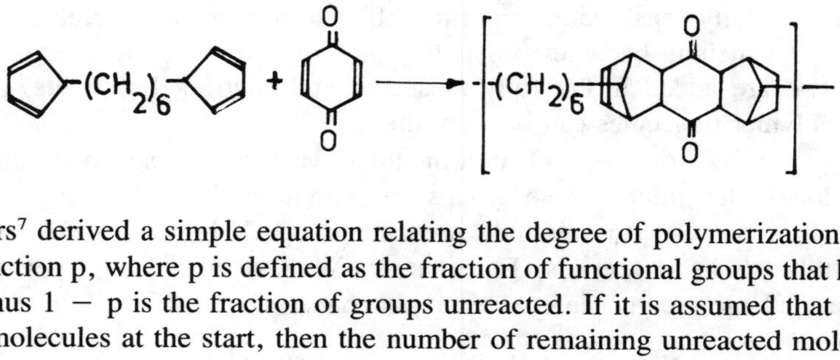

$$(13)$$

Carothers[7] derived a simple equation relating the degree of polymerization (DP) to the extent of reaction p, where p is defined as the fraction of functional groups that have reacted at time t. Thus $1 - p$ is the fraction of groups unreacted. If it is assumed that there are N_o number of molecules at the start, then the number of remaining unreacted molecules is

$$N = N_o(1 - p) \tag{14}$$

The average number of repeating units in all molecules at any stage in the reaction is the original number of molecules divided by the remaining number of molecules:

$$\overline{DP} = \frac{N_o}{N} \tag{15}$$

The expression for $\overline{DP}$ in terms of reaction conversion is known as the Carothers equation:

$$\overline{DP} = \frac{N_o}{N_o(1 - p)} = \frac{1}{1 - p} \tag{16}$$

According to this limiting equation, oligomers of $\overline{M} = 20$ to 50 are formed for conversions of 95 to 98%. Very few reactions in organic chemistry can be forced to and beyond 98%, yet this requirement is the first characteristic that should be considered when it is proposed to apply a new reaction to a step-growth polymerization. In order to obtain step-grown polymers of $\overline{M}$ equal to 200, the fractional conversion p should be 0.995.

The general equation for the formation of a linear polymer by the step reaction of bifunctional reactants A and B may be written as follows:

$$nA + nB \rightarrow A(BA)_{n-1}B \tag{17}$$

The probability of finding a repeating unit AB in the polymer chain is p and the probability of finding $n - 1$ of the repeating units is p^{n-1}. Since the probability of finding an unreacted molecule of A or B is $p - 1$, the probability (P_n) of finding a chain with n repeating units $(BA)_n$ is

$$P_n = (1 - p)p^{n-1} \tag{18}$$

Hence, the probability of the total number of repeating units $(BA)_n$ is

$$N_n = N(1 - p)p^{n-1} \tag{19}$$

where N is the number of molecules in the mixture after the reaction has occurred to the extent p. Since $N = N_o(1 - p)$, Equation 19 becomes

$$N_n = N_o(1 - p)^2 p^{n-1} \tag{20}$$

The corresponding weight-average molecular weight distribution W_n may be calculated from the relationship $W_n = nN_n/N_o$:

$$W_n = \frac{nN_o(1 - p)^2 p^{n-1}}{N_o} = n(1 - p)^2 p^{n-1} \tag{21}$$

The number-average molecular weight $\overline{M}_n$ and weight-average molecular weight $\overline{M}_w$ are as follows:

$$\overline{M}_n = \frac{mN_o}{N} \tag{22}$$

and

$$\overline{M}_w = \frac{m(1 + p)}{1 - p} \tag{23}$$

where m is the molecular weight of the -mer.

The index of polydispersity $\overline{M_w}/\overline{M_n}$ for the most probable molecular weight distribution is then

$$\frac{\overline{M_w}}{\overline{M_n}} = \frac{m(1 + p)/(1 - p)}{m/(1 - p)} = 1 + p \tag{24}$$

Thus, as mentioned above, when $p = 1$, i.e., for a total conversion, the index of polydispersity for the most probable distribution for step-reaction polymers is 2.

The limitation on $\overline{DP}$ is used to advantage when it is desired to obtain polymers of lower molecular weight by step-growth polymerizations. It is obvious that the value of p may be reduced by using an excess of one of the reactants or by adding a calculated amount of a monofunctional reactant (quenching the reaction before completion or adding a stoichiometric excess of one reactant is not economical).

If the number of monofunctional molecules present in the reaction mixture is N_1 in addition to N_o difunctional molecules A-B, the relationship (Equation 16) between the fractional conversion of either A or B groups in the $\overline{DP}$ must be modified to take into account the effect of this monofunctional, polymer-chain termination:

$$\overline{DP} = \frac{1 + N_1/N_o}{1 - p + N_1/N_o} \tag{25}$$

For example, if the monofunctional derivative is present to the extent of 1%, then Equation 25 predicts that a reaction conversion of 98% will yield a polymer with a $\overline{DP}$ of approximately 34 instead of 50 as seen above.

Assuming that the step-growth polymerization involves AA and BB monomers and that the B groups are present in excess, the stoichiometric imbalance defined by the ratio r

$$r = \frac{N_o^A}{N_o^B} \tag{26}$$

determines the number of unreacted functional groups after the reaction reaches the extent p, as shown below:

$$N^A = (1 - p)N_o^A \tag{27}$$

and

$$n^B = (1 - pr)N_o^B = (1 - pr)\frac{N_o^A}{r} \tag{28}$$

where N_o^A and N_o^B are the number of A and B functional groups, respectively, in the initial reaction mixture and N^A and N^B are the number of these groups at various stages of reaction.

The total number of chain ends is the sum of N^A and N^B, and the total number of chains, N_c, half the number of chain ends, is

$$N_c = \frac{1}{2}\left[(1 - p)N_o^A + (1 - pr)\frac{N_o^A}{r}\right] = \frac{1}{2}\left(1 + \frac{1}{r} - 2p\right)N_o^A \tag{29}$$

The total number of repeating units distributed among these chains, N_{ru}, is the number of monomer molecules present initially:

$$N_{ru} = \frac{1}{2}N_o^A + \frac{1}{2}N_o^B = \frac{1}{2}\left(1 + \frac{1}{r}\right)N_o^A = \frac{N_A}{2r}(1 + r) \tag{30}$$

Dividing the number of repeat units by the number of chains, one obtains the number-average degree of polymerization:

$$\overline{DP} = \frac{1 + r}{1 + r - 2p} \tag{31}$$

Considering Equation 30, for a 1% excess of BB, i.e., $r = 0.99$, a $\overline{DP}$ of 50 is attained at a reaction conversion of 98.5%.

The Carothers equation as originally derived[7] contained a factor, $\bar{f}$, to account for the effect of monomers having more than two functional groups on the degree of polymerization. In this case, $\bar{f}$ is the *average functionality*, i.e., the number of functional groups per monomer molecule for all types of monomer present:

$$\bar{f} = \frac{\Sigma N_i f_i}{\Sigma N_i} \tag{32}$$

where N_i and f_i are the number of molecules and the functionality of the i-th component in the reaction mixture, respectively. The original number of functional groups in the reaction mixture is then $N_o \bar{f}$ and the number of functional groups that have reacted is $2(N_o - N)$, N being the total number of molecules present in the reaction mixture at an extent of reaction p. Therefore

$$p = \frac{2(N_o - N)}{\bar{f} N_o} \tag{33}$$

Elimination of N from Equations 33 and 15 gives the modified Carothers equation:

$$\overline{DP} = \frac{2}{2 - p\bar{f}} \tag{34}$$

When the average degree of functionality is 2, this equation reverts to the previous form given by Equation 16. The modified Carothers equation demonstrates how sensitive step-growth polymerization is for the presence of small amounts of polyfunctional monomers which can act as branching or cross-linking sites in the growing chain.

For a fixed extent of reaction, the presence of multifunctional monomers in an equimolecular mixture of reactive groups increases the degree of polymerization. Conversely, for the same mixture, a lesser extent of reaction is needed to reach a specified $\overline{DP}$ with multifunctional reactants than without them. If the number of functional groups is unequal, this effect works in opposition to the multifunctional groups. For example, in Equation 17, if there is one trifunctional monomer in every ten original A monomer molecules, then $f = 2.1$ and a conversion of 95% will result in a polymer having a $\overline{DP}$ of 200 instead of 20, as mentioned above for a stoichiometric mixture of bifunctional partners.

Carothers attempted to correlate the concept of functionality with the structure of resulting polymers. He concluded that linear polymers are formed only when the monomers are bifunctional, and if the functionality is higher than 2 gelation takes place and a three-dimensional structure is formed.

Korshak[8] showed, however, that many tri-, tetra-, and higher functional monomers can form linear polymers. Korshak considered that there are three types of factors which determine the relation between monomer functionality and polymer structure. The first factor, illustrated by the reaction of glycerol ($f = 3$) with phthalic anhydride ($f = 2$), is determined by the monomer structure. The reactivity of functional groups is variable. Only at temperatures above 180°C are glyphthals formed. At lower temperatures, linear polyesters are formed since the polycondensation involves primarily the $-CH_2OH$ groups of glycerol, which are much more reactive than secondary $-CHOH$ hydroxyls.

The second type of factor refers to the influence of experimental conditions. The reactivity of functional groups might depend on the nature of the catalyst used, the proportion of monomers, the nature of solvent, and the temperature. For example, in the reaction of formaldehyde (f = 2) with phenol (f = 3), not all of the active hydrogens are equally active. Under acid conditions, the quinoid structure

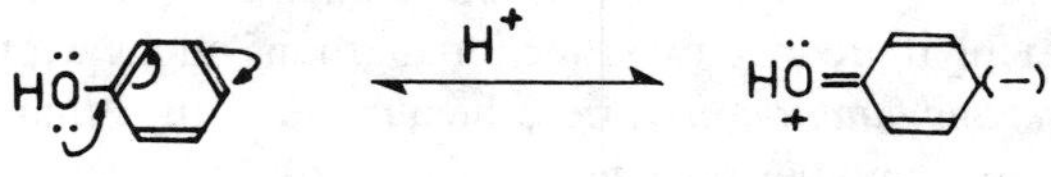

$$\ddot{H\ddot{O}} \underset{\cdot\cdot}{-} \hspace{-0.3em}\bigcirc\hspace{-0.3em} \xrightarrow{\quad H^+ \quad} \overset{+}{H\ddot{O}} \hspace{-0.3em}=\hspace{-0.3em}\bigcirc\hspace{-0.3em}-_{(-)} \tag{35}$$

is stabilized as a reaction intermediate favoring the *para* additions of the protonated for-maldehyde molecule. There are also indications that the reaction of one position alters the reactivity of others, so the reactivity depends on the extent of reaction as well. Low molecular weight polymers with chains capped by phenol repeat units (known as A-stage resins, novolacs, or resole prepolymers) are formed when the ratio of CH_2O to C_6H_5OH is less than unity. The reaction is either acid or base catalyzed, and branching is uncommon at this stage. When the reaction is carried out under base-catalyzed conditions and with a formal-dehyde/phenol ratio greater than unity, a final cross-linked polymer (called a C-state resin or resite) is formed.

The influence of solvents is illustrated by the reaction of 3,3′-diamino-4,4′-dihydroxy-diphenylmethane with carborane dicarboxylic acid: in common solvents, a three-dimensional product is obtained, while the addition of tributyl amine favors the formation of linear polymers.[8]

The third type refers to some favorable arrangements of functional groups in monomer molecules, and can be illustrated by the so-called "ortho effect".[9] This phenomenon explains the formation of linear polymers with heterocycles in the main chain from tri- and tetra-functional monomers. A typical example is the two-step preparation of an aromatic polyimide by the reaction of pyromellitic anhydride with aromatic diamines. In the first stage, polyamic acid in formed in aprotic solvents (dimethylformamide, dimethylsulfoxide):

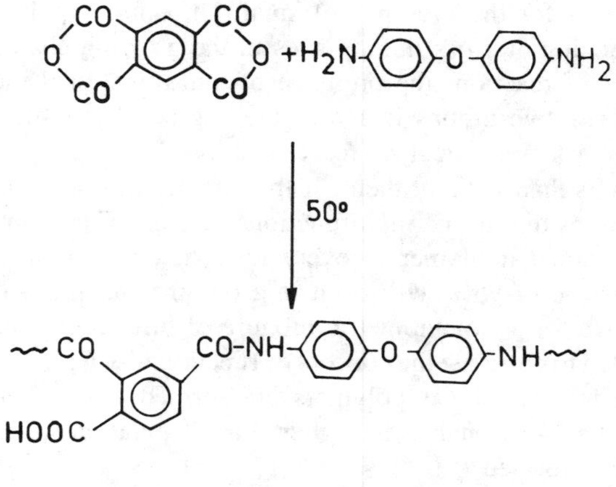

$$\text{(36)}$$

Polyamic acid

Bonding occurs predominantly at the *para* position; there is relatively little attack at the *meta* positions. In order to avoid cross-linking reactions, the solid content of the solution is restricted to 10 to 15% and the yield to 50%. In the second stage, water is eliminated at 300°C.

The step-growth reactions in the early stages of the process yield oligomers which react to form polymers. The course of further conversion depends not only on the amount of functional groups in monomers, but also on the arrangement and activity of functional groups in oligomer molecules. It is therefore necessary to extend the concept of functionality from monomers to oligomers. Moreover, side reactions may alter the original functionality of monomers or that of oligomers.

Consequently, the functionality should be defined as *possible functionality*, ϕ_{Po}, or theoretical functionality, i.e., the highest number of reactive functional groups derived from the molecular structure, or as *practical functionality*, ϕ_{Pr}, i.e., the number of functional groups capable of reacting under given conditions. The latter can change with reaction conditions (temperature, concentration of monomers, nature of catalyst, type of solvents, etc.).

Based on these definitions, it is possible to express the functionality of a given monomer in step-growth polymerizations in terms of *relative functionality*, ϕ_R, which is the ratio of ϕ_{Po} to ϕ_{Pr}. This new quantity characterizes the reactivity of both parent monomers and resulting oligomers. Usually, $\phi_{Po} \geq \phi_{Pr}$ and $\phi_R \geq 1$. When $\phi_{Pr} = 2$, the resulting polymers are linear and if $\phi_{Pr} \geq 3$, three-dimensional structures are formed.

Other types of functionality include the *number-average functionality*, $\overline{\phi}_n$, and the *weight-average functionality*, $\overline{\phi}_w$, which are defined by the following relations:

$$\overline{\phi}_n = \frac{\overline{\phi}_{Po1} + \overline{\phi}_{Po2} + \overline{\phi}_{Po3} + \cdots}{N_1 + N_2 + N_3 + \cdots} = \frac{\Sigma \phi_{Poi}}{\Sigma N_i} \tag{37}$$

and

$$\overline{\phi}_w = \frac{\phi_{Po1}^2 + \phi_{Po2}^2 + \phi_{Po3}^2 + \cdots}{\phi_{Po1} N_1 + \phi_{Po2} N_2 + \phi_{Po3} N_3 + \cdots} = \frac{\Sigma \phi_{Poi}^2}{\Sigma \phi_{Poi} N_i} \tag{38}$$

where P_{Poi} and N_i represent the possible functionality and the number of moles of species i, respectively.

III. MOLECULAR WEIGHT DISTRIBUTION

The molecular mass distribution functions described by Flory[10] shows that monomer molecules are the most numerous of all species at every value of p. On a weight basis, however, the proportion of oligomers is small and decreases as p increases. The maximum of the weight distribution curve occurs at n = −(1/lnp), a value which is close to the number-average degree of polymerization.

In the case of a polyamide produced from a dibasic acid and a diamine, two types of molecular species are formed—those with an odd number of structural units x in the chain which have end groups of the same chemical type (either acid or amine), and those with even values of x having one acid and one amine end group. If the reaction mixture contains equal amounts of diacid and diamine, the equation given for the simple ω-amino acid holds. When one component is in excess, only an average curve for W_n can be drawn. The validity of this molecular weight distribution has been challenged by Korshak,[11,12] who considered that the effects of hydrolysis and interchange are neglected in the derivation of Equation 20. Howard[13] has shown, however, that the new distribution function for linear condensation polymerization proposed by Korshak, i.e.,

$$W_x = \frac{\exp[-(\overline{x}_n)](\overline{x}_n)^{n-1}}{(x - 1)!} \tag{39}$$

has no sound theoretical basis.

The Korshak distribution (Equation 39) predicts a homogeneous polymer ($\overline{M}_w/\overline{M}_n \sim 1$) as a predominant species, with $\overline{DP}$ equal to the number-average value. A feature of the Flory distribution (Equation 20) is that monomers are the most numerous of all species, although contributing little on a weight-average basis.

These two dissimilar distribution functions for linear condensation polymers were checked by classical fractionation methods and by kinetic studies.[13] The experimental distribution curves fall into three classes:

1.   Those supporting Flory's most probable distribution
2.   Those in approximate agreement with Korshak's distribution
3.   Those of an intermediate form

According to Slonimsky,[14] the Flory distribution holds only in the absence of chain interchange reactions; under the influence of side reactions, the distribution gradually narrows until, at equilibrium, the Korshak distribution is obeyed.

Discussion of the theoretical molecular weight distribution has been made under the limiting conditions that all polymer molecules are linear. The formation of cyclic species will depend on the probability that the end groups of a linear species of the requisite length occupy an adjacent position in space relative to the probability that one or another of these end groups is close to a functional group of another molecule. At equilibrium, the weight fraction of cyclic species of x structural units (W_{rx}) formed in the condensation of a monomer of the type A-B (e.g., $H_2N{-}R{-}COOH$) is given by

$$W_{rx} = \frac{2BM_s(p'')^x}{cx^{3/2}} \tag{40}$$

where

$$B = \frac{(3/2\ n_s)^{3/2}}{2L^3N_A} \tag{41}$$

and c is the concentration (in g/ml), n_s the number of chain atoms per structural unit, L the effective bond length, N_A Avogadro's number, and p'' a revised extent of reaction defined by

$$(1 - p'') = \frac{1 - p}{1 - W_r} \tag{42}$$

where W_r is the weight fraction of all rings and p the extent of reaction of functional groups.[15]

Since

$$\sum_{x=1}^{\infty} N_x = N \tag{43}$$

then

$$W_r = \left(\frac{2BM_s}{c}\right) \sum_{1}^{\infty} \frac{(p'')^x}{x^{3/2}} \tag{44}$$

The weight fraction distribution of cyclic polymers decreases monotonically and is confined to very low values of x (oligomers). Because of ring formation, slight modification of the distribution for linear polymers (Equation 21) is necessary. Writing W_{Lx} as the weight fraction of linear species of DP $= x$, the value of W_{Lx} is given by:

$$W_{Lx} = (1 - W_r)(1 - p'')^2 \cdot x \cdot (p'')^{x-1} \tag{45}$$

and

$$\frac{x_w}{x_n} = (1 + p) - 2W_r$$

This revision is, however, of little importance since p'' is only slightly less than p and the area of the weight fraction distribution curve for linear species will be slightly reduced by introduction of the W_r value.

The step-growth mechanism is conveniently considered along with normal condensation polymers. Heterocyclic monomers, such as oxiranes, lactones, lactams, and siloxanes, are also involved in such a mechanism (i.e., step-addition polymerization).

A kinetic derivation of the distribution function for polymers obtained by the step-addition mechanism was originally given by Flory.[15] The mole fraction distribution is in this case

$$\frac{N_x}{N} = \frac{\exp(-d)d^{x-1}}{(x - 1)!} \tag{46}$$

where d represents the number of monomer units per macromolecule. The degree of polymerization x is defined as counting the initiation fragment as equivalent to one structural unit; hence

$$\bar{x}_n = d + 1 \tag{47}$$

and

$$W_x = \frac{d}{d + 1} \cdot \frac{x \cdot \exp(-d)d^{x-2}}{(x - 1)!} \tag{48}$$

IV. ORIGINS OF OLIGOMERS IN STEP-GROWTH REACTIONS

The following features should be taken into account when estimating the possibilities of step-growth reactions for the synthesis of oligomers:

1. The elementary acts of chain propagation are independent and, therefore, have a statistical character.
2. The chain propagation might be terminated in the course of the reaction due to the consumption of one of the monomers, or the reactivity of the active centers might decrease because of the conjugation effect and diffusion limitation.
3. The chain termination may be the result of both the introduction of compounds with $f \simeq 1$ and of chemical transformation of the functional groups by decarboxylation, dehydration, salt formation, etc.
4. The chain-growth process may be accompanied by ring formation, interchange, and reverse reactions (hydrolysis, ammonolysis, acidolysis, etc.). Side reactions, experimental conditions, the presence of certain additives, and the nature (structure) of monomers are determinant factors in the formation of oligomers in step-growth polymerization.[16-26]

By choosing the appropriate reaction conditions, the adverse processes may be reduced to a minimum (e.g., high initial concentration of reactants, removal of low molecular weight products, and low reaction temperature). Therefore, oligocondensation (oligoaddition) may be considered one of the most promising methods for the synthesis of oligomers.

The routes of obtaining synthetic oligomers by step-growth reactions are as follows:

1. The termination of oligocondensation before complete conversion of functional groups
2. The application of stoichiometric imbalance
3. The use of specific procedures: condensation telomerization[27] and metathetic polycondensation[28,29]

These procedures, together with some specific reactions such as the *cycloimidic effect* pointed out by Korshak[11] in the synthesis of polyamides and their implication in the process of internal or terminal functional groups,[12,13] are the main modes of formation of synthetic oligomers by polycondensation (polyaddition).

A. CARBONYL ADDITION-ELIMINATION REACTIONS

These reactions, utilized for the preparation of step-growth polymers, are principally those based on reactions of carboxylic and aldehydic functional groups. Most of the methods used for the synthesis of esters in organic chemistry have been investigated for the preparation of polyesters. Included among these are the following: direct esterification of a dibasic acid with a glycol or by self-condensation of an α-hydroxy acid; ester interchange (transesterification) of a diester of an aliphatic or aromatic dibasic acid with a glycol; esterification of an anhydride with a glycol; esterification of a dibasic acid with a diester of a glycol or bisphenol; and the Schotten-Baumann reaction between a diacid chloride and a glycol or bisphenol.

Poly(ethylene terephthalate), PET, is conveniently prepared in high molecular weight by the reaction of dimethyl terephthalate and ethylene glycol, using two successive ester interchange reactions. The first ester interchange reaction is an *in situ* monomer preparation step (Equation 49) which leads principally to the formation of bis(2-hydroxyethyl)terephthalate:

$$H_3C-O-\underset{\substack{\|\\O}}{C}-C_6H_4-\underset{\substack{\|\\O}}{C}-CH_3 \; + \; 2HO-(CH_2)_2-OH \xrightarrow{\;150-195°C\;}$$

$$HO(CH_2)_2-O-\underset{\substack{\|\\O}}{C}-C_6H_4-\underset{\substack{\|\\O}}{C}-O-(CH_2)_2OH \; + \; 2CH_3-OH \qquad (49)$$

Small amounts of linear oligomers up to approximately the tetramer are also formed in this process as a result of side reactions:

$$H_3C-O-\underset{\substack{\|\\O}}{C}-C_6H_4-\underset{\substack{\|\\O}}{C}-O-CH_3 \; + \; HO-(CH_2)_2-O-\underset{\substack{\|\\O}}{C}-C_6H_4-\underset{\substack{\|\\O}}{C}-O-(CH_2)_2-OH \rightleftharpoons$$

$$H_3C-O-\underset{\substack{\|\\O}}{C}-C_6H_4-\underset{\substack{\|\\O}}{C}-O-(CH_2)_2-O-\underset{\substack{\|\\O}}{C}-C_6H_4-\underset{\substack{\|\\O}}{C}-O-(CH_2)_2-OH \; + \; CH_3-OH \qquad (50)$$

and

$$2\,HO\text{-}(CH_2)_2\text{-}O\text{-}\underset{\underset{O}{\|}}{C}\text{-}C_6H_4\text{-}\underset{\underset{O}{\|}}{C}\text{-}O\text{-}(CH_2)_2\text{-}OH \; \rightleftharpoons$$

$$H\text{-}\left[\text{-}O\text{-}(CH_2)_2\text{-}O\text{-}\underset{\underset{O}{\|}}{C}\text{-}C_6H_4\text{-}\underset{\underset{O}{\|}}{C}\text{-}O\text{-}\right]_2\text{-}O\text{-}(CH_2)_2\text{-}OH \; + \; HO\text{-}(CH_2)_2\text{-}OH$$

$$+ \; HO\text{-}(CH_2)_2\text{-}OH \tag{51}$$

The second interchange reaction — the proper polymerization — is carried out well above the boiling point of ethylene glycol, which is continuously released during the process:

$$n\,H\text{-}\left[\text{-}O\text{-}(CH_2)_2\text{-}O\text{-}\underset{\underset{O}{\|}}{C}\text{-}C_6H_4\text{-}\underset{\underset{O}{\|}}{C}\text{-}\right]_x\text{-}O\text{-}(CH_2)_2\text{-}OH \xleftarrow{\;260°C\;}$$

$$\text{-}\left[\text{-}O\text{-}(CH_2)_2\text{-}O\text{-}\underset{\underset{O}{\|}}{C}\text{-}C_6H_4\text{-}\underset{\underset{O}{\|}}{C}\text{-}\right]_{nx}\text{-} \; + \; nx\,HO\text{-}(CH_2)_2\text{-}OH \tag{52}$$

The rates of the three possible ester interchange reactions between functional groups on either monomer molecule or polymer chain end groups are unequal: the reaction of two end groups is 1.8 times faster than the reaction of a monomer with an end group, and the latter process is 1.8 times faster than the exchange reaction between two monomer units.[30] This transesterification avoids, therefore, the necessity of having an exact, stoichiometric balance of functional groups at the start of the polycondensation. Slightly more than a 2:1 molar ratio of ethylene glycol to dimethyl terephthalate is used, and this ratio, along with the reaction conditions, determines the distribution of oligomers formed.

However, estimating the possibilities of the reactions of carboxylic acid derivatives for the formation of linear oligomers during industrial preparations and processing it is necessary to take into account that the chain propagation is terminated in the course of the reaction due to the consumption of one monomer or that the reactivity of the active centers decreases because of the *conjugation effects* and *diffusion limitations*. On the other hand, the chain termination can be realized owing to the introduction of *monofunctional compounds* or as a result of *chemical transformations* in functional groups (CO_2 or H_2O elimination, salt formation, etc.).

The amount of linear oligomers in PET may also be influenced by the nature of certain additives introduced into the system during synthesis or processing. In dope dyeing, the dyestuff is introduced into the polymerization system during synthesis. For this purpose, either *reactive or inert dyes* may be used.[31] In the first case, the dyestuff becomes a part of the main chain by combination with some suitable functional groups. In the second case, the dyestuff is dissolved in a polymer molten mass, and a physical blend of macromolecular product and dyestuff is obtained. In this case, the molecular weight distribution of polymer shows the systematic influence of the dyestuff. In the subsequent stages of polycondensation, the presence of dyestuff is accompanied by a systematic shift of the distribution curve of colored PET toward the range characteristic of low molecular weight species ($\overline{M}_n < 10^4$). This fact is determined by a supplementary degradation effect due to the attack of ester bonds by the dyestuff used (e.g., 1,4-aminoanthraquinone dyestuff).[32]

Condensation telomerization is a method based on chain termination by a monofunctional reactant that contains both a group taking part in polycondensation and a structural unit required for further reactions other than condensation.[33] The following general reaction can be written for the condensation telomerization:

$$n \text{ A–R–A} + n \text{ B–R'–B} + (n + 1)\underset{\underset{\text{X}}{|}}{\text{B–C=CH}_2} \xrightarrow{-\text{AB}} \underset{\underset{\text{X}}{|}}{\text{CH}_2\text{=C}}\text{–Z–(RZR'Z)}_n\text{–R–}\underset{\underset{\text{X}}{|}}{\text{C=CH}_2} \qquad (53)$$

where A = OH, NH$_2$; B = COOH, COOR″, Cl–CO, etc.; X = H, CH$_3$, halide, CN; Z = OCO, NH–CO; R, R′ = alkyl, aryl; and R″ = CH$_2$CH$_2$OH, CH$_2$CH$_2$NHR, etc.

Synthesis of reactive oligomers by this method may be realized using the oligocondensation of dicarboxylic acids with glycols or polyols with haloid anhydrides of dicarboxylic acids and by reesterification by glycols of mixtures containing esters of dicarboxylic and acrylic acids.[33]

The ratio between the dicarboxylic and monocarboxylic acids determines the $\overline{DP}_n$ value according to the following equation:

$$\overline{DP}_n = \frac{2DA}{MA} \cdot \frac{k_p}{k_t} \qquad (54)$$

where DA and MA are the number of dicarboxylic and monocarboxylic acid molecules, respectively, and k_p and k_t are the constants for propagation and termination, respectively.

Polymerizable oligomers with terminal reactive groups were prepared by reacting at low temperature bifunctional monomers (phthalic anhydride, ethylene glycol, etc.) in the presence of a monofunctional compound (methacrylic acid).[34-41] Depending on the molar ratio between difunctional monomer and monofunctional compound, reactive oligomers were obtained with various molecular weights.

Condensation telomerization has been also employed in the synthesis of oligothioacrylate esters by means of the interaction of various dicarboxylic acids with thioglycol and methacrylic acid or for the preparation of oligoacetylenic acrylates.[42,43]

The solvent used may also influence the amount and structure of the oligomers produced. Thus, during polycondensation of 5-hydroxy-methyl-2-furancarboxylic acid in pyridine, macrocyclic oligomers are formed; in hexane, the main products are linear oligoesters, and in imidazol, a complex mixture of polyester, macrocyclics, and linear oligoesters is obtained.[44] Formation of these compounds is described by the following reactions:

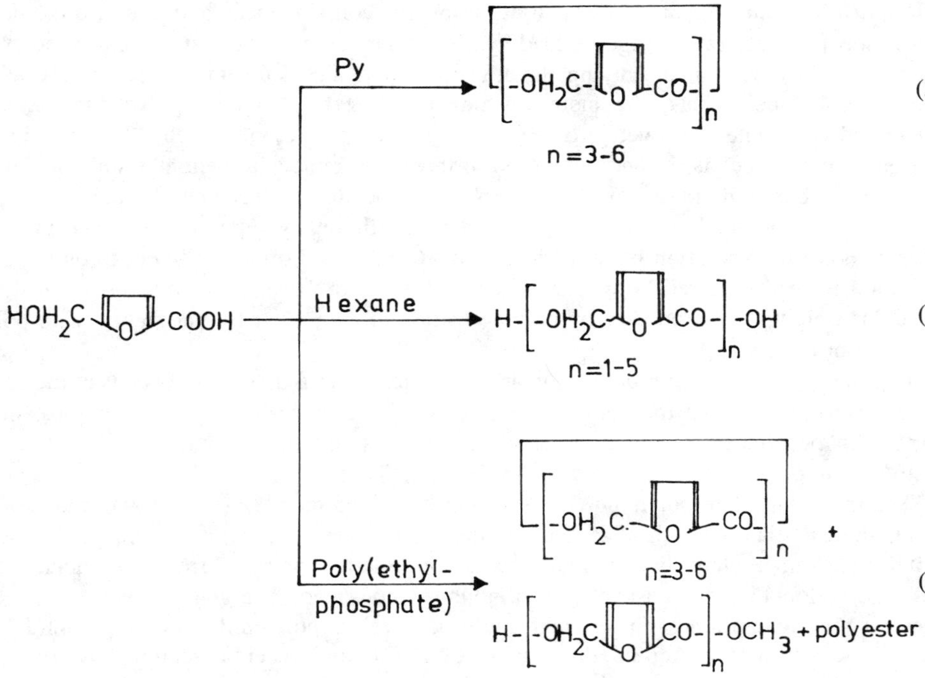

$$\xrightarrow{\text{Py}} \left[\left[-OH_2C-\underset{O}{\square}-CO-\right]_n\right] \qquad (55)$$

$$n = 3-6$$

$$HOH_2C-\underset{O}{\square}-COOH \xrightarrow{\text{Hexane}} H-\left[-OH_2C-\underset{O}{\square}-CO-\right]_n-OH \qquad (56)$$

$$n = 1-5$$

$$\xrightarrow[\text{phosphate)}]{\text{Poly(ethyl-}} \left[\left[-OH_2C-\underset{O}{\square}-CO-\right]_n\right] + \qquad (57)$$

$$n = 3-6$$

$$H-\left[-OH_2C-\underset{O}{\square}-CO-\right]_n-OCH_3 + \text{polyester}$$

From oxalic acid and ethylene glycol, linear oligoester and polymers may be obtained by low-temperature polycondensation.[45] Linear oligoesters containing alkyl perester functional groups were prepared by low-temperature condensation of glycols or polyether diols with 2,5-bis(*tert*-butyl peroxycarbonyl)terephthalic dichloride.[46] These oligomers can be used in the preparation of polyetheramides by condensation with 6-aminocaproic acid oligomers at various molar ratios.[47]

Oligoesters having melt anisotropy are prepared by reaction of oligo(ethylene terephthalate) with *p*-hydroxybenzoic acid in the presence of acylating agents[48,49] or by treating 4,4′-terephthaloyldioxybenzoylchloride with 5 meq of aliphatic diols.[50]

Polyesterification of carbonic acid and aromatic difunctional phenols leads to polycarbonates. Among them, the most important is the product formed by the reaction between phosgene and 2,2-bis(4-hydroxyphenyl)propane, which is better known by its trivial name of bisphenol A. The structural unit of this polyester is

$$\left[\begin{array}{c} CH_3 \\ | \\ -C-O-C_6H_4-C-C_6H_4-O- \\ \| \qquad\qquad | \\ O \qquad\qquad CH_3 \end{array} \right]_n \qquad (58)$$

The Schotten-Baumann reaction with phosgene (Equation 59) may be carried out either in an organic solvent or in an aqueous alkaline system:

$$\begin{array}{c} CH_3 \\ | \\ nHO-C_6H_4-C-C_6H_4-OH + nCOCl_2 \rightarrow \\ | \\ CH_3 \end{array} - \left[\begin{array}{c} CH_3 \\ | \\ -O-C_6H_4-C-C_6H_4-OC \\ | \qquad\qquad \| \\ CH_3 \qquad\qquad O \end{array} \right]_n - + 2nHCl \qquad (59)$$

High molecular weight aromatic polycarbonates contain linear and cyclic oligomers. These side products (i.e., oligomeric dihydric phenol carbonates) can be extracted with organic solvents without dissolving the high polymer. The solvents are preferably alkyl acetates or ketones. Thus, the mere washing of 50 g bisphenol A polycarbonate with 250 ml ethyl acetate dissolved 0.6 g (1.2%) of the sample, of which 70% was cyclic oligomeric polycarbonate.[51] The separation of these low molecular weight compounds with a semipermeable membrane avoids the problems associated with recovering the solvent.[52]

Oligomeric polycarbonates were prepared by continuously supplying a cooled mixture of aqueous alkaline solution of dihydric phenol, $COCl_2$, and organic solvent to one end of a reaction tower.[53,54] Cyclic oligocarbonates (DP = 2 to 16) can be transformed to high molecular weight polycarbonates by polymerization *in situ*[55,56] or by integration with polycarbonate processing.[57]

In general, *direct amidation* of a dicarboxylic acid with a diamine proceeds at the same rate as esterification, but the equilibrium is much more favorable for polymer formation. Despite this peculiarity, industrial polyamides contain significant amounts of low polymers, mostly cyclic oligomers.

The most important polyamide — Nylon-6.6 — is prepared by step-growth reaction of hexamethylene diamine with adipic acid. Other polyamides of practical interest are Nylon-6.10 from hexamethylene diamine and sebacic acid, Nylon-7 prepared from 7-aminoenanthic acid, and Nylon-11 prepared from 11-aminoundecanoic acid. The general synthetic procedures for the preparation of linear, high molecular weight polyamides are direct amidation of a dibasic acid with a diamine; direct amidation of an amino acid containing five or more carbon atoms between functional groups; amidation of a diester with a diamine; amidation of dinitriles with diamines or with carbonium ion-forming functional groups (the Ritter reaction); and, finally, amidation of a diacid chloride with a diamine (the Schotten-Baumann reaction).[58,59]

By direct extraction of industrial polyamides with appropriate solvents (e.g., methanol, dimethylformamide, and dioxan), low molecular weight fractions (1 to 11% by weight) were separated. In these fractions, the linear oligomers only amounted to 0.1%.[60,61] In the case of Nylon-6.6, linear oligoamides had a DP between 1 and 3.[62]

By oligomerization of 6-aminocaproic acid at 240 to 260°C in the presence of benzylic amines as terminating agent, a linear hexamer with a melting point of 198°C was obtained.[63]

Synthesis of a fully aromatic linear oligoamide by low-temperature interfacial polycondensation between terephthaloyl chloride and 1,4-phenylendiamine was reported. By this reaction, linear oligoamides were obtained with a chain length which consisted mainly of three to four monomer units.[64]

Linear oligomers are always "natural" components of all condensation polymers. They are formed through the same mechanism as that of formation of macromolecular compounds of similar structure. Cyclic oligomers, however, result from secondary reactions which are governed by specific mechanisms.

The distance between the two functional groups of the low molecular compound (monomer, dimer, trimer, etc.), i.e., the number of atoms that separate them, determines the probability of intramolecular collision of these groups and, therefore, that of cyclization. Since the step-growth polycondensation is second order while the cyclization is an intramolecular reaction of the first order, it follows that the competition between these two equilibrium processes is balanced by the Baeyer strain of the cyclooligomer.

The ring strain arises from:

1. Bond angle distortion (angle strain)
2. Bond stretching or compression; repulsion between eclipsed hydrogen atoms or substituents on neighboring ring atoms (conformational strain, bond torsion, bond opposition)

3.   Nonbonded interaction between atoms or substituents attached to different parts of the
       ring (transannular strain, compression of Van der Waals radii)

Severe distortion of bond angles is the major source of the high strain in three- and
four-membered rings. Bond opposition forces arising from eclipsed conformations are responsible for the strain in five-membered rings. Six-membered rings can adopt a strain-free
conformation and possess the highest stability. In medium-sized cyclooligomers, strain arises
primarily from nonbonded interactions and bond oppositions. In very large rings, containing
more than 12 to 15 atoms, transannular interactions can be avoided completely by arranging
the ring atoms into two almost parallel chains.

Poly(ethylene terephthalate) contains cyclic as well as linear oligomers, predominantly
cyclic oligomers. These species have been determined by solvent extraction or by chromatographic and spectroscopic methods.[65-71] The amount of these low molecular weight
species varies with the method of determination, degree of processing, and thermal history
of the polymer.[72-90] Some cyclic oligomer species contain one ethylene glycol linkage
($-O-CH_2-CH_2-O-$) and one diethylene glycol linkage ($-O-CH_2-CH_2-O-CH_2-CH_2-O-$).[91,92]

Heat treatment of polyester yarns induces the selective crystallization of the cyclic trimer
in the form of polygonal solids on the fiber surface at 200°C. The diffusion rate of cyclic
trimer is very rapid. Small quantities of the cyclic tetramer may also be found. The total
extractable oligomer content of the samples is reduced considerably after annealing at 200°C,
mainly due to solid-phase polymerization. Exposure to CH_2Cl_2 vapors also allows cyclic
oligomers to diffuse to the fiber surface, and larger cyclics migrate in appreciable quantities.
However, solvent-exposed samples, unlike the annealed fibers, exhibit on their surface
crystals of irregular shapes and sizes.[93]

Oligoesters were used in various domains of industry as solid coatings, textile auxiliaries,
lubricants, and adhesives.[94-111] Kinetic studies[65-80] and mathematical modeling[112,113] of polyesterification reactions were performed in order to establish the mechanism of cyclic oligomers formation.

On the basis of the Flory distribution for the weight fraction of molecular species with
size x (Equation 60)

$$W_x = xp^{x-1}(1 - p)^2 \qquad (60)$$

the proportion of oligomers is small, and decreases as p increases. Only linear oligomers
have been considered in this regard.[114]

The formation of *cyclic oligoesters* is caused by one of three specific mechanisms: (1)
the cyclization of the linear oligomers, (2) cyclodepolymerization proceeding from the chain
end, and (3) exchange-elimination reaction.[75,76] These three mechanisms are presented in
Figure 1.

Alcoholysis can also be involved in the formation of cyclic oligomers in polyesters.[80]

The amount of cyclic oligomers in polyesters is influenced by the extension of time, t,
for which the polyester is kept in a molten state, the molecular value of the polymer, and
the conversion p. On the basis of experimental data, Ha and Choun[80] proposed the following
reaction:

$$K_x = \frac{C_x}{p^x} \qquad (61)$$

where K_x is the molar cyclization equilibrium constant and C_x represents the molar concentration of cyclic oligomers. The amount of cyclic oligomers is related to t through an
exponential function:

$$C_x = t^{0.3} \qquad (62)$$

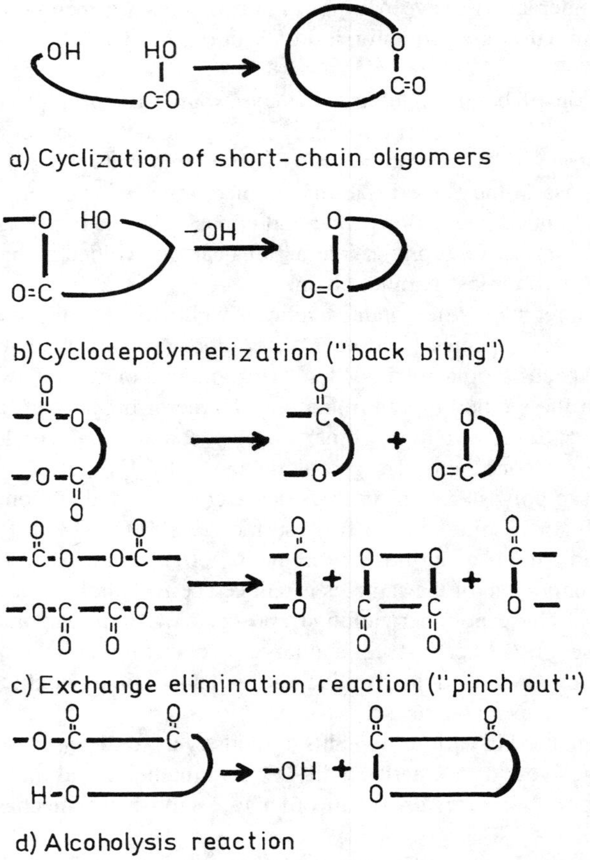

a) Cyclization of short-chain oligomers

b) Cyclodepolymerization ("back biting")

c) Exchange elimination reaction ("pinch out")

d) Alcoholysis reaction

FIGURE 1.   The mechanism of formation of cyclic oligoesters.

The polymerization of any lactam may be achieved by a series of repeated condensation or addition reactions.[115] It starts with the ring opening of the monomer (or activated monomer) by the initiator, e.g., H_2O:

$$H_2O + HN - CO \leftrightarrow H_2N \quad COOH \tag{63}$$

The growth of polymer chains then proceeds either through additions between terminal groups of two linear chains

$$H_2N\text{mm}CON - CO + H_2N\text{mm}CON - CO \rightleftharpoons$$

$$H_2N\text{mm}CO\text{–}NH \quad CO\text{–}NH\text{mm}CON - CO \tag{64}$$

or condensation between two linear molecules

$$H_2N\text{mm}COOH + H_2N\text{mm}COOH \rightleftharpoons H_2N\text{mm}CO\text{–}NH\text{mm}COOH + H_2O \tag{65}$$

or addition of monomer (or activated monomer) to the active end of the polymer molecule:

$$\text{mm}CON - CO + {}^\theta N \quad CO \rightleftharpoons A\text{mm}CON^\theta \quad CON \quad CO \tag{66}$$

Similar reactions of chains composed of more than one monomer unit lead to the following equilibria involving cyclic oligomers:

$$\left(\begin{array}{c} NH_2 \\ \\ COOH \end{array}\right. \rightleftharpoons H_2O + \left(\begin{array}{c} NH \\ | \\ CO \end{array}\right. \tag{67}$$

or

$$\left(\begin{array}{c} CO-NH\text{\small\textasciitilde} \\ \\ NH_2 \end{array}\right. \rightleftharpoons \left(\begin{array}{c} CO \\ | \\ NH \end{array}\right. + H_2N\text{\small\textasciitilde} \tag{68}$$

Transacylations between polymer molecules are classified as exchange reactions which do not alter the number of molecules and affect only the molecular weight distribution, e.g.,

$$\begin{array}{c} A\text{\small\textasciitilde}NH-CO\text{\small\textasciitilde}B \\ + \\ C\text{\small\textasciitilde}NH_2 \end{array} \rightleftharpoons \begin{array}{c} A\text{\small\textasciitilde}NH_2 \\ \\ C\text{\small\textasciitilde}NH-CO\text{\small\textasciitilde}B \end{array} \tag{69}$$

where A, B, and C represent fragments of initiating species.

` In homogeneous media, most of the transacylation reactions are reversible, and as soon as the first polymer (oligomer) amide groups are formed, the same kind of reaction can occur at both the monomer and polymer amide group. Unless the active species are steadily formed or consumed by some side reaction, a set of thermodynamically controlled *equilibria is established between monomer, cyclic as well as linear oligomers, and polydisperse linear polymers*. The existence of these equilibria is a characteristic feature of lactam polymerization and has to be taken into account in any kinetic treatment of the polymerization and analysis of polymerization products. The equilibrium fraction of each component depends on the ring size of the monomer, substitution and dilution, and temperature and catalyst concentration. For medium and large lactams, such as laurinlactam, the fraction of cyclic dimer (L_2) can be even higher than that of the monomer (L_1) or other cyclic oligomers, e.g., 0.33% (L_1), 0.94% (L_2), and 0.25% (L_3).[115a]

The conversion of caprolactam to linear macromolecules is also characterized by concomitant formation of low molecular weight cyclics. In this case, however, the fraction of the monomer at equilibrium is by far the highest, e.g., 8.5% (L_1), 0.7% (L_2), 0.43% (L_3), 0.22% (L_4), and 0.13% (L_5).[115]

The lower oligomers are soluble in water and C_1-C_3 alcohols, and cyclic oligomers have been determined in the extractable fraction. The fraction of cyclic oligomers remaining in the polymer increases with their molecular weight. Individual members ranging from cyclic dimer to cyclic nonamer were identified in polymer extracts.[116,117] The formation of these ring structures may be explained by general intra- or intermolecular transacylation and intermolecular transamidation, or direct cyclization of the corresponding linear oligomers (Equations 67 to 69). For the latter case, experimental findings did not appear to agree with cyclization theory.[118] Considerable research has been reported on the separation, preparation, and identification of cyclic structures in Nylon-6.[119-121]

During the polymerization of ε-caprolactam, about 3 to 10% (w/w) low molecular weight cyclic oligomers of caprolactam are formed.[122,123] The amount of cyclics in polyamide

TABLE 1
Concentration of Cyclic Oligomers (CO, % w/w) in
Equilibrium Caprolactam Polymers
$-[-HN(CH_2)_5CO-]_x-$[123]

	Reaction conditions			
	Hydrolytic polymerization		Anionic polymerization	
CO	250°C	250—180°C[a]	220°C	190—180°C[b]
x	(CO, %)	(CO, %)	(CO, %)	(CO, %)
1	7.80	1.82	6.15	1.92
2	1.13	0.43	0.90	0.29
3	0.78	0.14	0.58	0.20
4	0.59	0.06	0.43	0.14
5	0.45	0.02	0.34	0.11
6	0.34	0.01	0.27	0.07
$\overline{M}_n$ polymer	14 340	20 450	17 630	47 700

[a] Polymerized at 250°C and annealed for 48 h at 180°C.
[b] Polymerized at 190°C and annealed for 48 h at 180°C.

depends on the type of initiation, the reaction temperature, and the thermal history of the polymer (Table 1).

Formation of cyclic oligomers during the conversion of caprolactam to linear macromolecules may be described by the reequilibration between the *-meric cyclics and the y-meric linear chains, according to

$$-(M_y)- \rightleftharpoons -(M_{y-x}) + -(M_x)- \qquad (70)$$

where M denotes the monomeric unit $-HN(CH_2)_5CO-$.

The cyclization constant K_x is represented according to the interpretation given by Jacobson and Stockmayer[15a]

$$K_x = -[-(M_{y-x})-][-(M_x)-]/[-(M_y)-] = [-(M_x)-]/p^x \qquad (71)$$

where p is the extent of the reaction of the functional groups of the linear polymer molecules. It is defined by the relation

$$p = 1 - \frac{1}{\overline{DP}_n} = 1 - \frac{113}{\overline{M}_n} \qquad (72)$$

where 113 is the mass of the monomeric unit M. If $\overline{DP}_n \gg 1$, then $p^x \simeq 1$ and $K_x \simeq [-(M_x)-]$, where $[-(M_x)-]$ is given in mol/kg. For Nylon-6, with $\overline{M}_n$ in the range of 14,300 to 47,000, that had been obtained by polymerizing caprolactam at temperatures of 180 and 265°C, the value of K_x can be approximated by

$$\log K_x = \frac{A}{T} + B - 0.67C \qquad (73)$$

where C is the fraction of the polymer that has crystallized during the process of polymerization. T is the temperature in K, and values of A and B as a function of the ring size are listed in Table 2.[123]

Except for the monomer (x = 1), Equation 73 does not hold when the polymerization

TABLE 2
Values of Constants A and B as
a Function of Ring Size[123]

$[-HN(CH_2)_5CO-]_x$

x	A	B
1	−887	1.535
2	−833	0.291
3	−1128	0.519
4	−1170	0.351
5	−1073	0.045
6	−833	0.710

is carried out at temperatures above 220°C, and the resulting polymer then reequilibrates at temperatures below its melting point, with or without prior removal of fractions that are soluble in water or methanol. This parallels the observation that the concentration of cyclic oligomers in polymers quenched from the melt and then annealed at specific temperatures below the polymer melting point are different from those of polymers for which conversion equilibrium has been reached by polymerization of caprolactam at the corresponding temperatures (Table 1). Thus, it is not the mechanism or the particular process of polymerization that affects the oligomer concentration in polyamides, but different thermal histories. Different thermal treatment after equilibrium polymerization may give rise to different average chain conformation which, according to Semlyen, may in turn be responsible for different polymer compositions.[122] The author concluded that the high concentrations of cyclic monomer and dimer and the low concentrations of the higher rings that result from the ring-chain interconversion in the solid polymer (Table 1) is a consequence of a high concentration of low molecular weight chains in the amorphous regions and/or the failure of the linear macromolecules in these regions to obey Gaussian statistics.

The formation of cyclic structures may be favored by the polycondensation procedure. For example, only cyclic oligomers result from the interfacial polycondensation of chloroanhydrides of adipic acid or of phenylphosphonyl dichloride with N,N'-diethyleneamine. High-dilution polycondensation is also favorable for the synthesis of cyclic oligomers.[124,125]

B. ADDITION-ELIMINATION REACTIONS

Addition-elimination reactions of aldehydes have been applied to the synthesis of step-growth polymers through the reaction of a dialdehyde with a diamine. Polymers having repeating units with a Schiff base or azomethine structure have been prepared by this reaction, but unusually large ring structures and linear oligomers represent the main products.[126]

C. CARBONYL ADDITION-SUBSTITUTION REACTIONS

Polyacetals, phenol-, urea- and melamine-formaldehyde polymers as well as other types of polymers were obtained through carbonyl addition-substitution reactions. Phenol-formaldehyde polymers are prepared by either of two fundamentally different reactions, according to whether acidic or basic conditions are used.

Condensations of formaldehyde with compounds such as phenol, cresol, urea, and melamine involve so many steps that kinetic analysis is difficult. In the reaction with phenol in alkaline medium, the formation of methylol derivatives

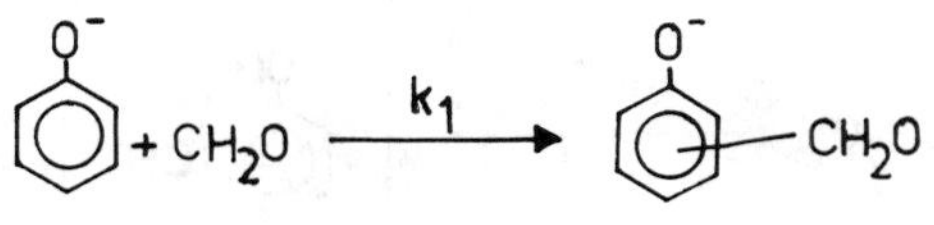

$$\text{(74)}$$

is relatively fast, compared to the coupling of aromatic rings:

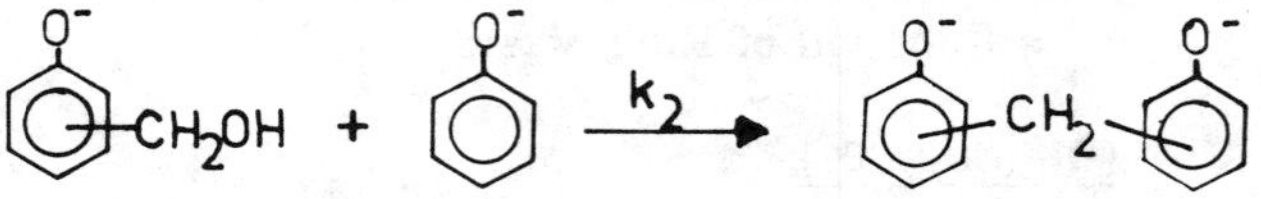

$$\text{(structure)} \quad (75)$$

The coupling reaction is the faster one in acidic medium, however. This difference is utilized to prepare low molecular weight phenolic oligomers using an alkaline catalyst. These prepolymers are moderately stable at low temperatures, and may be cross-linked by acidification. In the terminology of Baekeland, this step is termed an A-stage polymerization reaction and the product is an A-stage polymer. Bakelite A is a readily soluble hydroxymethyl phenol; it is produced at about 100°C and has one or more growing ends. A process carried out at 150 to 160°C produces the intermediate stage of partially condensed Bakelite B and, finally, at 160 to 200°C, the completely condensed Bakelite C is produced.

Ryabukhin[127] has proposed functional and kinetic relations for the condensation of phenol and formaldehyde in both acid and alkaline media. In alkaline systems, the concentration of reacted phenol x at any time was found to be in agreement with the equation

$$x = a - [(na - y)n^{-1}a^{(1-n)/n}]^n \qquad (76)$$

where a is the original concentration of phenol, y is the concentration of reacted formaldehyde, and n is the functionality of the phenol (n = 3 in alkaline medium). In acid medium, where n = 1, the data fit the equation

$$x = a - [(na - 2y)n^{-1}a^{(1-n)/n}]^n \qquad (77)$$

In alkaline media, the ratio of coefficients for the formation of mono-, di-, and trimethylolphenols had relative values of 1.0, 0.75, and 0.25, respectively.

For the reaction under acid conditions, it is convenient to think of the mechanism as involving an attack by a carbonium ion on an aromatic ring

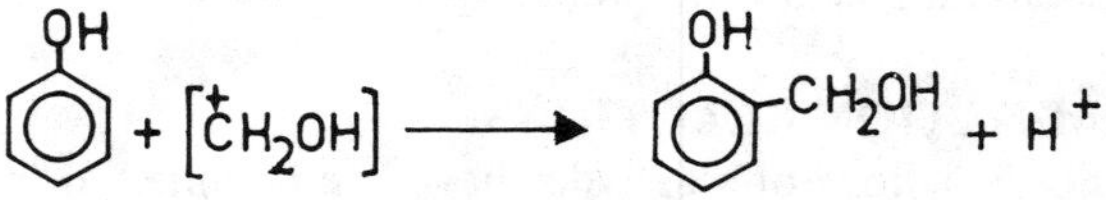

$$\text{(structure)} \quad (78)$$

while under basic conditions, the reaction is considered to involve the formation and addition of a quinone-type carbanion to formaldehyde:

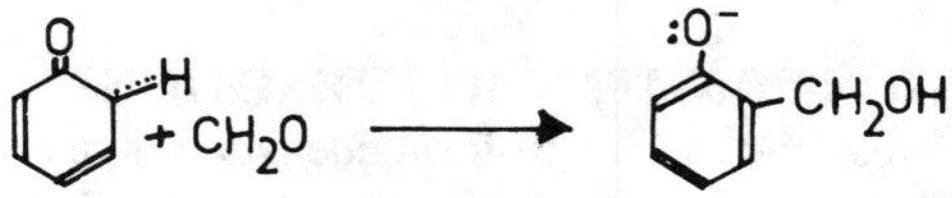

$$\text{(structure)} \quad (79)$$

The essential difference between the two types of A-stage reactions is that, under acidic conditions, the monomeric methylol derivatives of Reaction 78 are present only transiently in very small concentrations, while under basic conditions, the monomeric and dimeric methylol derivatives of Reaction 79 are stable. These compounds quickly condense to form low molecular weight poly(hydroxyphenyl methylene) structures:

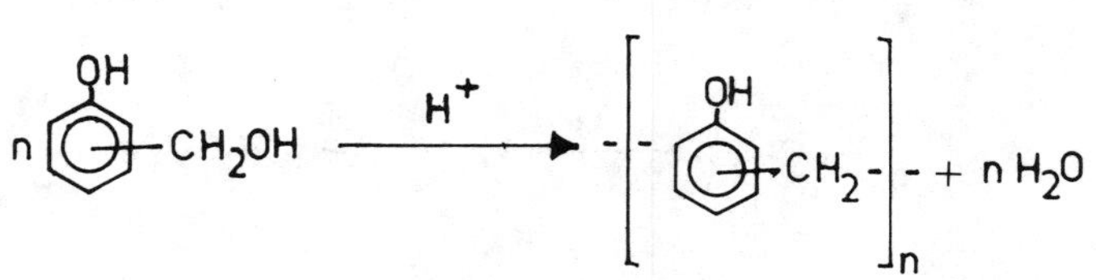

$$\text{(structure)} \quad (80)$$

These structures serve as prepolymers in applications of this step-growth polymerization reaction.

The mechanism of the base-catalyzed reaction of phenol with formaldehyde to form either o-methylolphenol or p-methylolphenol is similar to that of an aldol condensation.

The overall enthalpy of reaction — which is exothermic — is -50 kJ/mol and the activation energy is about 71 to 75 kJ/mol. Contrary to the acid-catalyzed reaction (vide infra), the p-site has only a slightly higher reactivity than the o-positions. Since there are twice as many o-sites as p-sites, chiefly o-substituted methylol groups are formed.

The rate of reaction is a function of pH. The ratio of ortho:para substitution is also strongly dependent upon pH, i.e., it decreases from 1.1 at pH 9.8 to 0.39 at pH 13.0.[128] It increased, however, along the series Mg < Ca < Sn < Ba < Li < Na < K. This effect was attributed to coordination of the reactants and chelate formation in the transition state.[129]

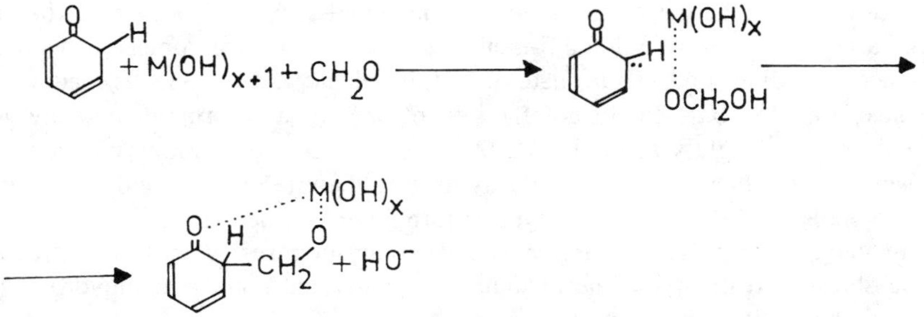

$$\text{(81)}$$

Recently, catalysis by cyclodextrin for a para-oriented attack of formaldehyde to phenol has been reported.[130] Cyclodextrins, cyclic oligomers of 6 to 8 glucose units, form inclusion complexes with various guest compounds.[131] β-Cyclodextrin exhibits a significant increase in selectivity and yield for the formation of (4-hydroxymethyl)phenol. At a concentration of 0.4 M of β-cyclodextrin, the para/ortho ratio is 5.3, corresponding to 84% selectivity for para, and the yield is 10.7 times that in its absence. Enhancement of the selectivity of α-cyclodextrin is minimal (70%) compared to that of the uncatalyzed reaction (63%). D-glucose, a noncyclic analogue of cyclodextrins, decreases this selectivity to 61%. These selectivities are indicative of the formation of an inclusion complex between phenol and β-cyclodextrin since the selective catalysis of the latter was observed only when the sodium hydroxide/phenol molar ratio was >1.[131]

Under the strongly alkaline conditions of prepolymer formation, the mononuclear methylolphenols obtained initially can undergo at least two different types of reactions to form a mixture of dinuclear and polynuclear phenolic compounds which is termed resole. By far the most important self-condensation reaction of methylolphenols under alkaline conditions is that which leads to the formation of diphenylmethanes (Reaction 82). The formation of dibenzyl ethers (Reaction 83) has also been suggested,

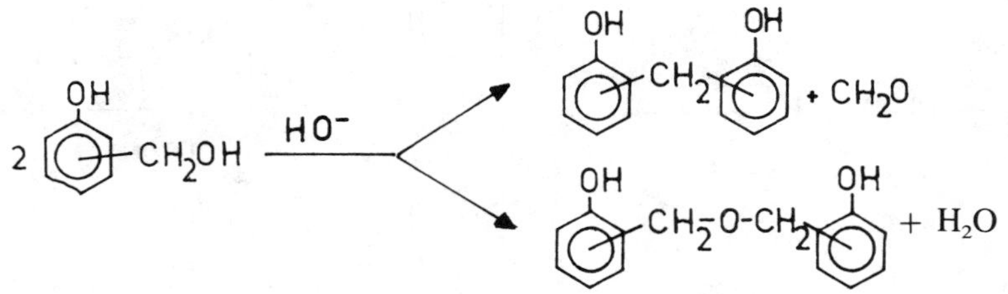

$$\text{(82)}$$

$$\text{(83)}$$

but this reaction probably does not occur to any significant extent under strongly alkaline conditions.[132] This choice is reflected by the corresponding energetic values of these structures, i.e., the activation energy for the formation of a $-CH_2-$ bridge is about 59 kJ/mol, while that of a $-CH_2-O-CH_2-$ bridge is 113 kJ/mol.

The average number of methylol groups on phenol nuclei in a particular resol mixture will depend upon a variety of factors, including the phenol/formaldehyde ratio, duration and temperature of reaction, type and concentration of catalyst, and the structure of the original phenol monomer. If these factors are varied, resoles of widely different structures are formed and, as a result, subsequent cross-linking reactions yield network polymers with significantly different properties. In the normal preparation of resoles, formaldehyde-to-phenol molar ratios between 1.5 and 3 are generally used.[132]

During the NaOH-catalyzed reaction of 4-*t*-butyl phenol with CH_2O, linear oligomers were formed in which the concentration of methylene bridges increased with temperature.[133] Condensation of very active phenols with formaldehyde to form resoles can be catalyzed by weak bases. Thus, resorcinol and 5-methylresorcinol can be oligomerized with CH_2O in the presence of sodium dodecilsulfonate or sodium oleate and in aqueous media.[134]

Resoles formed in base-catalyzed reactions of phenols and formaldehyde are generally neutralized or made slightly acidic before the second-stage cross-linking reaction is carried out. Cross-linking is then effected simply by heating the mixture of mono- and polynuclear methylol phenols, and the network polymers formed are termed *resites*.

Acid-catalyzed prepolymer (oligomer) formation supposes the following reactions in which the short-lived methylol intermediates are converted mainly to dihydroxydiphenylmethanes via benzylic carbonium ions:[135]

$$CH_2O + H^+ \leftrightarrow [\overset{+}{C}H_2OH] \tag{84}$$

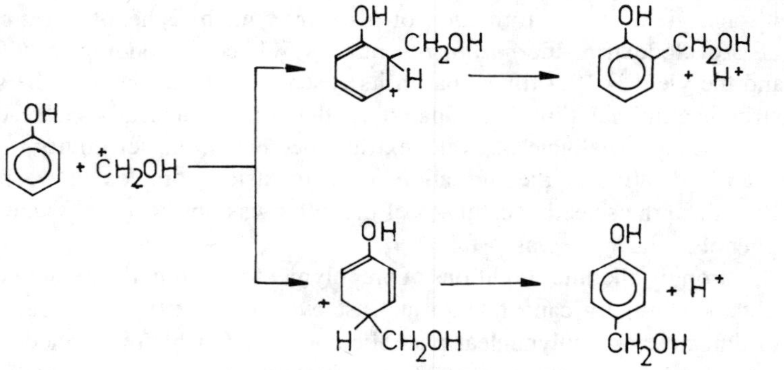

$$\tag{85}$$

$$\tag{86}$$

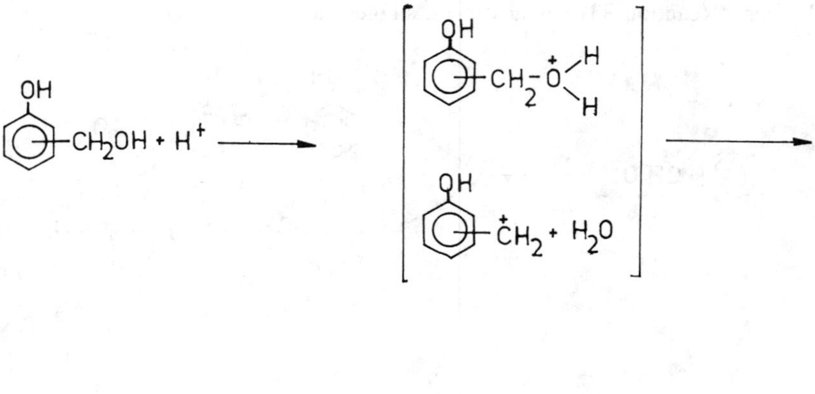

$$\tag{87}$$

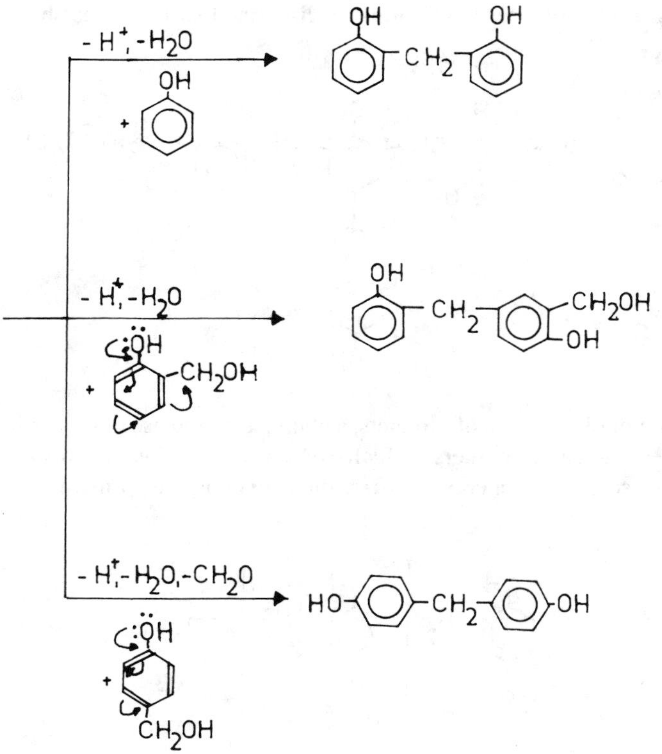

The acid-catalyzed reaction is exothermic to a considerable degree, with an overall reaction enthalpy of $\Delta H = -98.5$ kJ/mol. The addition reaction (addition of formaldehyde) contributes -20 kJ/mol and the condensation reaction (formation of methylene bridges) contributes -78.5 kJ/mol. The rates of addition and condensation reactions are in the ratio 1:42. The overall activation energy is about 84 to 110 kJ/mol.

Under strongly acidic conditions, both methylol substitution and methylene-bridge formation occur predominantly at the *para* position because of the electrostatic repulsion of the protonated phenolic hydroxyl group in attacking the *ortho* position. The *para* site in this acid-catalyzed reaction is about 2.4 times more reactive than the *ortho* site of the phenol. In general, then, *p*-methylol phenols are produced, but are not as attractive commercially as the *o,o'*-methylol-rich derivatives which have the reactive *para* sites available for curing reactions. The pH most favorable for the formation of *ortho* isomers is between 4 and 5. Continued methylation and methylene bridge formation by these reactions leads to a mixture of polynuclear compounds of considerable complexity which is termed *novolacs* ($\overline{M} = 600$ to 1500).

High *o,o'*-content novolacs occur at a moderately high H^+ concentration, however, because the ortho-methylol phenol formed as intermediate is rapidly stabilized by hydrogen bonding. The stability and, consequently, yield of this compound can be increased by the addition of chelate-forming metals (see above).

In order to prevent uncontrolled cross-linking in the first-stage preparation of novolac prepolymers, a slight excess of phenol is required. As a result, the prepolymers are of relatively low average degrees of polymerization (DP = 2 to 15) with broad molecular weight distribution. They are complex mixtures of linear and branched oligomers having methylene-bridged phenolic repeating units of ill-defined structure, as shown schematically in Structure 91:

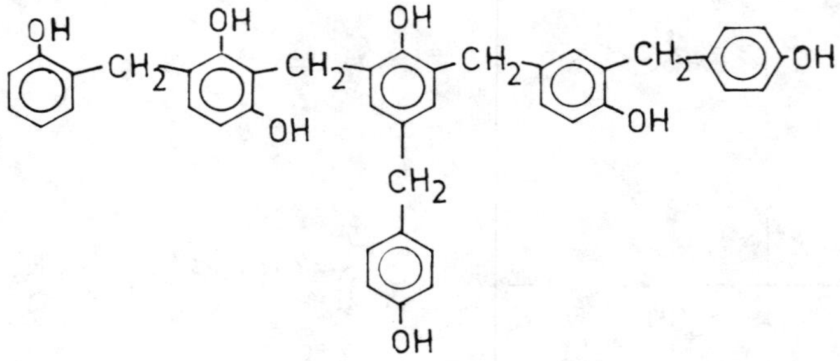

(91)

A much simpler distribution of products is obtained by the use of *para*-substituted phenols as the monomer. Linear oligomers, which were used as phenolic resin precursor, were prepared from 4-isopropenylphenol or 4-*tert*-butyl phenol and formaldehyde:[136]

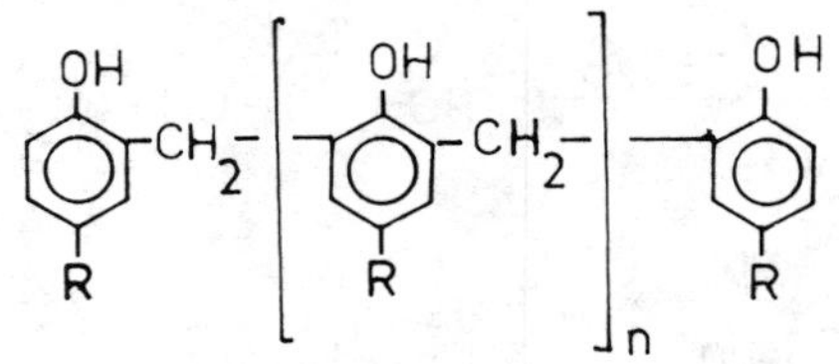

R	$i-C_3H_7$; $t-C_4H_9$
n	$O-4$

(92)

Besides these linear structures, Vicens[137] reported the formation of cyclic oligomers of the following type:

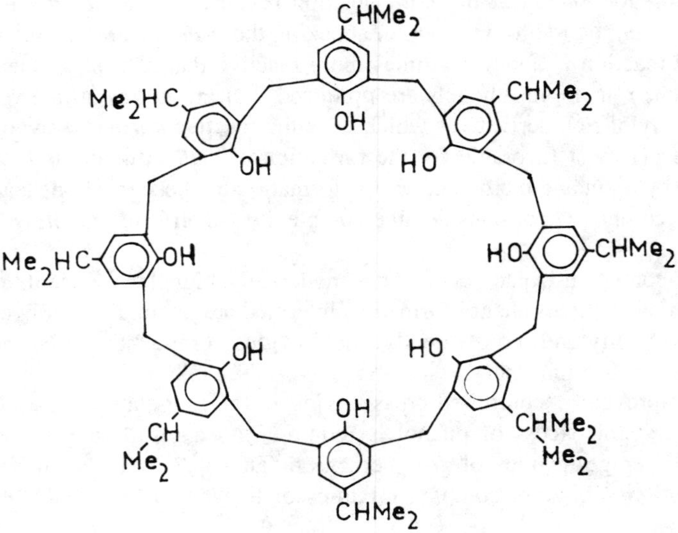

(93)

A much simpler distribution of products is obtained by the use of *para*-substituted phenols as the monomer. Linear oligomers, which were used as phenolic resin precursor, were prepared from 4-isopropenylphenol or 4-*tert*-butyl phenol and formaldehyde:[138]

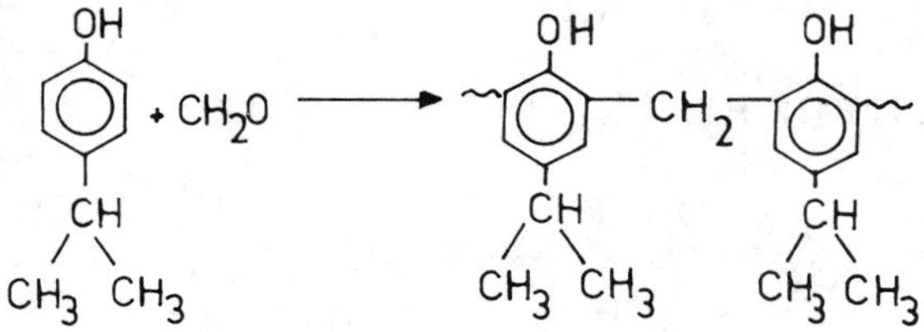

(94)

Novolacs free of cyclic compounds were obtained when 4-*tert*-butyl phenol was condensed with formaldehyde in aprotic nonpolar media.[139,140] Linear phenolic oligomers with narrow molecular weight distributions may also be prepared in the presence of Lewis acids as catalysts.[141]

Novolac resins (DP = 1 to 10) with a high degree of *ortho* substitution can be prepared by the metallic ion (Zn^{2+}, Mg^{2+}, and Al^{3+})-assisted reaction of phenols with paraformaldehyde.[139,140] These prepolymers are more readily cross-linked than randomly constructed novolacs obtained with protonic acids catalysts, apparently because of the much greater proportion of the more reactive *para* positions available for the second-stage reaction, i.e., these resins have been found to have much shorter curing times (while resoles are generally cured by heat without the addition of an auxiliary chemical cross-linking agent, novolacs require the assistance of such a compound).[142]

Recently, the Monte Carlo method was used for mathematical modeling of the synthesis of phenol-formaldehyde oligomers.[143] A series of differential equations for the kinetics of this condensation were derived for describing the formation of resoles and novolacs, especially in the early stages of polycondensation, by taking into account the functionality of all reacting species.

Higher aldehydes are used on a much lesser scale than formaldehyde for obtaining phenol-aldehyde resins. The base-catalyzed reaction of acetaldehyde and phenol does not yield resoles, since the aldehyde is most likely converted to aldol

$$CH_3CHO + HO^- \xrightarrow[-H_2O]{} {}^-CH_2CHO \xrightarrow{CH_3CHO}$$

$$\underset{\underset{CH_2CHO}{|}}{CH_3CHO^-} \xrightarrow[-\,{}^-CH_2CHO]{+CH_3CHO} \underset{\underset{OH}{|}}{CH_3CH-CH_2CHO} \qquad (95)$$

which further condenses with acetaldehyde to form 2,4-dimethyl-6-oxy-1,3-dioxane

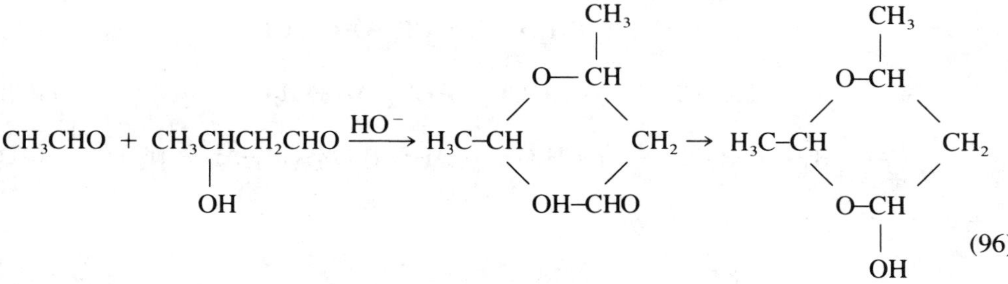

(96)

and higher oligomers, i.e., paraaldol (A), dialdol (B), and izoaldol (C):

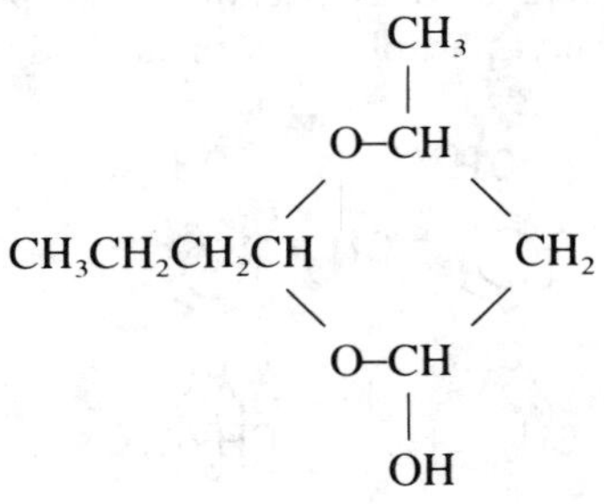

$$
\begin{array}{c}
\text{CH}_3 \\
|\\
\text{O–CH} \\
\diagup \qquad \diagdown \\
\text{CH}_3\text{CH}_2\text{CH}_2\text{CH} \qquad\qquad \text{CH}_2 \\
\diagdown \qquad \diagup \\
\text{O–CH} \\
|\\
\text{OH}
\end{array}
\qquad\qquad
\begin{array}{c}
\text{OH} \\
|\\
\text{CH} \qquad \text{CHO} \\
\diagup \quad \diagdown \ \diagup \\
\text{CH}_2 \qquad \text{CH} \\
|\qquad\qquad | \\
\text{CH} \qquad\quad \text{CH} \\
\diagup \ \diagdown \quad \diagup \ \diagdown \\
\text{H}_3\text{C} \qquad \text{O} \qquad \text{CH}_3
\end{array}
$$

(A) 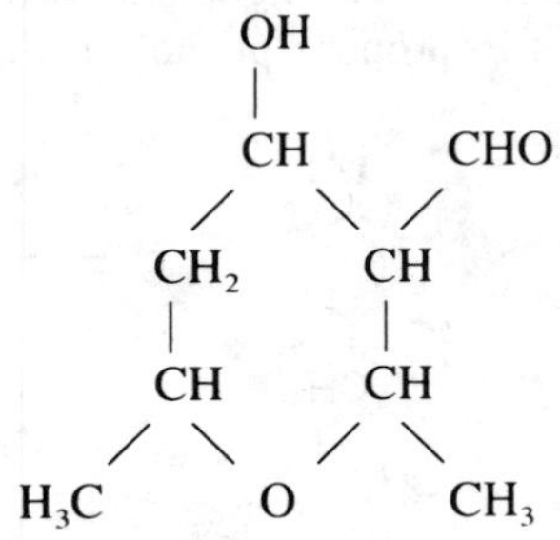   (B)

(97)

$$
\begin{array}{c}
\text{H}_3\text{C} \qquad \text{CH}_2 \qquad \text{O–CH–CH}_2 \qquad \text{O} \qquad \text{CH}_3 \\
\diagdown \ \diagup \quad \diagdown \ \diagup \quad | \qquad \diagdown \ \diagup \quad \diagdown \ \diagup \\
\text{CH} \qquad \text{CH} \ \ \text{CH}_3 \qquad \text{CH} \qquad \text{CH} \\
|\qquad\quad |\qquad\qquad\quad |\qquad\quad | \\
\text{O} \qquad\ \text{O} \qquad\qquad\quad \text{O} \qquad\ \text{CH}_2 \\
\diagdown \ \diagup \qquad\qquad\quad \diagdown \ \diagup \\
\text{CH} \qquad\qquad\qquad \text{CH} \\
\diagdown \qquad\qquad\qquad \diagup \\
\text{CH}_2\text{–CH}_2\text{–CH–O} \\
|\\
\text{CH}_3
\end{array}
$$

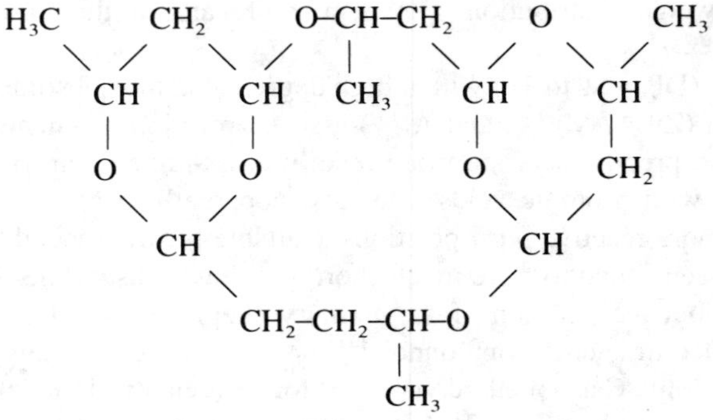

(C)

Under strongly acidic conditions (HCl, H_2SO_4), acetaldehyde reacts with phenols (1:1 molar ratio) to form novolac resins. Higher molar ratios of phenol:acetaldehyde are required when the aldehyde cyclic trimer, paraaldehyde, is used, and the characteristics of novolacs are dependent on the ratio used (Table 3).[144]

Novolac oligomers are also formed when acetaldehyde reacted under acidic conditions with cresols, 2,5-, 3,4-, or 3,5-xylenol, *p-tert-* or *p*-isobutylphenol, *p*-ethyl, *p*-octyl-, *p*-cumyl-, *p*-nonylphenol-, *p*-benzyl-, *p*-phenyl-, or *o*-allylphenol.[145] At the same time, other higher aldehydes can be used instead of acetaldehyde, such as propionaldehyde, iso-butyraldehyde, acrolein, benzaldehyde, and crotonaldehyde.[146-148]

Lower aliphatic aldehydes can also react with aromatic hydrocarbons or fractions of such hydrocarbons separated during oil processing to form oligomeric polycondensates. The following general scheme can be written for the acidic condensation of a generic aromatic hydrocarbon, Ar–H, with formaldehyde:

$$\text{Ar–H} + \text{CH}_2\text{O} \xrightarrow{\text{H}^+} \text{Ar–CH}_2\text{OH} \xrightarrow{\text{Ar–H}} \text{Ar–CH}_2\text{–Ar} + \text{H}_2\text{O} \qquad (98)$$

$$\text{Ar–CH}_2\text{OH} + \text{CH}_2\text{O} \rightarrow \text{Ar–CH}_2\text{–O–CH}_2\text{OH} \qquad (99)$$

$$2\,\text{Ar–CH}_2\text{OH} \rightarrow \text{Ar–CH}_2\text{–O–CH}_2\text{–Ar} + \text{H}_2\text{O} \qquad (100)$$

$$\text{Ar–CH}_2\text{–O–CH}_2\text{OH} + \text{Ar–CH}_2\text{OH} \rightarrow \text{Ar–(CH}_2\text{O)}_2\text{–CH}_2\text{–Ar} + \text{H}_2\text{O} \qquad (101)$$

TABLE 3
Characteristics of Oligomers Obtained by Condensation of CH_3CHO and C_6H_5OH in the Presence of HCl.[144] General Formula

$$H-[-C_6H_3(OH)-CH-]_n-C_6H_4OH$$
$$|$$
$$CH_3$$

Molar ratio C_6H_5OH/CH_3CHO		Softening range		
Feed	Oligomer	(0°C)	$\overline{M}_n$	n
4.00	2.00	123	214	1.00
3.00	1.68	39—63	270	1.47
1.50	1.33	75—103	457	3.07
1.25	1.27	99—123	633	4.50
1.00	1.14	128—144	940	7.00

Taking *m*-xylene as a model compound, xylene-formaldehyde oligomeric resins containing up to 16% oxygen were synthesized:[149,150]

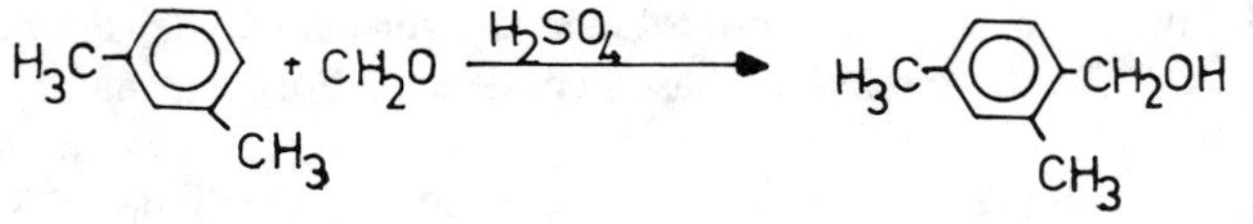

(102)

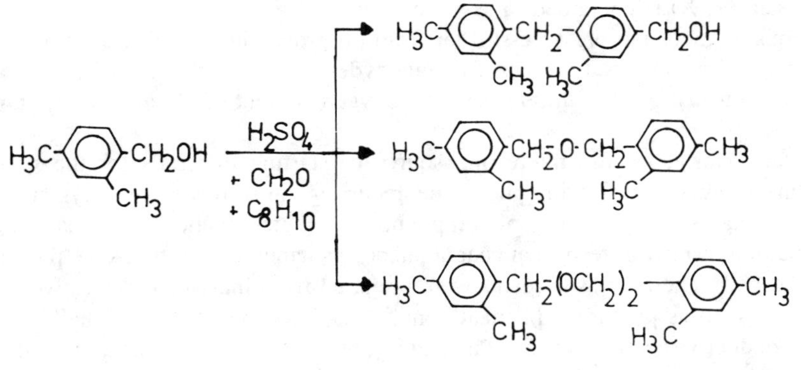

(103)

(104)

(105)

The following structure has been proposed for the product of the condensation between naphthalene and formaldehyde:[150]

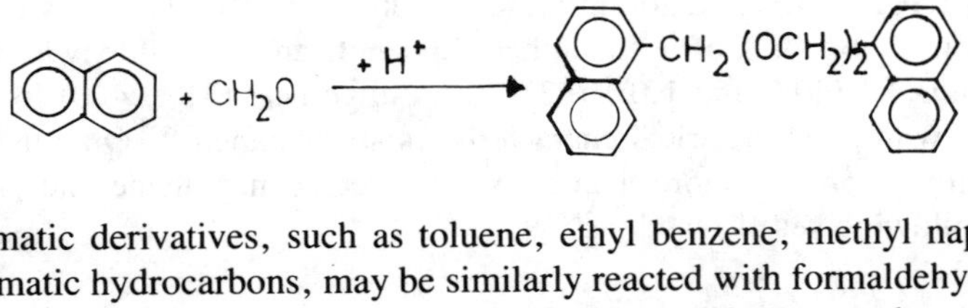

(106)

Other aromatic derivatives, such as toluene, ethyl benzene, methyl naphthalene, or a mixture of aromatic hydrocarbons, may be similarly reacted with formaldehyde under acidic conditions to form low oligomeric condensates.[151] Formaldehyde may be used in the form of formalin, paraformaldehyde, or trioxane, and the catalyst may be an aqueous solution of sulfuric acid (0.2 to 2 parts by weight vs. 1 part water) and C_1–C_2 monocarboxylic acids (1 part).

The following types of chemical structures were identified (IR, NMR) in the mixture of toluene-formaldehyde condensates:

1. Hydrocarbons, Ar–CH$_2$–Ar
2. Diaryl ethers, Ar–CH$_2$–O–CH$_2$–Ar
3. Acetals, Ar–CH$_2$–(OCH$_2$)$_n$–Ar, n = 1, 2, 3, . . .
4. Alcohols and glycols, Ar(CH$_2$OH)$_m$, m = 1, 2, 3, . . .
5. Methyl aryl ethers, Ar–CH$_2$–O–CH$_3$
6. Carbonyl derivatives, Ar–CHO

The distribution of these structures is determined by the concentration of the acid and purity of the aldehyde.

The promoting species seems to be the cationic intermediate Ar–CH$_2^+$. The presence of alcoholic HO- groups in aryl-formaldehyde oligomers opens the possibility for further reactions, known as *subsequent condensations*.[152,153] The reaction of *p*-xylylene glycol with 37% aqueous formaldehyde (formalin) in the presence of HCl as catalyst yielded polyacetalic oligomers which were similar to those obtained from xylene. When this glycol was reacted with phenol, dioxydiphenylmethane and *p,p'*-dioxydibenzylbenzene were obtained.

Since the compounds having etheric bonds are formed more easily, they predominate over diphenylmethane derivatives. Therefore, the reaction products have an oxygen content of 7 to 20%. For example, the resin obtained from toluene and formaldehyde (Polytol®) has an oxygen content of 8 to 12%, while the xylene-based condensate Polyxyl® has 12 to 18% oxygen ($\overline{M}$ = 300 to 600).

When the aldehyde is in excess (e.g., aromatic hydrocarbon/formaldehyde = $^1/_2$), most of the oxygen (75%) is contained in acetalic groups.[154]

Resinous extracts and pressure-still tar from oil processing are also used in industry as a raw material for condensation with formaldehyde.[155] Despite the higher extract/aldehyde feed ratios (up to $^1/_7$) used in these cases, the oxygen content of oligomers does not exceed 7%.[156]

Coal tar is an especially interesting source of starting materials for the synthesis of various linear prepolymers with phenylene recurring units linked by oxygen-containing bridges. It contains a large variety of compounds with heterocyclic and aromatic structures; of particular interest in this respect are naphthalene, acenaphthene, anthracene, phenanthrene, and pyrene.[157-168] Thermoplastic resins were prepared from individual derivatives separated from coal tar such as pyrene,[161] phenanthrene,[168] naphthalene,[159] and indene[160].

Photoconductive coatings were claimed to be obtained by the condensation of anthracene-rich coal tar fractions with CH$_2$O.[157]

The quality of novolac-type resins obtained from coal tar distillates is improved when phenol is added in a ratio of 1:1 to 2:1 (w/w) phenol:distillate. Formaldehyde usually is introduced as an aqueous solution (30 to 35%) in a molar ratio of 0.9:1.0 aldehyde:phenol, and the catalyst (HCl) in the reaction mixture amounts to 0.5 to 3% (by weight).

Condensation of mixtures of acenaphthene, phenol, and formaldehyde in molar ratios varying between 0.3:1.0:0.9 and 1.0:1.0:2.4 gives oligomers ($\overline{M}$ ~ 400) from which thermosetting resins with good electrical characteristics are obtained.[167] Oligomers contain 3.4 to 6.0% HO– groups and no more than 5 to 10% free acenaphthene and phenol. Kinetic studies of the following set of reactions[169]

$$\text{(107)}$$

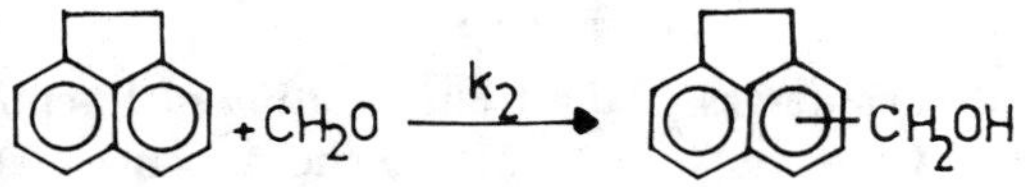

$$+ CH_2O \xrightarrow{k_2} \quad CH_2OH \tag{108}$$

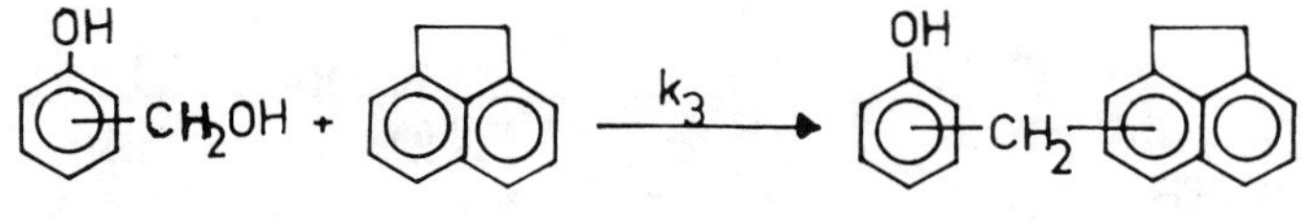

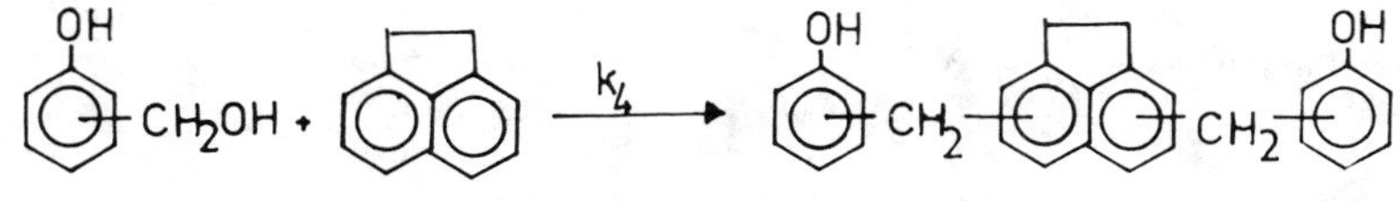

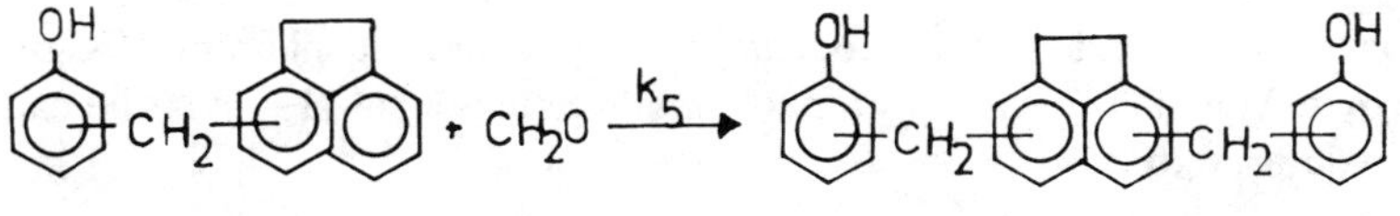

$$OH \quad CH_2OH + \quad \xrightarrow{k_3} \quad OH \quad CH_2 \tag{109}$$

$$OH \quad CH_2OH + \quad \xrightarrow{k_4} \quad OH \quad CH_2 \quad CH_2 \quad OH \tag{110}$$

$$OH \quad CH_2 \quad + CH_2O \xrightarrow{k_5} \quad OH \quad CH_2 \quad CH_2 \quad OH \tag{111}$$

$$OH \quad CH_2 \quad + \quad \xrightarrow{k_6} \quad OH \quad CH_2 \quad CH_2 \quad OH \tag{112}$$

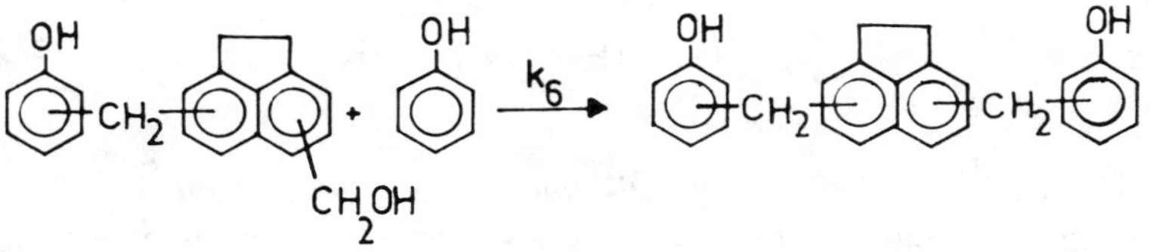

indicated that (1) in the presence of acenaphthene, Reaction 107 is much slower and the activation energy lower (Table 4), (2) $k_2 \ll k_1$, and (3) the probability of Reactions 111 and 112 is very low.

Some data have been reported for the reaction of formaldehyde with α-methylstyrene,[170] i.e., the Kriewitz-Prins reaction.[171] With a 40:60 water:dioxane mixture as solvent, the process was first order kinetically in each reactant, and the major reaction product was the olefin dimer 2,4-diphenyl-4-methylpentene-1. A direct relation between the logarithm of the rate coefficient and the acidity function H_o (Hammett) of the mineral acid catalyst (H_2SO_3) was found. In aqueous solutions of H_2SO_4 (0.5 to 3.0 mol/l), α-methylstyrene ($\sim 10^{-4}$ mol/l) is easily hydrated to dimethylphenylcarbinol and dimerization does not occur.

Methylol derivatives of unsaturated phenols obtained from formaldehyde and 4-isopropenyl- or 4-allylphenol can be oligomerized in the molten state without a catalyst by heating up to 200°C.[172] At 115°C, ether bridges are formed by the interaction of methylol groups which, at 140°C, give rise to methylenic bridges. Polymerization of isopropenyl groups starts at t = 150°C and of allyl groups at t = 190°C.

Polyformals, polyspiroacetals, and poly(thioacetals) can be obtained by the addition-substitution reaction of an aldehyde with alicyclic or long-chain aliphatic diols or by that between a dialdehyde and polyfunctional mercaptans. During the acid-catalyzed reaction of formaldehyde with a series of aliphatic glycols, the formation of five-, six-, and seven-membered rings prevented the preparation of polymers from glycols below 1,5-pentandiol. A large number of polythioacetals can be synthesized from polymethylene dithiols and aliphatic or aromatic monoaldehydes.[173] The molecular mass of these polycondensates is generally between 1000 and 1500 g/mol.

Higher aldehydes such as propionaldehyde, isobutyraldehyde, or acrolein are also active in the aforementioned reaction, but low oligomers are formed by aldol-type condensation if phenols or glycols are not present in the system to react with.[174,175]

TABLE 4
**Kinetic Parameters of the Condensation of Phenol with
Formaldehyde Catalyzed by H$^+$ in the Absence (A) and in
the Presence (B) of Acenaphthene[169]**

Parameter		Temperature					
		70°C		80°C		90°C	
		A	B	A	B	A	B
$k_1 \times 10^3$ l/mol/s		8.3	0.6	16.5	0.9	33.0	1.6
E_a, kcal/mol	A			16.80			
	B			12.00			
	A			−6.40			
logA	B			−4.54			

Amino resins are condensation products from compounds containing HN groups, which are joined by a Mannich reaction to a nucleophilic component H−Y via the carbonyl atom of an aldehyde or ketone:[176]

$$H-Y + =C=O + H-N= \rightarrow Y-\overset{|}{\underset{|}{C}}-N + H_2O \qquad (113)$$

The reaction is also called *α-ureidoalkylation*. Urea and melamine are used predominantly as NH group-containing compounds, followed to a lesser extent by the corresponding substituted and cyclic ureas, thiourea, guanidines, urethanes, cyanamides, acid amides, etc.

Originally, formaldehyde was exclusively employed as the carbonyl component, but more recently, higher aldehydes and ketones have also been used. However, the usefulness of the latter compounds is limited by aldol-type condensations, Cannizzaro reactions, enamine formation, and in some cases steric hindrance.

All acidic hydrogen compounds that possess a free electron pair at the condensation point can act as the *nucleophilic* partner. In this class of compounds are hydrogen halides, HO compounds such as alcohols, carboxylic acids, and hemiacetals; acidic HN compounds, e.g., amides, ureas, guanidines, melamines, urethanes, and primary and secondary amines; and acidic HS compounds such as mercaptans. In addition, it is possible to use all compounds that form a carbanion during proton donation (acidic HC compounds) or that tautomerize by prototropy, such as enolizing ketones. Besides the corresponding (with O_2N-, NC−, HOOC− groups, etc.) activated α-methylene group-containing substances (acidic HC compounds), analogous substituted aromatic compounds such as phenol, aniline, etc. are also suitable. Since urea itself as well as melamine (cyanourotriamide) can act as an acidic NH partner in such a reaction *(vide infra)*, the condensation between urea and formaldehyde and between melamine and formaldehyde will be discussed.

Urea-formaldehyde chemistry dates back to the end of the nineteenth century when methylene urea was isolated, but recognition of the importance of the polymeric products obtained from the reaction of urea and formaldehyde is attributed to John (1918).[177]

Application of the reaction of urea or melamine and formaldehyde to form step-growth network polymers is closely related to the procedure followed for the base-catalyzed phenol-formaldehyde polycondensation reaction in that the primary products of the first-stage reaction are methylol derivatives. Under the mildly basic conditions generally used for prepolymer formation in an aqueous medium, the methylol compounds are obtained in high yields and can be isolated as pure, crystalline compounds. For example, monomethylolurea melts at 111°C and dimethylolurea at 126°C, and both are water and alcohol soluble. N-methylol urea is stabilized by intramolecular hydrogen bonding:

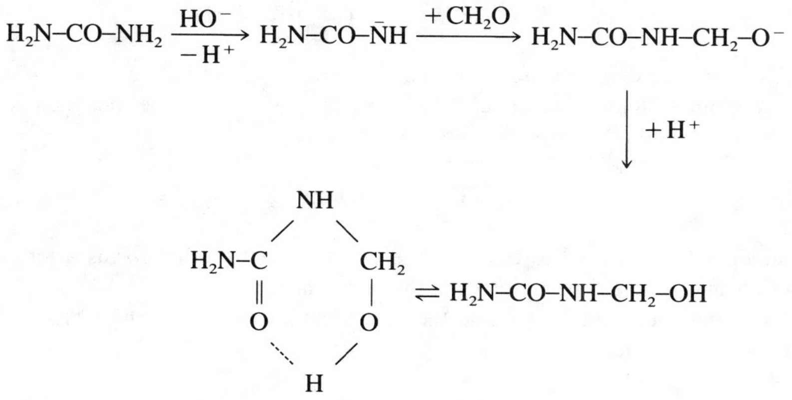

$$H_2N-CO-NH_2 \xrightarrow[-H^+]{HO^-} H_2N-CO-\bar{N}H \xrightarrow{+CH_2O} H_2N-CO-NH-CH_2-O^-$$

$$\downarrow +H^+$$

$$\begin{array}{c} NH \\ H_2N-C \quad\quad CH_2 \\ \| \quad\quad\quad\quad | \\ O \quad\quad\quad\quad O \\ \diagdown \quad\diagup \\ H \end{array} \rightleftharpoons H_2N-CO-NH-CH_2-OH \qquad (114)$$

If excess formaldehyde is added, then dimethylol urea can also occur, as well as tri- and tetramethylol urea, i.e., six possible compounds according to the following scheme:

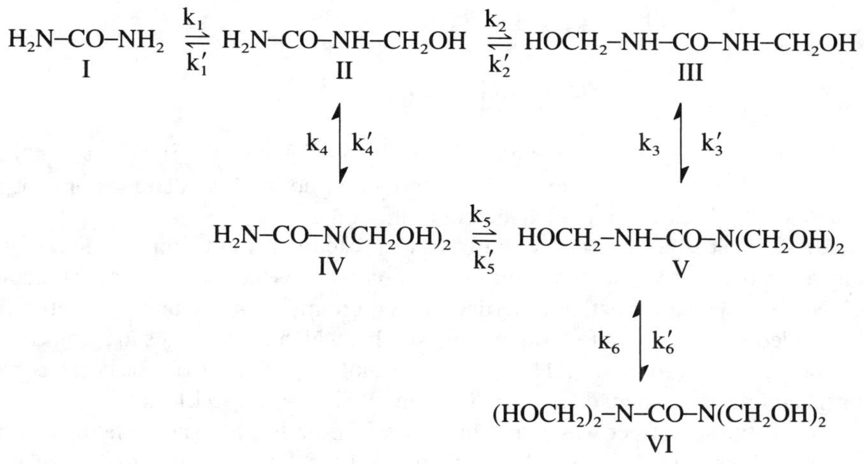

$$H_2N-CO-NH_2 \underset{k_1'}{\overset{k_1}{\rightleftharpoons}} H_2N-CO-NH-CH_2OH \underset{k_2'}{\overset{k_2}{\rightleftharpoons}} HOCH_2-NH-CO-NH-CH_2OH$$

I $\qquad$ II $\qquad$ III

$$k_4 \updownarrow k_4' \qquad\qquad\qquad k_3 \updownarrow k_3'$$

$$H_2N-CO-N(CH_2OH)_2 \underset{k_5'}{\overset{k_5}{\rightleftharpoons}} HOCH_2-NH-CO-N(CH_2OH)_2$$

IV $\qquad$ V

$$k_6 \updownarrow k_6'$$

$$(HOCH_2)_2-N-CO-N(CH_2OH)_2 \qquad (115)$$

VI

With melamine, in contrast to urea, two molecules of formaldehyde react with an NH_2, giving rise to nine possible methylol melamines.

The distribution of methylol derivatives fits the theoretical equation proposed by Aldersley and Gordon:[178]

$$n_{ij} = Nc_{ij}x^{(i+j)^2y^{i/2}}q^{i+j}(1-x)^{a-i-j} \qquad (116)$$

Here, n_{ij} is the mole fraction of the compound ij consisting of melamine (a = 6) or urea (a = 4) bearing i methylol groups belonging to secondary amino groups. N is the normalizer, and c_{ij} the combination factor, denoting the number of ways in which i + j formaldehydes can be placed to produce ij. Parameter q is a measure of the equilibrium degree of advancement of the reaction. The linear substitution parameter x reflects a term

$$\Delta G_x^+ = -2RT\ln x \qquad (117)$$

which is added i + j times to the free energy of adding a new methylol group to the derivative already bearing i + j methylols; thus:

$$\Delta G_{i+j} = \Delta G^* - 2(i + j)RT \ln x \tag{118}$$

where ΔG^* is the free energy of forming the monomethylol derivative (i = j = 0), or any link in the random system (without substitution effect). Similarly, the localized substitution parameter y reflects an additional free-energy term, i.e.,

$$\Delta G_y^* = -2RT \ln y$$

which is independently added to the free energy of a reaction that forms a tertiary amino group. Both x and y would be unity for a random reaction.

The following four equations relate the rate constants of Equation(s) 115 and the substitution parameters x and y:

$$k_2 = k_1^{1/2} + (y^{1/2}/4)x \tag{119}$$

$$k_2' = 2k_1'x^{-1} \tag{120}$$

$$k_3 = k_1 y^{1/2} x^2/2 \tag{121}$$

$$k_3' = k_1'(1 + 2y^{-1/2})x^{-2} \tag{122}$$

This kinetic treatment ignores compounds IV and VI (Equation 115) i.e., $k_4 = k_4' = k_5 = k_5' = k_6 = k_6' = 0$, since they might be formed in noticeable quantities only at high temperatures and high alkali and aldehyde concentrations.[179]

Melamine-formaldehyde chemistry conforms to a reaction pattern similar to that of urea-formaldehyde. In both cases, the dominant substitution effect is the localized, noncumulative one which counteracts the formation of tertiary amino groups. It was found[180] quantitatively in urea-formaldehyde to amount to a value of y = 0.17, which corresponds to a contribution to the free energy of activation of $-RT \ln y = 4.60$ kJ/mol (Equation 118). The corresponding results for melamine-formaldehyde are y = 0.5 and $-RT \ln y = 1.675$ kJ/mol.[180] The general (cumulative) substitution effect was found to be weak in both urea- and melamine-formaldehyde. It is stronger in urea (x = 0.71, $-2RT \ln x = 1675$ J/mol, Equation 117), as might be expected in that the pair of amino groups are separated by a single carbon atom in urea, while in melamine, two amino groups are separated by three atoms (x = 1.1, $-2RT \ln x = -419$ J/mol).

The methylolation of NH group-containing compounds with formaldehyde is first order with respect to the NH compound, formaldehyde, and catalyst; that is, Equation 123 holds:

$$v = k\left[HN\diagdown\diagup\right][HCHO][Cat] \tag{123}$$

Since a termolecular mechanism is improbable, it must be assumed that an associate is first formed, for example, in the bicarbonate ion-catalyzed reaction[181]

$$HCHO + HCO_3^- \rightleftharpoons HCHO \cdots HCO_3^- \tag{124}$$

with an equilibrium constant

$$K = \frac{HCHO \cdots HCO_3^-}{HCHO \quad HCO_3^-} \tag{125}$$

The rate-determining step is then the reaction between the associate and the NH compound, with a rate

$$v = k_r[HCHO \cdots HCO_3^-][HN] \tag{126}$$

The overall rate is then

$$v = k_r K \left[HN{\Big<} \right][HCHO][HCO_3^-] \tag{127}$$

For the same pK_a, HCO_3^-, $H_2PO_4^-$, and HPO_4^{2-}, catalysts give larger rate constants than CH_3COOH or R_3HN^+. The reason for this is that the former group of catalysts are bifunctional in that they are acidic in accepting protons and basic in donating protons.

Under acidic conditions, the N-methylol intermediates are extremely short lived since the effect of acids can very easily change them into a resonance-stabilized carbonium-immonium ion, e.g.,

$$R_3N{-}CO{-}NH{-}CH_2OH \underset{-H_2O}{\overset{+H^+}{\longleftrightarrow}} [R_2N{-}CO{-}NH{-}\overset{+}{C}H_2] \leftrightarrow [R_2N{-}CO{-}\overset{+}{N}H{=}CH_2] \tag{128}$$

Resonance-stabilized α-ureidoalkyl (carbonium-immonium) ion then reacts with a suitable nucleophilic reaction partner by electrophilic substitution reactions. Urea itself, as an acidic NH compound, can be the nucleophilic partner. A chain extension is obtaining according to

$$H_2N{-}CO{-}NH{-}\overset{+}{C}H_2 + H_2N{-}CO{-}NH_2 \rightarrow H_2N{-}CO{-}NH{-}CH_2{-}NH{-}CO{-}NH_2 + H_2^+ \tag{129}$$

Since the hydrogen in the NH group can likewise react further, acidification is a facile method for effecting second-stage, cross-linking reactions at relatively low temperatures. In contrast, under neutral conditions, the methylol derivatives are quite stable, but can be post-polycondensated in a controlled manner by heating in much the same way as that used for cross-linking resole prepolymers.

A broad spectrum of reaction pathways and reaction rates is available because of the differences in the nucleophilic potential of the NH_2 or NH groups, the electron distribution, and the resonance stabilization of the initial compounds. Reaction possibilities remain governed, besides the chemical constitution, by the alkylization rate, bonding capabilities of the carbonyl component, equilibrium position, and reactivity of the α-N-alkylol compounds.

Because the reaction is very sensitive to pH,[182,183] close control is maintained over the acidity of the reaction mixture by the use of buffers. In the absence of buffers, the pH of the initial alkaline reaction mixture tends to decrease, and may even fall below pH 7 as a result of the formation of acidic groups either through a Cannizzaro reaction or by oxidation of formaldehyde. The rate of the hydroxymethylation reaction increases rapidly from an initial pH of 8.7 to 2.7.[182] Under these conditions, a ratio of formaldehyde to urea of about 1.5:1.0 is generally used, which is essentially the same monomer ratio as that used in the preparation of resoles.

Hydroxymethylation as the first stage has been studied in a highly diluted solution without taking into account the consecutive reactions,[184-188] while investigations of oligo-condensation based on the determination of methylene bridge formation were carried out with nonuniform hydroxymethylmelamines by using aqueous/organic solvents in order to avoid de- and transmethylolation.[189-191]

Oligocondensation starts, however, immediately after formation of the initial hydroxy-methyl groups (Equation 130). It results in the formation of methylene bridges (Equation 131) and dimethylene ether bridges (Equation 132). Thus, these reactions proceed as parallel, consecutive, and competitive processes:[192]

$$R-NH_2 + CH_2O \xrightleftharpoons[k_{-i}]{k_i} R-NH-CH_2OH \tag{130}$$

$$R-NH-CH_2OH + H_2N-R \xrightleftharpoons[k_{-ii}]{k_{ii}} R-NH-CH_2-NH-R + H_2O \tag{131}$$

$$2R-NH-CH_2OH \xrightleftharpoons[k_{-iii}]{k_{iii}} R-NH-CH_2-O-CH_2-NH-R + H_2O \tag{132}$$

The effect of general acid-base catalysis on k_i is expressed by the minimum of this constant in the region near pH 7. At the formaldehyde/melamine molar ratio F/M = 3, the value of k_i was found to be three to five times higher than at F/M = 1. The demethylation constant k_{-i} increases with F and, independently of the pH value, passes a minimum at F/M = 1. The equilibrium constant $K = k_{-i}/k_i$ amounts to about 0.2 to 0.3 mol/l.[184,185,187]

The rate constant of methylene bridge formation k_{ii} increases as the pH value decreases and was found to be higher at F/M = 3 than at a molar ratio of F/M = 1. The rate constant of ether bridge formation k_{iii} passes a minimum at pH 8 and is higher at F/M = 1 than at a molar ratio of 3. The rate constant of ether cleavage k_{-iii} increases with pH and seems to be independent of F/M. Additionally, at pH = 7 to 10, the formation of methylene bridges by base-catalyzed scission of ether bridges (Equation 133 to 135) seems possible:

$$R-NH-CH_2-O-CH_2-NH-R' + HO^- \rightleftharpoons R-\bar{N}-CH_2-O-CH_2-NH-R' + H_2O \tag{133}$$

$$R-N-CH_2-O-CH_2-NH-R' \rightleftharpoons R-N{=}CH_2 + {}^-O-CH_2-NH-R' \tag{134}$$

$$R-N{=}CH_2 + \underset{\underset{R''}{|}}{H-N-R'} \rightleftharpoons R-NH-CH_2-\underset{\underset{R''}{|}}{N-R'} \tag{135}$$

where R,R' = amino resin residues and R'' = $-H$, $-CH_2OH$.

The higher activation energy of the melamine-formaldehyde reaction was registered for methylene bridge formation (76 kJ/mol), and the lowest for dimethylene ether bridge formation (27 kJ/mol) and scission (28 kJ/mol). For methylolation and demethylolation, activation energies were 40 and 46 kJ/mol, respectively.[192]

Stable organic soluble methylolmelamine oligomers (Equation 136) may be prepared by reacting these derivatives with alcohols such as butyl alcohol (Equation 137):

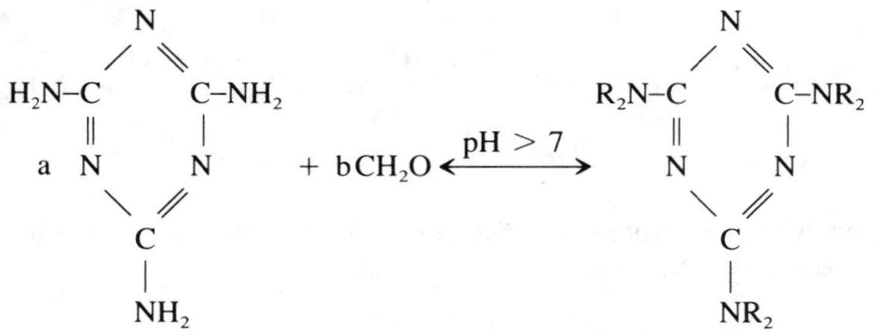

$$
\begin{array}{c}
\text{a}
\begin{array}{c}
\text{N} \\
\diagup \diagdown \\
\text{H}_2\text{N-C} \quad \text{C-NH}_2 \\
\| \qquad | \\
\text{N} \qquad \text{N} \\
\diagdown \diagup \\
\text{C} \\
| \\
\text{NH}_2
\end{array}
+ b\,\text{CH}_2\text{O}
\underset{}{\overset{\text{pH} > 7}{\longleftrightarrow}}
\begin{array}{c}
\text{N} \\
\diagup \diagdown \\
\text{R}_2\text{N-C} \quad \text{C-NR}_2 \\
\| \qquad | \\
\text{N} \qquad \text{N} \\
\diagdown \diagup \\
\text{C} \\
| \\
\text{NR}_2
\end{array}
\end{array}
$$

$$a \leq b; \quad R = -H \quad \text{or} \quad -CH_2OH \tag{136}$$

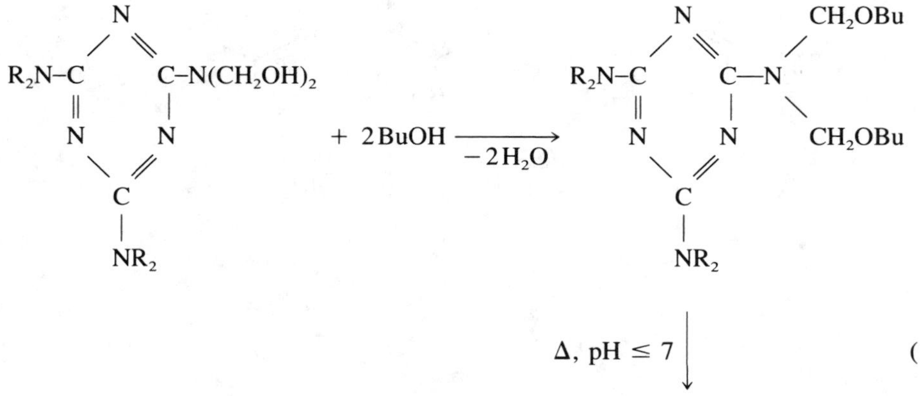

$$
\begin{array}{c}
\text{N} \\
\diagup \diagdown \\
\text{R}_2\text{N-C} \quad \text{C-N(CH}_2\text{OH)}_2 \\
\| \qquad | \\
\text{N} \qquad \text{N} \\
\diagdown \diagup \\
\text{C} \\
| \\
\text{NR}_2
\end{array}
+ 2\,\text{BuOH}
\xrightarrow[-2\text{H}_2\text{O}]{}
\begin{array}{c}
\text{N} \qquad \text{CH}_2\text{OBu} \\
\diagup \diagdown \quad \diagup \\
\text{R}_2\text{N-C} \quad \text{C-N} \\
\| \qquad | \quad \diagdown \\
\text{N} \qquad \text{N} \quad \text{CH}_2\text{OBu} \\
\diagdown \diagup \\
\text{C} \\
| \\
\text{NR}_2
\end{array}
$$

$$\Delta, \text{pH} \leq 7 \downarrow \tag{137}$$

$$
\begin{array}{c}
\text{N} \qquad \text{CH}_2\text{OBu} \\
\diagup \diagdown \quad \diagup \\
\text{\Large\sim\!\!\sim}\text{-NR-C} \quad \text{C-N} \\
\| \qquad | \quad \diagdown \\
\text{N} \qquad \text{N} \quad \text{R} \\
\diagdown \diagup \\
\text{C} \\
| \\
\text{RN} \\
\text{\}}
\end{array}
\tag{138}
$$

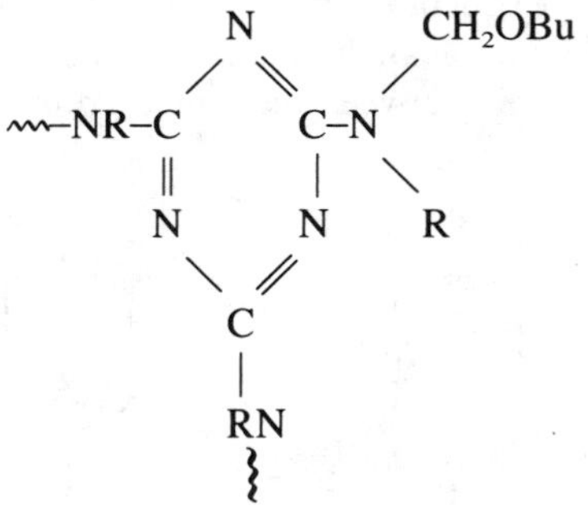

When these alkylated methylol-malamine compounds are heated under slightly acidic conditions (Equation 132), the ether linkages are slowly hydrolyzed and internuclear bridges are formed which involve primarily methylene groups formation.[193]

Synthesis parameters, e.g., pH, temperature, time, and monomer molar ratio, had various effects on the molecular weight distribution and oligomer distribution of melamine-formaldehyde resins. The average molecular weight of the reaction mixture in the first stages of the reaction was higher at monomer ratio 1:2.4 than at 1:1.8 due to a higher concentration of reactive groups. The average molecular weight of the oligomers had no effect on the properties of the molded end products.[194]

Three oligomers of increasing degree of condensation of melamine ($T_m = 350°C$) are reported in the literature:[195] melam, melem, and melom. The process can be schematically indicated as follows:

$$2n C_3H_6N_6 \xrightarrow[350°C]{-nNH_3} nC_6H_9N_{11} \xrightarrow{-nNH_3} (C_6H_6N_{10})_n \xrightarrow{-nNH_3} (C_6H_3N_9)_n \quad (139)$$

melamine   melam   melem   melom

While there is clear evidence that melam is the dimer of melamine in which two melamine rings are connected by a NH bridge:

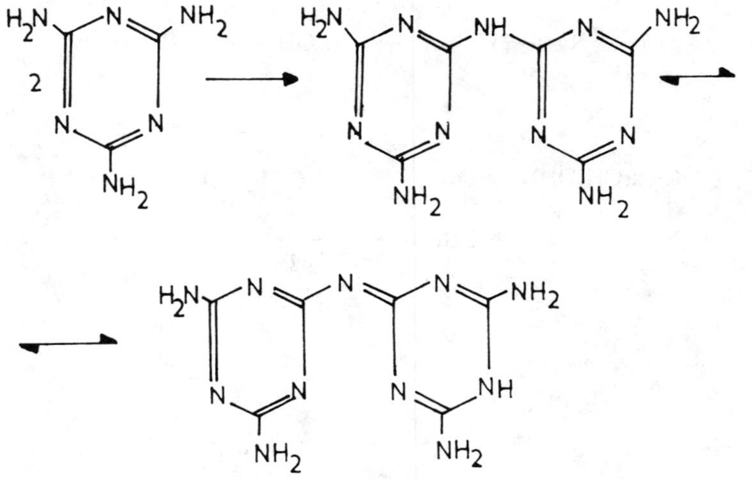

$$(140)$$

different structures have been proposed for melem, e.g., I (unlikely), II, and III (Equation 141). As far as melom is concerned, it is generally agreed that it is obtained by elimination of ammonia between the amine groups of melem.

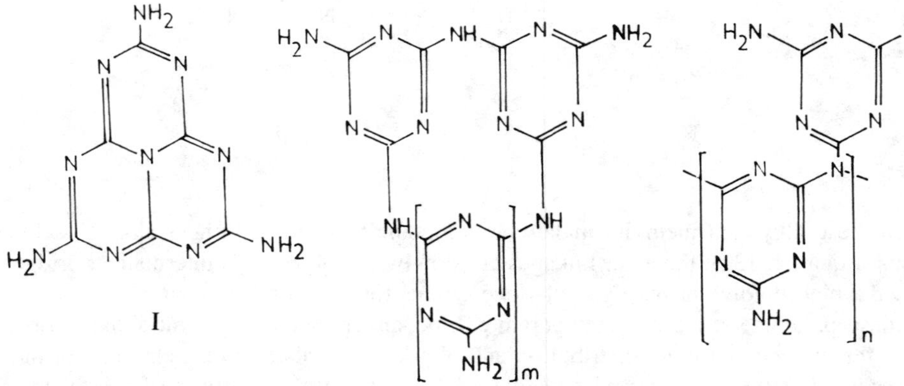

I   II   III

$$(141)$$

In spite of the uncertainty of the structure of melem and, consequently, melom, they can be identified by their IR spectra.

Attempts to prepare linear polymers by the acid-catalyzed step-growth polymerization reaction of thiourea (T) and formaldehyde (F), F/T = 1, have resulted in the formation of oligomers only:[196]

$$n\,H_2N-C-NH_2 \;+\; n\,CH_2O \rightarrow H-\left[\,-NH-C-NH-CH_2-\,\right]_n -OH \;+\; (n-1)H_2O \quad (142)$$

$$\overset{\|}{S} \qquad\qquad\qquad \overset{\|}{S}$$

Step-growth polymers which have been prepared by *nucleophilic substitution* include poly(alkylene polysulfides), low molecular weight prepolymers from the reaction of biphenols and epoxides, aliphatic and aromatic polyethers, polythioethers, polyamines, polyacetals, and low molecular weight hydrocarbon polymers from the Wurtz reaction.

The first linear poly(alkylene polysulfides) were prepared by Meyer[197] more than a century ago by two routes — by the reaction of bis(2-chloroethyl)sulfide with potassium sulfide and by reaction of the disodium salt of ethanedithiol with bromide. However, the commercial production of a polysulfide material started early in 1930 by the Thiokol Corporation and was based on the reaction of ethylene dichloride and sodium tetrasulfide.[98] The polymeric derivatives from this reaction were rubbery products of high molecular weight with attractive solvent resistance and physical properties.

In 1955, Thiokol Chemical Corporation started the industrial production of oligomeric polysulfides with −SH end groups. Simultaneously, these polysulfides were investigated in the U.S.S.R. American technologies for producing oligomeric polysulfides were thereafter applied in England and Japan, while the Soviet Union, East Germany, Poland, and Romania developed their own technologies for manufacturing these materials.[199]

In commercially significant organic polysulfides with the structural element $-(-RS_x-)-$, x is known as the sulfur grade or rank and represents the average number of sulfur atoms per structural element.

Organic polysulfides are synthesized from α,ω-dihalogen compounds and sodium polysulfide according to the general equation

$$n\,X-R-X \;+\; n\,Na_2S_x \;\rightarrow\; -(-RS_x-)_n- \;+\; 2\,NaX \quad (143)$$

Ethylene dichloride, bis(2-chloroethyl)formal, bis(2-chloroethyl)ether, and allyl or benzyl halides are suitable halogen compounds. The molecular weight of solid polymers is high, 2 to 5×10^5. The polymer properties depend first and foremost on the sulfur grade, i.e., a higher sulfur grade gives a better elastomer. Since alkylene polysulfide polymers, particularly those prepared from inorganic polysulfides higher than the disulfide, are composed of random mixtures of disulfide, trisulfide, tetrasulfide, and pentasulfide repeating units, x is generally some fractional value between two and four.

The flexible polysulfide chain is primarily responsible for the rubbery properties of these polymers. Monosulfide and disulfide polymers are hard, tough plastics, many of which crystallize readily. As the sulfur content is increased, the polymers become softer and more elastic.

For molecular weight reduction, the aqueous suspension of the polymer is reacted with a combination of sodium hydrosulfide and sodium sulfite.

$$R_nS_x-S_yR_n \;+\; NaSH \;+\; Na_2SO_3 \rightarrow R_mS_xNa \;+\; R_nS_{y+1}H \;+\; Na_2SO_3 \quad (144)$$

The molecular weight of the polymer may also be controlled by inclusion of a monofunctional halide such as butyl chloride, or a disulfide such as benzothiaryl disulfide. Both compounds react with mercaptide end groups by nucleophilic substitution to terminate the polymer chain. In this way, polysulfide oligomers (liquid thiokols) are formed. More than 90% of the alkylene polysulfide polymers manufactured in the world are of an oligomeric nature.

Presently, mainly 2,2′-dichlorodiethyl formal and the mixture of this compound with 2,2′-dichlorodiethyl ether or with 1,2-dichloromethane are used commercially. In these reactions, the formation of cyclic oligomers depends upon the rank and concentration of the inorganic polysulfide monomer and the structure of the organic dihalide monomer. In general, the rank of the oligomer will be close to, but somewhat lower than, that of the inorganic polysulfide monomer from which it is formed, depending upon the monomer ratio used in the polycondensation reaction.

Alkylene monosulfide polymers have been prepared by the reaction of dithiol salts with dihalides

$$n\,NaSRSNa\ +\ ClR'Cl \rightarrow -(-SRSR'-)_n- \ +\ 2\,NaCl \tag{145}$$

but these polycondensation reactions yield much lower molecular weight polymers ($\sim$5000).

Poly(thio-1,4-phenylene) deserves attention as an engineering plastic of excellent performance.

Generally, the formation of macromolecular compounds, poly(oxy-2,6-dimethyl-1,4-phenylene),[200] poly(1,4-phenylene sulfide),[201] and poly(1,4-phenylene)[202] are well-known reactions and can be considered as polycondensation reactions. However, Hartling and Lindberg[203] suggested the radical nature of the process of poly(1,4-phenylene sulfide) formation on the basis of the reaction of hydroquinone with elemental sulfur. In this case, when hydroquinone is used as monomer, the yield is considerably reduced in spite of the higher activation of the −OH group, compared with the activation by a chlorine substituent. Since hydroquinone is an inhibitor of radical reactions, it follows that the substitution of sulfur for chlorine in the present synthesis takes place through a radical mechanism.

When poly(p-phenylene sulfide) is prepared from 1,4-dichlorobenzene and disodium sulfide as starting materials, the mechanism of the reaction

$$n\,Cl-C_6H_4-Cl\ +\ n\,Na_2S \rightarrow -[-C_6H_4-S-]_n- \ +\ 2\,n\,ClNa \tag{146}$$

has usually been considered to be an addition-elimination reaction, the formation of the intermediate anion being the rate-determining step.

Experiments to limit the molecular weight of the products resulting from this reaction by employing an excess of one of the reactants yielded oligomers with a molecular weight of 2000. Application of equimolecular ratios gave relatively high molecular weight products, but the conversions were low. These results indicate that the formation of poly(thio-1,4-phenylene) does not follow a stepwise mechanism like a polycondensation. Therefore, an intermediate capable of maintaining a chain-carrying process has been postulated.[204] This is the radical cation which serves as a substrate for the attacking anions or as an electron acceptor in the transfer reaction with the anion. The radical cation is a highly reactive species when the chain is short. With an increase of the chain length, it becomes less reactive due to increasing conjugation. This is the reason why products with DP $\leq$ 20 are readily formed. The following reactions reveal the postulated mechanism:

Initiation

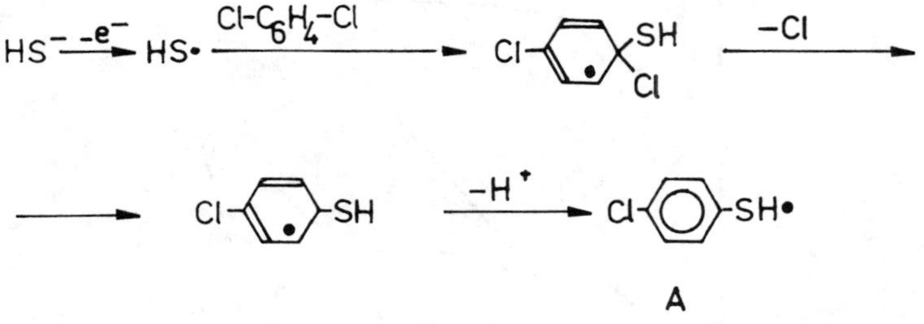

$$HS^- \xrightarrow{-e^-} HS\cdot \xrightarrow{Cl-C_6H_4-Cl} Cl-\underset{\cdot}{\bigcirc}\overset{SH}{\underset{Cl}{<}} \xrightarrow{-Cl}$$

$$\longrightarrow Cl-\underset{\cdot}{\bigcirc}-SH \xrightarrow{-H^+} Cl-\bigcirc-SH\cdot \qquad (147)$$

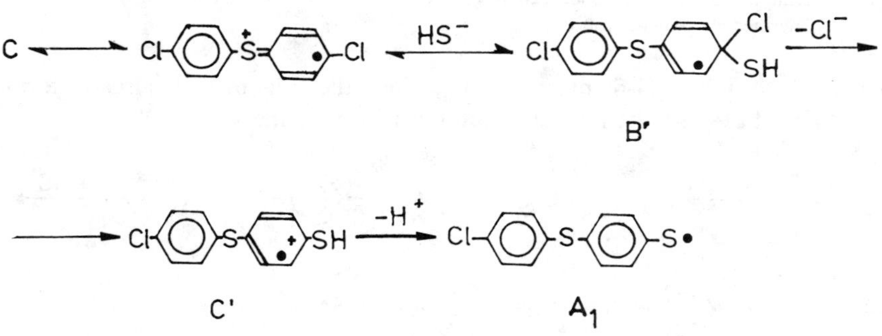

$$\mathbf{A}$$

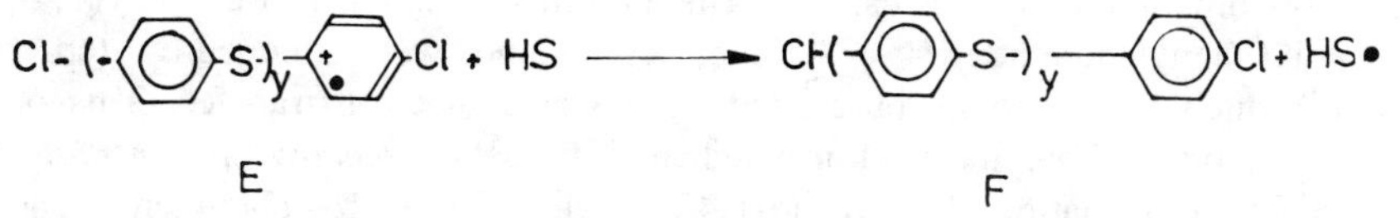

Propagation

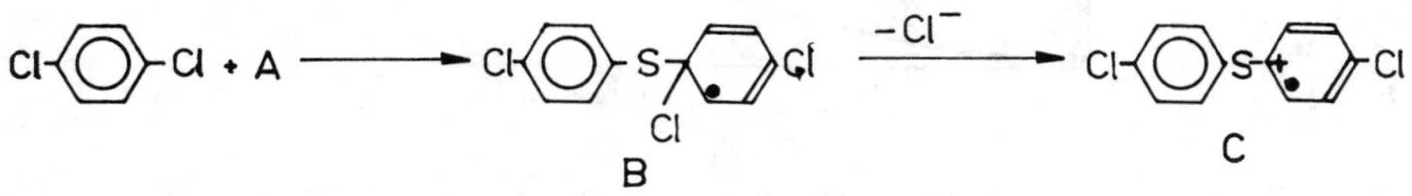

$$Cl-\bigcirc-Cl + A \longrightarrow Cl-\bigcirc-S-\underset{Cl}{\bigcirc}\cdot Cl \xrightarrow{-Cl^-} Cl-\bigcirc-S\overset{+}{\bigcirc}\cdot Cl \qquad (148)$$

$$\mathbf{B} \qquad\qquad\qquad \mathbf{C}$$

$$C \longleftrightarrow Cl-\bigcirc-\overset{+}{S}=\bigcirc\cdot-Cl \xleftarrow{HS^-} Cl-\bigcirc-S-\underset{\cdot}{\bigcirc}\overset{Cl}{\underset{SH}{<}} \xrightarrow{-Cl^-}$$

$$\mathbf{B'}$$

$$\longrightarrow Cl-\bigcirc-S-\overset{+}{\underset{\cdot}{\bigcirc}}-SH \xrightarrow{-H^+} Cl-\bigcirc-S-\bigcirc-S\cdot \qquad (149)$$

$$\mathbf{C'} \qquad\qquad\qquad \mathbf{A_1}$$

$$Cl-(-\bigcirc-S-)_y-\overset{+}{\underset{\cdot}{\bigcirc}}-Cl + HS^- \longrightarrow Cl-(-\bigcirc-S-)_y-\bigcirc-Cl + HS\cdot \qquad (150)$$

$$\mathbf{E} \qquad\qquad\qquad \mathbf{F}$$

Termination

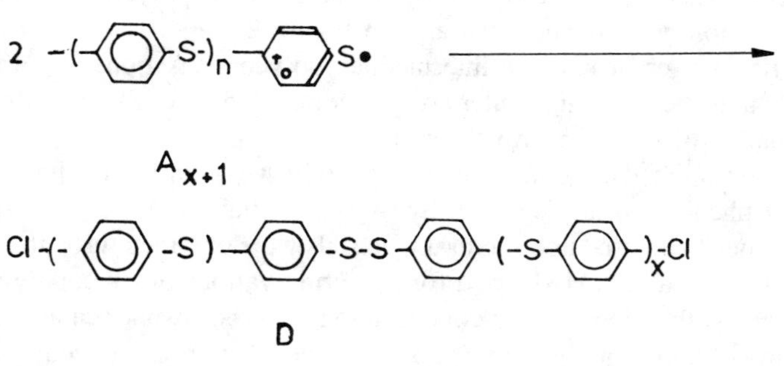

$$2 \ -(-\bigcirc-S-)_n-\overset{+}{\underset{\cdot}{\bigcirc}}-S\cdot \longrightarrow$$

$$\mathbf{A_{x+1}}$$

$$Cl-(-\bigcirc-S)-\bigcirc-S-S-\bigcirc-(-S-\bigcirc-)_x-Cl \qquad (151)$$

$$\mathbf{D}$$

The one-electron transfer, i.e., the formation of species HS·, initiates the chain. Propagation takes place only at high temperature, because only in this circumstance does the disulfide linkage cleave homolitically. Propagation is not restricted to the reaction of the

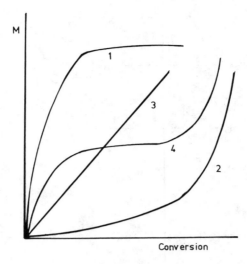

FIGURE 2.   The molecular weight-conversion relationship for (1) radical polymerization, (2) polycondensation, (3) living polymerization, and (4) reactive
intermediate polymerization.

sulfenyl radicals A with 1,4-dichlorobenzene, since the oligomers F already formed by
electron transfer of HS^- to the radical cation react in the same way.

$$A_{x+1} + F \xrightarrow{-Cl^-} Cl\text{-}(\langle O \rangle\text{-}S\text{-})_x\text{-}\langle O \rangle\text{-}S\text{-}(\text{-}\langle O \rangle\text{-}S\text{-})_y\text{-}\langle O \rangle\text{-}Cl \xrightarrow{+HS^-}$$

$$\longrightarrow Cl\text{-}(\langle O \rangle\text{-}S\text{-})_{\overline{x+y+1}}\text{-}\langle O \rangle\text{-}Cl + HS\bullet \qquad (152)$$

Other polyaromatic compounds can be formed by this mechanism, i.e., poly(1,4-phen-
ylene) and poly(phenylene oxide).[204,205] All these compounds were obtained by step-growth
processes, in which the reactions proceeded by a single electron transfer. This process
represents a new type of polymerization which may be called *reactive intermediate polycondensation*. In the formation of poly(thio-1,4-phenylene), poly(1,4-phenylene), and poly
oxy-1,4-(2,6-dimethylphenylene), the reactivity of end groups is dependent upon the chain
length.

Another characteristic feature of reactive intermediate polycondensation is the molecular
weight-conversion dependence illustrated in Figure 2.

In the first stages of reactive intermediate polycondensation, oligomers with DP $\sim$ 20
are formed, and the high molecular weight compounds are obtained only near the end of
the reactions, similar to step-growth processes.

Hydroxyterminated oligophenylene oxides with $\overline{M}_n \simeq 2800$ were prepared by the catalytic
oxidation of phenol and/or bisphenol by oxygen-containing gases.[206] Carborane-containing
oligophenylenes to be used as structural materials were obtained through a similar process.[207]

Macro-azo initiators (MAI) used for preparing various block copolymers via the radical
process were synthesized by polycondensation of azobiscyanopentanoyl chloride (ACPC)
with a series of diols or diamines[208] or by condensation of sebacoyl chloride (SC) with
azobisdiols (ABD) such as 2,2′-azobis(2-cyanopropanol), ACPO, 2,2′-azobis [2-methyl-*N*-
(2-hydroxyethyl)propionamide], AHPA, 2,2′-azobis [2-methyl-(1,1′-bis(hydroxymethyl)ethyl/
propionamide], ABHPA, and 2,2′-azobis [2-methyl-*N*-(1,1-bis(hydroxymethyl)-2-hydroxy-
ethylpropionamide], ABHHPA:[209]

$$\begin{array}{cc} CH_3 & CH_3 \\ | & | \\ R-C-N=N-C-R \\ | & | \\ CN & CN \end{array} + \begin{array}{cc} Cl-C-(CH_2)_8-C-Cl \\ \| & \| \\ O & O \end{array} \rightarrow MAI\ oligomers \qquad (153)$$

ABD   SC $-(-ABD-SC-)-$

where:

			$\begin{array}{c} CH_2OH \\ | \\ -NH-CO-C-CH_3 \\ | \\ CH_2OH \end{array}$	$\begin{array}{c} CH_2OH \\ | \\ -NH-CO-C-CH_2OH \\ | \\ CH_2OH \end{array}$
R	$-CH_2OH$	$-NH-CO-CH_2CH_2OH$		
ABD	ACPO	AHPA	ABHPA	ABHHPA

MAI obtained according to Equation 152 showed rather lower molecular weights, compared with those prepared from ACPC with a broad molecular weight distribution (MWD):

	MAI oligomer			
	$-(ACPO-SC-)-$	$-(AHPA-SC-)-$	$-(ABHPA-SC-)-$	$-(ABHHPA-SC-)-$
$\overline{M}_n$	1,600	2,300	1,800	1,800
M_w	2,800	10,400	19,600	3,500
MWD	1.75	4.50	10.90	1.90

Nucleophilic aromatic substitution was also used for preparation of maleimide-terminated amorphous aromatic polyether-polyketones by polymerization of bisphenol A with bis(*p*-chlorobenzoyl)benzene or 4,4′-difluorobenzophenone in the presence of *m*-aminophenol to form amine-terminated oligomers.[210] This reaction was followed by imidization of the end groups with maleic anhydride.

Polyimides contain the group $-CO-NR-CO-$. The basic member of this series, which is formally Nylon-1, arises from the spontaneous polymerization of isocyanic acid:

$$n\,H-N=C=O \xrightarrow[C_6H_6]{15°C} -(-NH-CO-)_n- \qquad (154)$$

Aromatic polyimides are usually high-temperature-stable materials. The step-growth polymerization is carried out in two stages. In the first, which is basically a step polyaddition process, polyamic acid is formed in aprotic solvents from dianhydrides and diamines:

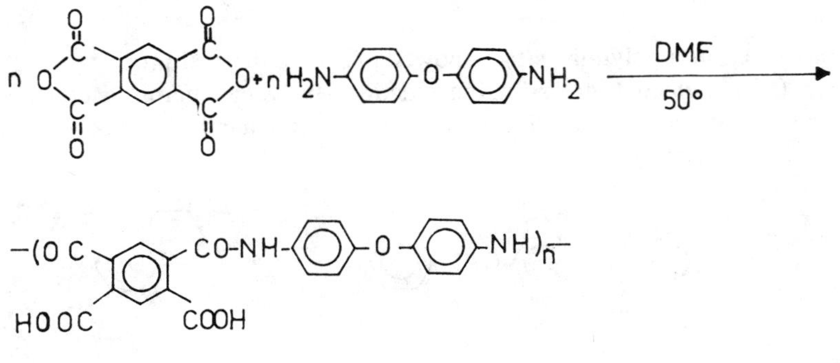

$$(155)$$

The molecular weight of the resulting acid polyamide can attain values up to $\overline{M}_n =$ 55,000 and $\overline{M}_w = 240,000$.[211] In the second stage, an intramolecular condensation, water is eliminated at high temperature:

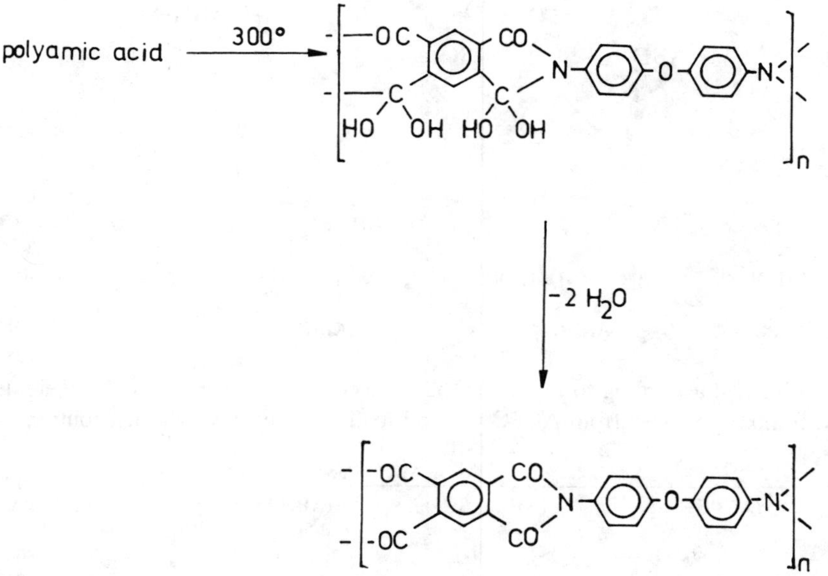

$$(156)$$

As a new cross-link for the thermally stable polyimides, the combination of internal acetylene and biphenylene end cap was found to give highly thermally stable polyimide after cure.[212] Biphenylene end-capped imide oligomers having internal acetylene groups in the backbone

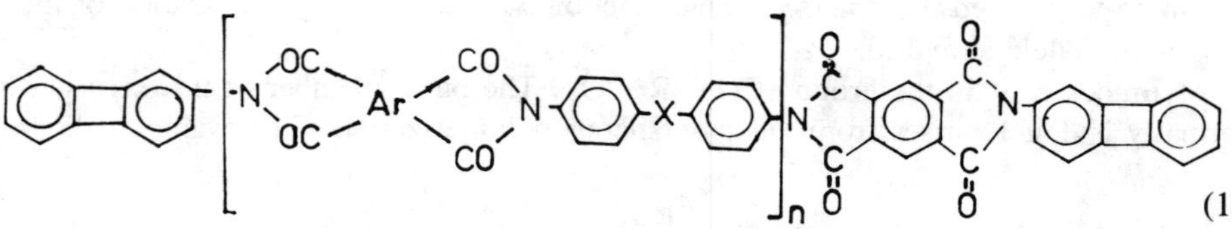

$$(157)$$

where X is $-C\,C-$ or $-O-$ and

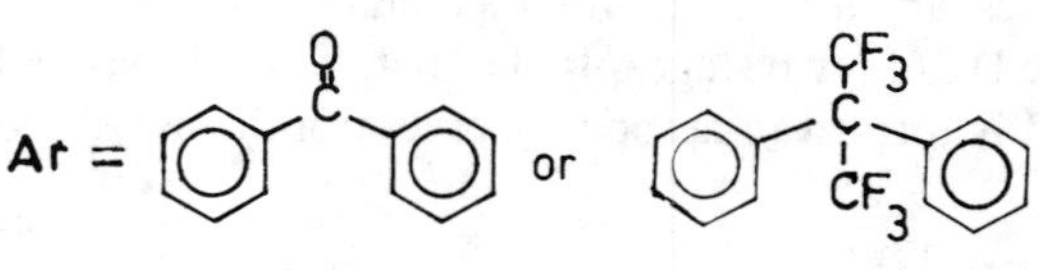

were melt-processed into polyimides that showed excellent thermal properties ($T_g \sim 400°C$ and $T_{dec}^{5\%} \sim 500°C$). Selective cross-linking reactions between internal acetylenes and biphenylene end caps of oligomers, with the formation of phenanthrene links

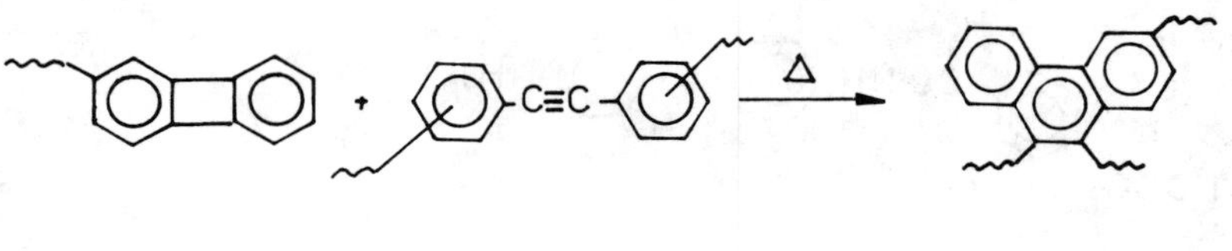

$$(158)$$

are considered the reason for enhanced thermal stability.[212] Solutions of polyamic acid containing internal acetylenes mixed with oligoamic acid solution (n = 5) having biphenylene end caps gave, after imidization and curing (400°C for 10 min), polyimides that are more stable ($T_{dec}^{5\%} \sim 550°C$) than those without biphenylene end caps.

The thermal stability of thermosetting resins (epoxy) was increased by incorporating heat-resistant imide groups into the resin network structure using an amine-terminated aliphatic polyimide oligomer:[213]

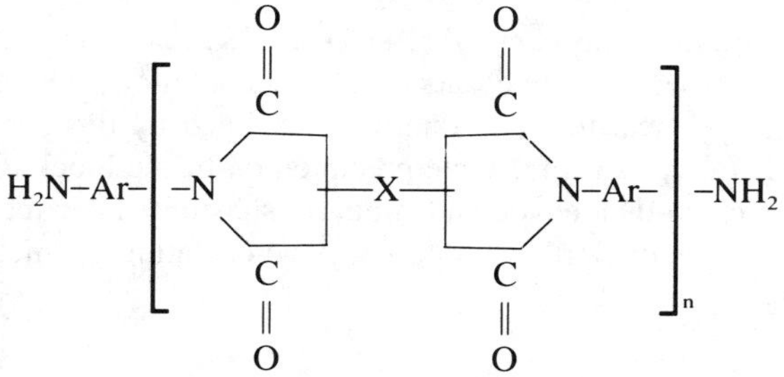

$$
H_2N-Ar-\left[-N \overbrace{\underset{\underset{O}{\overset{\parallel}{C}}}{\overset{\overset{O}{\overset{\parallel}{C}}}{\diamondsuit}}}-X-\overbrace{\underset{\underset{O}{\overset{\parallel}{C}}}{\overset{\overset{O}{\overset{\parallel}{C}}}{\diamondsuit}}N-Ar- \right]_n -NH_2 \qquad \overline{M}_n \cong 1000 \qquad (159)
$$

where Ar is an aromatic group and X is an aliphatic group. Cured products of the imide-oligomer/epoxy system can provide heat resistance at the level of bismaleimide, maintaining the excellent bonding strength of epoxy resin.

The preparation of oligomeric (diylidene) carbonothionic dihydrazides (n = 8 to 10) from equimolecular quantities of selected dialdehydes and carbonothionic dihydrazide takes place rapidly in acidic media (low pH, excess mineral acid) at 60 to 85°C.[214] Since mechanistic studies[215] indicate that the reaction of hydrazine derivatives with carbonyl compounds is pH dependent and optimum at an intermediate value, this oligomerization reaction is unusual among such reactions in operating at a low pH. The oligomers are insoluble in water, dilute acids, pyridine, DMSO, and DMF. The glyoxal oligomers, i.e., oligo[2,2'-(1,2-ethanediylidene)biscarbonothionic dihydrazone]

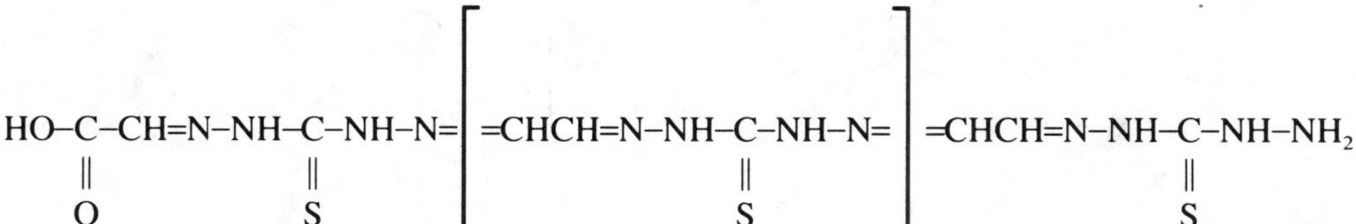

$$
\underset{O}{\overset{\parallel}{HO-C}}-CH=N-NH-\underset{S}{\overset{\parallel}{C}}-NH-N=\left[=CHCH=N-NH-\underset{S}{\overset{\parallel}{C}}-NH-N= \right]_8 =CHCH=N-NH-\underset{S}{\overset{\parallel}{C}}-NH-NH_2 \qquad (160)
$$

are soluble in dilute sodium hydroxide and give intensely colored solutions.

The terephthaldehyde oligomer, i.e., oligo[2,2'-(1,4-benzenediylidene)biscarbonothionic dihydrazide]

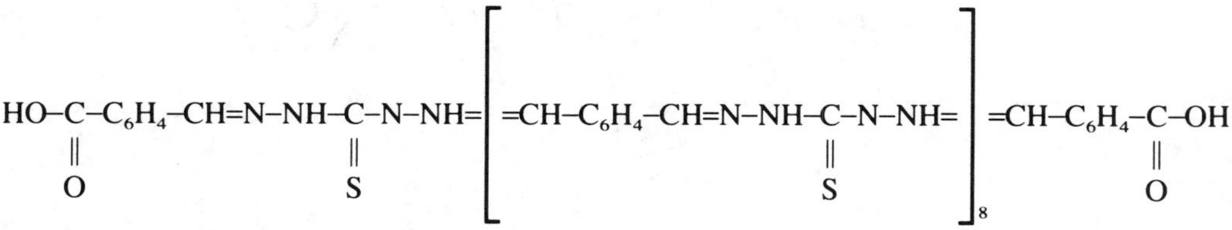

$$
\underset{O}{\overset{\parallel}{HO-C}}-C_6H_4-CH=N-NH-\underset{S}{\overset{\parallel}{C}}-N-NH=\left[=CH-C_6H_4-CH=N-NH-\underset{S}{\overset{\parallel}{C}}-N-NH= \right]_8 =CH-C_6H_4-\underset{O}{\overset{\parallel}{C}}-OH \qquad (161)
$$

is insoluble in aqueous base. In concentrated hydrochloric acid, the color of the insoluble product is yellow and it turns brilliant red in dilute sodium hydroxide. The conjugated unsaturation present in the sulfhydryl form functions as the chromophore.[215]

The enediol structure, $-C(OH)=C(OH)-$, characteristic of a variety of redox systems[216] can be obtained by condensation of dialdehyde derivatives.

Oligomers of two to nine units having a diazinediylethene-1,2-diol repeating unit have been prepared by a cyanide ion-catalyzed self-condensation of pyridazine-3,6-, pyrazine-2,5-, or pyrimidine-4,6-dialdehyde:[217]

$$OHC-C_4H_2N_2-CHO \xrightarrow{\text{cat.}} -C_4H_2N_2-C(OH)=[=C(OH)-C_4H_2N_2=]_n=C(OH)- \quad (162)$$

The oligomers, isolated as their potassium salts, are soluble in acid and base with an isoelectric point at pH 6.5. In base, they give dark red-brown solutions whose color, attributable to their semidone ion radical form, is rapidly discharged by oxygen (air) or by the addition of Fe^{3+} ion, which gives an immediate precipitation of the block chelate. The benzoin polymers from pyridine-2,6-dialdehyde and similarly substituted diazine types with their di-*ortho* (to N)-substituted enediol structures do not have chelating geometry of types 1 to 3 illustrated below in Formula(s) 163.

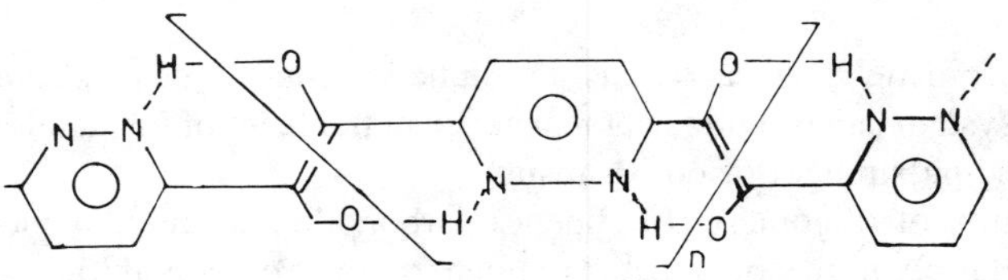

1

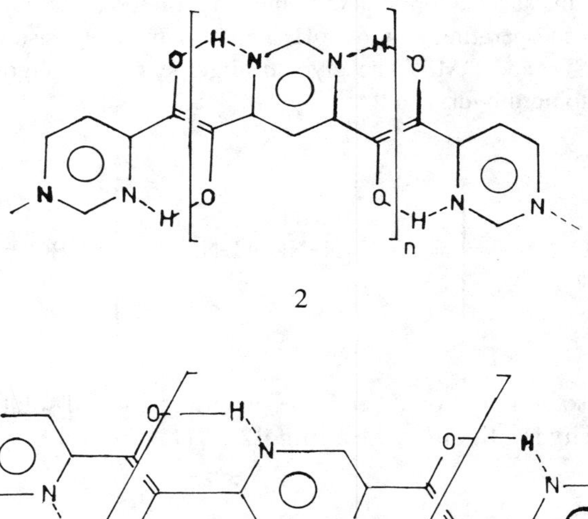

2

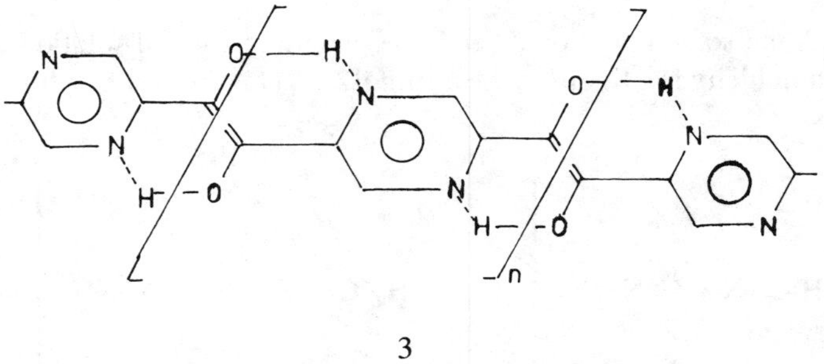

3   (163)

The oligomeric polyarylene obtained by oxidative polycondensation of aniline with NaOCl revealed that its chains consisted of equal amounts of quinonimine and phenylimine units and had $-NH_2$ end groups.[218]

Bromine containing oligomeric aromatic ethers[219] or carboxypiperazine units[220] were prepared by catalytic condensation of tetrabromobisphenol A or bis(chloromethyl)diphenyl ether with aromatic acids.

A series of polyethers containing only bromoalkane chain ends have been made by phase transfer-catalyzed polyetherification of 4,4'-dihydroxy-α-methylstilbene with 1,11-dibrom-undecane:[221]

$$n\,HO\text{–}C_6H_4\text{–}\underset{\underset{CH_3}{|}}{C}\text{=}CH\text{–}C_6H_4\text{–}OH \ + \ (n \ + \ 1)\,Br\text{–}(CH_2)_{11}\text{–}Br \xrightarrow{\ -2n\,HBr\ }$$

$$Br\text{–} (CH_2)_{11}\text{–}\left[\begin{array}{c} CH_3 \\ | \\ \text{–}O\text{–}C_6H_4\text{–}C\text{=}CH\text{–}C_6H_4\text{–}O\text{–}(CH_2)_{11}\text{–} \end{array}\right]_n\text{–}Br \qquad (164)$$

The low molecular products are completely crystalline. On increasing the DP, the polymers become monotropic nematics, and at higher molecular weights, enantiotropic nematics. The thermal stability of the mesophase increases with the polymer molecular weight up to about 10,000 and then remains constant. Both the enthalpy and entropy of melting and isotropization follow the same trend.

Chloroterminated oligo(phenylene sulfide) with glycidylthio end groups was obtained by the reaction of chloroterminated oligo(phenylene sulfide) with HSNa, followed by reaction with excess epichlorohydrin at 60°C in the presence of anhydrous NaOH to give a viscous, cinnamon-colored amorphous resin.[222]

Epoxy resins are prepared by a two-step polymerization sequence in which the first step is based on the stop-growth polymerization reaction of a bifunctional alkylene epoxide with a bifunctional nucleophile. The compounds prepared in the first step are essentially linear oligoethers with reactive end groups, obtained by using a large excess of the epoxide monomer. Technical epoxy resins can be synthesized by two different methods, both leading to products having the following structural formula:

$$CH_2\overset{O}{\overbrace{\diagup\ \diagdown}}CH\text{–}CH_2\text{–}\left[\begin{array}{c} CH_3 \qquad\quad OH \\ | \qquad\qquad\ | \\ \text{–}O\text{–}C_6H_4\text{–}C\text{–}C_6H_4\text{–}O\text{–}CH_2\text{–}CH\text{–}CH_2\text{–} \\ | \\ CH_3 \end{array}\right]_k \text{–}O\text{–}C_6H_4\underset{\underset{CH_3}{|}}{\overset{\overset{CH_3}{|}}{C}}C_6H_4\text{–}O\text{–}CH_2\overset{O}{\overset{|}{CH}}\text{–}CH_2$$

$$A \qquad (165)$$

In the so-called ''tuffy'' process, 2,2-bis(4'-hydroxyphenyl)-propane, which has been given the trivial name of bisphenol A because of its preparation from phenol and acetone, is reacted with a controlled excess of epichlorohydrin under alkaline conditions:

$$HO\text{–}C_6H_4\underset{\underset{CH_3}{|}}{\overset{\overset{CH_3}{|}}{C}}C_6H_4\text{–}OH \ + \ CH_2\overset{O}{\overbrace{\diagup\ \diagdown}}CH\text{–}CH_2\text{–}Cl \xrightarrow{\ NaOH\ } A \qquad (166)$$

This procedure leads to products containing both odd- and even-membered diepoxide oligomers of the type shown in Formula 165, where K = 0, 1, 2, 3, etc.

In the "fusion" or "advancement" process, the diglycidyl ether of bisphenol A is condensed with bisphenol A. The main constituents of the final resin are oligomeric diepoxides of Formula 167, where the significant values of k are even integers, since the process involves the base-catalyzed condensation of varying amounts of bisphenol A with a low molecular weight epoxy resin whose main constituent is the diglycidyl ether of bisphenol A:

$$\underset{\displaystyle CH_2}{\overset{\displaystyle O}{\diagup\diagdown}} - CH-CH_2-O-C_6H_4-\underset{\displaystyle CH_3}{\overset{\displaystyle CH_3}{\underset{|}{\overset{|}{C}}}}-C_6H_4-O-CH_2-CH \underset{\displaystyle CH_2}{\overset{\displaystyle O}{\diagup\diagdown}}$$

$$+ \ HO-C_6H_4-\underset{\displaystyle CH_3}{\overset{\displaystyle CH_3}{\underset{|}{\overset{|}{C}}}}-C_6H_4-OH \ \xrightarrow{\text{NaOH}} \ A \qquad (k = 0, 2, 4, 6) \qquad (167)$$

The product of the "tuffy" reaction and "advancement" process may be represented by the general formulas:

$$BB(AABB)_k; \quad k = 0, 1, 2 \text{ for "tuffy" process} \qquad (168)$$

and

$$BBAABB(AABBAABB)_j; \quad j = 0, 2, 4 \text{ for "advancement" process} \qquad (169)$$

In these formulas, BB represents the glycidyl group

$$(-CH_2-CH \underset{\displaystyle O}{\overset{\displaystyle \diagdown \diagup}{\underset{}{}}} CH_2)$$

when present as an end group, or a glyceryl group ($-CH_2-CHOH-CH_2-$) when present in the chain. When k = 0, BB represents the epichlorohydrin molecule. AA represents the nucleus of the bisphenol A molecule

$$-O-C_6H_4-\underset{\displaystyle CH_3}{\overset{\displaystyle CH_3}{\underset{|}{\overset{|}{C}}}}-C_6H_4-O- \qquad (170)$$

For k > 0, the molecular weight of the kth oligomer (or of the jth oligomer) is given by the following relations:

$$M_k = 340 + 284(k - 1) \qquad (171)$$

TABLE 5
Weight Percent of Oligomers Present
in a Resin Prepared by the
"Advancement" Process Having an
Epoxide Value of p = 2.35 eq/kg[223]

		Wt%	
J value	M	GPC	Theory
0	340	16.0	20.5
1	624	2—6	—
2	908	15.5	27.9
3	1192	6—8	—
4	1476	11.1	21.1

TABLE 6
Weight Percent of Oligomers Present in a
"Tuffy" Resin Having an Epoxide Value of
p = 2.63 eq/kg[223]

		Wt %	
k value	M	Theory	GPC
0	340	15.9	21.4
1	624	17.2	20.2
2	908	15.5	15.2
3	1192	12.9	11.4
4	1476	38.5	31.7

and

$$M_j = 340 + 568j \tag{172}$$

where $k = 1, 2, 3$, and $j = 0, 2, 4$.

The calculated and determined weight percent of oligomers present in "tuffy" and "advancement" resins are given in Tables 5 and 6, respectively.[223-226]

As may be seen from Table 5, the experimentally determined distribution differs considerably from that calculated theoretically. In the case of resins made by the "advancement" process, however, besides the oligomers with $j = 0$, 2, and 4, which are present as major components, smaller amounts of oligomers with $j = 1$, 3, 5, etc. may be found. Side reactions, such as cyclization during the chain growth and chain branching caused by the base-catalyzed addition of epoxide to the aliphatic hydroxyl groups of the $-CH_2-CH(OH)-CH_2-$ functional groups of the repeat unit of the polymer chain, are responsible for these anomalies.

A more recent process for the synthesis of epoxy compounds begins with olefins and NaOCl:

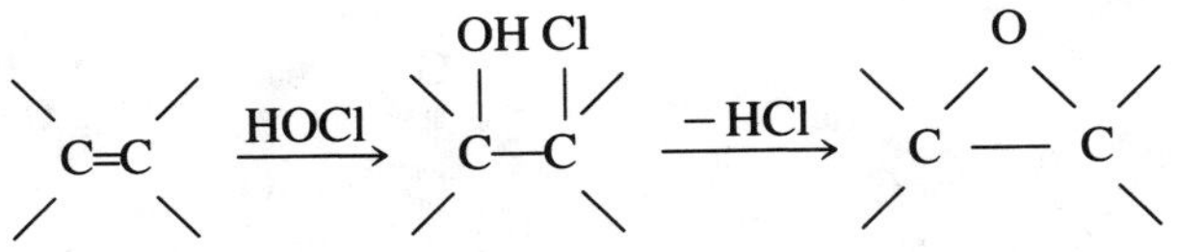

$$\backslash\!\!\diagup \!\!\diagdown C=C \diagup\!\!\backslash \xrightarrow{HOCl} \overset{OH\ Cl}{\underset{\diagup\ \ \backslash}{\backslash\!\!\diagup\!\!\mid\ \mid\!\!\diagup} C-C} \xrightarrow{-HCl} \overset{O}{\backslash\!\!\diagup\!\!\diagdown\!\!\backslash\!\!\diagup} C-C \tag{173}$$

where a double Walden inversion occurs:

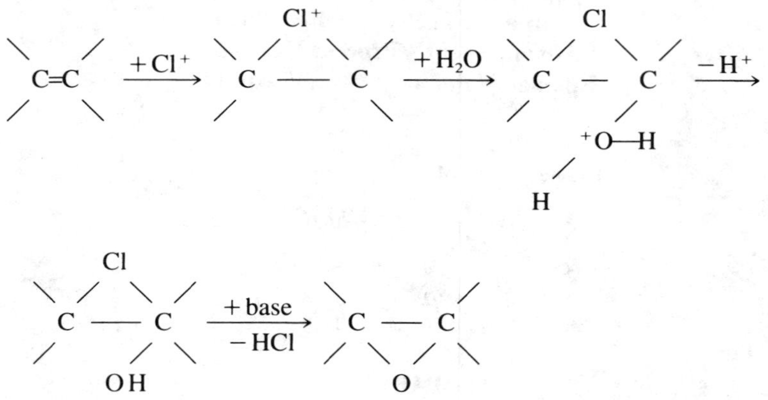

$$(174)$$

The direct epoxidation of olefinic bonds from hydrocarbon oligomers, i.e., oligobutadienes made anionically,[227] can be carried out with peracids, probably with the mechanism:[228]

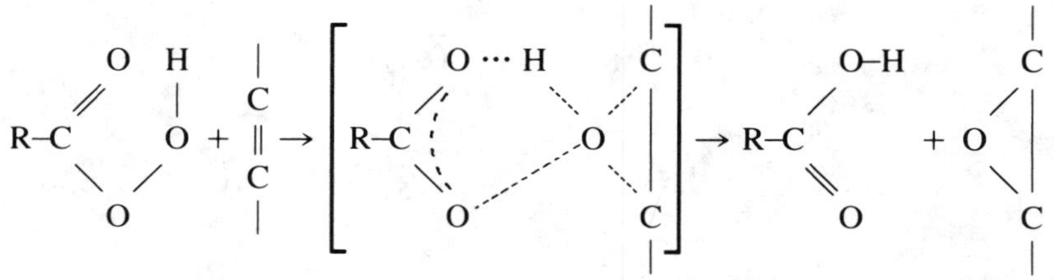

$$(175)$$

The peracids are frequently not added per se, but are produced *in situ*, e.g., from a mixture of acetic acid and hydrogen peroxide.[229]

Direct epoxidation enables polyepoxides to be produced, e.g., epoxidized plasticizers from the esters of unsaturated fatty acids, epoxides of low molecular weight poly(butadienes), and epoxides from compounds with many cycloaliphatic residues.[230,231]

In addition to epoxy resins, a large number of other types of polyethers have been prepared by condensation of alcohols and acetal interchange, Wurtz, and Wurtz-Fittig reactions.[232,233]

Dimerization of *t*-butyl (α-hydroxymethyl)acrylate, TBHMA,[234,235] produces an ether which is the precursor to the cyclopolymer of pyran or furan structure shown below, as determined from ^{13}C solution NMR spectra:

TBHMA

Dimer

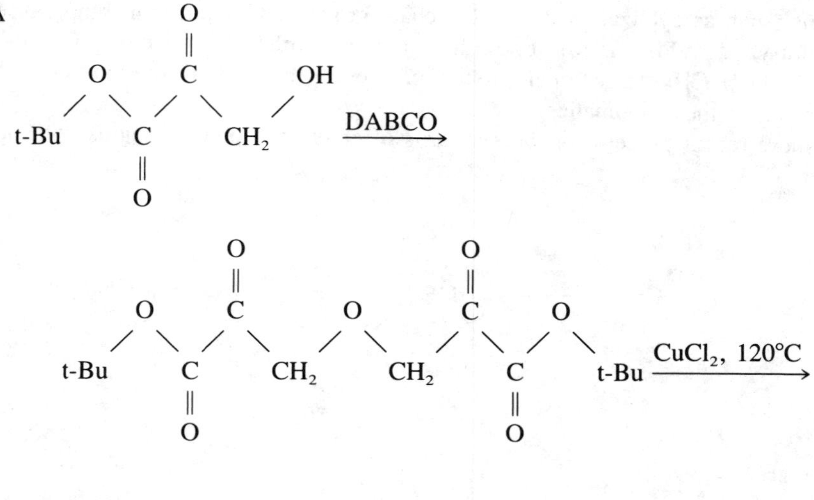

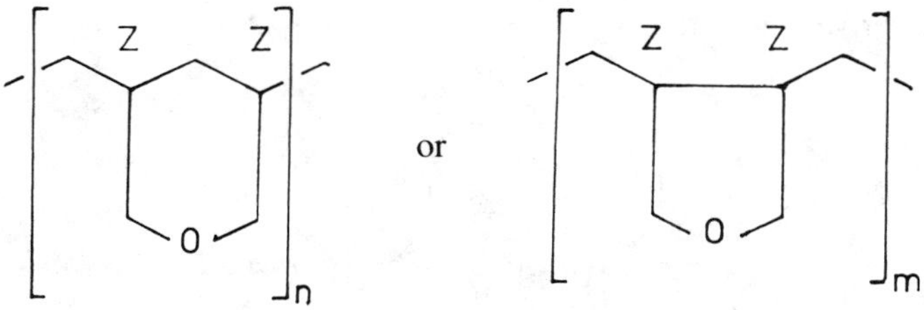

$$(176)$$

Condensation of methanol with formaldehyde catalyzed by acids results in methylal, which can be written as a formaldehyde dimer of linear structure:

$$2CH_3OH + CH_2O \xrightarrow{H^+} CH_3-[-OCH_2-]_2-H \qquad (177)$$

Since in the methylal oxidation

$$CH_3-OCH_2-OCH_3 + O_2 \rightarrow 3CH_2O + H_2O \qquad (178)$$

only 1 mol of H_2O forms for every 3 mol of CH_2O, the concentration of formaldehyde increases significantly ($>70\%$) over that obtainable with the classical methanol oxidation.[236]

Recently, a trioxane cyclic trimer of formaldehyde was developed using the highly concentrated aqueous formaldehyde solution obtained as shown above. Highly purified cyclic trimer was obtained by distillation,[237] since the synthesis was catalyzed by a noncorrosive solid acid[238] instead of the sulfuric acid commonly used in the commercial production of trioxane.

Poly(oxysiliconphthalocyanine)s have been synthesized with DP = 10 to 200 by the polycondensation of dihydroxysiliconphthalocyanine:[239,240]

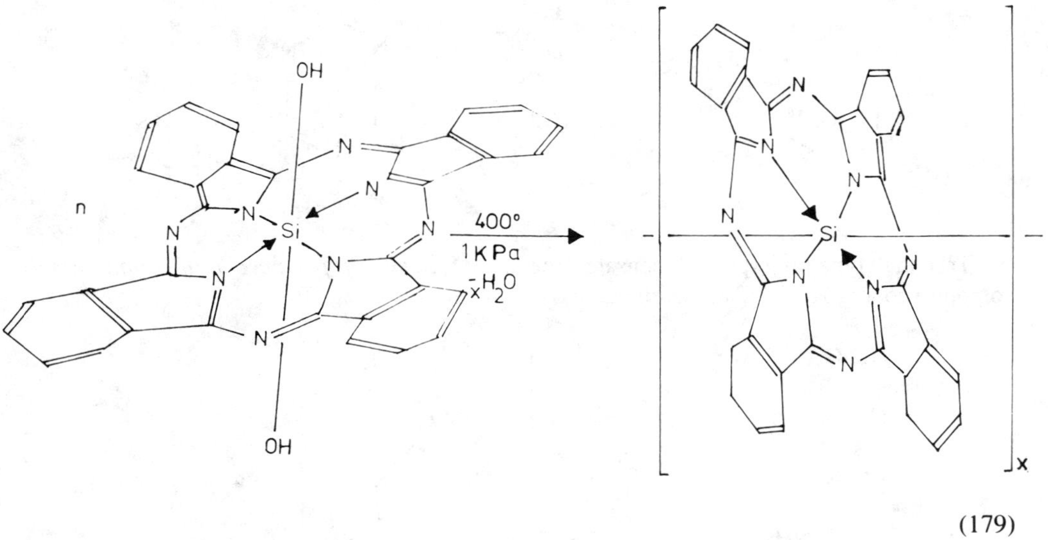

$$(179)$$

The electrical conductivity increases dramatically up to DP = 10. It seems that the function of the polymer chain is to bring about the desired packing of the macrocycles in stacks, but the actual chain length is of little consequence to the conductivity.

A series of *s*-triazine-containing aromatic oligoethers was prepared by polycondensation of sym-triazine derivatives with bisphenols:[241]

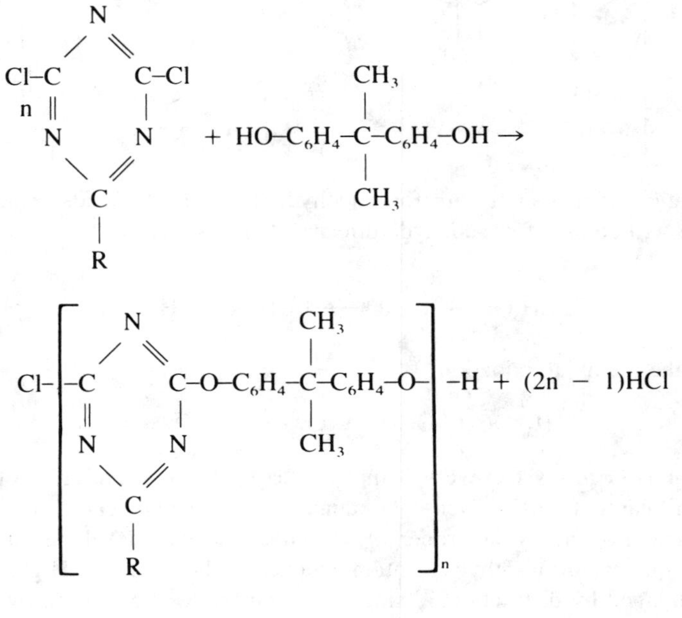

$$(180)$$

A S_N2 mechanism describes the process, which proceeds through a nucleophilic attack at the carbon atom of the triazinic cycle:

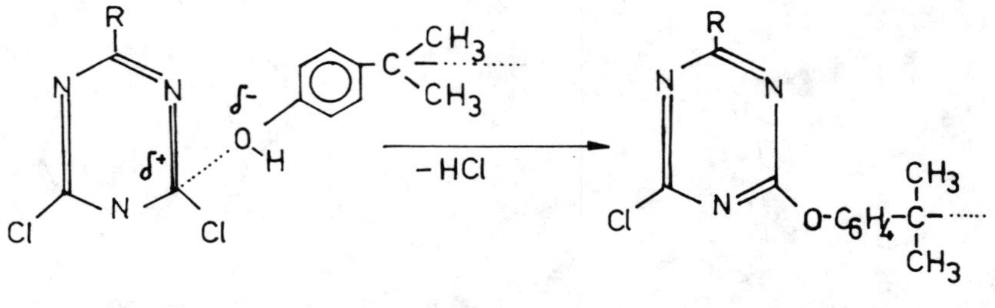

$$(181)$$

Triazine-containing polycondensates are also formed by *polyperetherification*, e.g., homopolycondensation of diphenoxyamino-sym-triazine

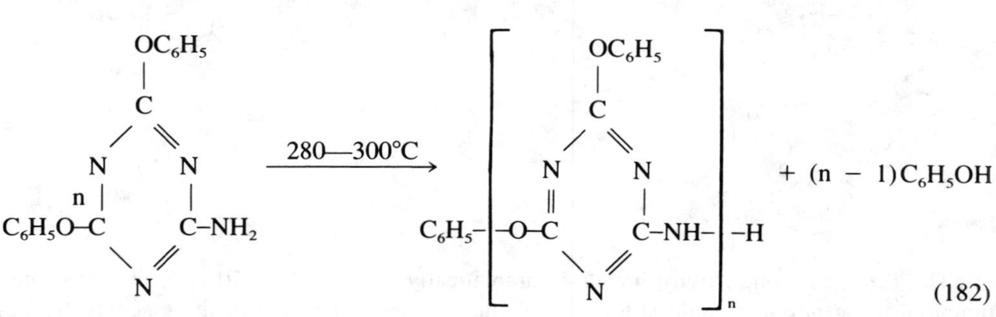

$$(182)$$

or the reaction between triazine-containing diesters and glycols

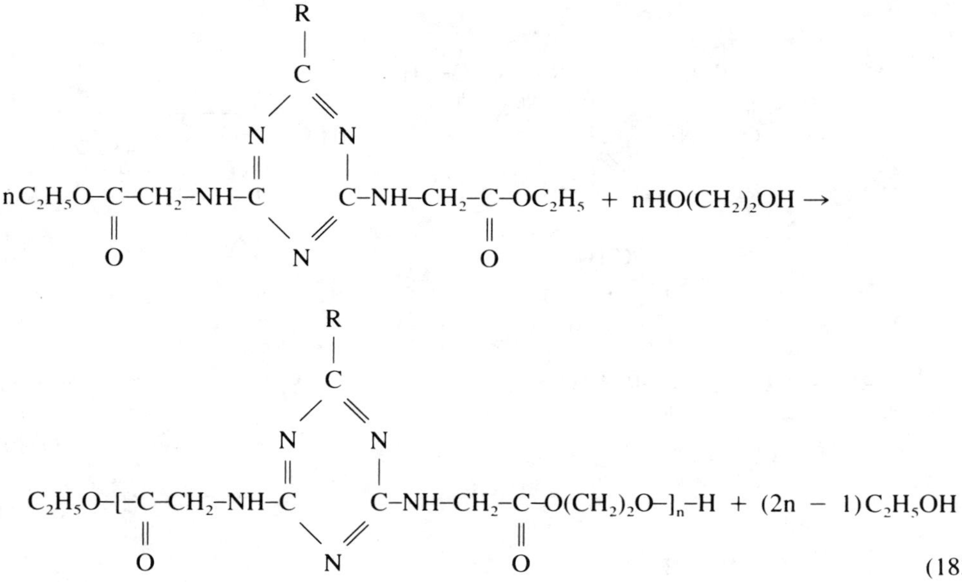

$$
nC_2H_5O-\underset{\underset{O}{\|}}{C}-CH_2-NH-C\underset{\diagdown N \diagup}{\overset{\overset{\overset{R}{|}}{\underset{\|}{C}}}{\overset{N\diagup\diagdown N}{}}}C-NH-CH_2-\underset{\underset{O}{\|}}{C}-OC_2H_5 \; + \; nHO(CH_2)_2OH \rightarrow
$$

$$
C_2H_5O-[\underset{\underset{O}{\|}}{C}-CH_2-NH-C\underset{\diagdown N \diagup}{\overset{\overset{\overset{R}{|}}{\underset{\|}{C}}}{\overset{N\diagup\diagdown N}{}}}C-NH-CH_2-\underset{\underset{O}{\|}}{C}-O(CH_2)_2O-]_n-H \; + \; (2n-1)C_2H_5OH
$$

(183)

or by *polyamination:*

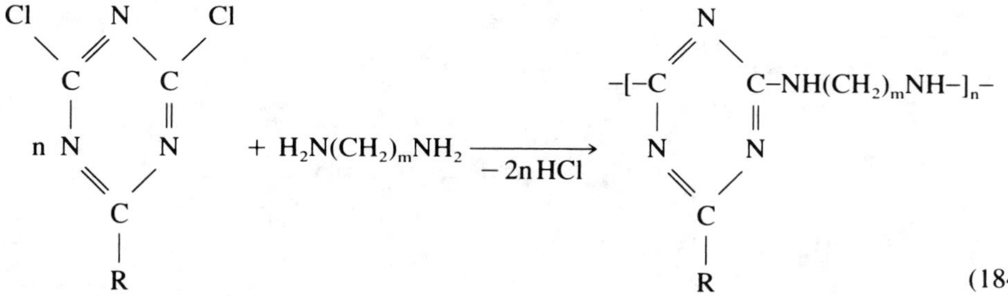

$$
n \; \begin{matrix} Cl \quad N \quad Cl \\ \diagdown \diagup\diagdown \diagup \\ C \quad\quad C \\ | \quad\quad\quad | \\ N \quad\quad N \\ \diagdown \diagup \\ C \\ | \\ R \end{matrix} \; + \; H_2N(CH_2)_mNH_2 \xrightarrow[-2nHCl]{} -[-C\underset{\diagup}{\overset{N}{\diagdown}}C-NH(CH_2)_mNH-]_n-
$$

(184)

polyamidation:

$$
nCl-\underset{\underset{O}{\|}}{C}-C_6H_4O-C\underset{\diagdown N \diagup}{\overset{\overset{\overset{OCH_3}{|}}{\underset{\|}{C}}}{\overset{N\diagup\diagdown N}{}}}C-OC_6H_4-\underset{\underset{O}{\|}}{C}-Cl \; + \; nH_2N-CH_2-NH_2 \xrightarrow[-2nHCl]{}
$$

$$
-[-\underset{\underset{O}{\|}}{C}-C_6H_4O-C\underset{\diagdown N \diagup}{\overset{\overset{\overset{OCH_3}{|}}{\underset{\|}{C}}}{\overset{N\diagup\diagdown N}{}}}C-OC_6H_4-\underset{\underset{O}{\|}}{C}-NH-CH_2NH-]_n-
$$

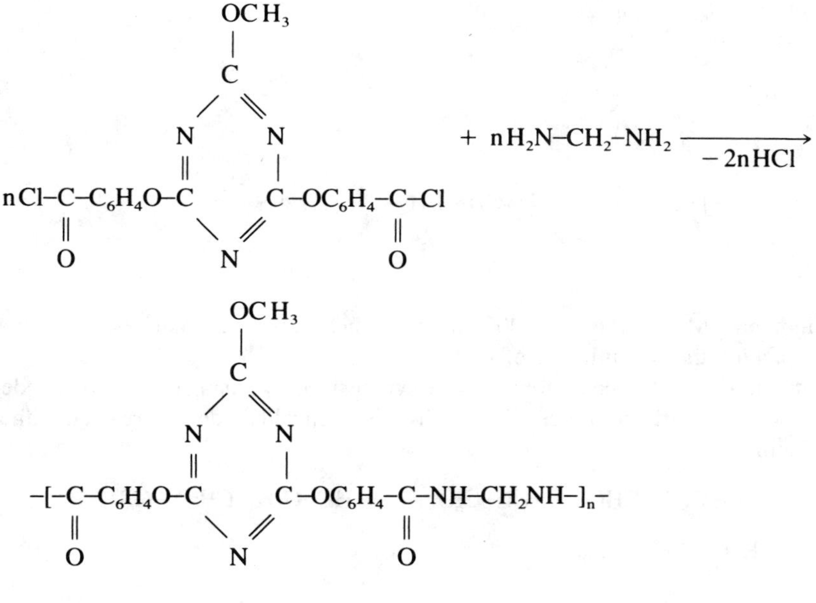

(185)

polyanhydrization:

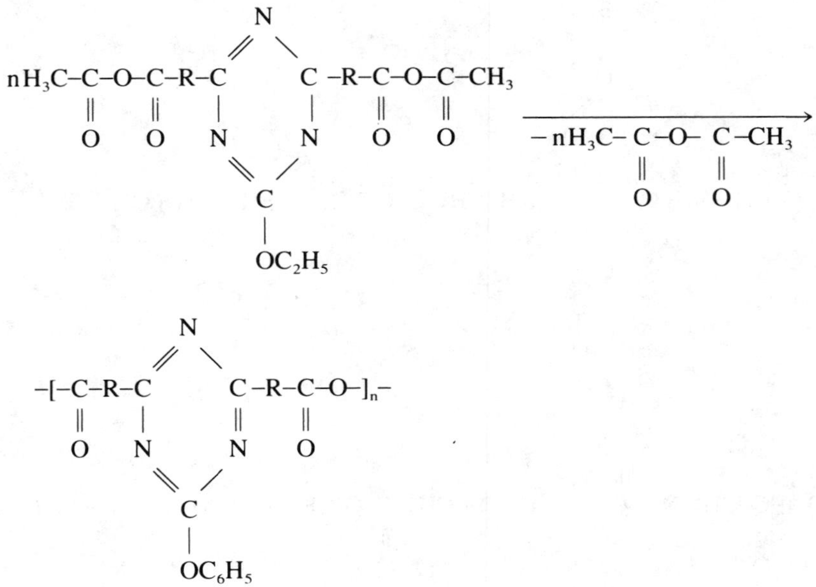

$$nH_3C-C-O-C-R-C \quad C-R-C-O-C-CH_3$$

(186)

and *aldol-* or *croton-type polycondensation:*[242]

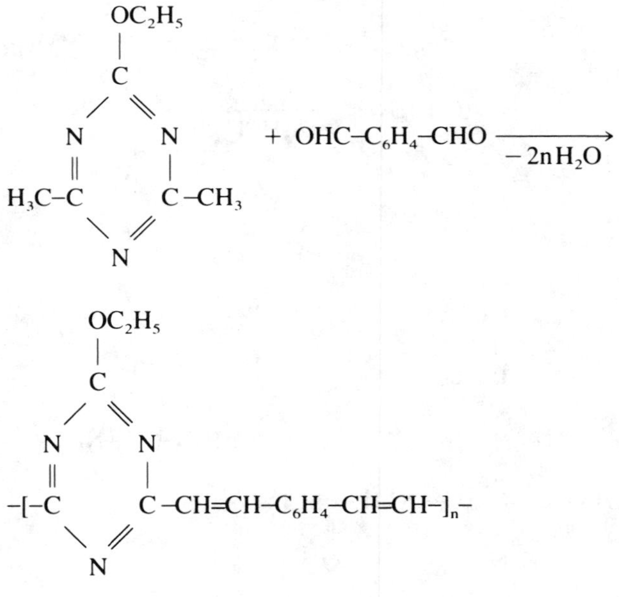

(187)

These oligomers ($\overline{M}_n$ = 1500 to 3000) have a much higher thermal resistance than their analogues lacking the triazinic cycle.

Discotic liquid crystalline oligomers were synthesized by Huang et al.[242] by condensation of phenols with cyanuric chlorides. Thus, 4-hydroxybenzaldehyde was reacted with a series of 4-alkylanilines

$$HO-C_6H_4-CHO + H_2N-C_6H_4-R \rightarrow HO-C_6H_4-CH=N-C_6H_4-R \qquad (188)$$

$$R = C_6H_{13}, C_{10}H_{21}$$

and the 4-hydroxybenzylidene-4'-alkylanilines formed were further condensed with cyanuric chloride to form 1,3,5-tris(4-benzylidene oxy-4'-alkylaniline) triazines:

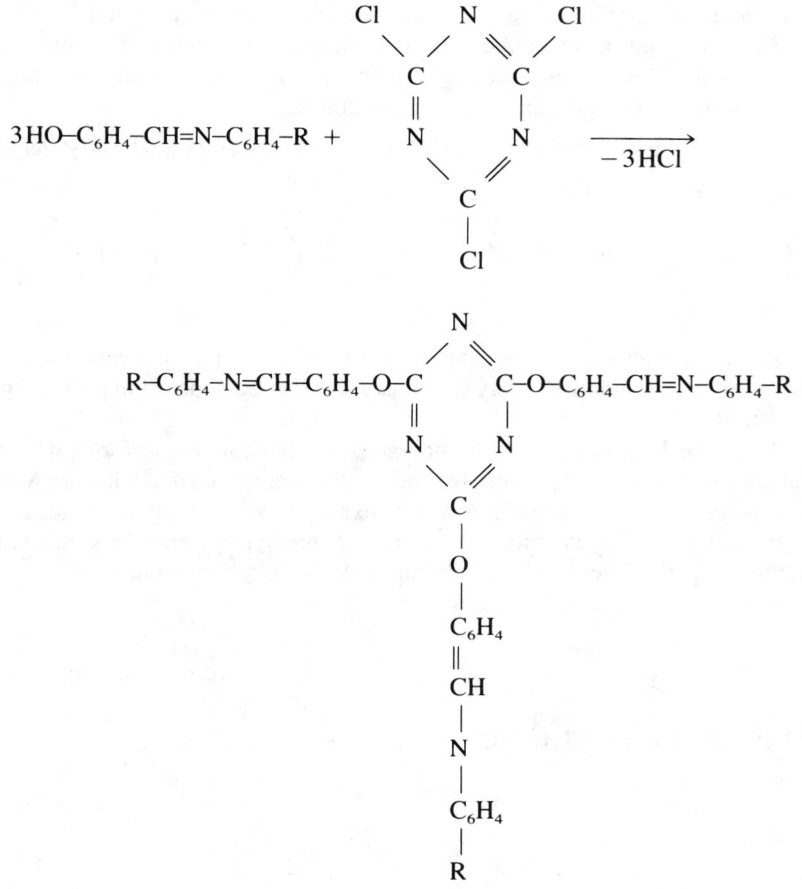

$$
3\,HO\text{--}C_6H_4\text{--}CH{=}N\text{--}C_6H_4\text{--}R \;+\; \underset{\displaystyle \begin{array}{c} Cl \\ | \\ C \\ \diagup\;\diagdown \\ N\qquad N \\ \| \qquad | \\ C \qquad C \\ \diagdown\;/\;\diagdown\;/ \\ Cl\quad N\quad Cl \end{array}}{} \;\;\xrightarrow[-\,3\,HCl]{}
$$

$$
R\text{--}C_6H_4\text{--}N{=}CH\text{--}C_6H_4\text{--}O\text{--}C \qquad C\text{--}O\text{--}C_6H_4\text{--}CH{=}N\text{--}C_6H_4\text{--}R
$$

(structure of triazine core with substituents C₆H₄–CH–N–C₆H₄–R)

(189)

These new discotic liquid crystals containing the aryloxy-*s*-triazine semiflexible core offer the advantage of processing the materials at the mesoporic phase with relatively low viscosity.

The two most important step-growth polymerization schemes which are based on *double-bond addition reactions* are the *ionic addition of diols to diisocyanates* (Reaction 190) and the *free-radical addition of dithiols to unconjugated diolefins* (Reaction 191).

$$
n\,HO\text{--}R\text{--}OH \;+\; n\,OCN\text{--}R'\text{--}NCO \;\rightarrow\; \text{--[--}ORO\text{--}CO\text{--}NH\text{--}R'\text{--}NH\text{--}CO\text{--]}_n\text{--} \tag{190}
$$

$$
n\,HS\text{--}R\text{--}SH \;+\; n\,CH_2{=}CH\text{--}R'\text{--}CH{-}CH_2 \;\rightarrow\; \text{--[--}SRS\text{--}CH_2\text{--}CH_2\text{--}R'\text{--}CH_2\text{--}CH_2\text{--]}_n\text{--} \tag{191}
$$

Reactive oligomers such as epoxy resins, isocyanate-terminated compounds, and urethane acrylates are extremely useful in the adhesive, coating, reaction injection molding, sealant, and composites industries. The application of reactive oligomers depends on their chemical structure, which affects the rheology, final physical characteristics, and curing speed of resin mixtures.

N-Cyanourea-terminated oligomers have been found to be very useful in preparing high molecular weight polymers for conversion into curing thermosets and for curing epoxy resins.[243] The preparation of *N*-cyanourea-terminated oligomers is simple, and the properties of the materials can be easily adjusted by structural change. Therefore, this chemistry is applicable to adhesives, sealants, and coatings. It was found that *N*-cyanourea compounds could be obtained directly from the reaction between isocyanate and cyanamide.[244] Using a diisocyanate as starting material, an *N*-cyanourea-terminated resin could be synthesized. Furthermore, by reacting a functional diol first with two molecules of a diisocyanate and then with two molecules of cyanamide, a structurally modified *N*-cyanourea oligomer can be prepared, as summarized by the following reactions:

$$HO-R-OH \ + \ 2OCN-R'-NCO \rightarrow OCN-R'-NH-\underset{\underset{O}{\|}}{C}-O-R-O-\underset{\underset{O}{\|}}{C}-NH-R'-NCO \quad (192)$$

$$(OCN-R'-NH-\underset{\underset{O}{\|}}{C}-O-)_2-R \ + \ 2H_2N-CN \rightarrow (NC-NH-\underset{\underset{O}{\|}}{C}-NH-R'-NH-\underset{\underset{O}{\|}}{C}-O-)_2-R \quad (193)$$

The *N*-cyanourea-terminated oligomers synthesized from a polycaprolactone diol, a diisocyanate, and cyanamide were very useful as adhesives and corrosion protection coatings for steel substrate.

Under basic conditions, cyanamide undergoes a dimerization and forms dicyandiamide at room temperature. Therefore, a monomer or an oligomer terminated with two *N*-cyanourea groups may react to form linear polymers. For example, at room temperature, a 2:1 molar ratio of diisocyanate and cyanamide dissolved in *N*-methyl pyrrolidone generated a linear polymer with a repeating segment *N,N'*-biscarbonyl-*N*-cyanoguanidine

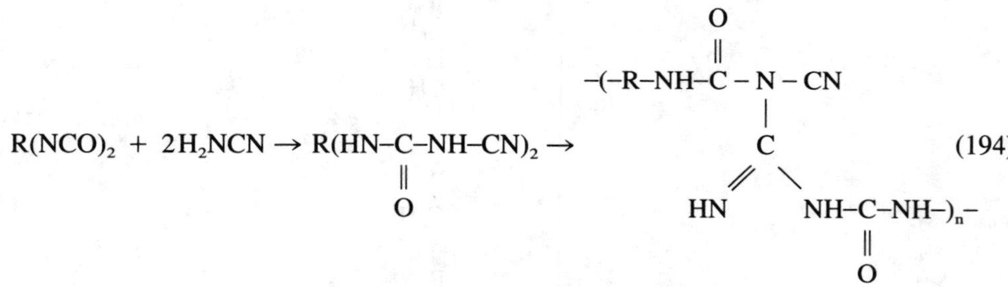

$$R(NCO)_2 \ + \ 2H_2NCN \rightarrow R(HN-\underset{\underset{O}{\|}}{C}-NH-CN)_2 \rightarrow$$

$$-(-R-NH-\underset{\underset{O}{\|}}{C}-N-CN$$

(194)

Oligomer I

where $R(NCO)_2$ is

- TDI, tolylene diisocyanate
- MDI, di(*p*-isocyanatephenyl)methane

Then, the slow ring closure reaction produces a structure like oligomer II, which is reactive toward protic derivatives (e.g., H_2O), forming a polymer (P) of ill-defined (cross-linked) structure:[244]

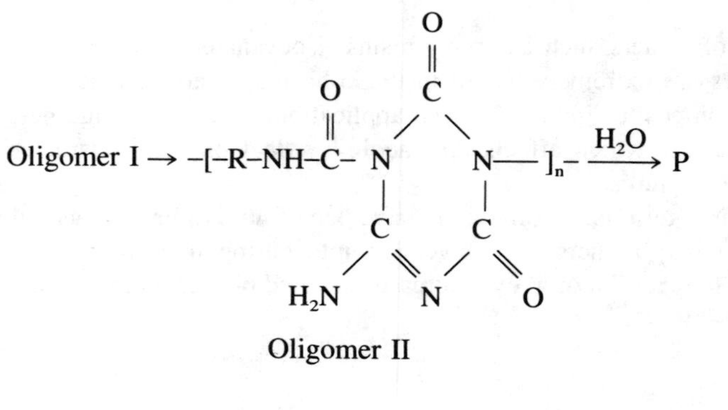

$$Oligomer \ I \rightarrow -[-R-NH-\underset{\underset{O}{\|}}{C}- \ N \cdots \cdots N-]_n- \xrightarrow{H_2O} P \quad (195)$$

Oligomer II

Cyclotrimerization of 2,4-tolylene diisocyanate and phenyl isocyanate results in soluble and fusible oligoisocyanurates,[245] and epoxyisocyanurate oligomers were prepared by polycyclotrimerization of 2,4-tolylene diisocyanate with hexamethylene diisocyanate at 100°C in the presence of 2 to 3% Co or Ni acetyl acetonate, followed by reaction with glycidol.[246-248]

In the presence of tertiary phosphines or peralkylated phosphoric triamides, isocyanates can be oligomerized at a desired degree of polymerization by addition of sulfonyl isocyates.[249,250] These oligomers are useful for the preparation of polyurethane coatings, lacquers, or polyurethane elastomers.[251,252]

Oligoesterdiols synthesized from the residues from the distillation of dimethylterephthalate or dimethylisophthalate are suitable for the preparation of rigid oligourethanes.[253] The physicomechanical properties of the oligomers depend upon the length of the oligoester units.[254] Oligourethane and oligomeric urea-type compounds prepared by Tsukube[255] and Tsukube and Araki[256] have specific binding interactions useful for biomimetic experiments.[257]

D. FREE-RADICAL COUPLING REACTIONS

These reactions form the basis of the preparation of arylene ether polymers, polymers containing acetylenic units, and arylene alkylidine polymers, by the following equations:

$$n R_3 - \underset{R_2}{\overset{R_1}{\bigcirc}} - OH + n[Ox] \longrightarrow -\left[\underset{R_2}{\overset{R_1}{\bigcirc}} - O-\right]_n + n[Ox]HR_3 \quad (196)$$

$$nCH \equiv C-C \equiv CH + n[Ox] \rightarrow -[-C \equiv C-C \equiv C-]_n- + n[Ox]H_2 \quad (197)$$

$$\underset{R}{\overset{R}{HC}} - \bigcirc - \underset{R}{\overset{R}{CH}} + 2nR' \longrightarrow -\left[\underset{R}{\overset{R}{C}} - \bigcirc - \underset{R}{\overset{R}{C}}\right]_n- + 2nR'H \quad (198)$$

These reactions require stoichiometric amounts of either an oxidizing agent, [Ox], or a derivative, R', capable of removing hydrogen atoms for monomer molecules to generate monomer and polymer end-group radicals.

Arylene ether polymers were prepared in 1916 by Hunter et al.[258] by ethyl iodide-induced decomposition of anhydrous silver 2,4,6-tribromophenoxide. The polymer had an estimated molecular weight between 6000 and 12,000. In more recent investigations, silver p-bromophenoxide was polymerized to a low molecular weight compound.[259]

The formation of arylene ether polymers by an oxidative coupling reaction has been shown to involve a gradual increase of molecular weight with a reaction conversion characteristic of a step-growth polymerization reaction. Polymerization of m-diethynylbenzene by a radical coupling reaction results in the formation of a soluble oligomer with a molecular weight of 7000.[260] The step-growth polymerization reaction has also been applied to the preparation of oligoxylylenes[261] and oligo(p-phenylene).[262]

Poly(p-phenylene), PPP, can be prepared by polycondensation of benzene with a heterogeneous system of Lewis acid-oxidant, such as $AlCl_3$-$CuCl_2$ with vigorous evolution of hydrogen chloride.[263] The polymer, an insoluble black powder, is branched but not cross-linked to a network. The molecular mass measured with the soluble sulfonate derivative does not exceed 8000 (DP = 100).

Treatment of benzene with $AlCl_3$ and CuCl (3:1 molar ratio) under O_2 gave a 187% yield of PPP based on the molecular quantity of CuCl without evolution of hydrogen chloride, indicating that the cuprous salt works as a catalyst:[264]

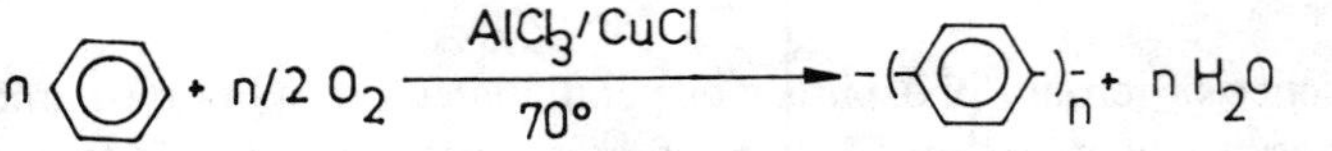

$$n \, \langle \bigcirc \rangle + n/2 \, O_2 \xrightarrow[70°]{AlCl_3/CuCl} -(\langle \bigcirc \rangle)_{\overline{n}} + n \, H_2O \qquad (199)$$

Fractions soluble in benzene accompanied the insoluble polymer (up to 46% at 60°C based on CuCl).

Polymerization of alkylbenzenes (e.g., toluene) was also achieved, using $AlCl_3$ and CuCl under oxygen. The polymers were of lower molecular weight and soluble in various organic solvents such as THF and $CHCl_3$.

One of the promising methods for the preparation of polyphenylene-type polymers is the polycondensation of mono- and diacetylaromatic compounds in the presence of a ketalization agent[265] or their ethyl ketals.[266] In the first stage, prepolymers with a rather low $\overline{M}_n$ and active acetyl and ketal end groups can be formed in benzene solution.[265-267] The first stage can be represented schematically as follows:

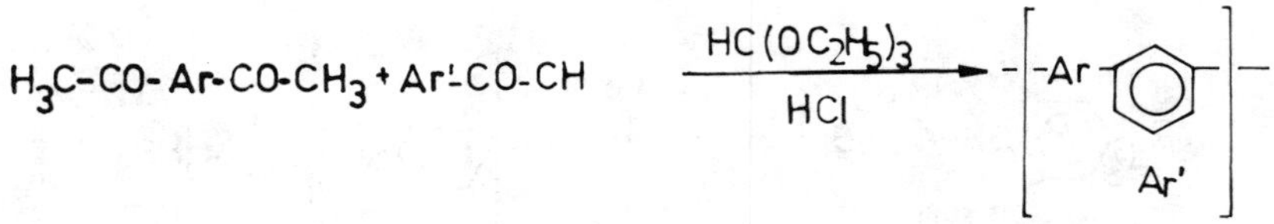

$$H_3C\text{-}CO\text{-}Ar\text{-}CO\text{-}CH_3 + Ar'\text{-}CO\text{-}CH \xrightarrow[HCl]{HC(OC_2H_5)_3} -\left[-Ar\text{-}\underset{Ar'}{\bigcirc}- \right]- \qquad (200)$$

and

$$H_3C(OC_2H_5)_2C\text{-}Ar\text{-}C(OC_2H_5)_2CH_3 + Ar\text{-}C(OC_2H_5)_2CH_3 \xrightarrow[-6C_2H_5OH]{HCl}$$

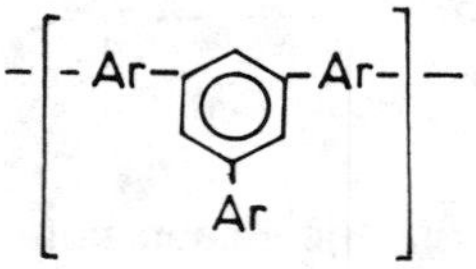

$$-\left[-Ar\text{-}\underset{Ar}{\bigcirc}\text{-}Ar\text{-} \right]- \qquad (201)$$

where Ar and Ar' can be

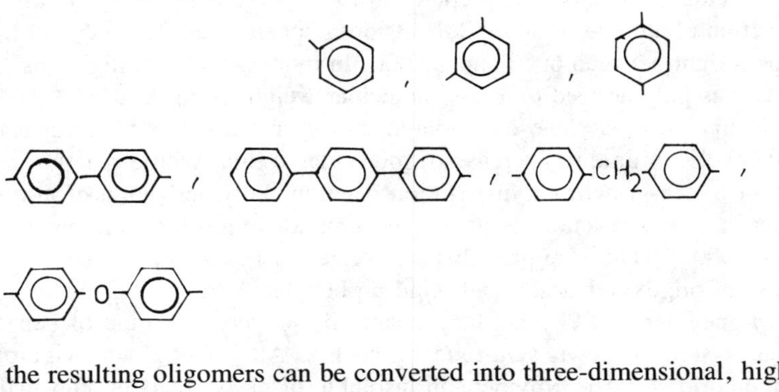

When heated, the resulting oligomers can be converted into three-dimensional, highly heat-resistant, and thermally stable polymers.

These reactions were exploited to obtain carborane-containing polymers.[268] The following mono- and diacetyl aromatic carboranes were condensed: 4,4'-diacetyldibenzyl-*o* (as well as *m* or *p*)-carborane and their ethyl ketal, 4-acetylbenzyl carboranes or their ketals. The oligomers had the structures:

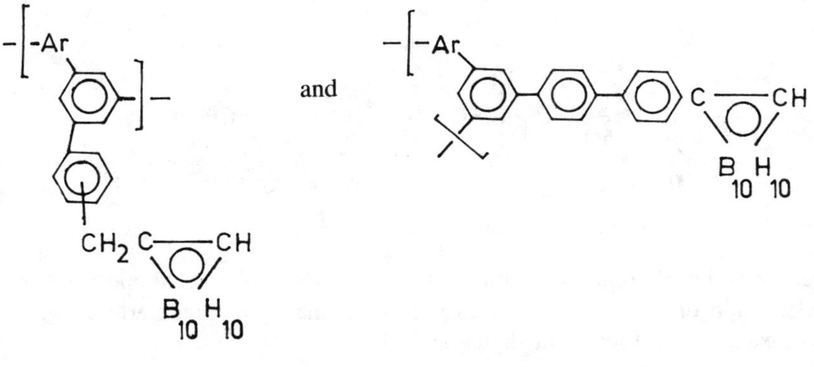

(202)

The molecular weight of oligomers was in the range of $\overline{M}_n$ = 1100 to 4600 (ebulioscopic) and the boron content was rather high (31 to 32.5%).

Aromatic oligocarboranes were also synthesized by radically promoted condensation of *o*-carborane at 200°C in solution (benzonitrile, naphthalene). An interaction between solute and solvent was noticed, and the structure of oligomers was dependent on the solvent used:[269]

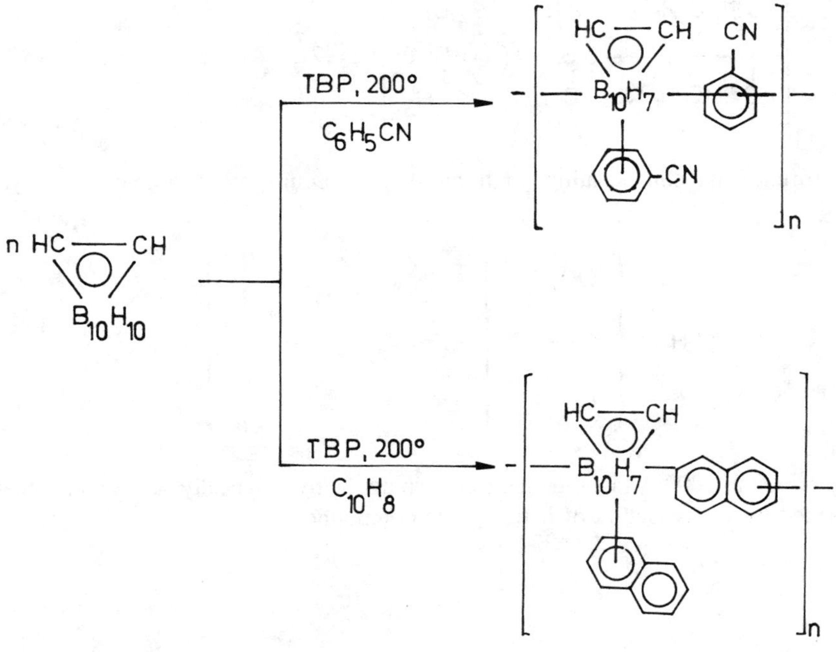

(203)

where TBP is *t*-butylperoxide (initiator) and n = 5 to 10.

Cohydrolysis of dimethyl(*o*-carboranylmethyl)methoxysilan and dimethyldichlorosilan results in oligomers ($\overline{M}_n \sim$ 650) with dimethyl(*o*-carboranylmethyl)siloxy end groups and $\overline{M}_w/\overline{M}_n$ = 1.38 to 2.40.[270,271]

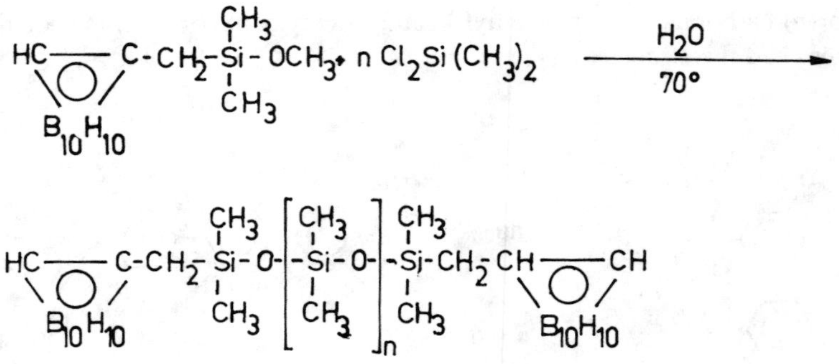

$$\text{HC} \underset{B_{10}H_{10}}{\overset{}{\bigtriangleup}} \text{C-CH}_2\text{-}\underset{CH_3}{\overset{CH_3}{\underset{|}{\overset{|}{Si}}}}\text{-OCH}_3 + \text{n Cl}_2\text{Si(CH}_3)_2 \xrightarrow[70°]{H_2O}$$

$$\text{HC}\underset{B_{10}H_{10}}{\overset{}{\bigtriangleup}}\text{C-CH}_2\text{-}\underset{CH_3}{\overset{CH_3}{\underset{|}{\overset{|}{Si}}}}\text{-O-}\left[\underset{CH_3}{\overset{CH_3}{\underset{|}{\overset{|}{Si}}}}\text{-O-}\right]_n\underset{CH_3}{\overset{CH_3}{\underset{|}{\overset{|}{Si}}}}\text{-CH}_2\text{CH}\underset{B_{10}H_{10}}{\overset{}{\bigtriangleup}}\text{CH} \tag{204}$$

n = 5 to 6.

Organometallic oligomers containing Ti, Si, or Ge in the backbone were obtained by polycondensation of specific derivatives with monomeric or oligomeric silandiols.[272] Oligotitanosiloxanes were formed in Equation 205.

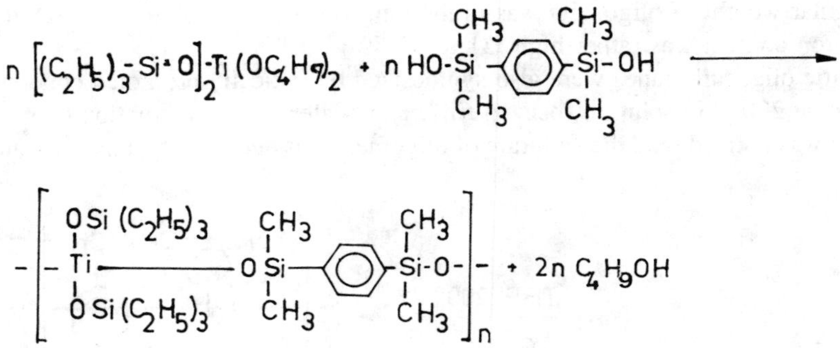

$$\text{n}\left[(C_2H_5)_3\text{-Si}\cdot\text{O}\right]_2\text{-Ti(OC}_4H_9)_2 + \text{n HO-}\underset{CH_3}{\overset{CH_3}{\underset{|}{\overset{|}{Si}}}}\underset{CH_3}{\overset{CH_3}{\underset{|}{\overset{|}{\bigcirc}}}}\text{Si-OH} \longrightarrow$$

$$-\left[\underset{\text{O Si}(C_2H_5)_3}{\overset{\text{O Si}(C_2H_5)_3}{\underset{|}{\overset{|}{\text{Ti}}}}}\text{---O-}\underset{CH_3}{\overset{CH_3}{\underset{|}{\overset{|}{Si}}}}\underset{CH_3}{\overset{CH_3}{\bigcirc}}\underset{CH_3}{\overset{CH_3}{\underset{|}{\overset{|}{Si}}}}\text{-O-}\right]_n + \text{2n C}_4H_9OH \tag{205}$$

Heterofunctional condensation of tetrabutoxygermanium gave oligomers of the type[273,274]

$$C_4H_9O\left\{\underset{OC_4H_9}{\overset{OC_4H_9}{\underset{|}{\overset{|}{\text{-Ge}}}}}\text{---O-}\left[\underset{OC_4H_9}{\overset{OC_4H_9}{\underset{|}{\overset{|}{\text{-Ge}}}}}\text{-(-O-Si-)}_m\text{-O-}\right]_p\right\}_n\text{-C}_4H_9 \tag{206}$$

with $\overline{M}_n$ = 400 to 5000, while oligomers containing the hydrolytically stable bond –Si–O–Ge– were formed by condensation of heterocyclic compounds.[275]

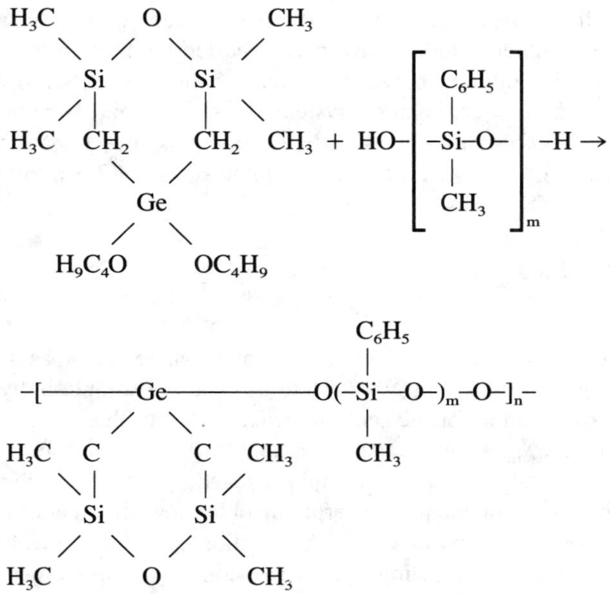

$$\text{(Structure 207 reactant and product diagrams)}$$

$$\overline{M}_n = 5000 \qquad (207)$$

A series of conjugated oligomers ($\overline{M}_n = 500$ to 2000) with aromatic or organometallic substituents were obtained by polymerization of acetylenic monomers or by polycondensation of corresponding aldehyde or ketones, as shown below:[276-278]

$$n\,H_3C–C=O \xrightarrow[-\,n\,H_2O]{\text{ZnCl}_2,\ 300°C} –(–HC=C–)_n–$$

with R substituent on each.

$$n\,HC–CH_2–R \xrightarrow[-\,n\,H_2O]{\text{ZnCl}_2,\ 300°C} –(–HC=C–)_n–$$

$$(208)$$

where R is

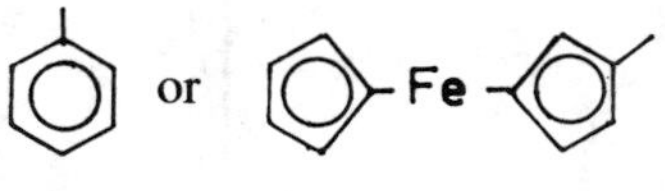

The polycondensation oligomers were accompanied by more complex insoluble products. Their magnetic (ESR) and electric (conductivity, E_a) properties characterized them as organic semiconductors.[277]

E. AROMATIC ELECTROPHILIC-SUBSTITUTION REACTIONS

This reaction has been applied to the preparation of poly(arylene alkylene) by the addition of benzyl alcohol to concentrated sulfuric acid.[279] The product can be separated into dioxane-soluble and -insoluble fractions. These fractions are white oligomers with an average DP of about 9 and 19 to 24, respectively. The reaction of benzyl alcohol with sulfuryl chloride results in a colored oligomer with DP = 10.

Polymers with molecular weights ranging from 3,500 to 12,000 are formed from the reaction of benzene and substituted benzenes with 1,2-dichloroethane. Poly(arylene alkylene) with a molecular weight of 2500 is obtained from methylene chloride and bromobenzene.[280]

Although organometallic catalysts have been commonly used for vinylic and dienic polymerizations, only a few synthetic studies have been reported in the field of step-growth reactions.[281-283] Dehalogenation coupling between aromatic halides with Ni(O) complexes results in poly(arylene)s having a π-conjugation system along the chain, such as the PPP.

Stirring a mixture of 2,5-dibromopyridine and Ni(O) complex in a 1:1 or 1:1.2 molar ratio in a solution gave yellow or yellowish-orange precipitates of poly(2,5-pyridinediyl):[281]

$$Br-\langle\bigcirc_N\rangle-Br + n\,NiLm \longrightarrow -(-\langle\bigcirc_N\rangle-)_{\overline{n}} + n\,Ni\,Br_2Lm \qquad (209)$$

Various Ni(O) complexes, NiLm, where Lm is the ligand, where usable as the reactant. The polymers are of an oligomeric nature (DP = 16 to 25) and are soluble in hydrochloric acid and formic acid. The spectrum in formic acid showed a relatively sharp π-π* absorption peak at about 370 nm, and the λ_{max} value was comparable to the λ_{max} of solid PPP.[282]

Palladium(O) complexes catalyze a variety of unique reactions out of which a new type of ring-opening polymerization is of interest in terms of obtaining oligomers. The studied monomers were vinyl cyclopropane derivatives.[283] The reaction mechanism involves a transfer of the activated proton from the growing end to the complexed monomer unit to be added, and the key intermediate is a Pd/π-allyl complex. The polymerization reaction also demands an initiator with an activated proton, e.g., diethyl malonate. Polymerization implies both vinylic and cycle atoms through a 1,5 connection, as shown below. The Pd(O) complex adds oxidatively to the monomer with the ring opening to generate the π-allyl complex, PdAC

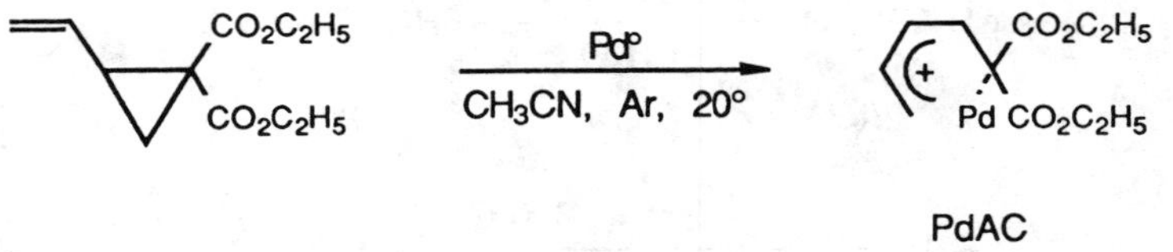

$$\overset{\qquad CO_2C_2H_5}{\underset{CO_2C_2H_5}{\bigtriangleup}} \xrightarrow[\text{CH}_3\text{CN, Ar, 20°}]{\text{Pd}^o} \overset{\qquad CO_2C_2H_5}{\underset{Pd\ CO_2C_2H_5}{(+)}} \qquad (210)$$

PdAC

which reacts with the initiator of diethyl malonate, followed by the reductive elimination of Pd(O):

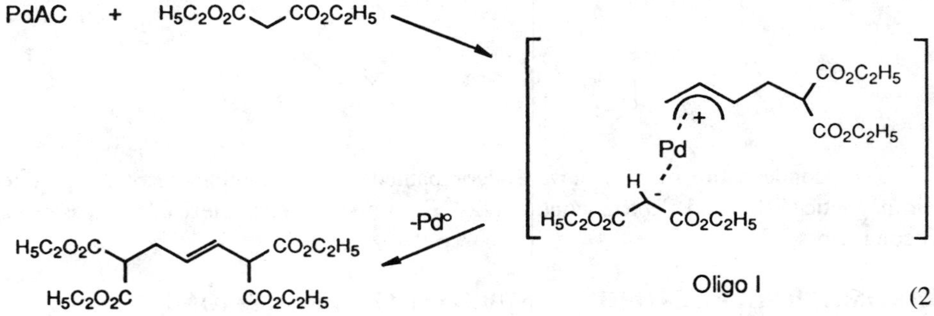

$$PdAC + H_5C_2O_2C\diagdown\diagup CO_2C_2H_5 \longrightarrow$$

$$\left[\begin{array}{c} \underset{Pd}{(+)} \diagup\diagdown\overset{CO_2C_2H_5}{\underset{CO_2C_2H_5}{\diagdown}} \\ H_5C_2O_2C\diagup\diagdown CO_2C_2H_5 \end{array} \right]$$

Oligo I

$$\xrightarrow{-Pd^o} \underset{H_5C_2O_2C}{H_5C_2O_2C}\diagup\diagdown\diagup\diagdown\overset{CO_2C_2H_5}{\underset{CO_2C_2H_5}{}} \qquad (211)$$

The product Oligo I still has an activated proton, which can react with another PdAC

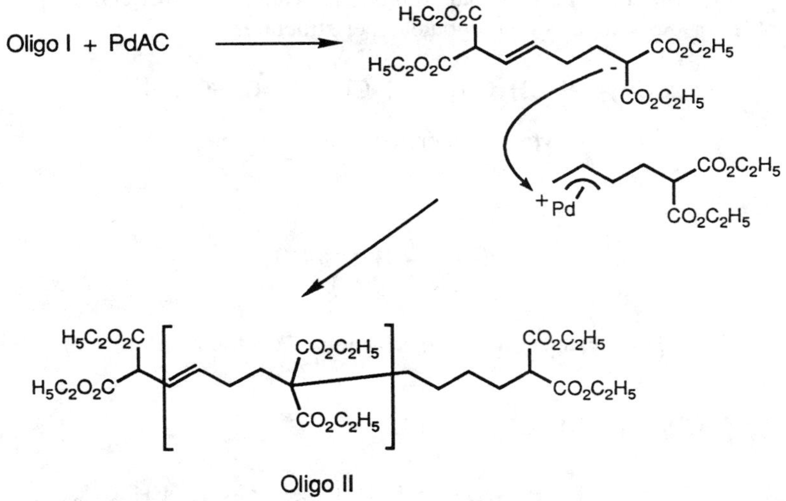

$$\text{Oligo I + PdAC} \longrightarrow$$

Oligo II

(212)

and the repetition of these reactions gives higher oligomers, up to DP $\sim$ 18:

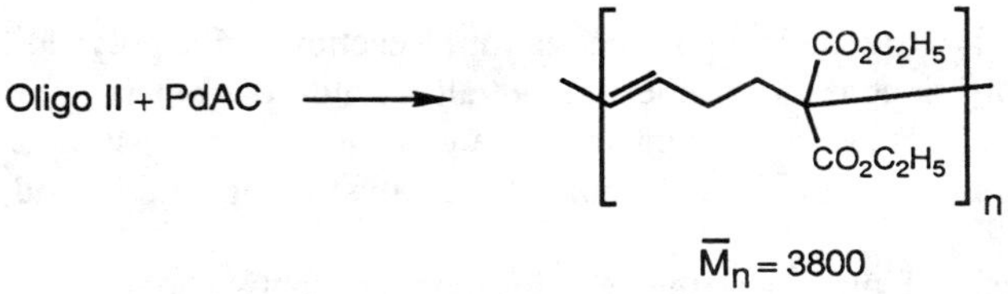

$$\text{Oligo II + PdAC} \longrightarrow$$

$$\overline{M}_n = 3800$$

(213)

This mechanism is supported by the fact that the isolated polymer could be used for the initiator. At the same time, the GPC curve was unimodal and shifted to the range of the higher molecular weight.

Synthesis of aromatic polyamides containing a poly(arylenevinylene) structure was also achieved in the presence of palladium catalysts.[284] Structures of this kind are generally difficult to obtain by the conventional Schotten-Baumann reaction. The reaction takes place between arylene bis(acrylamide) and aromatic dihalides in the presence of a base:

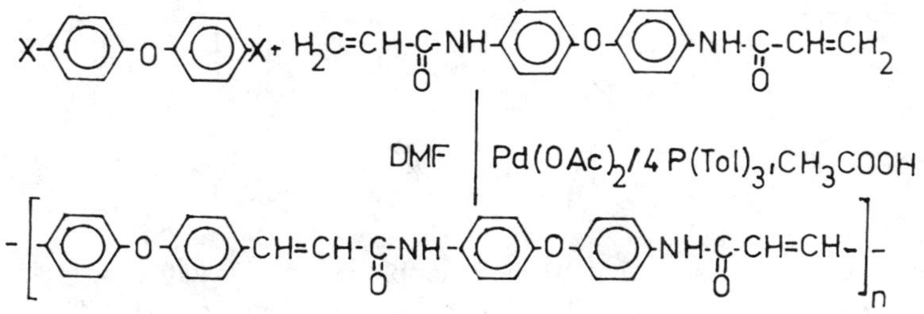

$$\text{X} \langle \bigcirc \rangle \text{-O-} \langle \bigcirc \rangle \text{X} + \text{H}_2\text{C=CH-C-NH} \langle \bigcirc \rangle \text{-O-} \langle \bigcirc \rangle \text{NH-C-CH=CH}_2$$

$$\text{DMF} \mid \text{Pd(OAc)}_2/4\,\text{P(Tol)}_3, \text{CH}_3\text{COOH}$$

$$-\left[\langle \bigcirc \rangle \text{-O-} \langle \bigcirc \rangle \text{CH=CH-C-NH-} \langle \bigcirc \rangle \text{-O-} \langle \bigcirc \rangle \text{NH-C-CH=CH-} \right]_n$$

(214)

where X = Br, I.

Oligomers were formed from the aromatic dibromide ether, whereas high molecular weight polyamides having the highest inherent viscosity of 0.84 dl/g were obtained using the diiodide as the aromatic dihalide. The polymers are soluble in DMF, NMP, DMSO, and pyridine and are photocross-linkable.

F. METATHETIC POLYCONDENSATION

The metathetic polycondensation of α,ω-dienes has considerably enlarged the possibilities of preparing unsaturated carbon chain oligomers. Thus, three different pathways appear to be available for the synthesis of polybutenamer structures:

$$nCH_2=CH-CH=CH_2 \rightarrow -[-CH_2-CH=CH-CH_2-]_n- \tag{215}$$

(1,4-type polymerization of butadiene)

$$n\begin{bmatrix} HC=CH \\ \\ (CH_2)_2 \end{bmatrix} \rightarrow -[-CH=CH-CH_2-CH_2-]_n- \tag{216}$$

(ring-opening polymerization of cyclobutene)

$$nCH_2=CH-CH_2-CH_2-CH=CH_2 \rightarrow$$

$$CH_2=CH-[-CH_2-CH_2-CH=CH-]_{n-1}-CH_2-CH_2-CH=CH_2 \tag{217}$$

(metathetic polycondensation of 1,5-hexadiene)

Reaction 215 is a true radical polymerization, Reaction 216 is polyaddition, and Reaction 217 is a true polycondensation which potentially yields oligomeric polyenes having two vinyl end groups.[285] Widely different results are obtained by Reaction 217, depending on the length of the monomeric hydrocarbon, the catalyst composition, and the reaction conditions.

The metathesis of olefins is the reaction converting two identical or different olefin molecules into two other olefins by formal permutation of the two double bonds in a four-centered array under the action of catalysts:[286]

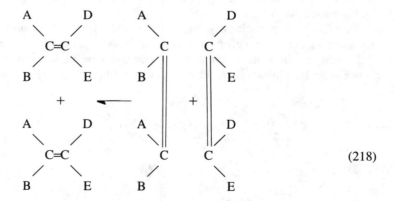

$$(218)$$

In general, when an unspecified alkene is employed in cometathesis, a cleavage of the acyclic polyene chain takes place. The process is called alkenolysis. Thus, unsaturated polymers can be converted, by alkenolysis, into fragments with a lower molecular weight:

$$-[-(CH_2)_n- = -]_m- \qquad -[-(CH_2)_n- = -]_{m-p}-R$$

$$+ \qquad \longrightarrow \qquad + \tag{219}$$

$$R- = -R \qquad -[-(CH_2)_n- = -]_p-R$$

The metathesis of cycloolefins (COs) takes place in a completely different way. Through the metathesis of the double bonds of two CO molecules, COs ought to yield cyclic dimers with two double bonds:

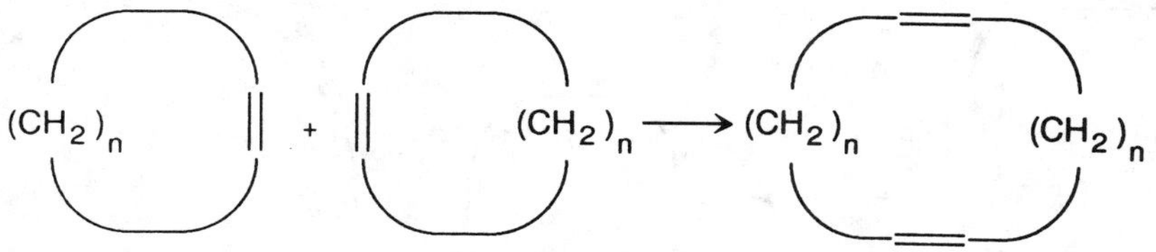

$$(220)$$

In most cases, however, in the presence of metathesis catalysts (usually based on Mo, W, or Re compounds), a ring-opening polymerization of CO takes place, with formation of polyalkenamers:

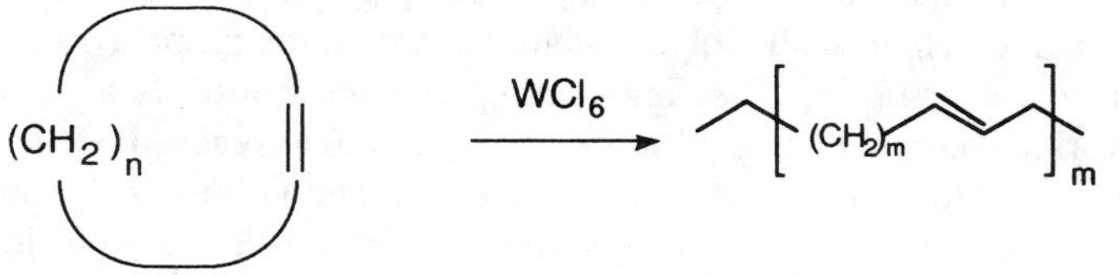

$$(221)$$

The presence of CO oligomers and unsaturated cycles along with polyalkenamers was interpreted as supporting the idea that the ring opening of CO proceeds by successive metathesis steps, initially forming dimers (Equation 221), then trimers, etc. Therefore, the metathesis of COs leads to the formation of cyclic and macrocyclic oligomers:

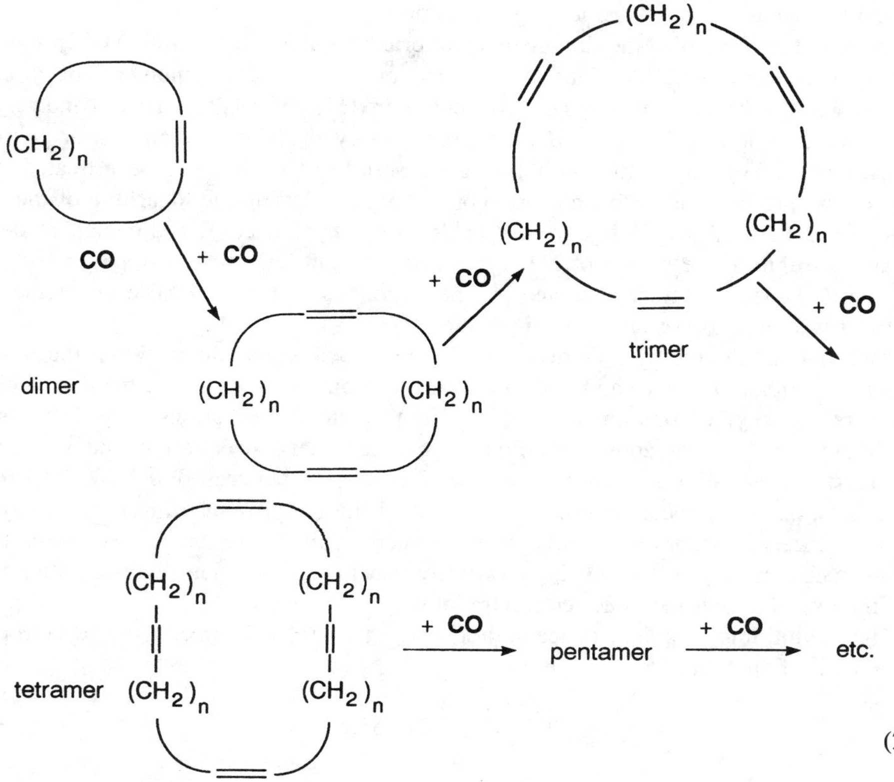

$$(222)$$

The macrocycles can be cleaved by intramolecular metathesis, with formation of COs and new macrocycles, a reaction related to the reverse of Equation 222.

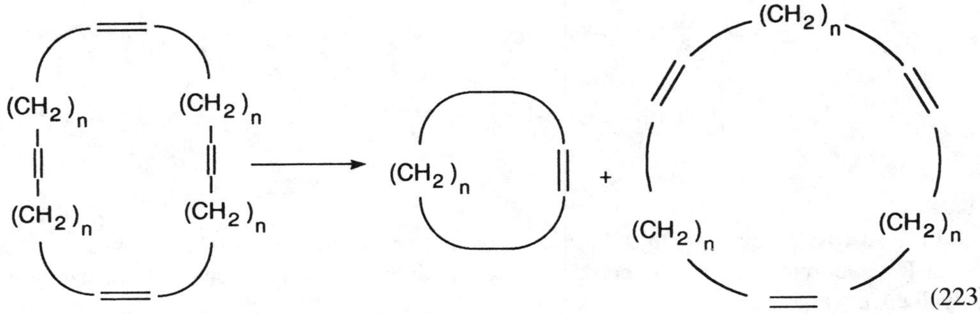

$$(223)$$

Recent data on the chain mechanism of ring-opening polymerization show that oligomeric products, obtained together with polyalkenamers, are more probably formed by cyclo-degradation of polyalkenamers. The ring-opening polymerization is a reversible process; under given working conditions, equilibrium conversion is reached. This equilibrium is determined by the stability of the monomer, oligomers, and polymers, by the characteristic thermodynamic parameters such as enthalpy and entropy, and by the reaction temperature:

$$\ln[M]_e = \frac{\Delta H_p}{RT} - \frac{\Delta S_p^\circ}{R} \qquad (224)$$

where $[M]_e$ represents the equilibrium concentration of the CO, ΔH_p the enthalpy variation during the polyaddition, ΔS_p° the standard entropy variation during the polyaddition, T the reaction temperature, and R the ideal gas constant.

Höcker et al.[287] studied the ring-chain equilibrium for the polymerization of cyclooctene and correlated the product distribution as a function of the initial monomer concentration. They showed that below a threshold CO concentration ($[M]_e$) of 0.1 M, practically no polymer was present, while at $[M] > 0.1$ M the presence of cyclic oligomers and linear polymers was evidenced. The ring-chain equilibrium for a series of COs has also been treated in the limits of the Jacobson and Stockmayer theory.[15a] When plotting the logarithm of the equilibrium constant, k_n, of the oligomers of cyclooctene, cyclodecene, and cyclopentadecene vs. the logarithm of the degree of polymerization, straight lines with a slope of $-{}^5/_2$ were obtained.[287] Deviations were observed for the oligomers with n > 4. The larger the ring, the lower was the position of the straight line.

An important thermodynamic feature of CO ring-opening polymerization is the *constant oligomer distribution* which can be obtained starting from the monomer, a fraction of higher oligomers, the polymer, or a single oligomer. The absolute concentrations of cyclic oligomers were found to be constant, above a minimum initial monomer concentration, and independent of catalyst concentration or temperature in a certain range between 0 and 100°C. Up to a minimum monomer concentration, which is called the *cutoff point* (threshold value), the monomer is generally converted into cyclic oligomers, while above this concentration, linear polyalkenamer is also formed. It is noteworthy that higher polymer/oligomer ratios result from the use of higher monomer concentrations.

The equilibrium oligomer concentration $[M_{oligo}]_e$ is related directly to the entropy of formation by Equation 225.

$$-\ln[M_{oligo}]_e = \Delta S^\circ / R \qquad (225)$$

Significantly, the plot of the entropy per carbon atom vs. the number of carbon atoms for cyclobutene, cyclooctene, cyclododecene, and cyclopentadecene resulted in a common curve.[287] Consequently, the number of double bonds has no effect on the entropy of formation or polymerization. Moreover, Chauvin et al.[288] found that the molecular weight distribution in both oligomer and polymer is constant whatever the starting material, i.e., CO, oligomer, polymer, or an oligomer-monomer mixture. The authors inferred that true thermodynamic equilibrium between cyclic oligomers and high molecular weight polymers is obtained without monomer participation:

$$\text{cycloolefin} \rightarrow [\text{oligomer-polymer}] \tag{226}$$

The existence of this equilibrium, which accounts for the "back-biting" process in the ring-opening polymerization similar to that encountered at the polymerization of cyclic formals,[289] was ascribed, in part, to conformational effects. In addition, no significant variations in the equilibrium composition or in the molecular weight of the polyalkenamer with temperature were observed at high conversions. The absence of such a variation of equilibrium oligomer concentration at low monomer content was attributed to the strain-free nature of the cyclooligomers. Therefore, the enthalpic term is negligible and the entropy governs the thermodynamic equilibrium (compare Equations 224 and 225).

Conformational changes of COs may also determine the oligomer distribution. For example, if during oligomerization the macrocyclic ring becomes twisted, it is possible to obtain *catenanes* or *knots*, depending on the number of half twists. The catenanes are formed when the macrocycles undergo n = 2 half twists and the knots, when the macrocycles undergo n = 3 half twists:

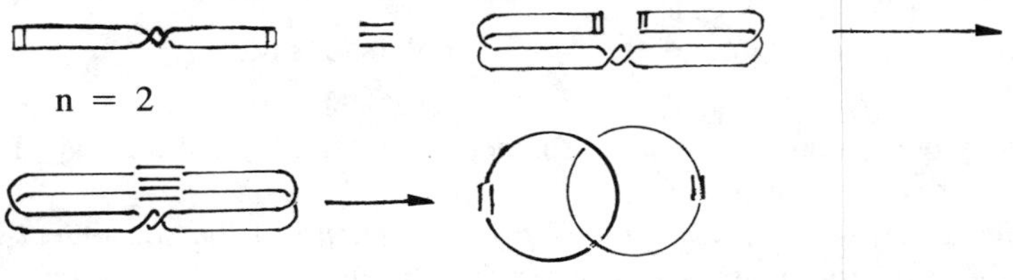

(227)

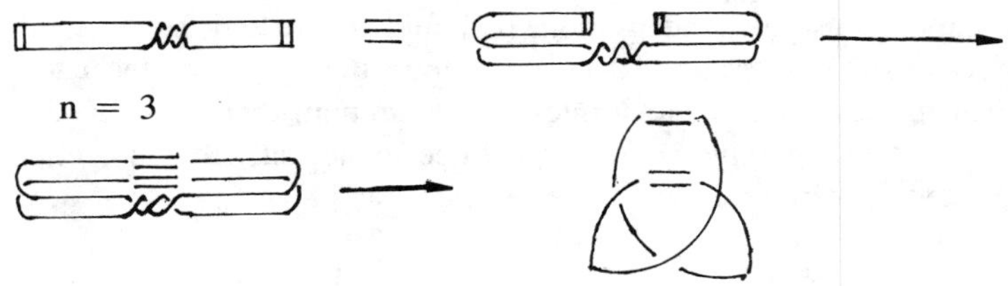

(228)

When the number of half twists is zero (n = 0), the macrocycles cleave into two COs

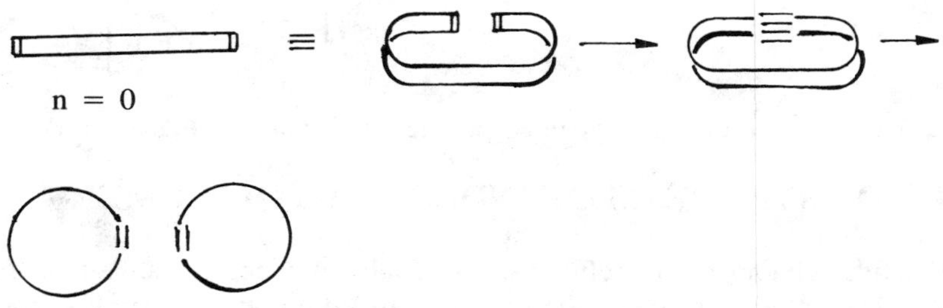

(229)

a reaction which represents the reverse of Equation 220. For one half twist (n = 1), a degenerate reaction occurs when the macrocycle is formed:

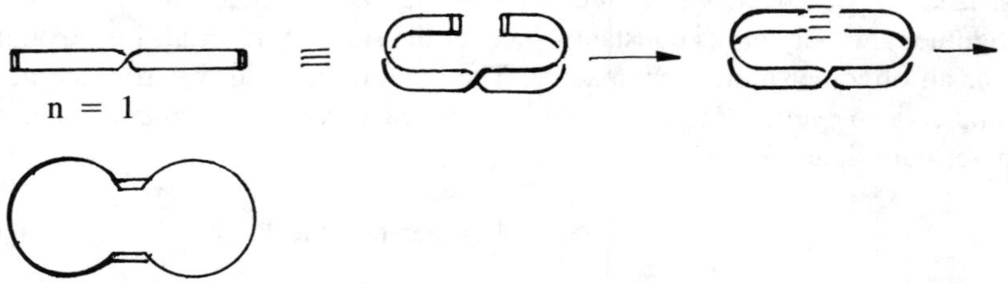

$$n = 1 \tag{230}$$

In this case, the double bonds merely change their position in the cycle.

Recently, Wagener et al.[290] reported a new type of metathesis reaction which resembles a stepwise polycondensation, since the polymerization would be driven by the removal of a small molecule:

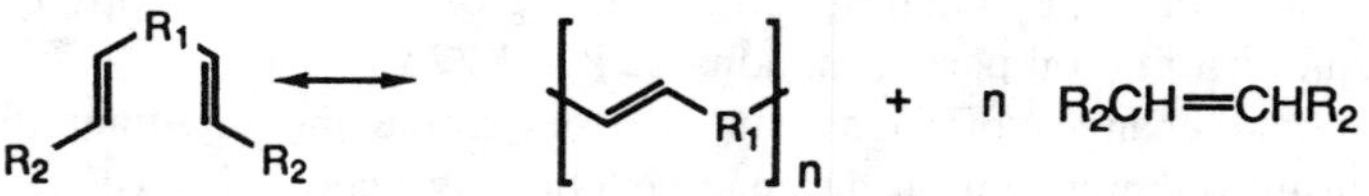

$$\text{(231)}$$

For example, acyclic diene metathesis polymerization of 1,9-decadiene

$$n CH_2{=}CH(CH_2)_6{-}CH{=}CH_2 \xrightarrow[\text{toluene}]{50°C, \text{ cat.}} {-}[{-}CH{=}CH{-}(CH_2)_6{-}]_n{-} \; + \; n CH_2{=}CH_2 \tag{232}$$

$$\text{polyoctenamer}$$

resulted in only two products — polyoctenamer ($\overline{M} \simeq 50{,}000$) and the gas ethylene, which is removed to drive the reaction.

In acyclic diene metathesis polymerization, the nature of the substituent R_1 would be restricted only by factors that govern intramolecular vs. intermolecular chemistry and by factors limiting the substituent propensity to poison the metathesis catalyst itself. No oligomer seems to accompany this polycondensation. It should be stressed, however, that the type of metathesis encatenation is strongly dependent upon the nature of the catalyst used. For example, when the tungsten hexachloride-ethyl aluminum dichloride catalyst system was employed for metathesis polymerization of 1,9-decadiene, only low molecular weight polymers were formed,[291] while a Lewis acid-free catalyst, i.e.

$$W(CH{-}t{-}C_4H_9)[N{-}2,6{-}C_6H_3(i{-}C_3H_7)_2] \left[{-}O{-}C \begin{smallmatrix} CH_3 \\ \diagup \\ \diagdown \\ (CF_3)_2 \end{smallmatrix} \right]_2$$

promoted the formation of a highly stereospecific polyoctenamer (92% *trans*).[290]

V. SYNTHESIS OF INDIVIDUAL OLIGOMERS

Well-defined oligomers are useful as standard substances for the calibration of analytical methods, and homologous series of oligomers are needed for physicochemical investigations

TABLE 7
Physical Data of Linear Oligoesters

General formula		Mol wt	mp	Ref.
H–[G–T]$_n$–G–H	n = 1	254	109	298
	n = 2	446	173	298
	n = 3	638	200	298
	n = 4	830	213	298
	n = 5	1023	218	298
HO–T–[G–T]$_n$–OH	n = 1	358	360	297
	n = 2	550	285	297
	n = 3	742	275	60
	n = 4	934	254	60
	n = 5	1127	233	60
H–[G–T]$_n$–OH	n = 1	210	178	299
	n = 2	402	200	298
	n = 3	594	220	298

in order to establish the true relationships between the structure, molecular weight, and physical parameters of these compounds.

Synthesis of individual oligomers with the desired DP values is possible only by the coupling of monofunctional reaction partners. In order to obtain higher yields and avoid undesired reactions, the couplings are carried out with the activation of one of the monofunctional compounds. For example, activation of the carboxyl group is carried out by transformation into acid chloride or azide, as well as to a mixed anhydride or active ester.

As it is not possible to discuss here the synthesis of each individual oligomer, special reference is given to the review articles by Zahn and co-workers[60,292,293] and Rothe.[294] The synthesis of oligomers may be carried out either in a stepwise manner, i.e., by adding one base unit at a time, or by the combination of fragments comprised of lower oligomers to form the higher oligomer desired.

Condensation of ethylene glycol with terephthalic acid, along with the expected macromolecular compound, also induces the formation of various oligomers.

Three series of linear oligomers are possible: ester diols, ester dicarboxylic acids, and ester-hydroxy acids. For simplicity, the abbreviation method proposed by Zahn and Krzikalla[295] was used, i.e., H–[G]–H and HO–[T]–OH for the glycol and terephthalic acid, respectively. After this abbreviation method, the above series of oligomers may be represented by the following formulas: H–[G–T]$_n$–G–H (ester diols), HO–T–[G–T]$_n$–OH (ester dicarboxylic acids), and H–[G–T]$_n$–OH (ester hydroxy acids). These three series of oligoesters were obtained by step-by-step synthesis, isolation from industrial polymer, and the duplication method.[296]

Step-by-step synthesis has been systematically used by Zahn et al.[297,298] in order to obtain the well-defined linear oligoester. The physical characteristics of linear oligoesters are listed in Table 7.[299]

Cyclic oligoesters of terephthalic acid and ethylene glycol may be synthesized by reacting the dichlorides of oligoester dicarboxylic acids with oligoester diols in dilute solutions:

$$Cl–[–CO–C_6H_4–CO–O–(CH_2)_2–O–]_m–CO–C_6H_4–CO–Cl \; +$$

$$H–[–O–(CH_2)_2–O–CO–C_6H_4–CO–]_n–O–(CH_2)_2–OH \xrightarrow[-2HCl]{base}$$

$$\overline{[–O–(CH_2)_2–O–CO–C_6H_4–CO–]_{m+n+1}} \qquad (233)$$

TABLE 8
Physical Data of Cyclic Oligoesters

General formula		Mol wt	mp	Ref.
$[-G-T-]_n$	n = 2	384	229	300
	n = 3	576	317—320	302
	n = 4	709	328	303
	n = 5	961	264	303
	n = 6	1153	306	301
	n = 7	1345	238	301

By the proper choice of initial reactants, Repin and Papanikolau[300] and Zahn and Repin[301] succeeded in synthesizing the whole series of cyclic oligoesters from dimer to heptamer. Characterization of these oligomers has been carried out by chromatographic methods.[302,303] The physical characteristics of the cyclic oligoesters are listed in Table 8.

It was also interesting to synthesize a series of well-defined and uniform oligomers as models for poly(butylene terephthalate). Evaluation of the physical properties of these well-defined oligomers is of importance if one wants to understand the properties of poly(butylene terephthalate). The synthesis of these oligomers has been realized by Hässlin et al.[303-305] and Meraskentis and Zahn[306] using the following reactions:

$$Bz-T-B-H \ + \ Cl-[T-B]_n-TCl \rightarrow Bz-[T-B]_{n+2}-T-Bz \rightarrow Cl-[T-B]_{n+2}-T-Cl \quad (234)$$

$$THP-B-T-B-H \ + \ Cl-[T-B]_n-T-Cl \rightarrow THP-[B-T]_{n+3}-B-THP \rightarrow$$
$$H-[B-T]_{n+3}-B-H \quad (235)$$

$$Bz-T-Cl \ + \ H-[B-T]_n-B-H \rightarrow Bz-[T-B]_{n+1}-H \quad (236)$$

$$Cl-[T-B]_{n+2}-T-Cl \ + \ THP-B-H \rightarrow \text{transition from } -T-Cl \text{ to } B-H \text{ end group} \quad (237a)$$

$$H-[B-T]_{n+3}-B-H \ + \ Bz-T-Cl \rightarrow \text{transition from } -B-H \text{ to } -T-Cl \text{ end group} \quad (237b)$$

where THP = tetrahydropyrane, B = $-O-(CH_2)_4-O-$, T = $-CO-C_6H_4-CO-$, and Bz = $-O-CH_2-C_6H_5$.

Using Reactions 234 to 237, the four series of oligoesters have been obtained. The physical characteristics of these oligomers reveal that their melting behavior is strongly influenced by the degree of polymerization and the nature of the end groups. From X-ray and IR data, it can be concluded that oligoesters of this type have structures similar to the α-form of poly(butylene terephthalate).[307] Physical properties of well-defined oligoesters also have been studied in the case of poly(propylene terephthalate).[308]

Recently, a new route for the synthesis of monodisperse oligoester diols from common monomers has been developed.[309] The essential part of this method is the use of monosilylated diols, as outlined in the following reactions:

$$HO-R-OH \ + \ 2 \, \text{(phthalic anhydride)} \rightarrow HOOC-C_6H_4-COO-R-OOC-C_6H_4-COOH \quad (238)$$
$$I$$

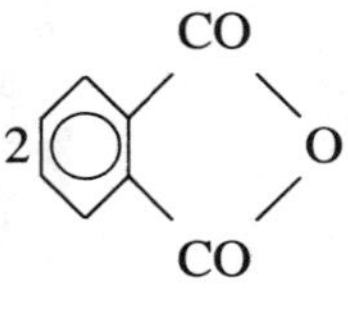

$$HO-R-OH \ + \ Cl-Si(CH_3)_2-C(CH_3)_3 \xrightarrow{NaH} HO-R-O-Si(CH_3)_2-C(CH_3)_3 \quad (239)$$
$$II$$

$$\text{I + II} \xrightarrow[\text{carbodiimide}]{\text{dicyclohexyl-}}$$

$$(CH_3)_2C-Si(CH_3)_2-(-O-R-OOC-C_6H_4-CO-)_2-O-R-O-Si(CH_3)_2-C(CH_3)_3 \quad (240)$$

III

$$\text{III} \xrightarrow{\text{deprotection}} H-(O-R-OOC-C_6H_4-CO-)_2-O-R-OH + HO-Si(CH_3)_2-(CH_3)_3 \quad (241)$$

where $R = -CH_2-C(CH_3)_2-CH_2-$ or $-(CH_2)_4-$.

The key to the above synthesis is the deprotection procedure, because the evaluated methods broadened the molecular weight distribution mainly through the cyclization processes.

For the synthesis of oligoamides, the carboxyl group is usually activated by transformation into acid chloride or azide, as well as a mixed anhydride or active ester. Activation of the amine group of oligomers has also been used. Higher homologues are formed by selective removal of one protecting group, activation, and coupling of the intermediate products.

Zahn and co-workers have pioneered the synthesis of well-defined oligoamides.[310-313]

Linear oligoamides may be represented by the following general formulas: $H-[B-A]_n-BH$, $HO-A-[B-A]_n-OH$, and $H-[B-A]_n-OH$, where in the case of oligoamides based on adipic acid and hexamethylene diamine, $A = -CO-(CH_2)_4-CO-$ and $B = -HN-(CH_2)_6-NH-$.

Well-defined oligoamides with aminic end groups were prepared by the stepwise method using the following reactions:

$$2NC-(CH_2)_5-NH_2 + Cl-CO-(CH_2)_4-CO-Cl \xrightarrow[-HCl]{Py}$$

$$NC-(CH_2)_5-NH-CO-(CH_2)_4-NH-(CH_2)_5-CN \xrightarrow[CH_3OH]{H_2, PtO_2}$$

$$H_2N-(CH_2)_5-NH-CO-(CH_2)_4-CO-NH-(CH_2)_6-NH_2 \quad (242)$$

The same method has been used for the preparation of individual oligoamides with carboxylic end groups. The following reactions were used:

$$2H_5C_2OOC-(CH_2)_4-CO-Cl + H_2N-(CH_2)_6-NH_2 \xrightarrow[-HCl]{Py}$$

$$H_5C_2OOC-(CH_2)_4-CO-NH-(CH_2)_6-NH-CO-(CH_2)_4-COOC_2H_5 \xrightarrow[\substack{-2C_2H_5OH \\ -2KCl}]{+KOH, HCl}$$

$$HOOC-(CH_2)_4-CO-NH-(CH_2)_6-NH-CO(CH_2)_4-COOH \quad (243)$$

TABLE 9

Physical Data of Nylon-6-Type Oligoamides

General formula	Mol wt	mp	Ref.
H–[B–A]$_n$–BH[a]			
n = 1	230	141	312
n = 2	460	230	312
n = 3	625	241	312
HO–A–[B–A]$_n$–OH			
n = 1	260	197	312
n = 2	485	205	312
n = 3	650	218	312
H–[B–A]$_n$–OH			
n = 1	244	190	312
n = 2	470	208	312
n = 3	636	245	312
n = 4	923	245	312
n = 5	1145	247	312

[a] A = –CO–(CH$_2$)$_4$–CO–; B = –HN–(CH$_2$)$_6$–NH–.

Oligoamides with aminic and carboxylic end groups were obtained by the following reactions:

$$NC–(CH_2)_5–NH_2 + Cl–CO–(CH_2)_4–COOC_2H_5 \xrightarrow{-HCl}$$

$$NC–(CH_2)_5–NH–CO–(CH_2)_4–COOC_2H_5 \xrightarrow[\substack{-2C_2H_5OH \\ -KCl}]{+H_2,\ CH_3OH}$$

$$H_2N–(CH_2)_6–NH–CO–(CH_2)_4–COOC_2H_5 \xrightarrow[-KCl]{+KOH,\ HCl}$$

$$H_2N–(CH_2)_6–NH–CO–(CH_2)_4–COOH \tag{244}$$

Various oligoamides with the desired DP have been obtained by Reactions 242 to 244, especially those of the Nylon-6 type. The physical characteristics of these compounds are listed in Table 9. The synthesis of Nylon-6 oligomers is also possible at temperatures well above 0°C by the mixed-anhydride method.[314]

Linear oligoamides of Nylon-4, bearing various end groups, were prepared in a stepwise fashion using the *p*-nitrophenyl active ester-coupling method.[315] The formulas and melting points of the synthesized oligomers are shown in Table 10.

Linear oligoamines of the Nylon-6,10 type were obtained by the reduction of oligoamides with lithium-aluminum hydride.[316] The formulas and physical data of the synthesized oligoamides are shown in Table 11.

Linear oligoamides of the Nylon-6 type were prepared by the following reactions:[317]

$$Z–[Cap]_3–OH \xrightarrow[N(C_2H_5)_3]{Cl–CO–C_2H_5} Z–[Cap]_3–COOC_2H_5 \xrightarrow{H–[Cap]_m–OR}$$

$$Z–[Cap]_{3+m}–OR \xrightarrow{HBr} H–[Cap]_{3+m}–OH \tag{245}$$

TABLE 10
Physical Data of End-Capped Nylon-4 Oligomers with the General Formula X–[–HN–(CH$_2$)$_3$–CO–]$_n$–Y[315]

n	M	X	Y	mp (°C)
1	358.35	C$_6$H$_5$CH$_2$CO–	–O–C$_6$H$_4$–NH$_2$	82
2	336.39	C$_6$H$_5$CH$_2$CO–	–OCH$_3$	88.5—89
3	421.50	C$_6$H$_5$CH$_2$CO–	–OCH$_3$	139.5—140
3	329.40	CH$_3$CO–	–OCH$_3$	147—149
3	406.49	C$_6$H$_5$CH$_2$CO–	–NH$_2$	200—202
3	407.40	C$_6$H$_5$CH$_2$CO–	–OH	—
3	272.35	H–	–NH$_2$	138—144
4	506.61	C$_6$H$_5$CH$_2$CO–	–OCH$_3$	173—175
4	414.52	CH$_3$CO–	–OCH$_3$	177—178
4	399.49	CH$_3$CO–	–NH$_2$	210—214
4	441.57	CH$_3$CO–	–NH–C$_3$H$_7$	212—216

TABLE 11
Formulas and Physical Data of Nylon-6,10-Type Oligomers[316]

n	Structure[a]	M	mp (°C)
1	H–[–A–]–OH	272	64—67
2	H–[–A–]$_2$–OH	530	70—72
3	H–[–A–]$_3$–OH	786	85—86
1	HO–(CH$_2$)$_{10}$–[–A–]–OH	430	99—100
2	HO–(CH$_2$)$_{10}$–[–A–]$_2$–OH	686	90—93
3	HO–(CH$_2$)$_{10}$–[–A–]$_3$–OH	942	80—84
1	H–[–A–]–NH–(CH$_2$)$_6$–NH$_2$	372	70—72
2	H–[–A–]$_2$–NH–(CH$_2$)$_6$–NH$_2$	628	86—88
3	H–[–A–]$_3$–NH–(CH$_2$)$_6$NH$_2$	984	81—85

[a] A = [–HN–(CH$_2$)$_6$–NH–(CH$_2$)$_{10}$–]–OH.

where R = H, Cap = –HN–(CH$_2$)$_5$–CO–, and Z = C$_6$H$_5$CH$_2$COO–. The physical characteristics of these oligoamides are listed in Table 12.

The synthesis of cyclic oligoamides is based on the known methods of peptide chemistry. In these methods, the carboxyl end group is usually activated by transformation into acid chloride or azide (''azide method''), as well as a mixed anhydride or active esters. Activation of amino groups has also been used to obtain cyclic oligoamides.

The azide method is based on the following schematic reactions:[318-324]

$$Z–[Cap]_n–OH \; + \; HCl \cdot H–[Cap]_m–OCH_3 \xrightarrow[\mathrm{N(C_2H_5)_3}]{\mathrm{Cl–CO–CH_3}} Z–[Cap]_{n+m}–OCH_3$$

$$\xrightarrow{\mathrm{N_2H_4/H_2O}} Z–[Cap]_{n+m}–NHNH_2 \xrightarrow{\mathrm{HBr}} H \cdot Br \cdot H—[Cap]_{n+m}–NH \cdot NH_3Br$$

$$\xrightarrow{\mathrm{NaNO_2/HCl}} HBr \cdot H–[Cap]_{n+m}–N_3 \xrightarrow{\mathrm{NaOH}} \boxed{[Cap]_{n+m}} \quad (246)$$

where Cap = –HN–(CH$_2$)$_5$–CO–, Z = C$_6$H$_5$CH$_2$–CO–Cl, n = 2, 3, and m = 4, 6.

TABLE 12
Melting Temperatures of
Linear Oligoamides of Nylon-6

Structure	n	Melting point
H–[Cap]$_n$–OH	9	208—211
	10	212—213
	11	209—212
	12	211—213
Z–[Cap]$_n$–OH	9	195
	10	200
	11	205
	12	208
Z–[Cap]$_n$–OBz^a	5	169
	6	186
	7	194
	8	198
	9	203
	10	206

^{a} Bz = –CH$_2$C$_6$H$_5$.

Cyclic oligoamides can also be prepared by activation of amino groups.[325] The following reactions are used in this procedure:

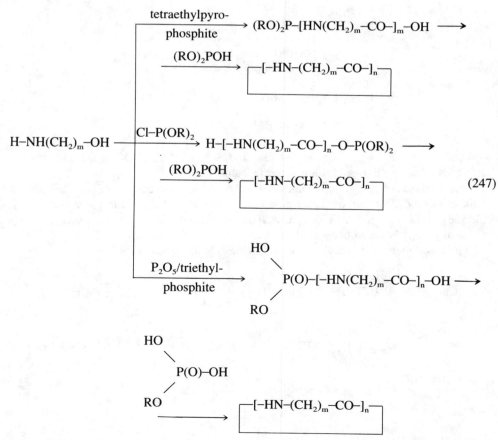

$$(247)$$

where m = 1, n = 2, 3, and R = alkyl.

TABLE 13
Melting Points of Cyclic Oligoamides of
ε-Aminocaproic Acid and
ω-Aminoundecanoic Acid

Structure	m	n	Melting point (°C)
┌─[Cap]$_m$─┐ [a]	2	—	348
	3	—	245
	4	—	261
	5	—	253
	6	—	260
┌─[Und]$_n$─┐ [b]	—	2	186
	—	3	183
┌─[Cap$_m$Und]$_n$─┐	1	1	187—189
	2	1	181
	3	1	191
	1	2	193—195

[a] Cap = −HN−$(CH_2)_5$−CO−.
[b] Und = −HN−$(CH_2)_{10}$−CO−.

The reaction between the acid chloride of dicarboxylic acids and diamines can also be used for the preparation of cyclic oligoamines.[326,327] Through the above general methods, various cyclooligoamines were obtained.[328-331] The physical characteristics of these compounds are listed in Table 13.

Only a few studies have been reported on oligourethanes. Isolation of the cyclic oligourethan of hexamethylene diisocyanate and 1,4-butanediol has been claimed by Bayer[332] and linear oligomers with DP = 2 to 9 have been reported by Kern and co-workers[333-336] and Zahn and Dominik.[337] These authors established the duplication method for the preparation of oligourethans of diisocyanates and diols.

Oligoethylurethanes, oligotrimethylurethanes, and oligopentamethylurethanes have been synthesized by the stepwise procedure using 2-acetoxyethyl isocyanate, 3-acetoxypropyl-isocyanate, and 5-acetoxypentyl isocyanate, respectively.[338-340] The method involves an addition of ω-acetoxyalkyl isocyanate to a preceding hydroxy-terminated oligomer, followed by elimination of the terminal acetyl group, giving a corresponding hydroxy-terminated oligomer. The physical characteristics of these well-defined oligomers are listed in Table 14.

Because it is desirable to control the combined thermoplastic processing properties and elastomeric mechanical properties of the urethanic thermoplastic elastomers, model urethane *hard-segment* oligomers from 4,4'-methylene bis(phenylisocyanate) — MDI — and 1,4-butanediol, and from piperazine and 1,4-butanediol, have been prepared and reacted with oligo(tetramethylene oxide) *soft-segment* to form well-defined elastomers.[341,342] By mixing the hard-segment oligomers, it was concluded that the crystalline fraction of the elastomers consists of the eutectics of the short segments and mixed crystals of hard segments of greater than three repeat units.[342]

As a model for the soft blocks of both polyurethanes and copoly(ether)s, a series of homologous α,β-dihydroxyoligo(tetramethylene oxide)s with n = 1 to 14, the corresponding phenyl isocyanate end-capped bis(phenyl urethane)s, and α-hydro-ω-chloro-oligo(tetramethylene oxide)s have been prepared by two methods.[343] The lower homologues with n ≤ 7 were prepared by the reaction of either a diol or chloroalcohol, obtained by the

TABLE 14
Physical Characteristics of Oligourethanes

n	General formula	M	Melting point (°C)
	H–[–O–(CH$_2$)$_2$–NH–CO–]$_n$–OCH$_3$		
1		119.1	0
2		206.2	62—63
3		293.3	101.5—102.5
4		380.4	129.0—130.0
5		467.4	152.0—153.0
6		554.5	168.5—169.5
	H–[–O–(CH$_2$)$_3$–NH–CO–]$_n$–OCH$_3$		
1		133.2	Oily
2		234.3	Oily
3		335.4	76.5—79.0
4		436.5	106.5—108.5
5		537.6	126.0—128.0
	H–[–O–(CH$_2$)$_5$–NH–CO–]$_n$–OCH$_3$		
1		161.2	15
2		290.4	73.0—75.0
3		419.5	99.0—105.0
4		548.7	112
5		677.8	116
6		807.0	122.0—125.0
7		936.2	131.0—134.5
8		1065.3	139.0—140.5

cationic ring-opening of THF, with 2,3-dihydrofuran, followed by cleavage of the resulting bis- or monoacetal. The cleavage reaction resulted in a chain extension of from 0 to 2 units, and the mixture of oligomers was separated by fractional distillation. The higher homologues were prepared by a phase-transfer reaction between the appropriate α-acetal-ω-chloro-oligo(tetramethylene oxide) and a diol. The acetal was then cleaved, resulting in a mixture of the mono- and bis-substituted diols, which were separated first by distillation to remove the shorter oligomers and then by preparative gel permeation chromatography (GPC). Although the thermal behavior of these oligomers depends on whether there is an odd or even number of repeat units and on the end group, the melting point of all extrapolate to that of pure crystalline poly(tetramethylene)oxide.[344]

Oligo(tetramethylene terephthalate)s have been studied as models for the hard segments of segmented poly(ether ester)s. A series of well-defined and uniform oligomers with up to seven repeat units was prepared by a route involving protection and deprotection of butanediol with tetrahydropyran groups, and of terephthaloyl chloride and benzyl groups.[345-347] Compatibility studies demonstrated that the oligoesters do not crystallize and that the melt of the lower component is not a good solvent for the higher-melting component. The addition of oligo(tetramethylene terephthalate)s to block copoly(oxytetramethylene-b-tetramethylene terephthalate) induced different phase behavior in the copolymer, resulting in higher melting points in the blend than in the pure oligomer or copolymer.[347] These results confirm that only part of the ester hard segments crystallize in the block copolymers, which fractionates the segments of different length.[348-351] The triblock copoly(ether ester)s in which the middle block was oligo(tetramethylene oxide) with n = 2 to 6 and the end blocks were oligo(tetramethylene therephthalate) also crystallized in a lattice similar to that of the corresponding oligoester.[348] These models were prepared by reaction of 1 mol of the α-β-dihydroxyoligo(tetramethylene terephthalate) in which the other end was protected by a benzyl group.

Well-defined segmented block copolymers were constructed of oligomethylene chains as the soft segments and the condensation products of bisphenol A with terephthalic acid as the hard segments.[349,350] *Hard-soft* diblock copolymers and *hard-soft-hard* and *soft-hard-soft* triblock copolymers were prepared by specific stepwise chain-lengthening sequences of the appropriately protected monomers and/or telechelic oligomer blocks.[349] Benzyl and *t*-butyl groups were used to protect ester and phenol functionalities, and tetrahydropyranyl groups to protect aliphatic hydroxide functionalities.

Thermal stability and thermal behavior with respect to phase transformations in the solid and liquid state have been described for these segmented block copolymers of uniform chain length and defined structure.[350]

A highly chemoselective polyamidation of multifunctional dicarboxylic acids and diamine without protection of the acylation-sensitive groups has been achieved by the use of diphenyl(2,3-dihydro-2-thioxo-3-benzoxazolyl)phosphonate, DPP, as activating agent.[352] A series of model compounds was prepared to obtain information regarding the reaction conditions necessary for polymer formation and to aid in polymer identification. The reaction of 5-substituted isophthalic acids with aniline in the presence of activating agent DPP was carried out in *N*-methylpyrrolidone at room temperature, and gave the desired diamides without formation of byproducts:

$$\text{HOOC}-\bigcirc-\text{COOH} + 2\,\text{H}_2\text{N}-\bigcirc \longrightarrow \bigcirc-\text{NH-CO}-\bigcirc-\text{CONH}-\bigcirc \qquad (248)$$

where X is HO–, H_2N–, or HOOC–.

In order to investigate the relationship between ionic conduction in polymer electrolytes and the polymer structure, poly[γ-methoxy-oligo(ethylene oxide)-L-glutamate]s, PM_nEOG, where chosen as model compounds.

The ionic conduction in polymeric electrolytes must correlate closely with the solvated ion radius, which may be controlled by the segment size involved in the cooperative rearrangement. PM_nEOG is expected to have a rigid α-helix main chain and flexible side chains attached to the outside of the main chain, so that the motion of the backbone and that of the pendent chain is decoupled. The polymeric models were synthesized by the ester exchange reaction of poly(γ-methyl-L-glutamate) with oligo(ethylene oxide)monomethyl-ethers, where the number of ethylene oxide units (n) was 1, 2, 3, and 7:

$$-(-\text{CO}-\text{CH}-\text{NH}-)- \ + \ \text{HO}-(-\text{CH}_2\text{CH}_2\text{O}-)_n-\text{CH}_3 \rightarrow$$

$$\begin{array}{c} | \\ \text{CH}_2 \\ | \\ \text{CH}_2 \\ | \\ \text{C} \\ /\!\!/ \quad \backslash \\ \text{O} \qquad \text{OCH}_3 \end{array}$$

$$-(-\text{CO}-\text{CH}-\text{NH}-)-$$

$$\begin{array}{c} | \\ \text{CH}_2 \\ | \\ \text{CH}_2 \\ | \\ \text{C} \\ /\!\!/ \quad \backslash \\ \text{O} \qquad \text{O}-(\text{CH}_2\text{CH}_2\text{O}-)_n-\text{CH}_3 \end{array} \qquad (249)$$

It was shown that the dependence of ionic conductivity became larger when the side-chain length was longer. This result implies that the ion transport was mediated by the side-chain motion and that the solvated ion radius became larger when the side-chain length was longer.[353]

A search of the literature before the discovery of the complexing power of 2,3,11,12-dibenzo-1,4,10,13,16-hexaoxacyclooctadeca-2,11-diene[354] disclosed only several references to cyclic polyethers.[355-363] Since 1967, a nearly inestimable number of original publications, reviews, and monographs, the topic of which is research of cyclic oligoether properties, have been published.

Pedersen, working for du Pont, wanted to synthesize the bisphenol derivative by the following reaction:

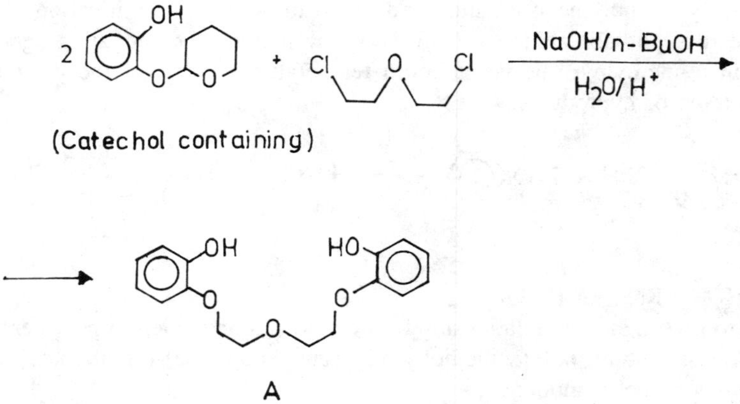

(Catechol containing)

(250)

A

"Luckily — that is what one has to say today — pyrocatechol derivative was not pure when applied. There was still free catechol present. That is the reason why, when working up the reaction product beside the target compound the cyclic oligoether was found in small yield (0.1%)."[364]

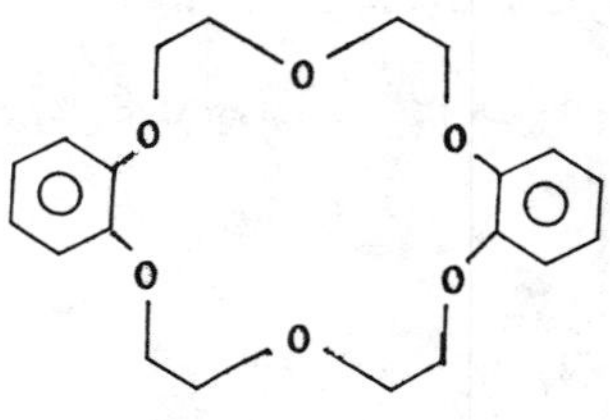

(251)

Pedersen noticed the cyclic compound at once because of its behavior concerning solubility. While only a small amount of the substance was solved in pure methanol, on addition of sodium salts, it was surprisingly soluble. This observation led the discoverer of this new type of substance to make a bold statement: "It seemed clear to me now that the sodium ion had fallen into the hole in the center of the molecule."[364] The fact that only a short time previously it had been known that certain natural products, so-called ionophores, e.g., nonactin and valinomycin,

nonactin   valinomycine   (252)

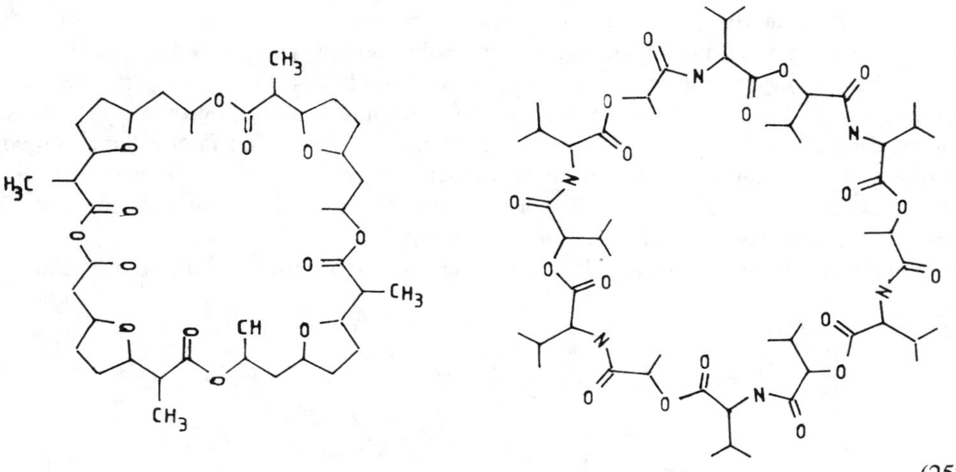

can enclose ions of alkali metals and transport them in biological membranes, led Pedersen to parallel these compounds with macroring cycloethers. This was to become fertile soil for extensive dreams. On such an occasion, Pedersen created the term *crown*. "My excitement, which had been rising during this investigation, now reached its peak and ideas swarmed in my brain. I applied the epithet "crown" to the first member of this class of macrocyclic polyethers because its molecular model looked like one and with it, cations could be crowned and uncrowned without physical damage to either, just as the heads of royalty."[364]

The discovery of cyclic polyethers was an enormous advance in chemistry due to the various applications of these compounds in separation science, biology,[365] and polymer science.[366] As a consequence of this research, it became apparent that such substances, or derivatives thereof, could render membranes selectively permeable to substrate species, i.e., induce selective transport processes. Thus, like molecular recognition and catalysis, transport represents one of the functional features displayed by cycloethers.

Crown ethers may be considered to be physical catalysts.[367] These compounds operate a translocation on the substrate, like a chemical catalyst operates a chemical transformation.

In order to determine the general connections between the structure and properties of these compounds, chemists started to vary all possible parameters of the classical crown ether ring, such as ring size, nature and number of donor sites, or molecular flexibility. Finally, additional bridges were introduced, or the ring was opened. Crown ethers are subdivided into three large groups: *podands, coronands,* and *cryptands.*

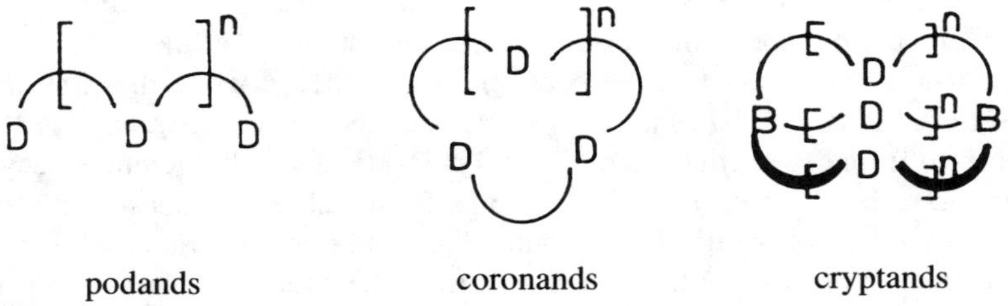

podands   coronands   cryptands   (253)

As it is not possible to discuss here the synthesis and properties of each crown ether, the reader is referred to the review articles by Molson,[368] Vögtle and Weber,[369,370] Izatt and Christianson,[371] and Medved et al.[372]

The term "crown", which was originally meant as a joke, was retained and soon a separate system of notation was developed for the new type of compound: *crown ether.*

The nomenclature of crown ethers represents, in order: the number and kind of hydrocarbon rings, the total number of atoms in the polyether ring, the class name (crown), and the number of oxygen atoms in the polyether ring. The placements of the hydrocarbon rings and the oxygen atom are as symmetrical as possible in most cases, and the exceptions are indicated by *asym*. The ring size is given in square brackets. The family term ''crown'' comes next. The number of donor sites is given last. Additional substituents or sites of condensation are placed first. For the compound with the formula shown in Structure 251, the classical Pedersen notation is dibenzo 18 crown-6.

For the preparation of cyclic polyethers, Pedersen has proposed five different methods:[354]

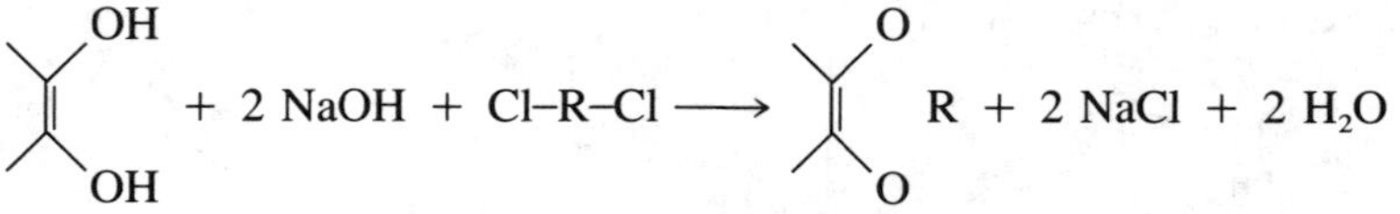

$$\text{(diol)} \; + \; 2\,NaOH \; + \; Cl\text{--}R\text{--}Cl \; \longrightarrow \; \text{(product)} \; R \; + \; 2\,NaCl \; + \; 2\,H_2O \qquad (254)$$

$$\text{(catechol derivative)} \; + \; 2\,NaOH \; + \; Cl\text{--}T\text{--}Cl \; \longrightarrow \; \text{(product)} \; + \; 2\,NaCl \; + \; 2\,H_2O \qquad (255)$$

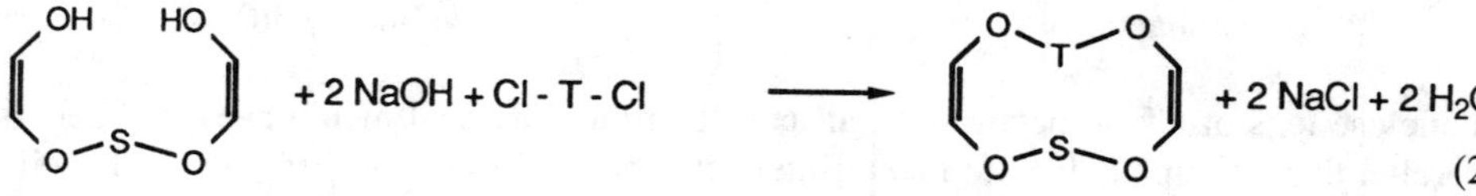

$$2\;\text{(diol)} \; + \; 4\,NaOH \; + \; 2\,Cl\text{--}U\text{--}Cl \; \longrightarrow \; \text{(product)} \; + \; 4\,NaCl \; + \; 4\,H_2O \qquad (256)$$

$$2\;\text{(Cl--V diol)} \; + \; 2\,NaOH \; \longrightarrow \; \text{(product)} \; + \; 2\,NaCl \; + \; 2\,H_2O \qquad (257)$$

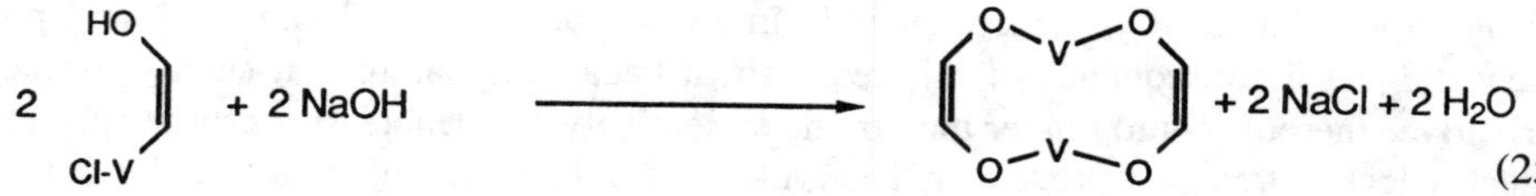

The last method consists of the hydrogenation of the benzo compound to the 1,2-cyclohexyl derivatives in *p*-dioxane using ruthenium dioxide catalyst. In the above reactions, R, S, T, U, and V represent divalent organic groups which may or may not be identical. The physical characteristics of the crown ethers obtained by Pedersen[354] are listed in Table 15.

Macromolecules with unusual architectures (i.e., *rotaxanes, cascade molecules,* and *calixarenes*) have been prepared in the last decade mainly by step-growth polymerization.[373-376] For example, the condensation polymerization of appropriate monomers in the presence of macrocyclic solvents is operationally the simplest approach to polyrotaxanes.[377]

The first mention in print of the idea of a stable union of a linear molecule threaded through a cyclic one appears to have been made in 1961,[378] while the first experimental report was a brief note in 1967, which also suggested the name *rotaxane*.[379] A ''hooplane'' was synthesized from decamethylene diol, trityl chloride, and a 28-membered acyloin bound to a polymeric resin.[380] At 120°C, C_{25}-C_{29} rings form stable rotaxanes, while rings larger than C_{29} allow passage of the trityl end group.[381] A number of generic types of polyrotaxane architectures may be envisioned, as shown in the schematic structures 1 to 4 of Scheme 258.

TABLE 15
Structure and Melting Point of Crown Ethers Obtained by Pedersen[354]

No.	compound	M	Melting point (°C)
1	2	3	4
1		180	67—69
2	a = benzo	220	44—45.5
3	a = cyclohexyl	230	Below 26
4	a = benzo; R = H	268	79
5	a = benzo; R = t-butyl	324	43.4—44.5
6	a = 2,3-naphtho	318	117—119
7	a = cyclohexyl; R = H	274	Below 26
8	a = cyclohexyl; R = t-butyl	3301	Below 26
9	a = 2,3-decalyl	328	Below 26
10	a = benzo; R = H	312	43—44
11	a = benzo; R = t-butyl	368	35—37
12	a = 2,3-naphtho	362	110—111.5
13	a = cyclohexyl; R = H	318	Below 26
14	a = cyclohexyl; R = t-butyl	374	Below 26
15	a = benzo; R = H	300	150—152
16	a = benzo; R = t-butyl	412	149—152
17	a = cyclohexyl; R = H	312	153.5—155.5
18	a = cyclohexyl; R = t-butyl	424	—
19		384	139—141
20		496	137—138

TABLE 15 (continued)
Structure and Melting Point of Crown Ethers Obtained
by Pedersen[354]

No. compound		M	Melting point (°C)
1	2	3	4
21	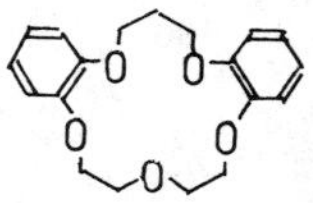	330	117—118
22	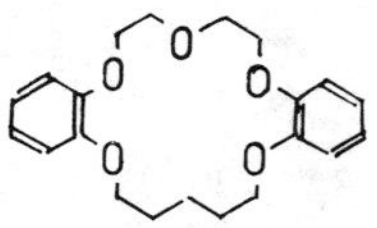	358	157—158
23	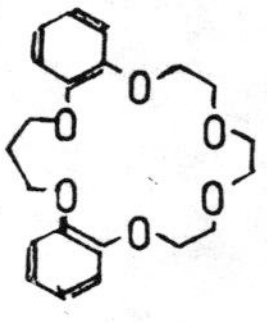	374	84.5—86
24	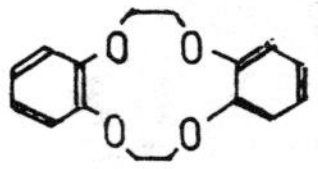	316	113.5—115
25	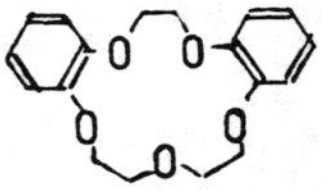	272	208—209
26	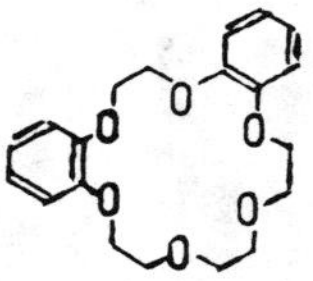	360	117—118
27	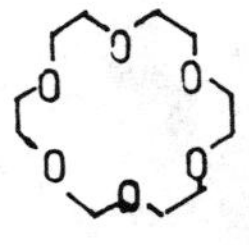	264	39—40
28	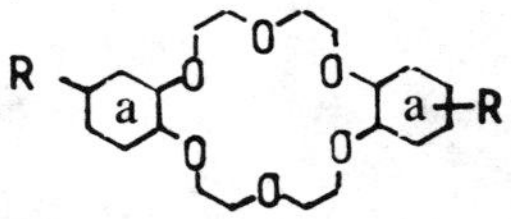 a = benzo; R = H	360	164
29	a = benzo; R = *t*-butyl	472	135—137
30	a = 2,3-naphtho	460	244—246
31	a = cyclohexyl; R = H	372	68.5—69.5
32	a = cyclohexyl; R = *t*-butyl	484	Below 26
33	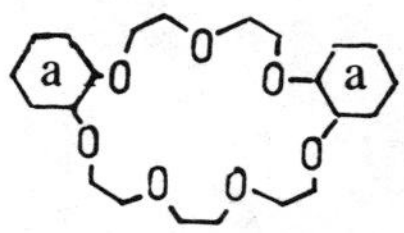 a = benzo	404	106.5—107.5

TABLE 15 (continued)
Structure and Melting Point of Crown Ethers Obtained
by Pedersen[354]

1	2	M 3	Melting point (°C) 4
	No. compound		
34	a = cyclohexyl	416	Below 26

35	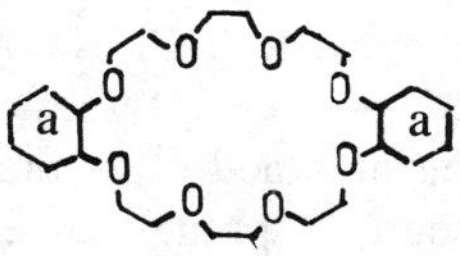	448	113—114
	a = benzo		
36	a = cyclohexyl	460	Below 26
37	a = 2,3-naphtho	548	190—190.5

38	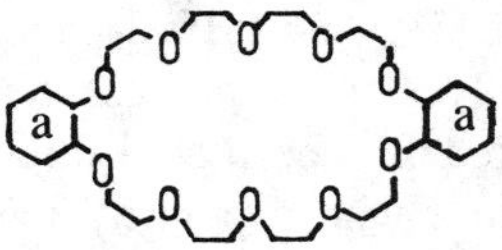	536	106—107.5
	a = benzo		
39	a = cyclohexyl	548	Below 26

40	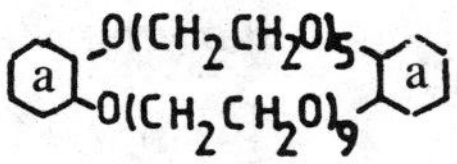	800	Below 26
	a = benzo		
41	a = cyclohexyl	812	Below 26

42	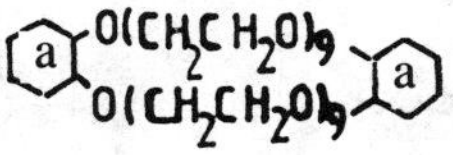	416	82—83
	a = benzo		
43	a = cyclohexyl	472	125—127

| 44 | 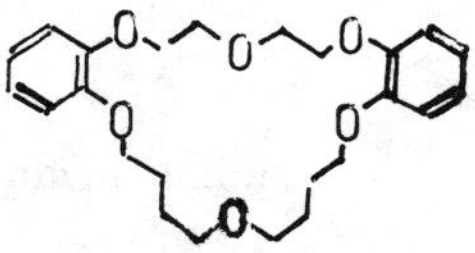 | 408 | 190—192 |

| 45 | 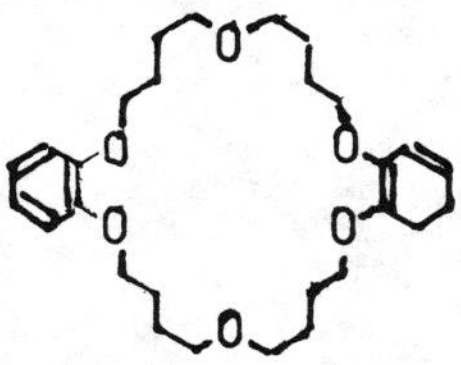 | 422 | 147—149 |

| 46 | 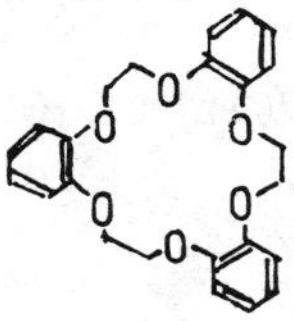 | 452 | 98.5—100 |

| 47 | 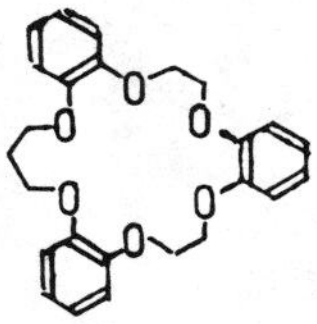 | 544 | 150—152 |

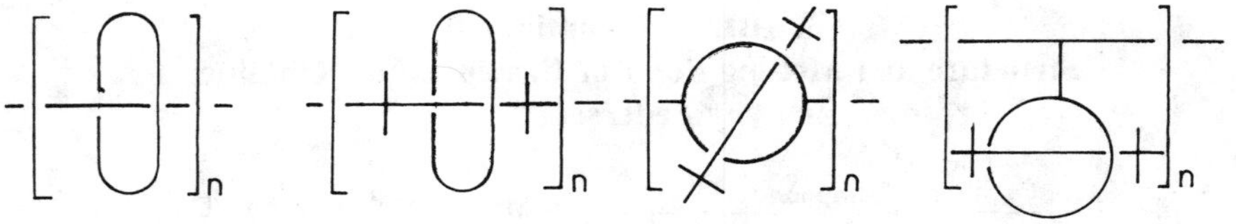

1   2   3   4   (258)

The first syntheses of polymeric rotaxanes appear to be those of Agam and Zilkha,[382] who equilibrated oligomeric ethylene glycols (DP = 3 to 22) with crown ethers (15-, 30-, 44-, and 58-membered rings) and then used the obtained oligorotaxanes to form macromolecular urethanic polyrotaxanes by the addition of naphthalene-1,5-diisocyanate.

A step-growth polymerization in the presence of a suitable macrocycle seems to be the simplest way to obtain polyrotaxanes. Thus, Lecavalier et al.[377] synthesized an oligomeric heterorotaxane by the combination of a polar macrocycle (i.e., 30-crown-10 ether) with a nonpolar polyester backbone. Sebacoyl chloride (SC) and 1,10-decanediol (DD) were chosen as monomers because they afford a nonpolar polyester and possess the optimum chain length to thread the macrocycle via a statistical approach. The model oligoester (DP $\cong$ 23) was prepared using a 1% excess of diol in diglyme, whose polarity is comparable to that of crown ethers. As the blocking group, the 3,3,3-triphenyl propionyl chloride (TPPC) was used:

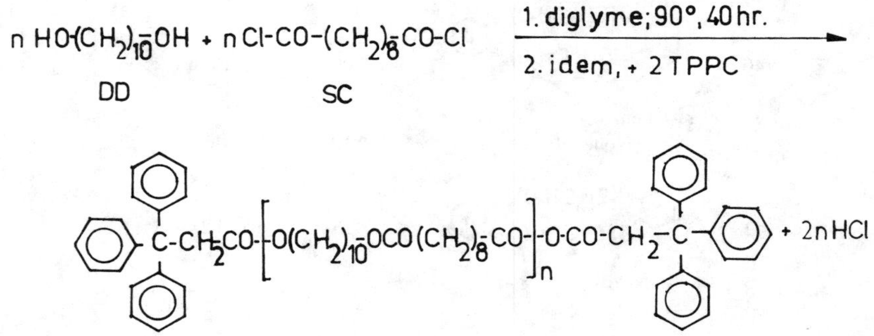

M_n = 8500

The oligorotaxane (DP = 15) was prepared in a similar manner, except that diglyme was substituted by 30-crown-10:

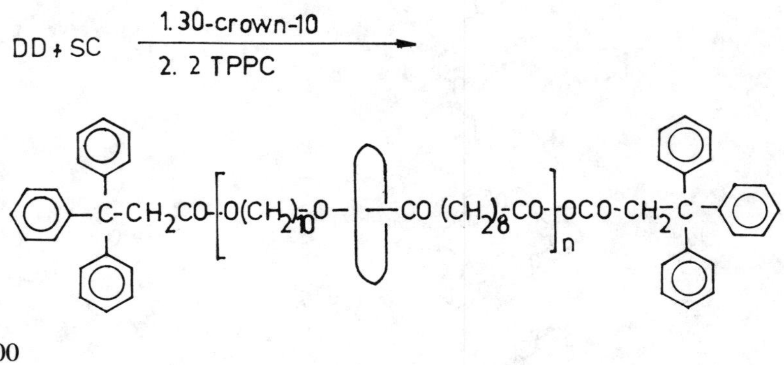

M_n = 6000

Interestingly, it has been observed that the coligative molecular weights (vapor pressure osmometry [VPO] and cryoscopy) of rotaxanes are not the sums between those of linear and cyclic components. This was rationalized in terms of the large ring size and the consequent degree of independence of motion of the two components.[383]

Multiarmed compounds have long alkyl chains with terminal ionic groups, attached to a core molecule, and behave like crown ethers, cryptands, or micelles. *Hexapus, octopus, tentacle, turbine,* and *polypode* molecules are examples of such compounds in which the core molecule is benzene:[384]

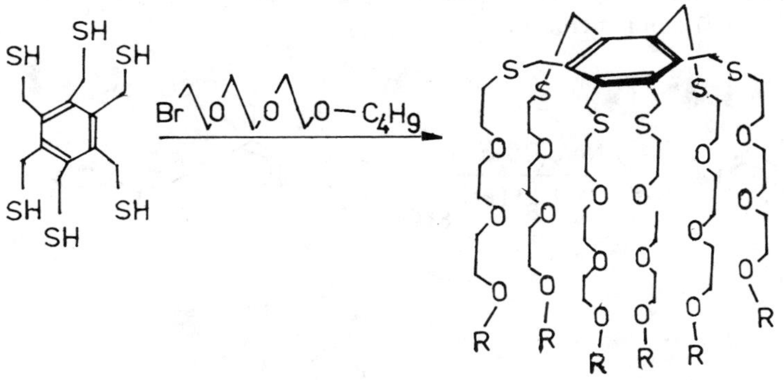

Octopus molecule (1)

$R = C_4H_9$

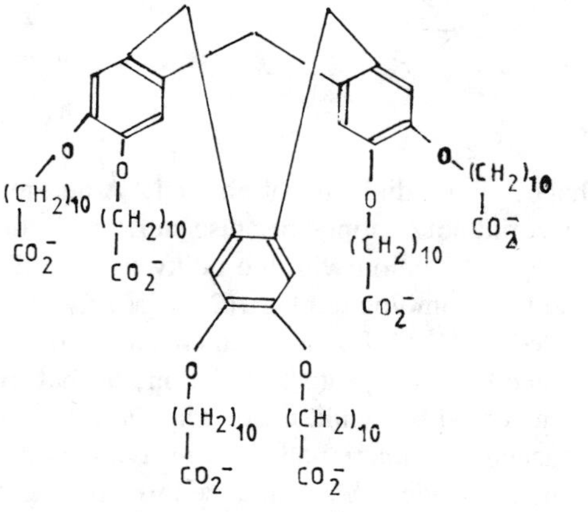

Hexapus molecule (2)

Tentacle molecule (5)

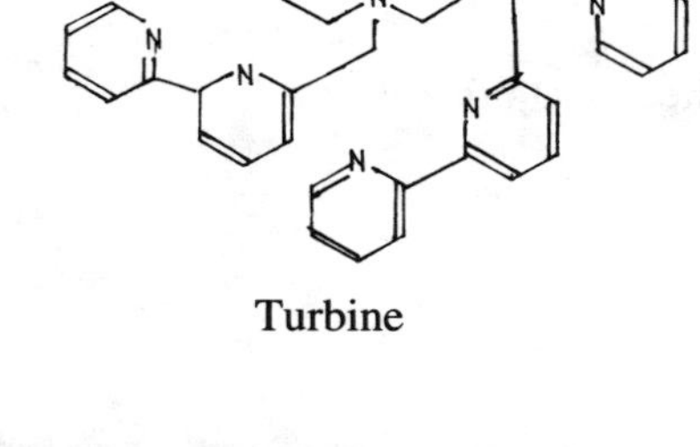

Turbine

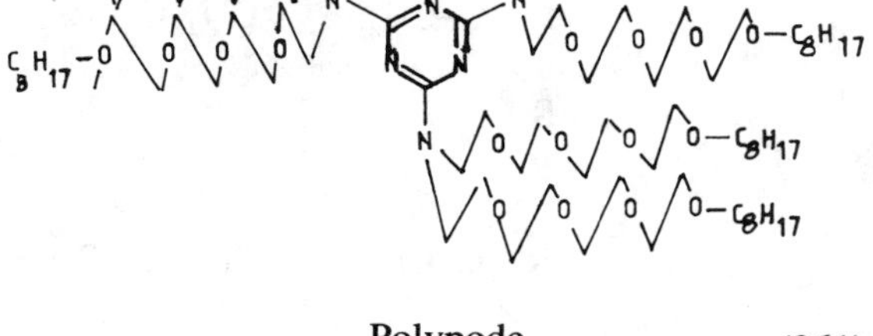

Polypode

(261)

The number of arms can be increased with a mathematical growth progression (e.g., 1 → 3 → 9 → 27) by using a new synthetic methodology named cascade synthesis, which is a stepwise procedure, to yield the multibranched polymers. Two types of cascade molecules are of special interest: *arborols* and *dendrimers.*[385] The word *arborol* derives from *arbor,* the Latin word for tree. Similarly, the word *dendrimer* derives from *dendron,* the Greek word for tree. These new types of compounds have a spherical polyfunctionalized surface over a lipophilic core and would be an ideal model compound of unimolecular micelles.

The first arborols were monodirectional, made by alkylating an alkyl halide (e.g., 1-bromopentane) with triethylmethane-tricarboxylate, then amidating the triester with tris(1), tris(hydroxymethyl)-aminomethane.

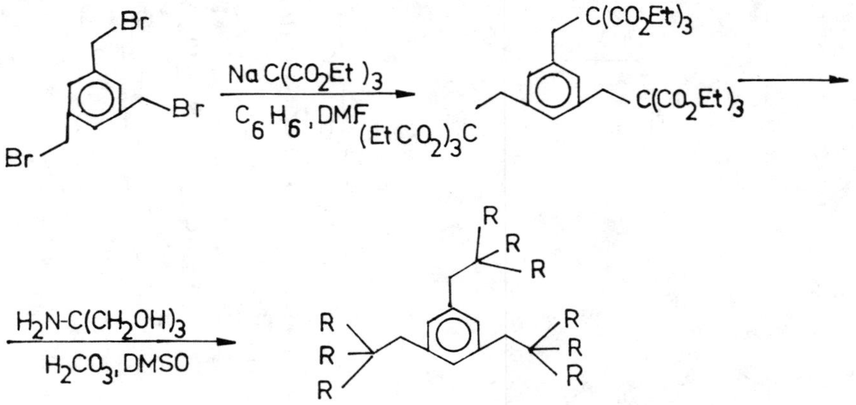

(262)

where R = $-CO-NH-C(CH_2OH)_3$.

The resulting molecule was spherical in shape, with a diameter of about 12 Å and nine hydroxyl terminal groups. However, the process could not continue because steric inhibition at the hydroxymethylene carbon prevented a second alkylation with the bulky triester. The solution was to synthesize a new compound, the bis- homologue of tris(2), tris(3-hydroxypropyl)-aminomethane, as shown in Reaction 263.[375,385-388] For a tridimensional arborol, the two-carbon extension provided the "room" needed for repeated alkylation, so that true "cascade molecules" with high molecular weights could be synthesized. The *directionality* refers to the number of dense cascade spheres attached to a functionalized core. For example, a tetradirectional arborol might consist of four such cascade spheres attached to a neopentyl core, as shown below:

```
HOH2C     CH2OH    BrH2C     CH2Br
    \     /            \     /
       C         →        C        →
    /     \            /     \
HOH2C     CH2OH    BrH2C     CH2Br

 R                  R
  \                /
R - C-CH2     CH2-C - R
  /      \   /     \
 R         C        R
 R                  R      tetradirectional arborol
  \      /   \     /
R - C-CH2     CH2-C - R
  /                \
 R                  R
```

(263)

A model synthesis of a monodirectional, multibranched arborol oligomer is given below:

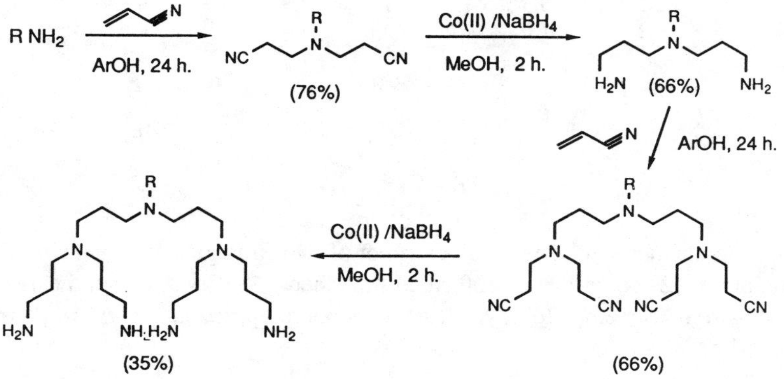

$$R\ NH_2 \xrightarrow[\text{ArOH, 24 h.}]{} \text{(76\%)} \xrightarrow[\text{MeOH, 2 h.}]{\text{Co(II) /NaBH}_4} \text{(66\%)}$$

(264)

A schematic drawing is shown in Structure 265:[384]

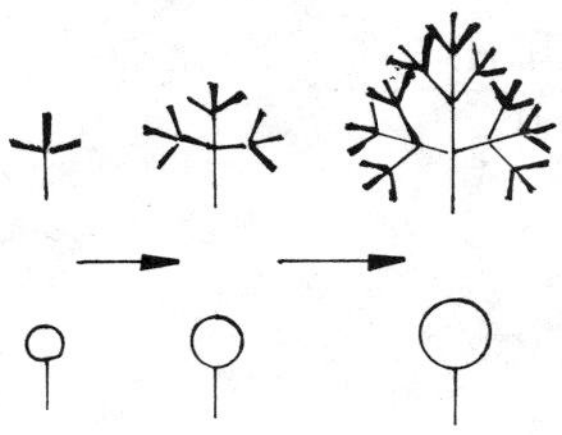

(265)

Dendrimer synthesis is also a stepwise process.[374,389-397] The resulting micronetworks were referred to as possessing *starburst topology,* and the molecular entities, constructed by reiterative synthetic steps, were coined starburst dendrimers. The dimensional projections of these dendrimers, A to E (Structure 266), illustrate the concentric construction of branches upon branches (generation, G = 1, 2, 3, 4, . . .) to produce radial arrays of branch junctures and terminal functionality around a so-called initiator core, I.

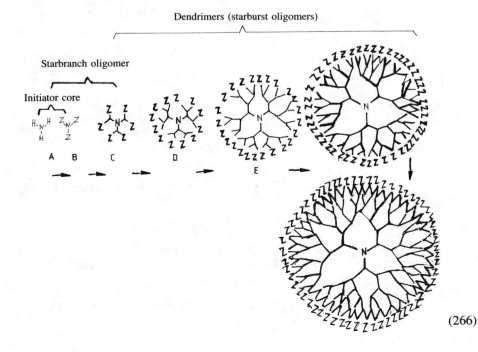

(266)

Z	Generation					
–COOMe	0.5	1.5	2.5	3.5	4.5	5.5
–NH$_2$	1.0	2.0	3.0	4.0	5.0	6.0
	A	B	C	D	E	
	(A)$_n$	(B)$_n$	(C)$_n$	(D)$_n$	(E)$_n$	

Starburst Polymers

Presently, dendrimer synthesis involves either of two strategies:[395] (1) *in situ* construction of branch points or (2) coupling of preformed branched moieties. The synthesis of polyamidoamines shown in Scheme 267 typifies the general approach for *in situ* branch-point construction.[374,389-394]

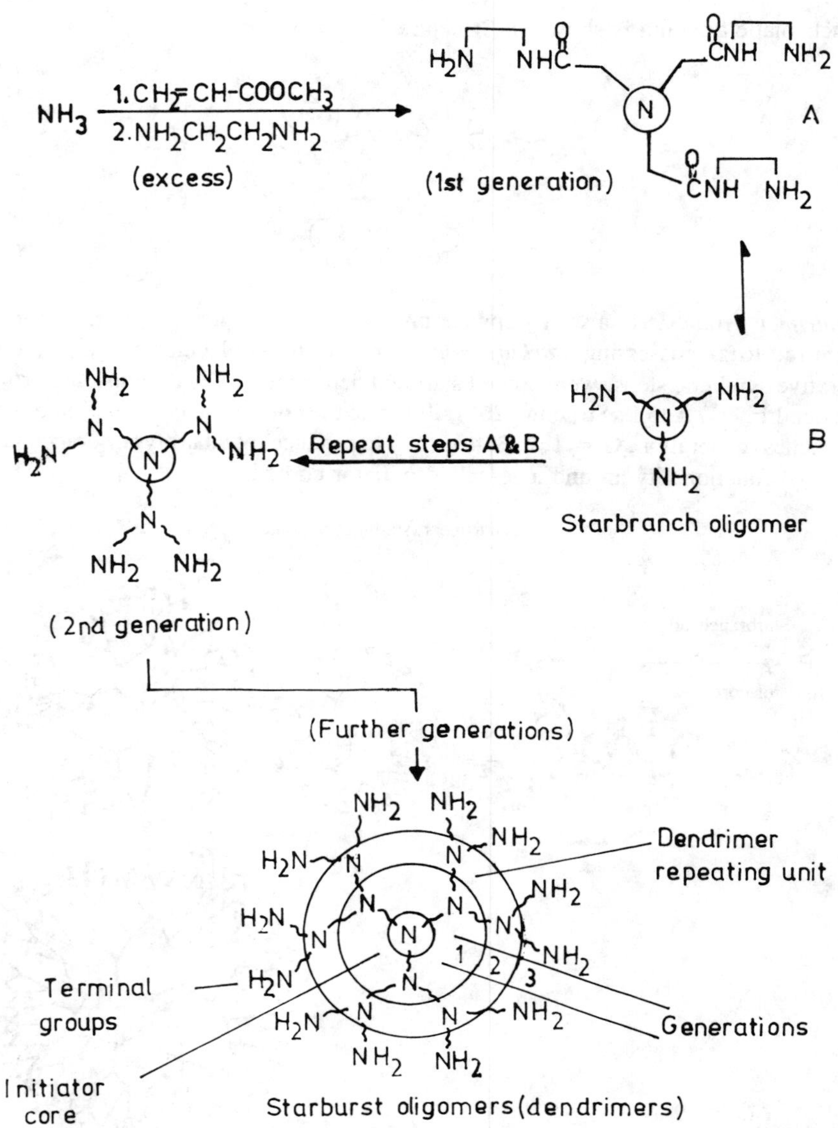

Starburst oligomers (dendrimers) (267)

In this synthesis, ammonia was used as the core. Exhaustive Michael addition of an amine with excess methyl acrylate was followed by exhaustive amidation of the resulting ester with excess diamine. Polyamidoamines (PAMAMs) have been synthesized as far as the 10th generation. A variety of initiator cores have been used in this series, such as ammonia with a multiplicity of three and ethylene diamine with a multiplicity of four. The shape of the core, as well as its multiplicity (N_c) and that of the repeating unit (N_r), determine the general shape and ultimately the size of the dendrimer. The number of terminal units (N_t), and therefore of functionalities, in each generation can be calculated by the following formula:

$$N_t = N_c(N_r)^G \tag{268}$$

If ammonia is used as the initiator core ($N_c = 3$), the progression is 3, 6, 12, etc. up to 1536 for the 10th generation ($G = 10$). With increasing size, the molecules tend to assume a spheroidal shape. In contrast, if ethylene-diamine is used as the initiator core ($N_c = 4$), the progression is 4, 8, 16, etc. These molecules tend to become ellipsoidal.

The second strategy of synthesis, i.e., coupling of preformed branched units, was used to prepare starburst polyether. The general concept is outlined in Scheme 269.[380,395-397]

[Core] - (OH)$_n$

    conversion to reactive function

[Core] - X$_n$

    n Y - cyclic reaction with masked synthon

[Core] - (- O - [Cyclic])$_n$

    deprotection

[Core] - [- O - {R} -(OH)$_m$ -] $_n$

    unmasked 1st generation

    to second generation $\tag{269}$

In this approach, the building block for the next generation contains, in addition to the reactive functionality, multiple functionalities that are masked in a cyclic structure. A bicyclic orthoester structure is used to mask three hydroxyl functions. Specifically, pentaerytritol (PE) is the initiator core and 4-(hydroxymethyl)-2,6,7-trioxabicyclo[2.2.2]octane, HTBO, is the building block for the consecutive generation. In this situation ($N_c = 4$ and $N = 3$), a highly branched, compact starburst molecule is obtained.

The 2nd (G = 2) and 3rd (G = 3) generation were synthesized in a completely analogous manner, as shown in Scheme 270.

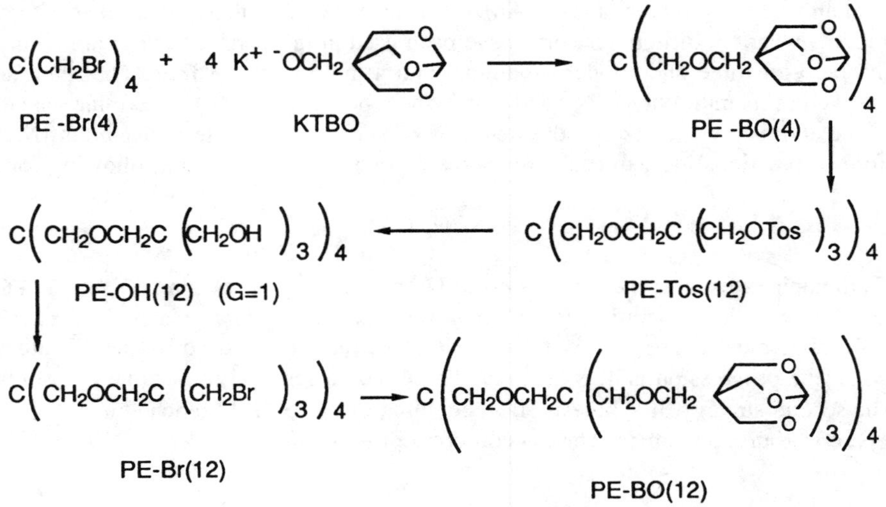

PE -Br(4)   KTBO   PE -BO(4)

$$C\{CH_2OCH_2C[CH_2OCH_2C(CH_2OH)_3]_3\}_4 \rightarrow PE\text{-}Tos(36) \rightarrow PE\text{-}Br(36) \rightarrow PE\text{-}BO(36) \rightarrow$$

$$PE\text{-}OH(36)$$
$$(G = 2)$$

$$C\{CH_2OCH_2C\{CH_2OCH_2C[CH_2OCH_2C(CH_2OH)_3]_3\}_3\}_4 \tag{270}$$

$$PE\text{-}OH(108)$$
$$(G = 3)$$

All intermediates were solid and were completely characterized. The 4th generation (G = 4), namely, PE–OH(324), is a *forbidden molecular structure* which is beyond the starburst-limited generation.[395,396]

The structure of the hydroxyl intermediates of the first and second generation were confirmed by fast-atom bombardment mass spectroscopy.[397] The analytical methods for structure verification of the other dendrimers have been described in References 398 and 399.

In size and in shape, starburst dendrimers are characterized by covalent connectivity that extends over colloidal dimensions. In addition, the stepwise methods used in the synthesis of these "precision macromolecules" can in principle be readily adapted to the introduction of radial variations in structure and properties within the polymer. A full exploitation of the starburst topology (e.g., as, *inter alia*, sequestering agents, catalysts, and vehicles for controlled delivery) requires a thorough understanding of the structural and dynamic properties of these dendritic chains. ^{13}C NMR relaxation was used to probe local motion in a series of PAMAM starburst dendrimers.[398] The results are interesting in that the relaxation time T_1 for each of the terminal carbon signals (f) decreases continuously from generation 1 to generation 11, while the relaxation times of the other methylene carbons decrease initially but level off at generation 3. The longer T_1 of carbon f suggests increased segmental mobility for this carbon, compared with the internal methylene groups, and the continuous decrease in T_1 with molecular weight indicates that the segmental motion is increasingly restricted at the surface of the higher-generation dendrimers.

Calixarenes, the cyclic oligomers depicted in Scheme 271, are synthesized from phenols and formaldehyde.[376] A variety of new host molecules from calixarenes (calix = goblet in Latin) have been synthesized.[400] They acted as cyclic ionophores

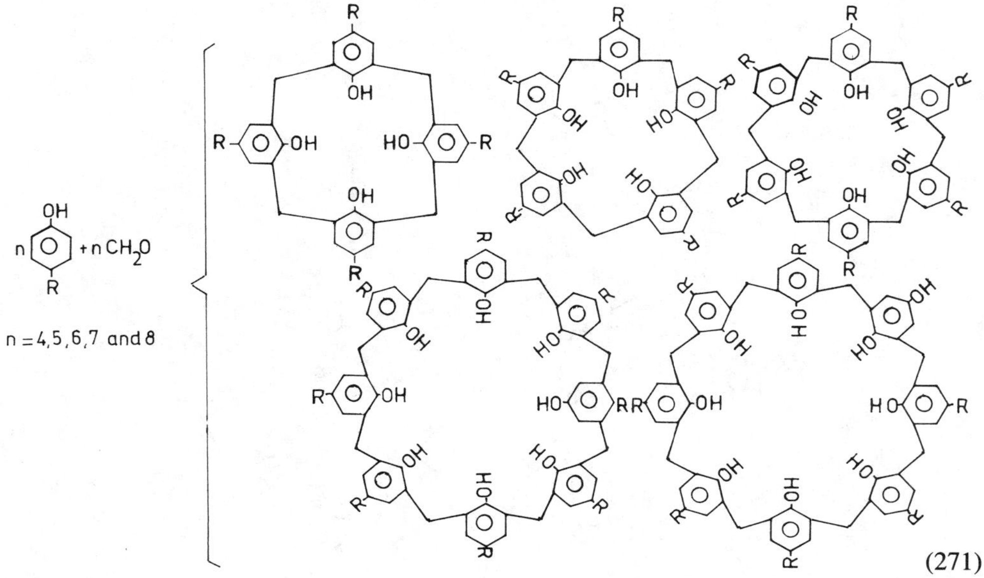

$$n \underset{R}{\overset{OH}{\bigcirc}} + n\,CH_2O$$

n = 4,5,6,7 and 8

(271)

with crown-like functions by introducing ionophoric groups, as shown in Scheme 272.

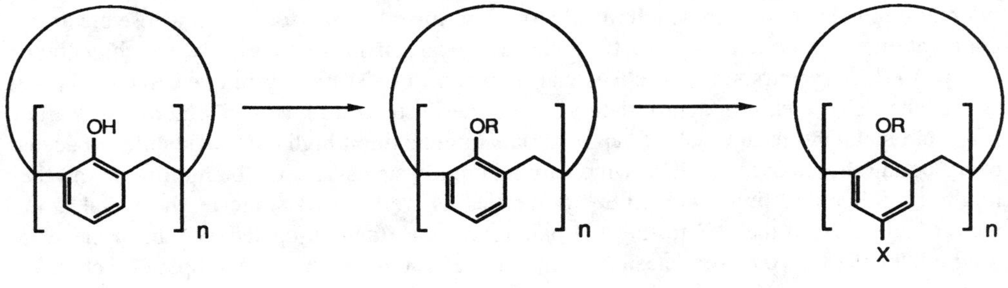

where: R = –H, X = –SO₃Na, –CH₂NH₂, –N=Ar, –CH₂Cl

R = –CH₃, X = –COAr, –Br

R = –CH₂COOH, X = –SO₃Na   (272)

In particular, calix[5]arene and calix[6]arene oligomers with carboxylate groups acted as "super uranophiles", exhibiting the record-breaking uranyl affinity (K_{uranyl} = $10^{19.2}$ M^{-1}) and uranyl selectivity (e.g., $K_{uranyl}/K_{Me}n+$ = $10^{17.5}$, where K_{uranyl} and $K_{Me}n+$ are the stability constants for uranyl and metallic ion, respectively; here, M^{n+} = M_g2+; Scheme 273).[376,401,402]

Inclusion of organic guest molecules in solution was affected for the first time by using water-soluble calixarenes which had either anionic, cationic, or neutral hydrophilic groups. Chiral, water-soluble calixarenes changed their conformation upon the inclusion of guest molecules, which was conveniently detected by the circular dichroism (CD) spectra and provided several lines of important information for the host-guest-type complex formation.[403-406]

Through these studies, it was demonstrated that the moderate rigidity and remaining conformational freedom are important characteristics of complex formation with calixarene-based host molecules.

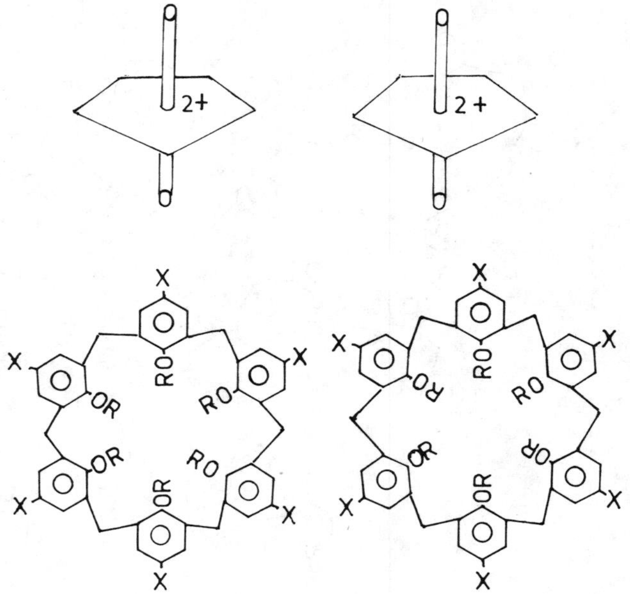

(273)

Finally, one should also mention the polymeric *helicene*[407] and *belt-like macropolycycles*[408] as new classes of oligomers of particular structure and characteristics. The fundamental idea in obtaining helicenes — in which metal atoms are linked by hydrocarbons whose electronic conjugation is unbroken — is to twist the hydrocarbon into helices. By attaching bulky groups to their precursors, the helices can be made to twist mainly in one direction. If the size of the helix is chosen appropriately, a polymeric structure forms in which hydrocarbon rings and metal atoms alternate. Such materials might exhibit high thermal stability (because every bond other than the C–H's, which are inherently strong, should be multiple), and they might exhibit interesting conductivity properties as well, for if some of the metal atoms were oxidized from the $+2$ to the $+3$ oxidation state, the conjugated system of electrons might delocalize the electron vacancy from one metal to the next. An example of a chemical representation is structure H shown below (Scheme 274).

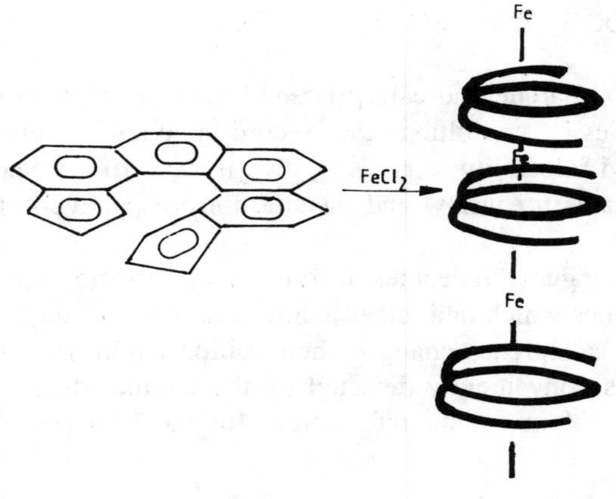

(274)

This is a helicene dianion capped at both ends by unsaturated five-membered rings. It should, with transition metal halides (e.g., ferrous chloride), give helical polymers.

Although neplanar, the pitch of the helix should attenuate the overlap of adjacent *p*-orbitals by only a miniscule amount, and oxidation should then allow electrons to delocalize throughout. A prominant feature of this helical structure is its chirality. If the hydrocarbon used to prepare it were of the optical pure antipodes, an electron translating from one metal atom to the next would twist, and always in the same direction. The material would be a molecular solenoid. Its optical, magnetic, and conductive properties should be unique.[407]

The synthesis of the pentalene dianion many years ago by Katz et al.[409] suggested that its combination with transition metal halides might yield new structures in which hydrocarbon rings and metal halides alternate:

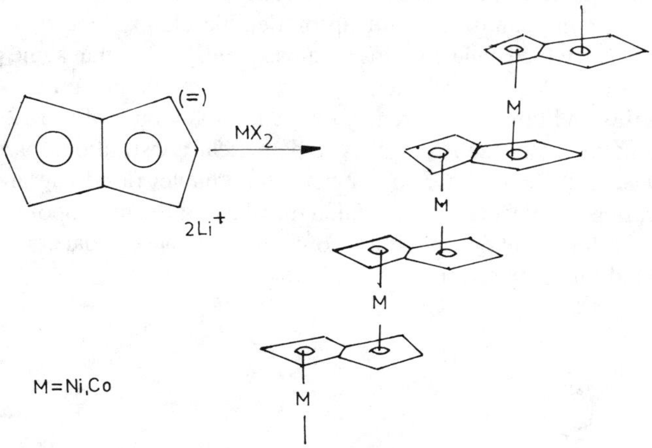

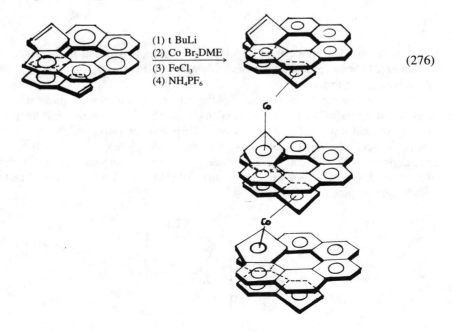

However, the materials were only dimers (n = 2).[410] Quite recently, Katz et al.[411] synthesized the oligomeric cobaltocenium salt shown in Scheme 276 with $\overline{M}_n \sim 1900$ which, according to elemental analysis, corresponds to 3.19 helicene units and 2.13 cobalt atoms.

Accordingly, the fast-atom bombardment mass spectrum had as the only prominent peaks at high mass those corresponding to four nine-membered rings (144 carbons) and three cobalts (n = 4), three nine-membered rings and two cobalts (n = 3), and two nine-membered rings and one cobalt (n = 2). The oligomeric mixture exhibits high optical activity: at 100% enantiomeric excess, $[\alpha]_D = -26,000°$. Comparison of the molar elipticities and rotational strengths of the oligomer and the monomer (i.e., the values were about seven times larger for the oligomeric cobaltocenium salt than for the monomer one) indicated that they increase with the number of units in the oligomer, and suggested that the optical activity of larger polymeric species could be enormous.

By the mid 1970s, Pedersen, Lehn and Cram had established — as mentioned above — the art of molecular recognition with wholly synthetic compounds designed and constructed to act as molecular receptors toward a wide range of chemical species.[413] The legacy of macrocyclic polyethers to the supramolecular chemist was a necklace-like design feature in which large ring compounds are built up of flexible chains, one bond at a time, often around rigid templates which can be removed subsequently.[413] In that same period, Diederich and Staab[414] had succeeded in synthesizing *kekulene,* a cyclic hexamer, $(C_8H_4)_6$, shown in Scheme 277. In this and other cycloarenes, cyclic homologation of the 12 benzenoid nuclei occurs in the plane of the benzene rings. A hypothetical constitutional isomer of kekulene, called [12]*cyclacene,*[208] should have the 12 benzenoid nuclei fused laterally in a polyacene-like manner, forming a molecular belt in which the plane of the macropolycycle is orthogonal to the mean plane of the bent benzene rings, now obliged to adopt boat conformation (Scheme 278). When spread out in two dimensions (Scheme 279),

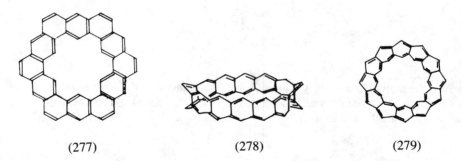

(277)   (278)   (279)

the constitutional formula of [12]cyclacene suggests that it can be regarded as two anti-aromatic [24] annulenes, joined to each other by single bonds between every alternate carbon atom around the annulene rings.

Whatever the electronic character of the [n]cyclacene oligomers, their status was heightened by the prediction that their condensed phases might show high-temperature ferromagnetism and warm superconductivity.[415] Perhydro derivatives of the cyclacenes (Scheme 280) — made up of cyclohexan-1,4-diene rings — are called *beltenes.* Force-field calculations[416] showed that the strain energy per macro ring unit decreases monotonically from n = 3 to n = 12, and predicted that [9]beltene (i.e., n = 9) should be the optimum molecular receptor for acetylene.

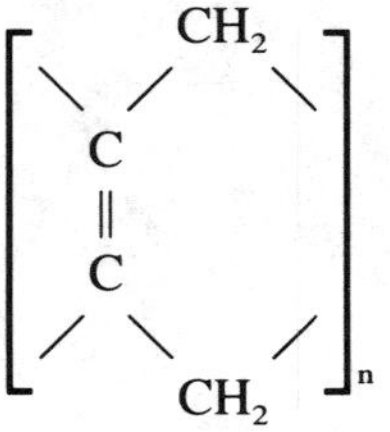

$$\left[\begin{array}{c} \diagdown \diagup \;\; CH_2 \;\; \diagup\diagdown \\ C \\ \| \\ C \\ \diagup\diagdown \;\; CH_2 \;\; \diagdown\diagup \end{array} \right]_n$$

Since Diels-Alder reactions afford six-membered rings via a mechanism in which two bonds are formed more or less simultaneously with high regioselectivity, complete stereo-specificity (*cis*-addition) and high stereoselectivity (i.e., *endo* vs. *exo* configurations in bicyclic systems), a point should be reached where an intramolecular cycloaddition between a diene and a dienophile in the same molecule will lead to macropolycyclization. Stoddart[412] took advantage of these particularities of Diels-Alder reactions to synthesize belt-like cy-clooligomers. The challenged reside in selecting a bisdiene and a bisdienophile that work and go together to complete a molecular loop. The sequential cycloadditions are shown in Scheme 280.[408,417] *Syn*-isomer of 1,4; 5,8-diepoxy-1,4,5,8-tetrahydroanthracene (Structure 280a) was the building block of choice as far as the bisdienophile was concerned.

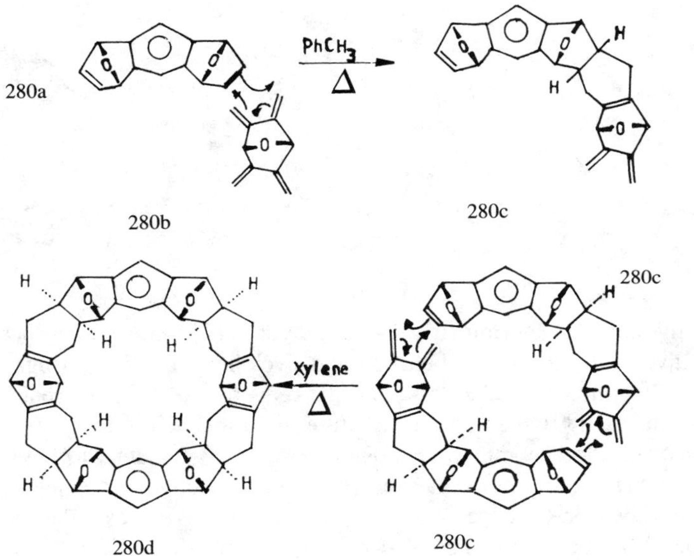

(280)

The building block of choice for the bisdiene was 2,3,5,6-tetramethylene-7-oxabicy-clo[2.2.1]heptane (Structure 280b). On heating the 1: adduct (Structure 280c) at 150°C, the loop was closed and the macrocycle (Structure 280d) was made. The systematic name for Structure 280c is *rel*-(1R,4S,4aS,7aR,8R,10S,10aS,13aR,14R,17S,17aS,aOaR,21R,-23S,23aS,26aR)-1,4:6,25:8,23:10,21:12,19:14,17-hexaepoxy-1,4,4a,5,6,7,7a,8,10,-10a,11,12,13,13a,14,17,17a,18,19,20,20a,21,23,23a,24,25,26,26a-octacosahydro-2,16:3,15-dimethanoundecane, or *kohnkene,* a trivial name honoring its maker (Franz Kohnke).[418]

Deoxygenation of Structure 280d gave dideoxykohnkene, Structure 281a, which was

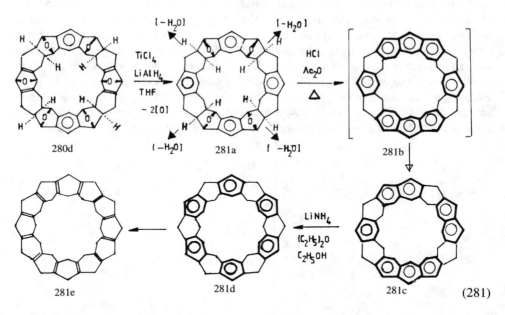

(281)

thereafter dehydrated to a symmetrical dodecahydrol[12]cyclocene (Structure 281b) — a stable derivative with alternating benzene and cyclohexy-1,4-diene rings — called, conservatively, [12]*collarene*.[419] The last step, the reduction of [12]collarene, would result in [12]beltene (Scheme 280). The crystal structure of kohnkene disclosed that its cavity shape and size is just right to accommodate a phenyl ring. This feature was exploited as the first application for these new molecular receptors: the selective piezoelectric quartz crystal detection of nitrobenzene was demonstrated by using kohnkene as a detector coating.[408] At the same time, it has been shown by X-ray measurements that a "free" water molecule can be trapped in the middle of ideoxykohnkene (Structure 281a).[419] The water molecule is disordered, with its hydrogens pointing principally toward the pairs of diagonal methine hydrogens (e.g., 1 and 7), but still > 2.7Å away from any potential interactive sites within the hydrophobic cavity.

Oligomers with precise dimensions and forms, such as molecular belts (large and small), molecular cages (small and large), and molecular strips (linear and coiled), not to mention molecular nets and stacks, all are part of the uncharted territory of unnatural product synthesis ready to become a reality.[411,420] Direct condensation and telomerization, and controlled degradation or partial decomposition of high polymers seem to be promising routes for the synthesis of these and other well-defined oligomers.

REFERENCES

1. **Simionescu, C. I. and Bulachovschi, V.,** *Treatise of Macromolecular Chemistry,* Vol. 3, EDP, Bucharest, 1976.
2. **Saunders, A. and Dobinson, F.,** The Kinetics of Polycondensation Reactions, in *Comprehensive Chemical Kinetics,* Vol. 5, *Non-Radical Polymerisation,* Bamford, C. H. and Tipper, C. F. F., Eds., Elsevier, Amsterdam, 1976, chap. 7.
3. **Seymour, R. B. and Carraher, Ch. E., Jr.,** *Polymer Chemistry,* Marcel Dekker, New York, 1988, chap. 7.
4. **Morton, M.,** *Anionic Polymerization: Principles and Practice,* Academic Press, New York, 1983, chap. 1.

5. **Berlin, A. A. and Matveyeva, N. G.**, Progress in the chemistry of polyreactive oligomers and some trends in its development. I. Synthesis and physico-chemical properties, *J. Polym. Sci. Macromol. Rev.*, 12, 1, 1977.

6. **Hiemenez, P. C.**, *Polymer Chemistry*, Marcel Dekker, New York, 1984, chap. 5.

7. **Carothers, W. H.**, Polymers and polyfunctionality, *Trans. Faraday Soc.*, 32, 39, 1936.

8. **Korshak, V. V.**, Dependence of polymer structure on monomer functionality, *Acta Polym.*, 34, 603, 1983.

9. **Korshak, V. V.**, *Thermostable Polymers*, Nauka, Moscow, 1972, 211 (in Russian).

10. **Flory, P. J.**, Random reorganization of molecular weight distribution in linear condensation polymers, *J. Am. Chem. Soc.*, 64, 2205, 1942.

11. **Korshak, V. V.**, Über den reaktions Mechanismus der Polykondensation, *Faserforsch. Textiltech.*, 6, 241, 1955.

12. **Korshak, V. V. and Zamyatina, I. V.**, High molecular weight compounds. XVI. The polydispersity of polyamide, *Izv. Akad. Nauk S.S.S.R.*, 59, 909, 1948 (in Russian).

13. **Howard, G. J.**, The molecular weight distribution of condensation polymers, in *Progress in High Polymers*, Robb, J. C. and Peaker, F. W., Eds., Heyword, London, 1961, 187.

14. **Slonimski, G. N.**, Mutual solubility of polymers and properties of their mixtures, *J. Polym. Sci.*, 30, 410, 1958.

15. **Flory, P. J.**, Molecular size distribution in ethylene oxide polymers, *J. Am. Chem. Soc.*, 62, 1561, 1940.

15a. **Jacobsen, H. and Stockmayer, W. H.**, Intramolecular reaction in polycondensation. I. The theory of linear systems, *J. Chem. Phys.*, 18, 1601, 1950.

16. **Iamanis, I. and Adelman, M.**, Significance of oligomerizations in the transesterification of dimethyl terephthalate with ethylene glycol, *J. Polym. Sci. Polym. Chem. Ed.*, 14, 1945, 1976.

17. **Iamanis, I. and Adelman, M.**, Two models for the kinetics of the transesterification of dimethyl terephthalate with ethylene glycol, *J. Polym. Sci. Polym. Chem. Ed.*, 14, 1961, 1976.

18. **Krumpolc, M. and Malek, J.**, Esterification of benzene carboxylic acids with ethylene glycol, *Makromol. Chem.*, 171, 69, 1973.

19. **Nondek, L. and Malek, J.**, Esterification of benzene carboxylic acids with ethylene glycol, *Makromol. Chem.*, 185, 2211, 1977.

20. **Rosenfeld, J. C. and Pierrone, F.**, Kinetic studies of the transesterification of neopentyl glycol, dimethyl terephthalate and the polycondensation of neopentylterephthalate oligomers, *Polym. Prepr.*, 17(2), 524, 1976.

21. **Tang, A.-C. and Yao, K.-S.**, Mechanism of hydrogen ion catalysis in esterification. II. Studies on the kinetics of polyesterification reactions between dibasic acids and glycols, *J. Polym. Sci.*, 35, 219, 1959.

22. **Griehl, W. and Jäger, P.**, Formation of oligomers in polyesters, *Faserforsch. Textiltech.*, 26, 378, 1975 (in German).

23. **Birley, A. V., Kyriakos, D., and Dawkins, J. W.**, Unsaturated polyesters of terephthalic acid, *Polymer*, 19, 1290, 1978; *Polymer*, 21, 632, 1980.

24. **Lazarowa, R. and Dimov, K.**, Kinetic of the polycondensation of poly(ethylene terephthalate) with low molecular weight, *Angew. Makromol. Chem.*, 55, 1, 1976.

25. **Lubimowa, G. V., Trofimova, G. N., and Novicov, O. D.**, Synthesis of oligoesteracrylates. Adipic acid-diethylenglycol-methacrylic acid system, *Vysokomol. Soedin. Ser. B*, 22, 182, 1980.

26. **Schmit, F. and Rossbach, V.**, Catalytic esterification of carboxylic groups in poly(ethylene terephthalate), *Polym. Bull.*, 2, 491, 1980.

27. **Berlin, A. A., Kefeli, T. Ya., and Suchareva, L. A.**, Investigation of the structure and properties of crosslinked polymers based on oligo carbonate methacrylates in film formation, *J. Macromol. Sci. Chem.*, 11, 977, 1977.

28. **Dall'Asta, G., Stigliani, G., Greco, A., and Motta, L.**, Concurrent oligomerization and Friedel-Crafts reactions in diolefin metathesis, *Chim. Ind. (Milan)*, 55, 142, 1973.

29. **Dall'Asta, G., Stigliani, G., Greco, A., and Motta, L.**, Concurrent oligomerization, isomerization and Friedel-Crafts reactions in olefin metathesis, *Polym. Prepr.*, 13, 910, 1972.

30. **Challa, G.**, The formation of poly(ethylene therephthalate) by ester interchange. I. The polycondensation equilibrium, *Makromol. Chem.*, 38, 105, 1960.

31. **Maréchal, E.**, Polymeric dyes—synthesis, properties and uses, *Prog. Org. Coat.*, 10, 251, 1982.

32. **Uglea, C. V. and Stan, V.**, The characterization of dope dyed poly(ethylene terephthalate), *Br. Polym. J.*, 14, 39, 1982.

33. **Berlin, A. A. and Matveyeva, N. G.**, The progress in the chemistry of polyreactive oligomers and some trends of its development. II. Specific features of network formation of oligomers and properties of network polymers, *J. Polym. Sci. Macromol. Rev.*, 15, 107, 1980.

34. **Berlin, A. A., Kefeli, T. Y., Fillipovskaya, U. M., and Sivergin, U. M.**, Condensation, telomerization, and synthesis of a new type of unsaturated polyesters, *Vysokomol. Soedin.*, 1, 951, 1959.

35. **Berlin, A. A., Popova, G. L., and Isaeva, E. F.**, Polymerization and properties of mixed polyesters of the acrylic series, *Dokl. Akad. Nauk S.S.S.R.*, 126, 83, 1959.

36. **Berlin, A. A., Kefeli, T. Y., and Korolev, G. V.,** Polymerizable oligomers, *Khim. Prom. (Moscow),* 80, 870, 1962.
37. **Berlin, A. A. and Bogdanov, I. F.,** The chemistry and technology of synthetic high molecular compounds. I. Polymerization of ethylene glycol dimethacrylate, *Zh. Obshch. Khim.,* 17, 1699, 1947.
38. **Berlin, A. A., Dabagova, A. K., and Rodionova, E. F.,** Three dimensional polymerization of allyl ethers and mixed allyl ethers of methacrylic esters of glycols, *Sb. Statei Obshch. Khim. S.S.S.R.,* 2, 1560, 1953.
39. **Berlin, A. A., Kefeli, T. Y., Fillippovskaya, U. M., and Sivergin, U. M.,** Synthesis and basic properties of polyether acrylates of different degrees of polymerization, *Vysokomol. Soedin.,* 2, 411, 1960.
40. **Berlin, A. A., Korovin, L. P., and Sumin, L. G.,** Oligomer homologous composition of oligoester acrylates, *Vysokomol. Soedin. Ser. B,* 11, 424, 1969.
41. **Berlin, A. A., Kefeli, T. Ya., and Suchareva, L. A.,** Investigation of the structure and properties of crosslinked polymer based on oligocarbonate methacrylates in film formation, *J. Macromol. Sci. Chem.,* 11, 977, 1977.
42. **Rabie, A. M.,** Synthesis and characterization of some oligothioacrylate esters and their polymers, *Eur. Polym. J.,* 8, 687, 1972; *Eur. Polym. J.,* 8, 841, 1972.
43. **Rabie, A. M., Nosseir, M. H., and Faltus, B. M.,** Synthesis and study of polymerization and copolymerization of some oligounsaturated esters, *Eur. Polym. J.,* 11, 177, 1975.
44. **Hirai, H.,** Oligomers from hydroxymethylfuran carboxylic acid, *J. Macromol. Sci. Chem.,* 21, 1165, 1984.
45. **Aleksnis, A., Deme, Dz., and Surna, J.,** Synthesis of oligoesters and polyesters from oxalic acid and ethylene glycol, *J. Polym. Sci. Polym. Chem. Ed.,* 15, 1855, 1977.
46. **Fedorova, V. A., Donchak, V. A., and Puchin, V. A.,** Deposited Doc., CPSTL Kph-D8, 1981 (in Russian); *Chem. Abstr.,* 98, 107855, 1983.
47. **Ahn, J. J. and Jao, G.,** A study on preparation of polyesteramide from oligoester and oligoamide, *Han'guk Somyn Konghakhoechi,* 23, 123, 1986 (in Korean); *Chem. Abstr.,* 105, 153622r, 1986.
48. **Nozawa, S., Nakamura, M., and Homa, M.,** Japanese Patents 60,221,422 and 85,221,422; *Chem Abstr.,* 207905k, 1986.
49. **Wang, D. Z. and Jones, F. N.,** Synthesis of cross-linkable heterogeneous oligoester diols by direct esterification with p-hydroxybenzoic acid, *ACS Symp. Ser.,* 367, 335, 1988.
50. **Dimian, A. F. and Jones, F. N.,** Model cross-linkable liquid crystal oligoester diols as coating binders, *Polym. Mater. Sci. Eng.,* 56, 640, 1987.
51. **Rosenquist, N. R., Howery, R. W., Pyles, R. A., and Mulvey, P. J.,** European Patent Appl. 227,109,1985; *Chem. Abstr.,* 107, 23792n, 1987.
52. **Pyles, R. A. and Mulvey, P. J.,** European Patent Appl. 228,699,1985; *Chem. Abstr.,* 107, 199v, 1987.
53. **Okayama, H., Kitachi, T., and Ohara, O.,** Japanese Patent 62,167,321, 1987; *Chem. Abstr.,* 107, 23759lj, 1987.
54. **Okayama, H., Kitachi, T., and Ohara, O.,** Japanese Patent 62,267,324, 1987; *Chem. Abstr.,* 109, 38492u, 1988.
55. **Rosenquist, N. R.,** German Patent 3,638,260, 1987.
56. **Rosenquist, N. R.,** German Patent 3,638,323, 1987.
57. **Brunelle, D. J., Evans, T. L., Shannon, T. G., and Williams, D. A.,** European Patent Appl. 162,379,1985; *Chem. Abstr.,* 104, 130502s, 1986.
58. **Tessier, M. and Maréchal, E.,** Synthesis and structural study of α-amino and α,ω-diaminooligoamides, *Eur. Polym. J.,* 22, 877, 1986.
59. **Fuchs, H., Brend, U., and Buchackert, H.,** German patent 3,524,394, 1987.
60. **Zahn, H. and Gleitsmann, G. B.,** Oligomere und Pleinomere von synthetischen faser bildenden Polymeren, *Angew. Chem.,* 75, 772, 1963.
61. **Zahn, H., Roedel, H., and Kunde, J.,** Oligoamide aus α-Aminoundecansaure, *J. Polym. Sci.,* 36, 539, 1959.
62. **Zahn, H., Miro, P., and Schmidt, F.,** Über cyclische Oligoamide aus Nylon, *Chem. Ber.,* 90, 1411, 1957.
63. **Antos, K., Hodul, P., Bartos, L., Benicky, M., and Krajiciki, L.,** Czechoslovakian Patent 210,495, 1979.
64. **Cum, G., Gallo, R., Febo, S., and Spadori, A.,** Oligomeric 1,4-phenyleneterephthalamide. Synthesis by interfacial polycondensation and structural properties, *Angew Makromol. Chem.,* 138, 111, 1986.
65. **Ross, S. D., Coburn, E. R., Leach, W. A., and Robinson, W. B. J.,** Isolation of a cyclic trimer from poly(ethylene terephthalate) film, *J. Polym. Sci.,* 13, 406, 1954.
66. **Giuffria, Ruth,** Microscopic studies of mylar film and its low molecular weight extract, *J. Polym. Sci.,* 49, 427, 1961.
67. **Gootfried, W. and Wenk, H.,** Cyclic oligomers in polyesters from diols and aromatic dicarboxylic acids, *Angew. Makromol. Chem.,* 112, 59, 1983.
68. **Toraj Industries, Inc.,** Japanese Patent 57,191,370, 1981; *Chem. Abstr.,* 144995, 1983.

69. **Ryn, D. and Ha, W. S.,** Cyclic oligomers, *Hanguk Somyn Konghakhoeki,* 22, 391, 1985; *Chem. Abstr.,* 51144y, 1986.
70. **Kiselev, V. V., Anfinogentov, A. A., Malych, V. A., Dorochov, I. N., and Kefasov, V. V.,** Optimization of bifunctional monomer polycondensation, *Dokl. Akad. Nauk S.S.S.R.,* 294, 122, 1987.
71. **Takeda, M. and Tanabe, K.,** Japanese Patent 61, 216, 203, 1986; *Chem. Abstr.,* 106, 34308d, 1987.
72. **Derminot, J., Agege, R., and Jaquemart, J.,** Polyester oligomers. Migration determined by thermal treatments, *Bull. Sci. ITF,* 5, 215, 1976.
73. **Derminot, J.,** Polyester oligomers. Qualitative and quantitative analysis, *Bull. Sci. ITF,* 5, 355, 1976.
74. **Berari, D., Halip, V., Secure, V., and Homotescu, J.,** Study of PET oligomers and their influence upon processing of TEROM® fibres and yarn, *Mater. Plast. (Bucharest),* 10, 243, 1973 (in Romanian).
75. **Goodman, I. and Nesbitt, R. F.,** The structures and reversible polymerization of cyclic oligomers from poly(ethylene terephthalate), *Polymer,* 1, 384, 1960.
76. **Goodman, I. and Nesbitt, B. F.,** The structures of reversible polymerization of cyclic oligomers from poly(ethylene terephthalate), *J. Polym. Sci.,* 48, 423, 1960.
77. **Peebles, L. H., Jr., Hufmann, M. V., and Ablett, C. T.,** Isolation and identification of the linear and cyclic oligomers of poly(ethylene terephthalate) and the mechanism of cyclic oligomer formation, *J. Polym. Sci. Part A,* 7, 479, 1969.
78. **Cooper, D. R. and Semlyen, A.,** Equilibrium ring concentration and the statistical conformations of polymer chains. XI. Cyclic in poly(ethylene terephthalate), *Polymer,* 14, 185, 1973.
79. **Burzin, K., Haltrup, W., and Feinauer, R.,** Cyclic esters in poly(ethylene terephthalate), *Angew, Makromol. Chem.,* 74, 93, 1978.
80. **Ha, W. S. and Choun, Y. K.,** Kinetic studies on the formation of cyclic oligomers in poly(ethylene terephthalate), *J. Polym. Sci. Polym. Chem. Ed.,* 17, 2103, 1979.
81. **Gueris, C. and Meybeck, J.,** The study of poly(ethylene terephthalate) degradation. I. Preparative chromatographic separation of oligomers, *Bull. Soc. Chim. Fr.,* 6, 2163, 1971.
82. **Seidel, B.,** Spectroscopic studies of linear and cyclic oligomers in PET, *Z. Elecktrochem.,* 62, 214, 1958.
83. **Dorman-Smith, V. A.,** The separation of some PET oligomers, *J. Chromatogr.,* 29, 265, 1967.
84. **Greiswald, P. and Wuntke, K.,** Column and paper chromatography of PET oligomers, *Plaste Kautsch.,* p. 274, 1968.
85. **Ito, E. and Okajima, S.,** Studies on cyclic tris(ethylene terephthalate), *J. Polym. Sci. Polym. Lett. Ed.,* 7, 483, 1969.
86. **Lomakin, S. M., Assejewa, R. M., and Saikov, G. J.,** Thermodynamische stabilität ungersättigter oligoester, *Plaste Kautsch.,* 1980, 611.
87. **Hudgins, W. R., Theuer, K., and Mariani, T.,** Separation of PET oligomers by high performance liquid chromatography and thin-layer chromatography, *J. Appl. Polym. Sci. Appl. Polym. Symp.,* 14, 145, 1978.
88. **Zaborski, L. M., II,** Determination of polyester prepolymer oligomers by high performance liquid chromatography, *Anal. Chem.,* 49, 1167, 1977.
89. **Mori, S., Iwasaki, S., Furukawa, M., and Takeuki, T.,** Thin-layer chromatography of lineaar oligomers of ethylene terephthalate, *J. Chromatogr.,* 62, 109, 1971.
90. **Takyama, H.,** Japanese Patent 62,53,478, 1987; *Chem. Abstr.,* 107, 200324y, 1987.
91. **Shiono, S.,** Separation and identification of PET oligomers by gel permeation chromatography, *J. Polym. Sci. Polym. Chem. Ed.,* 17, 4123, 1979.
92. **Repin, H. and Papanikolau, E.,** Synthesis and properties of cyclic di(ethylene terephthalate), *J. Polym. Sci. Part A,* 7, 3426, 1969.
93. **Cimecioglu, A. L., Zerotnian, S. H., Alger, K. W., Colling, M. J., and Easton, G. C.,** Properties of oligomers present in poly(ethylene terephthalate), *J. Appl. Polym. Sci.,* 32, 4719, 1986.
94. **Tanaka, H. and Yoshimi, H.,** Japanese Patent 61,126,128, 1986.
95. **Jones, F., Lu David, D. L., Pechacek, J., and Kangus, S.,** Synthesis of telechelic oligoesters having low polydispersity using dicyclohexylcarbodiimide, *Ind. Eng. Chem. Prod. Res. Dev.,* 25, 385, 1986.
96. **Vinayuk, D. K.,** Indian Patent 158,902, 1987.
97. **Vinayuk, D. K.,** Indian Patent 158,901, 1987.
98. **McCoy, D.,** Canadian Patent 1,132,606, 1987.
99. **Royc, W.,** U.S. Patent 4,659,778, 1987.
100. **Nguyen, H. A. and Maréchal, E.,** Synthesis of reactive oligomers and their use in block polycondensation, *J. Macromol. Sci. Rev. Macromol. Chem. Phys.,* C28,187, 1988.
101. **Marchessault, K. H. and Faure, A. J.,** Synthesis of configurationally controlled oligomers of poly-α-hydroxy butyrate, *ACS Polym. Prepr.,* 25(1), 87, 1974.
102. **Iwakura, Y., Iwata, K., Matsuo, S., and Tohara, A.,** Oligoesters. IV. The synthesis of optically active oligo(L-γ-hydroxy isovalertes), *Makromol. Chem.,* 146, 33, 1971.
103. **Schnell, D.,** The kinetics of oligomerization, *Angew. Makromol. Chem.,* 39, 131, 1954.
104. **Berlin, A. A. and Rabin, A. M.,** The synthesis of oligoesters, *Vysokomol. Soedin. Sec. B,* 15, 336, 1973.
105. **Alimov, A. P., Kogan, L. K., Gusevi, V. K., and Shchepkina, R. S.,** Production, properties and use of neopentyl polyol ester, *Vses. Nauchno Issled. Inst. Prerab. Nefti,* 42, 41, 1982.

106. **Maslosh, V. Z., Myakukhina, V. T., Mishura, L. B., Kravtsov, A. I., and Bukhan'ko, A. I.**, Russian Patent 1,333,675, 1987.
107. **Saam, J. C. and Chou, Y. J.**, U.S. Patent 4,355,154, 1982.
108. **Kirpichnikov, P. A.**, Synthesis and chemical transformation of reactive oligomers, *J. Polym. Sci. Polym. Symp.*, No. 67, 192, 1980.
109. **Teijin, Ltd.**, Japanese Patent 57,154,483, 1982; *Chem. Abstr.*, 98, 73770j, 1983.
110. **Jones, M., Benet, E., and Corey, J. G.**, European Patent Appl. 162,651, 1985; *Chem. Abstr.*, 104, 130809x, 1986.
111. **Vilenskii, A., Kercha, Yu. Yu., Kiviralik, S., and Fomina, S. A.**, Thermal properties of cooligoesters, in *Fizikochim. Poliuretanov*, Khimia, Moscow, 1981, 93 (in Russian).
112. **Yamada, T. and Imamura, Y.**, A mathematical model for computer simulation of a direct continuous esterification process between terephthalic acid and ethylene glycol, *Polym. Eng. Sci.*, 28, 385, 1988.
113. **Choi, K. J.**, A modeling of semibath reactors for melt transesterification of dimethylterephthalate with ethylene glycol, *Polym. Eng. Sci.*, 27, 1703, 1987.
114. **Flory, P. J.**, *Principles of Polymer Chemistry*, Cornell University Press, Ithaca, NY, 1953.
115. **Sebenda, J.**, Lactams, in *Chemical Kinetics*, Bamford, C. H. and Tipper, C. F. H., Eds., Elsevier, Amsterdam, 1976.
115a. **Mori, S., Furukawa, M., and Takeuki, T.**, Reduction gas-chromatographic determination of cyclic monomer and oligomers in polyamides, *Anal. Chem.*, 42, 661, 1970.
116. **Zahn, H., Kunde, J., and Heidemann, G.**, Uber cyclische oligoamide aus ε-aminocapronsäure, *Makromol. Chem.*, 43, 220, 1961.
117. **Hermans, P. H., Heikens, D., and Van Valden, P. F.**, On the mechanism of the polymerization of ε-caprolactam. II. The polymerization in the presence of water, *J. Polym. Sci.*, 30, 80, 1958.
118. **Semplyen, J. A. and Walker, G. R.**, Equilibrium ring concentrations and the statistical conformations of polymer chains. III., *Polymer*, 10, 597, 1969.
119. **Arai, Y.**, Kinetics of hydrolitic polymerization of ε-caprolactame. III. The formation of cyclic dimer, *Polymer*, 22, 273, 1981.
120. **Feldman, R.**, Cyclic oligomers in polyamides, *Angew. Makromol. Chem.*, 34, 9, 1973.
121. **Munoz-Escalona, A. and Oteyza, M.**, Sintesis de los oligomeros diamino de los nylon 6.6 and nylon 6.5, *Rev. Plast. Mod.*, 34, 695, 1977.
122. **Andrews, J. M., Jones, F. R., and Semlyen, J. A.**, Equilibrium ring concentrations and the statistical conformations of polymer chains. XV. *Polymer*, 15, 420, 1974.
123. **Reimschuessel, H. C.**, Nylon 6, *J. Polym. Sci. Macromol. Rev.*, 12, 65, 1977.
124. **Harris, D. M., Jenkins, R. L., and Nielsen, M. L.**, Phosphorous-containing polymers by a two-phase condensation, *J. Polym. Sci.*, 35, 540, 1959.
125. **Cleaver, C. S. and Pratt, B. C.**, Polyamides from 2,2'-bis-5(4H)-oxazolone, *J. Am. Chem. Soc.*, 77, 1572, 1955.
126. **Michel, R. H. and Murphey, W. A.**, Polymers from the condensation of dihydrazides with dialdehydes and diketones, *J. Appl. Polym. Sci.*, 7, 617, 1963.
127. **Ryabukhin, A. G.**, Kinetics of reaction of phenol with formaldehyde in different media, *Vysokomol. Soedin. Sec. A*, 11, 2562, 1969.
128. **Peer, H. G.**, The reaction of phenol with formaldehyde. II. The ratio of *ortho-* and *para-*hydroxymethyl-phenol in the base catalysed hydroxymethylation of phenol, *Rec. Trav. Chim. Pays-Bas*, 78, 851, 1959.
129. **Lenz, R. W.**, *Organic Chemistry of High Polymers*, Interscience, New York, 1967, chap. 5.
130. **Komiyama, M.**, Catalysis by cyclodextrin for *para*-oriented attack of formaldehyde to phenol, *Polym. J.*, 20, 439, 1988.
131. **Bender, M. L. and Komiyama, M.**, *Cyclodextrin Chemistry*, Springer-Verlag, Berlin, 1978.
132. **Lenz, R. W.**, *Organic Chemistry of High Polymers*, Interscience, New York, 1967, chap. 6.
133. **Casiroghi, G., Cornia, M., Sartori, G., Casnati, G., Bocchi, V., and Andretti, G. D.**, Selective step-growth phenol aldehyde polymerization. I. Synthesis, characterization and X-ray analysis of linear all-*ortho* oligonuclear phenolic compounds, *Makromol. Chem.*, 183, 2611, 1982.
134. **Ivanova, E. A., Erkova, L. N., and Lazarevich, S. Ya.**, Study of condensation of resorcinol and 5-methylresorcinol with formaldehyde in the presence of emulsifier, *Zh. Prikl. Khim. (Moscow)*, 59, 2377, 1986 (in Russian).
135. **Gutsalyuk, V. G.**, *Arenephenol-Aldehyde Oligomers* Nauky, Alma-Ata, 1986, chap. 5 (in Russian).
136. **Perrin, R. and Lamartine, R.**, Preparation of well-defined phenolic resins from pure precursors, *Makromol. Chem., Makromol. Symp.*, 9, 69, 1987.
137. **Vicens, J., Pilat, T., Ganiet, D., Lamartine, R., and Perrin, R.**, Synthesis and characterization of precursor of phenolic resins produced from 4-isopropylphenol, *C. R. Acad. Sci. Ser. 2*, 302, 15, 1986.
138. **Yoshimura, Y., Sase, S., Suzuki, H., Ishida, S., and Nakamoto, Y.**, Japanese Patent 61,221,212, 1985; *Chem. Abstr.*, 107, 24291g, 1987.

139. **Casiroghi, G., Cornia, M., Ricci, G., Balduzzi, G., Casnati, G., and Dario, G.**, Selective step-growth phenol-aldehyde polymerization. II. Synthesis of 4-*tert*butylphenol novolak resins and related open chain oligomeric compounds, *Makromol. Chem.*, 184, 1363, 1983.

140. **Kuchesenskoi, G. K., Mazaleva, A. P., Makatkin, A. V., and Siding, M. I.**, Reactions of methylol derivatives in the system of phenol-formaldehyde oligomers, *Lukokras. Mater. Ikm. Primen.*, No. 6, p. 14, 1986 (in Russian).

141. **Yoshimura, Y. and Sase, S.**, Japanese Patent 61,221,217, 1985; *Chem. Abstr.*, 107, 24307i, 1987.

142. **Negulescu, I. I. and Uglea, C. V.**, *Synthetic Oligomers*, CRC Press, Boca Raton, FL, in preparation.

143. **Kumskov, V. N. and Iufenov, A. M.**, Mathematical model of the synthesis of phenol-formaldehyde oligomers, *Plast. Massy*, No. 9, 7, 1987 (in Russian).

144. **Chaisky, V. Ya., Karlynsky, L. E., and Lupatov, L. F.**, *Chemical Products from Coking of Coal from East S.S.S.R.*, Vol. 1, Sverdlovsk, 1963.

145. U.S. Patent 3,674,723, 1972.

146. U.S. Patent 3,474,065, 1969.

147. U.S. Patent 3,385,824, 1968.

148. **Andrews, P. K.**, British Patent, 1,541,023, 1979.

149. **Wegler, R.**, Condensation of aromatic compounds with formaldehyde. A new group of reactive formaldehyde resins, *Angew. Chem.*, 60, 88, 1948.

150. **Gutsalyuk, V. G., Neviskyi, V. M., and Safronova, A. S.**, *Arenephenolaldehyde Oligomers*, Nauka, Alma-Ata, 1986, chap. 1 (in Russian).

151. U.S. Patent 3,372,147, 1968.

152. French Patent 2,362,873, 1978.

153. French Patent 1,052,049, 1954.

154. **Moshchynskaya, N. K.**, *Polymeric Materials Based on Aromatic Hydrocarbons and Formaldehyde*, Technica, Kiev, 1970, 250 (in Russian).

155. **Bayarstanova, Zh. Zh. and Erdenova, Sh. E.**, *Heavy Oil Stocks and Polymers Based on Them*, Nauka, Alma-Ata, 1984, 228 (in Russian).

156. Japanese Patent 62,530,381, 1987; *Chem. Abstr.*, 107, 178192e, 1987.

157. **Simionescu, C. I., Dumitrescu, S., Rotstein, M., and Negulescu, I. I.**, Romanian Patent 71,086, 1980.

158. **Chaisky, V. Ya., Karlynsky, L. E., and Lupatov, L. F.**, *Chemical Products from Coking of Coal from East S.S.S.R.*, Vol. 3, Sverdlovsk, 1965.

159. **Chaisky, V. Ya., Karlynsky, L. E., and Lupatov, L. F.**, *Chemical Products from Coking of Coal from East S.S.S.R.*, Vol. 4, Sverdlovsk, 1967.

160. **Chaisky, V. Ya., Karlynsky, L. E., and Lupatov, L. F.**, *Chemical Products from Coking of Coal from East S.S.S.R.*, Vol. 5, Sverdlovsk, 1967.

161. **Shoryghina, N. V.**, Pyrene-phenol-formaldehyde resins, *Plast. Massy*, No. 9, p. 19, 1967.

162. **Shoryghina, N. V., Ninin, V. K., and Zhylina, N. V.**, Thermoreactive arene-phenol-formaldehyde oligomers, *Plast. Massy*, No. 4, 34, 1973.

163. **Zhuravleva, I. I. and Akutin, M. S.**, Method for obtaining phenol- and naphthalene-phenol-formaldehyde oligomers, *Plast. Massy*, No. 7, 40, 1981.

164. French Patent 2,372,182, 1978.

165. British Patent 1,466,641, 1977.

166. **Kostitsyn, B. A. and Tretyakov, A. S.**, *Chemical Industry in Ukraina*, Vol. 4, Naukova Dumka, Kiev, 1970, 10 (in Russian).

167. **Jedlinski, Z. J.**, Novel coal-based monomers and thermally stable polymers, in *IUPAC Macro '83*, Vol. 1, Plenary and Invited Lectures, Bucharest, 1983, chap. 1.

168. **Herbert, L. D., Robinson, J. G., and Marr, G.**, Preparations and properties of phenenthrene-formaldehyde resins, *Br. Polym. J.*, 13, 154, 1981.

169. **Gutsalyuk, V. G., Neviskyi, V. M., and Safronova, A. S.**, *Arenephenolaldehyde Oligomers*, Nauka, Alma-Ata, 1986, chap. 4 (in Russian).

170. **Gaillard, J., Hellin, M., and Causemont, F.**, Etude de la réaction de Prins. XIII. Systéme α-methylstyrène-formol; cinetique en phase aqueuse homogène, *Bull. Soc. Chim. Fr.*, 9, 3360, 1967.

171. **Vatsuro, K. and Mishchenko, G.**, *Réactions Organiques Classées par Auteurs*, Editions Mir, Moscow, 1981, 513.

172. **Golovaneva, N. M., Sergeev, V. A., Shitikov, V. K., and Komarova, L. I.**, Methylol derivatives of unsaturated phenols and preparation of oligomers and polymers from them, *Nov. Issled. Obl. Pr-va Primen. Fenol Smol Ionitov*, 4, 1985 (in Russian); *Chem. Abstr.*, 104, 149455a, 1984.

173. **Cohen, S. M. and Lavin, E.**, Polyspiroacetal resins. I. Initial preparation and characterization, *J. Appl. Polym. Sci.*, 6, 503, 1962.

174. **Uglea, C. V., Romanţov, A., Georgescu, Florentina, and Bălănescu, G.**, Romanian Patent 63405, 1976.

175. **Uglea, C. V., Georgescu Florentina, Romanţov, A. and Bălănescu, G.**, Romanian Patent 63406, 1976.

176. **Elias, H. G.**, *Macromolecules*, Vol. 2, *Synthesis and Materials*, Plenum Press, New York, 1977, sect. 28.2.

177. **John, H.**, British Patent 151016, 1918.

178. **Aldersley, J. W. and Gordon, M.**, Polyaddition and Polycondensation. Substitution effects in polycondensation systems, *J. Polym. Sci. Part C*, 16, 4567, 1969.

179. **Ito, K.**, On the methylolated products of urea with formaldehyde, *Kogyo Kagaku Zasshi*, 64, 382, 1961.

180. **Gordon, M., Alliwell, A., and Wilson, T.**, *Symposium on the Chemistry of Polymerization Processes*, Monogr. 20, Society of Chemical Industry, London, 1965, 187.

181. **Petersen, H.**, Kinetik und Katalyse bei Aminoplastkondensation, *Chem. Ztg.*, 95, 692, 1971.

182. **Volkodaeva, U. V., Akutin, M. S., Andrianov, B. V., Potekhina, E. S., Yacobson, B. V., and Ghebychev, B. S.**, Russian Patent 1,301,832, 1987.

183. **de Jong, J. I. and de Jong, J.**, Kinetics of the formation of methylene linkages in solutions of urea and formaldehyde, *Rec. Trav. Chim.*, 72, 139, 1953.

184. **Okano, M. and Ogata, Y.**, Kinetics of the condensation of melamine with formaldehyde, *J. Am. Chem. Soc.*, 74, 5728, 1952.

185. **Gordon, M., Aliwell, A., and Wilson, T.**, Kinetics of the addition stage in the melamine-formaldehyde reaction, *J. Appl. Polym. Sci.*, 10, 1153, 1966.

186. **Sato, K.**, Thermodynamics of the hydroxymethylation of melamine and urea with formaldehyde, *Bull. Chem. Soc. Jpn.*, 40, 724, 1967.

187. **Sato, K. and Ouchi, S.**, Studies on formaldehyde resins. XVI. Hydroxymethylation of melamine with formaldehyde in weakly acidic region and without additional catalyst, *Polym. J.*, 10, 1, 1978.

188. **Tomita, B.**, Melamine-formaldehyde resins: molecular species distributions of methylolmelamines and some kinetics of methylolation, *J. Polym. Sci. Polym. Chem. Ed.*, 15, 2347, 1977.

189. **Sato, K.**, Condensation of a methylolmelamine (with a low degree of methylation) in acidic media, *Bull. Chem. Soc. Jpn.*, 40, 2963, 1967.

190. **Sato, K.**, Condensation of methylolmelamine, *Bull. Chem. Soc. Jpn.*, 41, 7, 1968.

191. **Sato, K. and Naito, T.**, Studies on melamine resin. VII. Kinetics of the acid-catalyzed condensation of di- and trimethylolmelamine, *Polym. J.*, 5, 144, 1973.

192. **Nastky, R., Dietrich, K., Reinisch, G., Rafler, G., and Gojewski, H.**, The initial stage of the reaction of melamine with formaldehyde, *J. Macromol. Sci. Chem.*, A23, 579, 1986.

193. **Schildknecht, C. E.**, *Polymer Processes*, Interscience, New York, 1956, 304.

194. **Herma, H., Dietrich, K., Schulze, C., and Terge, W.**, Melamine-formaldehyde resins. Effect of conditions of synthesis on molecular weight, oligomer distribution and molding properties, *Plaste Kautsch.*, 33, 324, 1986.

195. **Costa, L., Camino, G., and Martinasso, G.**, Thermal behavior of malamine salts, *ACS Polym. Prepr.*, 30(1), 531, 1986.

196. **Staudinger, H., Krasig, H., and Wetzel, G.**, Über die Konstitution von harnstoff thio- resp. thioharstoff-formaldehyde Kondensation. II. Mitteilung über makromolekulare Verbindungen, *Makromol. Chem.*, 20, 1, 1956.

197. **Meyer, V.**, Über thiodiglycol Verbindungen, *Chem. Ber.*, 19, 3259, 1886.

198. **Gaylord, N. G.**, Ed., *Polyethers*, Part III, Interscience, New York, 1962, 43.

199. **Averko-Antonovichi, L. A. and Smisclova, R. A.**, *Polysulfide Oligomers*, Khimia, Leningrad, 1983, 4.

200. **Hay, A. S., Blanchard, H. S., Enders, G. F., and Eustance, J. W.**, Polymerization by oxidative coupling, *J. Am. Chem., Soc.*, 81, 6335, 1959.

201. **Edmonds, J. T., Jr. and Hill, H. W., Jr.**, U.S. Patent 3,354,129, 1967.

202. **Kovacic, P. and Kyriakis, A.**, Polymerization of aromatic nuclei. II. Polymerization of benzene to poly-p-phenyls by $AlCl_3$-$CuCl_2$, *Tetrahedron Lett.*, 1962, 467.

203. **Hartling, B. and Lindberg, J. J.**, The formation of poly(phenylene sulfide)s and the substituent effect on reactions of substituted chlorobenzenes with sulfur, *Makromol. Chem.*, 179, 1707, 1978.

204. **Koch, W. and Heitz, W.**, Models and mechanism of the formation of poly(thio-1,4-phenylene), *Makromol. Chem.*, 184, 779, 1983.

205. **Koch, W., Risse, W., and Heitz, W.**, Radical ions as chain carriers in polymerization reactions, *Makromol. Chem. Suppl.*, 12, 105, 1985.

206. **Tache, P. and Freitag, D.**, German Patent 3,529,093, 1987.

207. **Korshak, V. V., Teplyakov, M. M., Khodina, A. I., and Kalinin, V. N.**, Russian Patent 919,326, 1982.

208. **Ueda, A. and Nagai, S.**, Block copolymers derived from azobiscyano-pentanoic acid. IV. Synthesis of a polyamide-polystyrene copolymer, *J. Polym. Sci. Polym. Chem. Ed.*, 22, 1611, 1984.

209. **Ueda, A., Agari, Y., and Nagai, S.**, Synthesis of a new macro-azo-initiator for block copolymerization, in *Prepr. IUPAC Macro '88*, Kyoto, Japan, 1988, 5.4-5.16, 203.

210. **Lyle, G. D., Mahanty, D. K., Cecere, J. A., Wu, S. D., Senger, J. S., Chen, P. H., Kilic, S., and McGrath, J. E.,** Synthesis, curing and physical behavior of maleimide terminated poly(arylene ether), *Int. SAMPLE Symp. Exhib.,* 33, 1080, 1988.

211. **Simionescu, C. I. and Negulescu, I. I.,** *Treatise of Chemistry of Macromolecular Compounds,* Vol. 4, EDP, Bucharest, chap. 9 in press (in Romanian).

212. **Tekeiki, T., Date, H., Takayama, T. and Stille, J. K.,** Imide oligomers containing acetylenic and biphenylene groups: preparation, curing and thermal properties, in *Prepr. IUPAC Macro '88,* Kyoto, Japan, 1988, 4.7-4.16, 462.

213. **Adachi, T., Saito, X., Kamio, K., and Shibata, M.,** Characterization and application of organosoluble amine-terminated polyimide oligomer, in *Prepr. IUPAC Macro '88,* Kyoto, Japan, 1988, 3P-08b, 650.

214. **Wiley, R. H.,** Oligomeric carbobothionic dihydrazides from dialdehydes and carbonothionic dihydrazide, *J. Macromol. Sci. Chem.,* A25, 231, 1988.

215. **Knipe, A. C. and Watts, W. E.,** *Organic Reaction Mechanisms,* John Wiley & Sons, New York, 1981, 6.

216. **Hesse, G.,** *Houben Weyl Methoden der Organische Chemie,* Vol. 6, 4th ed., 1978, 217.

217. **Wiley, R. H.,** Poly(di-1,2-diazinediyl-ethene-1,2-diol)s, *J. Macromol. Sci. Chem.,* A24, 1183, 1987.

218. **Gadzhiev, G. G., Kasanov, F. Kh., Seridov, M. A., and Ragimov, A. V.,** Oligoaniline structure, *Dokl. Akad. Nauk Az. S.S.R.,* 43, 50, 1987 (in Russian).

219. **Fuhr, K., Mueller, F., and Eicher, T.,** German Patent 3,504,189, 1986.

220. **Heitz, W. and Schwalm, R.,** German Patent 3,542,230, 1987.

221. **Percec, V. and Nava, H.,** Functional polymers and sequential copolymers by phase transfer catalysis. XXV. Transformation of a monotropic mesophase into an enantiotropic one by increasing the molecular weight of the polymer and by copolymerization, *J. Polym. Sci. Part A,* 25, 405, 1987.

222. **Sergeev, V. A., Nedel'kin, V. J., Andrianova, O. B. I., and Novikov, V. U.,** α,ω-Oligophenylene sulfide diepoxide, *Vysokomol. Soedin. Ser. B,* 28, 317, 1986.

223. **Batzer, H. and Zahir, S. A.,** Studies in the molecular weight distribution of epoxide resins. I. Gel permeation chromatography of epoxide resins, *J. Appl. Polym. Sci.,* 19, 585, 1975.

224. **Batzer, H. and Zahir, S. A.,** Studies in the molecular weight distribution of epoxide resins. II. Chain branching in epoxide resins, *J. Appl. Polym. Sci.,* 19, 601, 1975.

225. **Batzer, H. and Zahir, S. A.,** Studies in the molecular weight distribution of epoxide resins. III. Gel permeation chromatography of epoxide resins subject to postglycidylation, *J. Appl. Polym. Sci.,* 19, 609, 1975.

226. **Batzer, H. and Zahir, S. A.,** Studies in the molecular weight distribution of epoxide resins. IV. Molecular weight distribution of epoxide resins made from bisphenol A and epichlorohydrin, *J. Appl. Polym. Sci.,* 21, 1814, 1977.

227. **Negulescu, I. and Ionescu, E.,** Romanian Patent 53410, 1964.

228. **Schwartz, W. W. and Blumbergs, J. H.,** Epoxidations with *m*-chloroperbenzoic acid, *J. Org. Chem.,* 29, 1976, 1944.

229. **Negulescu, I.,** Aspects regarding the epoxidation of certain liquid polymers, *Mater. Plast.,* 6, 263, 1969 (in Romanian).

230. **Elias, H. G.,** *Macromolecules,* Vol. 2, *Synthesis and Materials,* Plenum Press, New York, 1977, sect. 26.2.3.

231. **Negulescu, I. and Mociarov, V.,** Romanian Patent 51730, 1966.

232. **Dabas, I., Dotrek, B., Stary, S., Lidarik, M., and Makovski, L.,** Czechoslovakian Patent 210,234, 1982.

233. **Heitz, W., Wirth, Th., Peters, R., Strabe, G., and Fischer, E. V.,** Synthese und ligenschaften molekulareinherthicher n-Paraffine bis zum $C_{140}H_{282}$, *Makromol. Chem.,* 162, 63, 1962.

234. **Mathias, L. J., Kuseforgen, S. H., and Kress, A. D.,** Functional methacrylate monomers. Simple synthesis of alkyl-α-(hydroxymethyl)-acrylates, *Macromolecules,* 20, 2326, 1987.

235. **Warren, R. M. and Mathias, L. J.,** *t*-Butyl(α-hydroxymethyl)-acrylate: a precursor for a potential anti-cancer cyclopolymer, *ACS Polym. Prepr.,* 30(1), 235, 1989.

236. **Iwaisako, T., Chohno, M., Masamoto, J., Ohtake, J., and Kawamura, M.,** Development of methylol process for production of highly concentrated aqueous formaldehyde solution, *ACS Polym. Prepr.,* 30(1), 170, 1989.

237. U.S. Patent 4,493,752, 1985.

238. **Iwaisako, T., Masamoto, J., Yoshida, K., Matsusaki, K., Kagawa, K., and Nagahaia, H.,** Development of advanced process for manufacturing acetal copolymer, *ACS Polym. Prepr.,* 30(1), 241, 1989.

239. **Orthmann, E., Enkelmann, V., and Wegner, G.,** Synthesis and electrochemical doping of phthalocyanatopolysilixane, *Makromol. Chem. Rapid. Commun.,* 4, 687, 1983.

240. **Orthmann, E. and Wegner, G.,** Catalysis of the polycondensation of dihydrosiliconphthalocyanine, *Makromol. Chem. Rapid Commun.,* 7, 243, 1986.

241. **Pogosyan, G. M., Pankratov, V. A., Zaphystonyi, V. N., and Matsoyan, S. G.,** Polytriazines, in *Izv. A. N. Armyan S.S.S.R.*, Erevan, U.S.S.R., 1987, chap. 2 (in Russian).

242. **Huang, S. J., Feldman, A. J., and Cercena, J. L.,** Polymeric and monomeric aryloxy-S-triazines, *ACS Polym. Prepr.*, 30(1), 348, 1989.

243. **Lin, S. C.,** U.S. Patent 4,379,728, 1983.

244. **Lin, S. C.,** N-Cyanourea-terminated resins, *ACS Polym. Prepr.*, 25(1), 112, 1984.

245. **Sergeev, V. A., Shitikov, V. K., Chizhova, N. V., and Korshak, V. V.,** Cyclotrimerization of 2,4-tolylene diisocyanate and phenyl isocyanate, *Vysokomol. Soedin. Ser. A*, 24, 2352, 1982.

246. **Sorokin, M. F., Shode, L. G., and Nogteva, S. I.,** Synthesis of epoxyisocyanurate oligomers and some properties of coatings, *Lakokras. Mater. Ikh. Primen.*, No. 6, 11, 1982.

247. **Pernikos, R., Lozdina, B., Apsite, B. K., Karlivans, V., and Barkane, R.,** Russian Patent 973,551, 1982.

248. **Giogun, S. M., Boronina, G. P., Colisman, I. I., Gulyaliva, N. A., Ubvotseva, U. V., Bogdanov, A. P., Prainin, L. B., and Erabiko, L. B.,** Oligoepoxyurethanes as modifiers for polymer materials, *Plast. Massy*, No. 7, 41, 1988.

249. **Richter, R., Mueller, H., Kubitza, W., Erybert, T., and Mennicken, G.,** German Patent 3,432,081, 1986.

250. **Van Geenen, A. A. and Vincent, J. A. J. M.,** European Patent Appl., 77,105, 1983.

251. **Eremeeva, T. V. and Karyakina, M. J.,** Gelling during formation of polyurethane coatings, *Lakokras. Mater. Ikh. Primen.*, 4, 13, 1983.

252. **Sorokin, M. S., Shade, L. G., Onosova, L. A., and Gonseva, O. F.,** Preparation of urethane oligomers, deposited doc., VINITI 3538, 1981.

253. **Troev, K., Todorov, K., and Borisov, G.,** Synthesis of rigid polyurethane foams from oligoester alcohols based on the residues from distillation of crude dimethyl terephthalate and recrystallization of the dimethyl terephthalate/dimethyl isophthalate fraction, *J. Appl. Polym. Sci.*, 28, 2491, 1983.

254. **Degtyaseva, A. A., Shrubovich, V. A., and Tkach, V. P.,** Effect of the thermal orientation of linear aliphatic polyester-polyurethanes on their properties, *Izv. Vyssh. Uchebn. Zaved. Khim. Khim. Tekhnol.*, 29, 94, 1986.

255. **Tsukube, H.,** Noncyclic crown-type polyether polymers for transport of alkali cations and amino acid anion, *J. Polym. Sci. Polym. Chem. Ed.*, 20, 2989, 1982.

256. **Araki, T. and Tsukube, H.,** New type urea- and thiourea-oligomers: role of chemical specificity for transport and radical forming photo-reduction, in *Prepr. IUPAC Macro Florence 1980*, Vol. 4, 1980, 135.

257. **Maruyama, K., Tsukube, H., and Araki, T.,** An artificial oligomer carrier for transport of organic substrates, *JCS Chem. Commun.*, 1222, 1980.

258. **Hunter, W. H., Olson, A. O., and Daniels, E. A.,** Über eine katalysche Zersetzung gewisser Phenol-silbersalze, *J. Am. Chem. Soc.*, 38, 1761, 1916; *Chem. Zentralbl.*, 87, 1133, 1916.

259. **Staffin, G. O. and Price, C. C.,** Polyethers. IX. Poly(2,6-dimethyl-1,4-phenylene oxide), *J. Am. Chem. Soc.*, 82, 3632, 1960.

260. **Hay, A. S.,** Oxidative coupling of acetylenes, *J. Org. Chem.*, 25, 1275, 1960; *J. Org. Chem.*, 27, 3320, 1962.

261. **Lunk, H. E. and Youngman, E. A.,** *J. Polym. Sci. Part A*, 3, 2983, 1963.

262. **Josefowicz, M. and Buvet, R.,** Charactérisation spectrophotométrique d'une melange d'oligophenylenes, *C. R.*, 253, 1801, 1961.

263. **Kovacic, P. and Wu, C.,** Cross-linking on unsaturated polymers with dimaleimides, *J. Polym. Sci.*, 47, 45, 1960.

264. **Toshima, N., Kanaka, K., and Hirai, H.,** New method for preparation of (p-phenylene) and its derivatives, *Prepr. IUPAC Macro '88*, Kyoto, Japan, 6.3-07, 1988, 152.

265. **Korshak, V. V., Tplyakov, M. M., and Šergeev, V. A.,** New method for synthesis of polyphenylene-type-polymers and investigation of polycondensation of aromatic diacetyl derivatives, *Dokl. Akad. Nauk S.S.S.R.*, 208, 1360, 1973.

266. **Korshak, V. V., Teplyakov, M. M., and Chebotarev, V. P.,** Polycondensation of ethyl ketals of aromatic acetyl compounds—a new method of synthesis of polyphenylenes, *Vysokomol. Soedin. Ser. A*, 16, 497, 1974.

267. **Teplyakov, M. M. and Tschebotarjow, A. V.,** Polycondensation von Ketalen. Eine neue Methode der Polyphenylene-Synthese, *Acta Chim. (Budapest)*, 81, 281, 1973.

268. **Korshak, V. V., Teplyakov, M. M., Gelashvili, Ts. L., Komarov, S. D., Kalinin, V. N., and Zhakhakin, L. I.,** Analysis and investigation of polymers on the basis of 4,4'-diacetyldibenzylcarboranes, *J. Polym. Sci. Polym. Lett. Ed.*, 17, 115, 1979.

269. **Kabakii, Yu. A., Valetskii, P. M., and Pshenychkin, P. A.,** Free radical initiated interaction between o-carbonate and aromatic compounds, *Dokl. Akad. Nauk S.S.S.R.*, 280, 1180, 1985.

270. **Izmailov, B. A., Kalinin, V. N., and Myakushev, V. D.,** Hydrolytic polymerization of (carboranyl-methyl)organochlorosilans, *Zh. Obshch. Khim.*, 50, 1588, 1980.

271. **Izmailov, B. A., Kalinin, V. N., and Myakushev, V. D.,** Hydrolytic copolycondensation of carboraneacylcooxysilane with trimethylchloro- and dimethyldichlorosilane, *Zh. Obshch. Khim.,* 53, 1807, 1983.

272. **Zhdanov, A. A. and Levitskii, M. M.,** Organosilicon polymers: polymers with inorganic backbones, in *Progress in Synthesis of Elemento-Organic Polymers,* Korshak, V. V., Ed., Nauka, Moscow, 1988, chap. 6 (in Russian).

273. **Kapustina, A. A., Avilova, T. P., and Bykov, V. T.,** Synthesis of bis(triethyl-silyloxy)dibutoxygermanium, *Zh. Obshch. Khim.,* 46, 1655, 1976.

274. **Avilova, T. P., Bykov, V. T., and Kapustina, A. A.,** Synthesis and investigation of polygermanosiloxanes, *Vysokomol. Soedin. Ser. B,* 17, 325, 1978.

275. **Kapustina, A. A.,** Synthesis and investigation of polygermano-organosiloxanes, Ph.D. Thesis, Institute of Organic Chemistry, Acadamy of Science, Irkutsk, U.S.S.R., 1979, 27.

276. **Simionescu, C. I. and Negulescu, I. I.,** Polimeri organometalici, *Mater. Plast. (Bucharest),* 18, 294, 1978 (in Romanian).

277. **Sheats, J. H., Carraher, Ch. E., Jr., and Pittman, C. V., Jr., Eds.,** *Metal-Containing Polymeric Systems,* Plenum Press, New York, 1985, 511.

278. **Paushkin, Ya. M., Vyshnykova, T. P., Lunin, A. F., and Nizova, S. A.,** *Polymeric Organic Semiconductors,* Khimia, Moscow, 1971, 224 (in Russian).

279. **Shriner, R. L. and Berger, A.,** Condensation products of benzalcohols. Polybenzyle, *J. Org. Chem.,* 6, 305, 1941.

280. **Kolesnikov, G. S., Korshak, V. V., and Susanova, T. V.,** Synthesis of polyarylenealkyls. I. Polycondensation of methylene chloride with halogenoderivatives of benzene, *Izvest. Akad. Nauk. S.S.S.R. Otd. Khim. Nauk,* 1957, 1478.

281. **Yamamoto, T., Ito, T., and Kubota, K.,** Preparation of poly(2,5-pyridinediyl) by dehalogenation polycondensation of 2,5-dibromopyridine with reducing metal. A soluble π-conjugated polymer with a large degree of polymerization, in *Prepr. IUPAC Macro '88,* Kyoto, Japan, 1988, 5.3-5.28, 150.

282. **Yamamoto, T., Hayashi, T., and Yamamoto, A.,** A novel type of polycondensation utilizing transition metal-catalyzed O–C coupling. I. Preparation of thermostable polyphenylene type polymers, *Bull. Chem. Soc. Jpn.,* 51, 2091, 1978.

283. **Suzuki, M., Sawada, M., Yoshida, S. and Saegusa, T.,** A new ring-opening polymerization catalyzed by Palladium (O) complex, in *Prepr. IUPAC Macro '88,* Kyoto, Japan, 1988, 2.3-2.17, 85.

284. **Yoneyama, M., Tanaka, M., Kakimoto, M., and Imai, Y.,** Synthesis of poly(arylenevinylene) analogs by palladium catalyzed polycondensation of aromatic dibromides with divinyl compounds, in *Prepr. IUPAC Macro '88,* Kyoto, Japan, 1988, 4.4-4.19, 203.

285. **Dall'Asta, G., Stigliania, G., Greco, A., and Motta, L.,** Concurrent oligomerization, isomerization and Friedel-Crafts reactions in diolefin metathesis, *Chim. Ind. (Milan)* 55, 142, 1973.

286. **Drăguţan, V., Balaban, A. T., and Dimonie, M.,** *Olefin Metathesis and Ring-Opening Polymerization of Cyclo-Olefins,* John Wiley & Sons, New York, 1985, chap. 1.

287. **Höcker, H., Reif, L., Reimann, W., and Riebel, K.,** Equilibrium oligomer concentration of the metathesis products of cycloolefins, *Rec. Trav. Chim.,* 96, M47, 1977.

288. **Chauvin, N., Commerenc, D., and Zalorowski, D.,** Equilibrium oligomeric concentration in the polymerization of 1,5-cyclooctanene, *Rec. Trav. Chim.,* 96, M131, 1977.

289. **Dreyfuss, P. and Dreyfuss, M. P.,** Polymerization of cyclic ethers and sulfides, in *Chemical Kinetics,* Bamford, C. H. and Tipper, C. T. H., Eds., Elsevier, Amsterdam, 1976, chap. 4.

290. **Nel, J. G., Wagener, K. B., Boncella, J. M., and Duttweiler, R. P.,** Acyclic metathesis polymerization. The synthesis of higher *trans* polyoctenamer, *ACS Polym. Prepr.,* 30(1), 283, 1989.

291. **Lindmark-Hamberg, M. and Wagener, K. B.,** Acyclic metathesis polymerization of 1,9 decadiene, *Macromolecules,* 20, 2449, 1987.

292. **Zahn, H., Rathgeber, P., Rexroth, E., Krzikalla, R., Lauer, W., Miro, P., Spoor, H., Schmidt, F., Seidel, B., and Hildebrand, D.,** Oligomere von Polyamid- und Polyester-typ, *Angew. Chem.,* 68, 229, 1956.

293. **Zahn, H.,** Oligomere vom Polyamid- und Polyester-typ in Textilindustrie und Forshung, *Z. Textilind.,* 66, 928, 1964.

294. **Rothe, M.,** Physical data of oligomers, in *Polymer Handbook,* 2nd ed., Brandrup, J. and Immergut, E. H., Eds., John Wiley & Sons, New York, 1975, VI-1.

295. **Zahn, H. and Krzikalla, R.,** Synthesis of the homogeneous linear oligoesters of poly(ethylene terephthalate) types, *Makromol. Chem.,* 23, 31, 1957.

296. **Fordyce, R. and Hibbert, H.,** Studies on reactions relating to carbohydrates and polysaccharides. LVII. The synthesis of 90 membered oxyethylene glycols, *J. Am. Chem. Soc.,* 61, 1910, 1939.

296a. **Fordyce, R. and Hibbert, H.,** Studies on reactions relating to carbohydrates and polysaccharides. LVIII. The relation between chain length and viscosity of the polyethylene glycols, *J. Am. Chem. Soc.,* 61, 1912, 1939.

296b. **Lovell, E. L. and Hibbert, H.,** Studies on reactions relating to carbohydrates and polysaccharides. LIX. The precipitability of pure hemicolloidal polyoxyethylene glycols, *J. Am. Chem. Soc.,* 61, 1916, 1939.

297. **Zahn, H. and Seidel, B.,** Weitere Oligoester der Terephthalsäure mit Glycol, *Makromol. Chem.,* 29, 70, 1959.

298. **Zahn, H., Borstlop, C., and Volk, G.,** Synthesis of oligoester of poly(ethylene terephthalate) type, *Makromol. Chem.,* 64, 18, 1963.

299. **Zahn, H. and Krzikalla, R.,** Synthesis of the homogeneous linear oligoester of poly(ethylene terephthalate) type, *Makromol. Chem.,* 23, 31, 1957.

300. **Repin, H. and Papanikolau, E.,** Synthesis and properties of cyclic di(ethylene terephthalate), *J. Polym. Sci. Sect. A,* 7, 3426, 1969.

301. **Zahn, H. and Repin, H.,** Synthese linear homolgen Reihe von Cyclischen Äthylen Terephthalaten, *Chem. Ber.,* 103, 3041, 1970.

302. **Uglea, C. V., Popa, A., Popa, M., and Popa, N.,** Synthesis and properties of poly(ethylene terephthalate) oligomers, *Abstr. IUPAC Macro '83,* Section 1, Bucharest, September, 1983, 505.

303. **Hässlin, H. W., Dröscher, M., and Wegner, G.,** Structure and properties of segmented poly(etheresters). III. Synthesis of defined oligomers of poly(butylene terephthalate), *Makromol. Chem.,* 181, 301, 1980.

304. **Hässlin, H. W., Dröscher, M., and Wegner, G.,** Melting and mutual compatibility of oligo(tetramethylene terephthalate) as models for poly(ether esters), *Makromol. Chem.,* 179, 1373, 1978.

305. **Hässlin, H. W. and Dröscher, M.,** Structure of oligo(butylene terephthalate), *Polym. Bull.,* 2, 769, 1980.

306. **Meraskentes, E. and Zahn, H.,** Synthesis of cyclic tris(ethylene terephthalate), *J. Polym. Sci. Part A,* 4, 1890, 1966.

307. **Pilati, F., Manaresi, P., Fortunato, B., Munari, A., and Passalaqua, V.,** Formation of poly(butylene terephthalate): secondary reactions studied by model molecules, *Polymer,* 22, 1566, 1981.

308. **Birley, A. W., Dawkins, J. V., and Kiriakos, D.,** Unsaturated polyesters based on terephthalic acid. III. Characterization of poly(propylene terephthalate) prepolymer by gel permeation chromatography, *Polymer,* 21, 632, 1980.

309. **Gang-Fung, C. and Jones, F. N.,** Toward synthesis of monodisperse oligoesters using silyl ether protecting groups. Cyclization during deprotection, *ACS Prepr.,* 29(2), 256, 1958.

310. **Zahn, H. and Kunde, J.,** Synthesis of cyclic oligoamides from ε-aminocaproic acid, *Ann.,* 618, 158, 1958 (in German).

311. **Zahn, H. and Kunde, J.,** Heptacyclo-ε-aminocapryl, *Angew. Chem.,* 69, 713, 1957 (in German).

312. **Zahn, H. and Lauer, W.,** Upon the linear oligoamide from adipic acid and hexamethylene diamine, *Makromol. Chem.,* 23, 85, 1957.

313. **Zahn, H., Kush, P., and Shah, J.,** Oligo-diamine und Oligo-aminosäure vom Nylon 6.6-typ, *Kolloid Z.,* 216, 298, 1967.

314. **Kush, P. and Blessing, E.,** Zur Synthese von Oligomeren des Nylon-6,6 nach der Anhydridmethode bei höheren Temperature, *Makromol. Chem.,* 167, 69, 1973.

315. **Bacskai, R.,** Synthesis and thermal stability evaluation of end-capped nylon-4 oligomers, *ACS Polym. Prepr.,* 24(2), 63, 1983.

316. **Zahn, H. and Gleitsmann, G. B.,** Linear oligoamides of nylon-6,10 type, *Makromol. Chem.,* 60, 45, 1963.

317. **Zahn, H. and Hildebrnad, D.,** Nonakis-, Dekakis-, Undekakis- and Dodekakis-ε-aminocaproamide, *Chem. Ber.,* 92, 1963, 1969.

318. **Rothe, M.,** Cyclic amides in polycaprolactame, *J. Polym. Sci.,* 30, 227, 1958.

319. **Zahn, H. and Kunde, J.,** Heptakis-cyclo-ε-aminocaproyl, *Angew. Chem.,* 69, 713, 1957.

320. **Zahn, H. and Kunde, J.,** Octakis- and nonakis-cyclo-ε-aminocaproyl, *Angew. Chem.,* 70, 189, 1958.

321. **Zahn, H., Kunde, J., and Heidemann, G.,** About the cyclic oligoamides of ε-aminocaproic acid, *Makromol. Chem.,* 53, 220, 1961.

322. **Zahn, H., Kunde, J., and Heidemann, G.,** Synthesis of cyclic oligoamides from ε-aminocaproic acid, *Justus Liebigs Ann. Chem.,* 618, 158, 1958.

323. **Zahn, H. and Pieper, W.,** Cyclic oligomer of ε-aminocaproic and aminoundecanoic acids, *Angew. Chem.,* 95, 1069, 1962.

324. **Zahn, H. and Kunde, J.,** Cyclic oligoamides of ω-aminoundecanoic acid, *Chem. Ber.,* 94, 2470, 1961.

325. **Rothe, M., Rothe, I., Bünig, H., and Schvensee, K. D.,** The cyclization of peptides by activation of aminic group, *Angew. Chem.,* 71, 700, 1959.

326. **Stetter, H. and Marx, J.,** The synthesis of macrocyclic amides, *Ann.,* 607, 59, 1957.

327. **Zahn, H. and Schmidt, F.,** Mono- and bis-cyclohexamethylene-adipamide, *Chem. Ber.,* 92, 1381, 1959.

328. **Hall, H. K.,, Jr.,** Structural effects on the polymerization of lactams, *J. Am. Chem. Soc.,* 80, 6404, 1962.

329. **Glover, I. G., Smith, R. B., and Rapoport, H.,** Amide-amide reaction via cyclols, *J. Am. Chem. Soc.,* 87, 2003, 1965.

330. **Zahn, H., Miro, P., and Schmidt, P.,** Über cyclische Oligoamide aus Nylon, *Chem. Ber.,* 90, 1411, 1957.

331. **Huigen, R., Brade, H., Walz, H., and Glogger, I.**, Die eigenschaften aliphatscher Lactame und die *cis-trans*-Isomerie der Saureamidgruppe, *Chem. Ber.*, 90, 1437, 1957.

332. **Bayer, O.**, Bemerkungen zu der Abhandlung von Th. Lieser und Karl Macura: Kunstliche organische Hochpolymere, *Ann.*, 549, 286, 1941.

333. **Kern, W. and Thoma, W.**, Methods of formation of macromolecular substances by polymerization, I., *Makromol. Chem.*, 11, 10, 1953.

334. **Kern, W. and Thoma, W.**, Methods of formation of macromolecular substances by polymerization, II., *Makromol. Chem.*, 16, 89, 1955.

335. **Kern, W. and Rauterkus, K. J.**, Synthesis of macromolecules of uniform size. IV. Basis of the duplication procedure, *Makromol. Chem.*, 28, 221, 1958.

336. **Kern, W., Kalsch, H., Raiterkus, K., and Sutter, H.**, Synthese mehrerer Polymer homologer Reichen von Diol-oligo-urethanen. Synthese von Macromolekeln enheitlicher Grösse. V., *Makromol. Chem.*, 44/46, 78, 1961.

337. **Zahn, H. and Dominik, M.**, Lineare Oligomere aus Hexamethylendiisocyanat und Butandiol-(1,4), *Makromol. Chem.*, 44, 290, 1961.

338. **Iwakura, Y., Hayashi, K., and Iwata, K.**, Linear oligomers of polypentamethylurethan, *Makromol. Chem.*, 89, 214, 1965.

339. **Iwakura, Y., Hayashi, K., and Iwata, K.**, Linear oligoethyleneurethans. II. Oligourethanes, *Makromol. Chem.*, 95, 217, 1966.

340. **Iwakura, Y., Hayashi, K., and Iwata, K.**, Linear oligotrimethyleneurethans. III. Oligourethans, *Makromol. Chem.*, 98, 13, 1966.

341. **Qin, H. Y., Macosko, C. W., and Wellinghoff, S. T.**, Synthesis and characterization of model urethane compounds, *Macromolecules*, 18, 553, 1985.

342. **Eisenbach, C. D., Nefzger, H., Baumgartner, M., and Grunter, C.**, Struktur und Eigenschaften von Modellverbindungen für segmentierte Polyurethanelastomere, *Ber. Bunsenges. Phys. Chem.*, 89, 1190, 1985.

343. **Bill, R., Droscher, M., and Wegner, G.**, Structure and properties of segmented poly(ether ester)s and polyurethanes. V. Synthesis of uniform oligo(oxytetramethylene)s, *Makromol. Chem.*, 182, 1033, 1981.

344. **Bill, R., Droscher, M., and Wegner, G.**, Oligo(oxytetramethylene)s and their derivatives: models for segmented poly(ether ester)s and polyurethanes, *Makromol. Chem.*, 179, 2993, 1978.

345. **Hässlin, H. W., Droscher, M., and Wegner, G.**, Melting and mutual compatibility of oligo(tetramethylene terephthalate)s as models for poly(ether esters), *Makromol. Chem.*, 179, 1373, 1978.

346. **Hässlin, H. W., Droscher, M., and Wegner, G.**, Structure and properties of segmented polyether-esters. III. Synthesis of defined oligomers of poly(buthylene terephthalate), *Makromol. Chem.*, 181, 301, 1980.

347. **Hässlin, H. W. and Droscher, M.**, Structure and properties of segmented poly(ether esters). IV. Compatibility of oligo(tetramethylene terephthalate)s with oligoethers and block copoly(ether ester)s, *Makromol. Chem.*, 181, 2357, 1980.

348. **Schmidt, F. C. and Droscher, M.**, Structure and properties of segmented poly(ether ester)s. VII. Model triblock copolyesters of monodisperse block length, *Makromol. Chem.*, 184, 2669, 1983.

349. **Seliger, H., Bitar, M. B., Nguyen-Trong, H., Marx, A., Roberts, R., Kruger, J. K., and Unruh, H. G.**, Segmented block copolymers of uniform chain length and defined structure. I., *Makromol. Chem.*, 1450, 1984.

350. **Kruger, J. K., Marx, A., Roberts, R., Unruh, H. G., Bitar, M.B., Trong, H. N., and Seliger, H.**, Segmented block copolymers of uniform chain length and defined structure, *Makromol. Chem.*, 185, 1469, 1984.

351. **Percek, V. and Pugh, C.**, Oligomers, in *Encyclopedia of Polymer Science and Engineering*, Vol. 10, 2nd ed, Mark, H. F., Bikales, N. B., Overberger, C. G., and Menges, H., Eds., John Wiley & Sons, New York, 1988, 432.

352. **Ueda, M., Kameyama, A., Hashimoto, K., and Honma, K.**, Chemoselective polyamidation, *Prepr. IUPAC Macro '88*, Kyoto, Japan, 1988, 2.4-2.16, 162.

353. **Watanabe, M., Aoki, S., Sanui, K., and Ogata, N.**, Ionic conduction in polymer electrolytes based on poly[γ-methoxyoligo(ethylene oxide)-L-glutamate]s, *Prepr. IUPAC Macro '88*, Kyoto, Japan, 1988, 3.8-09, 517.

354. **Pedersen, C. J.**, Cyclic polyethers and their complexes with metals salts, *J. Am. Chem. Soc.*, 89, 7017, 1967.

355. **Luttringaus, A. and Irmengard, S.**, Zur Struktur der Losungen. I. Einfluss de Losungmittels auf die Gestalt beweelicher Paraffinketten, ermittelt aus ring Schlussversuchen, *Makromol. Chem.*, 18/19, 511, 1956.

356. **Luttringaus, A.**, Über vielgliedrige Ringsysteme. VIII. Über eine neue Anwendung des verdungs Prinzips, *Ann.*, 528, 155, 1937.

357. **Luttringaus, A.**, Über vielgliedrige Ringsysteme. X. Neuartige Dioxibenzol- und Dioxynaphthalinäther, *Ann.*, 528, 181, 1937.

358. **Luttringaus, A.,** Über vielgliedrige Ringsysteme. XI. Über die Gestalt des Diphenyl- und des Diphenyl-methanmoleküls, *Ann.,* 528, 211, 1937.

359. **Luttringaus, A.,** Über vielgliedrige Rinsysteme. XII. Über den valenz Winkel des Sauerstoffatoms in Derivaten des Diphenylethers, *Ann.,* 528, 223, 1937.

360. **Adams, R. and Whitchill, L. N.,** Many-membered compounds by direct synthesis from two ω,ω'-bi-functional molecules, *J. Am. Chem. Soc.,* 63, 2073, 1941.

361. **Ackman, R. G., Brown, W. H., and Wright, G. I.,** Condensation of methyl ketones with furan, *J. Org. Chem.,* 20, 1955, 1947.

362. **Down, J. L., Lewis, J., Moore, B., and Wilkinson, G. V.,** Solubility of K and Na-K alloy in ethers, *Proc. Chem. Soc.,* 1957, 209.

363. **Down, J. L., Lewis, J., Moore, B., and Wilkinson, G. V.,** The solubility of alkali metals in ethers, *J. Chem. Soc.,* 1959, 3767.

364. **Pedersen, C. J.,** Macrocyclic polyethers for complexing metals, *Aldrich Chimica Acta,* 4, 1, 1971.

365. **Lehn, J. M.,** Cryptates: macropolycyclic inclusion complexes, *Pure Appl. Chem.,* 49, 857, 1977.

366. **Percec, V. and Rodenhouse, R.,** Liquid crystal polyethers containing dibenzo-18-crown-6 structural units, *Polym. Prepr.,* 29(2), 296, 1988.

367. **Lehn, J. M.,** Chemistry of transport processes—design of synthetic carrier molecules, in *Physical Chemistry of Transport Membrane Ion Motion,* Spach, G., Ed., Elsevier, Amsterdam, 1983, 181.

368. **Melson, G. A., Ed.,** *Coordination Chemistry of Macrocyclic Compounds,* Plenum Press, New York, 1979.

369. **Vögtle, F. and Weber, E., Eds.,** *Host Guest Complex Chemistry,* Vol. 3, Current Topics in Chemistry 121, Springer-Verlag, New York, 1984.

370. **Vögtle, F. and Weber, E., Eds.,** *Biomimetic and Bioorganic Chemistry,* Vol. 1-3, Current Topics in Chemistry 128, 132, 136, Springer-Verlag, New York, 1985.

371. **Izatt, R. M. and Christiansen, J. J., Eds.,** *Progress in Macrocyclic Chemistry,* Vol. 3, John Wiley & Sons, New York, 1987.

372. **Medved, S. S., Belova, N. A., Gracheva, L. J., and Kostenko, T. J.,** Basic uses of alkylene-oxide-based crown ethers and their acrylic analogs, *Plast. Massy,* No. 3, 7, 1986 (in Russian).

373. **Maciejewski, M.,** Polymerization and copolymerization of some monomers as adducts with β-cyclodextrins, *J. Macromol. Sci. Chem.,* A13, 77, 1979.

374. **Tomalia, D. A., Baker, H., Dewald, J., Hall, M., Kallos, G., Martin, S., Raeck, J., Ryder, J., and Smith, O.,** A new class of polymers: starburst-dendritic macromolecules, *Polym. J.,* 17, 117, 1985.

375. **Newkome, G. R., Yao, J., Baker, G. R., and Gupta, V. K.,** Cascade molecules. A new approach to micelles a [27]-arborol, *J. Org. Chem.,* 50, 2003, 1985.

376. **Arimura, T., Shinkai, S., and Matsuda, T.,** Synthesis and molecular recognition abilities of functionalized calixarenes, *J. Synth. Org. Chem. Jpn.,* 47, 523, 1989.

377. **Lacavalier, P. R., Eugen, P. T., Shen, Y. X., Yoardar, S., Ward, T., and Gibson, H. W.,** Monomeric and polymeric heterorotaxanes, *ACS Polym. Prepr.,* 30(1), 189, 1989.

378. **Frish, H. L. and Wasserman, E.,** Chemical topology, *J. Am. Chem. Soc.,* 83, 3789, 1961.

379. **Schill, G. and Zollenkopf, H.,** Directed synthesis of catenane compounds. VII. Conversion of triansa compound into a catenane compound, *Chem. Ber.,* 100, 2021, 1967.

380. **Harrison, I. T. and Harrison, S.,** The synthesis of a stable complex of a macrocycle and a threaded chain, *J. Am. Chem. Soc.,* 89, 5723, 1967.

381. **Harrison, I. T.,** The effect of ring size on threading reactions of macrocycles, *J. Chem. Soc. Chem. Commun.,* 231, 1972.

382. **Agam, G. and Zilkha, A.,** Synthesis of a catenane by a statistical double-stage method, *J. Am. Chem. Soc.,* 98, 5214, 1976.

383. **Agam, G., Graiver, D., and Zilkha, A.,** Studies on the formation of topological isomers by statistical methods, *J. Am. Chem. Soc.,* 98, 5206, 1976.

384. **Arai, S.,** Synthesis of a cascade molecule, *J. Synth. Org. Chem. Jpn.,* 47, 62, 1989.

385. **Worthy, W.,** New families of multibranched macromolecules *Chem. Eng. News,* 66, 19, 1988.

386. **Newkome, G. R. and Baker, G. R.,** Two directional cascade molecules: synthesis and characterization of [9]-n[9] arborols, *J. Chem. Soc. Chem. Commun.,* 1986, 752.

387. **Newkome, G. R. and Yao, Z. Q.,** Cascade molecules. Synthesis and characterization of a benzene [9]-arborol, *J. Am. Chem. Soc.,* 108, 849, 1986.

388. **Newkome, G. R. and Gupta, V. K.,** A convenient synthesis of tetrakis(2-bromoethyl)-methane, *J. Org. Chem.,* 52, 5480, 1987.

389. **Tomalia, D. A., Baker, H., and Dewald, J.,** Dendritic macromolecules. Synthesis of starburst dendrimers, *Macromolecules,* 19, 2466, 1986.

390. **Tomalia, D. A. and Dewald, J. R.,** U.S. Patent 4,507,466, 1985.

391. **Tomalia, D. A. and Dewald, J. R.,** U.S. Patent 4,558,120, 1985.

392. **Tomalia, D. A. and Dewald, J. R.,** U.S. Patent 4,568,737, 1986.

393. **Tomalia, D. A. and Dewald, J. R.,** U.S. Patent 4,587,329, 1986.

394. **Tomalia, D. A. and Dewald, J. R.,** U.S. Patent 4,631,337, 1986.
395. **Wilson, L. R. and Tomalia, D. A.,** Synthesis and characterization of starburst dendrimers, *ACS Polym. Prepr.,* 30(1), 115, 1989.
396. **Padias, A. B. and Hall, H. K.,** Starburst polyether dendrimers, *J. Org Chem.,* 52, 5305, 1987.
397. **Padias, A. B., Hall, H. K., Jr., and Tomalia, D. A.,** Starburst polyethers, *ACS Polym. Prepr.,* 30(1), 119, 1989.
398. **Meltzer, D. A., Tirrell, D. A., Jones, A. A., Inglefeld, P. T., Downing, D. M., and Tomalia, D. A.,** ^{13}C NMR relaxation in poly(amido amine) starburst dendrimers, *ACS Polym. Prepr.,* 30(1), 121, 1989.
399. **Smith, B. P., Martin, J. S., Hall, J. M., and Tomalia, D. A.,** in *Applied Polymer Analysis and Characterization,* Mitchell, J. Jr., Ed., Hanser, New York, 1986, 375.
400. **Arimura, T., Shinkai, S., Matsuda, T., Hirata, Y., Satoh, H., and Manabe, O.,** Fries rearrangement in calixarene esters: a new entry for the synthesis of p-substituted calixarenes, *Bull. Chem. Soc. Jpn.,* 61, 3733, 1988.
401. **Shinkai, S., Horeishi, H., Veda, K., and Manabe, O.,** A new hexacarboxylate uranophile derived from calix[6]arene, *J. Chem. Soc. Chem. Commun.,* 1987, 233.
402. **Shinkai, S., Koreishi, H., Ueda, K., Arimura, T., and Manabe, O.,** Molecular design of calixarene-based uranophiles which exhibit remarkably high stability and selectivity, *J. Am. Chem. Soc.,* 109, 6371, 1987.
403. **Shinkai, S., Mori, S., Koreishi, H., Tsubaki, T., and Manabe, O.,** Hexasulfonated calix[6]arene derivatives: a new class of catalysts, surfactants and host molecules, *J. Am. Chem. Soc.,* 108, 2409, 1986.
404. **Shinkai, S., Tsubaki, T., Sone, T., and Manabe, O.,** A new synthesis of p-nitrocalix[6]arene, *Tetrahedron Lett.,* 26, 3343, 1985.
405. **Shinkai, S., Arimura, H., Satoh, H., and Manabe, O.,** Chiral calixarene, *J. Chem. Soc. Chem. Commun.,* 1495, 1987.
406. **Arimura, T., Edamitsu, E., Shinkai, S., and Manabe, O.,** Calixarenes, *Chem. Lett.,* 2269, 1987.
407. **Katz, T. J., Sudhakar, A., Yang, B., and Nowick, J. S.,** Synthesis of conjugated optically active polymetallocenes, *J. Macromol. Sci. Chem.,* A26, 309, 1989.
408. **Stoddart, F.,** Molecular lego, *Chem. Br.,* 24, 1203, 1988.
409. **Katz, T. J., Rainmuth, W. H., and Smith, D. E.,** Electrolytic reduction of cyclooctatetraene, *J. Am. Chem. Soc.,* 84, 802, 1962.
410. **Katz, T. J. and Acton, N.,** Bis(pentalenylnickel), *J. Am. Chem. Soc.,* 94, 3281, 1972.
411. **Katz, T. J., Acton, N., and McGinnis, J.,** Sandwiches of iron and cobalt with pentalene, *J. Am. Chem. Soc.,* 94, 6205, 1972.
412. **Stoddart, J. F.,** Molecular lego, *Nature (London),* 334, 10, 1988.
413. **Pedersen, C. J.,** Crown ethers, *Angew. Chem. Int. Ed. Engl.,* 27, 1021, 1988.
414. **Diederich, F. and Staab, H. A.,** Benzoide versus annulenoide Aromazität: Synthese und Ligenschaften des Kekulens, *Angew. Chem.,* 90, 383, 1978.
415. **Kivelson, S. and Chapman, O. L.,** Polyacene and a new class of quasi-one-dimensional conductors, *Phys. Rev. B, Condens. Mater.,* 28, 7236, 1983.
416. **Alder, R. W. and Sessions, R. B.,** Force field calculations on molecular belts built from cyclohexa-1,4-diene rings, *J. Chem. Soc. Perkin Trans.,* 2, 1849, 1985.
417. **Kohnke, F. H.,** Ein hexaepoxy octacosanhydro 12 Cyclacen, *Angew. Chem.,* 99, 947, 1987.
418. **Kohnke, F. H.,** *Angew. Chem. Int. Ed. Engl.,* 26, 892, 1987.
419. **Aston, P. R., Isaacs, N., Kohnke, F. H., Slawin, A., Spencer, C. M., Stoddart, J. F., and Williams, D. J.,** En route to 12 collarene, *Angew. Chem.,* 100, 981, 1988.
420. **Ellwood, P., Mathias, J. P., Stoddart, J. F., and Kohnke, F. H.,** Stereoelectronically programmed molecular "Lego" sets, *Bull. Soc. Chim. Belg.,* 97, 669, 1988.

Chapter 1.3

IONIC OLIGOMERIZATION

I. BASIC PRINCIPLES OF IONIC POLYMERIZATION

Ionic polymerizations are propagated by growing macroions. A distinction can be made between cationic encatenation

$$\text{wwwM}_n^+ + M \rightleftharpoons \text{wwwM}_{n+1}^+ \tag{1}$$

and anionic encatenation

$$\text{wwwM}_n^- + M \rightleftharpoons \text{wwwM}_{n+1}^- \tag{2}$$

In real systems, however, more than one type of growing species generally occurs, since the ions can exist as free ions, ion pairs, or ionic associates of three, four, or more ions. A rapid dynamic equilibrium often occurs between two or more of the ionic forms:[1]

$$R\text{--}X \rightleftharpoons R^{\delta+} - X^{\delta-} \rightleftharpoons R^+X^- \rightleftharpoons R^+/S/X^- \rightleftharpoons R^+ + X^- \tag{3}$$

	contact ion pairs	solvent-separated ion pairs		free ions

polarization	ionization	dissociation

Additionally, free ions, solvated ion pairs, and contact ion pairs can also be in equilibrium with the corresponding ionic associates. For example:

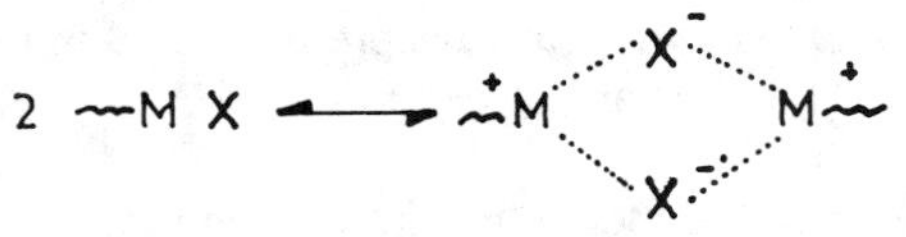

$$2 \ \text{~M X} \rightleftharpoons \text{~M}^+ \begin{matrix} X^- \\ \diamond \\ X^- \end{matrix} M^+\text{~} \tag{4}$$

The overall driving force for the ionic polymerization of olefinic monomers is the same as that in free radical processes, i.e., the reduction in free energy associated with the loss of unsaturation in forming polymeric chains. This fall in free energy arises from a large negative enthalpy change, which more than compensates for the accompanying unfavorable entropy change. With cyclic monomers, the thermodynamic driving force arises from the relief of ring strain in forming linear chains, although there is a smaller enthalpy change (negative) and entropy change (positive).[2]

Both modes of ionic polymerization are described by the same vocabulary as the corresponding steps in the free radical mechanism for chain-growth polymerization. However, initiation, propagation, transfer, and termination are quite different in many ways between anionic and cationic mechanisms.

Moreover, the ionic ring-opening polymerization is a distinct category of polymerization, sharing common features with both step-growth and addition polymerizaton.[3] For example, step-growth polymer chains also continue to grow through the entire duration of the ring-opening reaction. In step-growth polymerization, a reaction between groups on any molecular species is possible; only monomers react with growing chains in ring-opening polymerizations. The latter feature also characterizes addition polymerization.

Initiation of ionic polymerization usually involves the transfer of an ion or an electron to or from the monomer, with the formation of an ion pair. It is thought that the counterion, or *gegenion,* of this pair stays close to the growing chain end throughout its lifetime, particularly in media of low dielectric constant. The actual polymerization-starting species reacts with the monomer in the start step. Two cases can be distinguished: (1) transfer of two electrons with the formation of a bond between the polymerization starter and the monomer molecule and (2) transfer of one electron without formation of a bond.[1] In the two-electron mechanism, the starting species is always joined to the monomer unit:

1.   A cation adds on to a monomer molecule and forms a monomer cation:

$$H^+ \ + \ CH_2{=}C(CH_3)_2 \rightarrow H{-}CH_2\overset{+}{C}(CH_3)_2 \tag{5}$$

2.   An anion adds on to a monomer molecule and forms a monomer anion:

$$R^- \ + \ CH_2{=}\underset{\underset{C_6H_5}{|}}{CH} \ \rightarrow R{-}CH_2{-}\underset{\underset{C_6H_5}{|}}{\bar{C}H} \tag{6}$$

3.   A neutral molecule adds on to a monomer molecule and forms a zwitterion:

$$R_3N \ + \ \underset{\underset{CH_2{-}O}{|\quad|}}{CH_2{-}CO} \rightarrow R_3\overset{+}{N}CH_2CH_2COO^- \tag{7}$$

The *two-electron* transfer is not a transfer reaction in terms of macromolecular terminology, where the concept of *transfer* is reserved for a reaction where an active species reacts with another molecule, with the formation of an active species from this molecule.

On the other hand, the *one-electron* transfer is concerned with a genuine electron transfer reaction. In the first step, a radical ion is always produced and in the second, a dimerization to diions generally occurs. The mechanism can imply:

1.   An electron transfer from a donor, e.g., from an anion, to the monomer molecule:

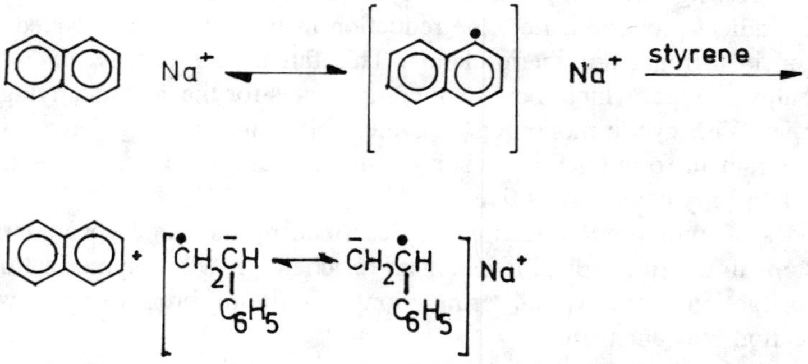

$$ \tag{8}$$

The radical anion dimerizes:

$$2\underset{\underset{C_6H_5}{|}}{\overset{\bullet}{C}H_2{-}\bar{C}H} \ \rightarrow \underset{\underset{C_6H_5}{|}}{\bar{C}H}{-}\underset{\underset{C_6H_5}{|}}{CH_2{-}CH_2}{-}\underset{\underset{C_6H_5}{|}}{\bar{C}H} \tag{9}$$

2. An electron transfer from a monomer to an acceptor, e.g.,

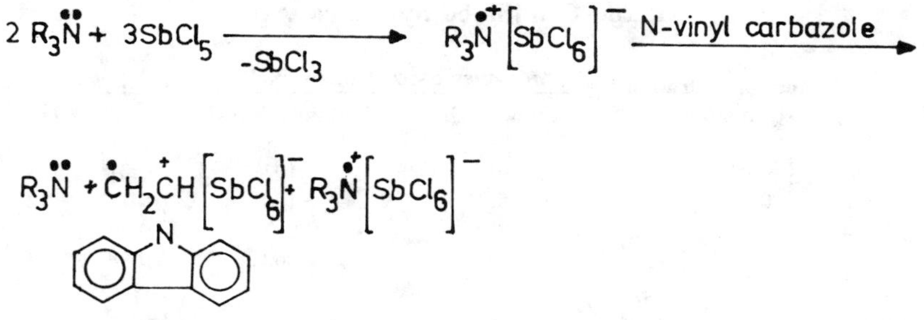

$$2\ R_3\ddot{N} + 3SbCl_5 \xrightarrow[-SbCl_3]{} R_3\overset{\bullet+}{N}\left[SbCl_6\right]^- \xrightarrow{\text{N-vinyl carbazole}}$$

$$R_3\ddot{N} + \overset{\bullet}{C}H_2\overset{+}{C}H\left[SbCl_6\right]^- + R_3\overset{\bullet+}{N}\left[SbCl_6\right]^- \tag{10}$$

where R is an electrodonating radical, e.g., Br–C$_6$H$_4$–.

The monomeric radical cation presumably dimerizes to the di(N-vinylcarbazole) cation.

In the two-electron mechanism, macroions are produced for the start reaction; with the one-electron mechanism, there are diions. The polymer chain grows by a further addition of monomer molecules to these monoions or diions.

The equilibrium

$$\text{wwww}M_i^* + M \leftrightarrow \text{wwww}M_{i+1}^* \tag{11}$$

between the addition to and elimination from an active growing chain $\text{ww}M_n^*$ of a monomer M is independent of the reaction path. It is therefore also independent of the mechanism and, hence, the chemical nature of the growing end (anion, cation, or radical). The equilibrium constant of the reaction is given by

$$K_e = \frac{[\text{wwww}M_{i+1}]_e}{[M]_e[M_i^*]_e} \tag{12}$$

If the degree of polymerization and therefore i and the molecular weight are very large, then the concentrations of growing chains i and i + 1 will be practically identical

$$[\text{wwww}M_i^*] \simeq [\text{wwww}M_{i+1}^*] \tag{13}$$

and consequently

$$K_e = 1/[M]_e \tag{14}$$

The obtaining of high molecular polymers normally requires low concentrations of initiator. However, with higher initiator concentrations, both the degree of polymerization (\overline{DP}) and the monomer concentration at equilibrium decrease (Table 1). This effect, observed because the assumption from equation 13 no longer applies, is taken as an advantage for the preparation of low molecular polymers (oligomers). In this case, the equilibrium reaction between initiator and monomer begins to assume a major role.

In the absence of transfer and termination reactions (i.e., living polymerization) and as long as the initiator is not incorporated into the chain, the number of monomer units per initiator molecule, $\overline{X}_n$, is at equilibrium

$$\overline{X}_n = j\ \frac{[M]_o - [M]_e}{[I]_o - [I]_e} \tag{15}$$

where j = 1 for growing macroions and i = 2 for growing diions.

TABLE 1
Equilibrium Bulk Polymerization of Lauryl Lactam (M = 197)
at 280°C Initiated by Lauric Acid[1]

Initiator concentration $[I]_o$, mol/kg	Monomer concentration, mol/kg		$\dfrac{[I]_o}{[M]_o}$	$\overline{DP}$
	Initial $[M]_o$	Equilibrium $[M]_e$		
0.049	5.010	0.107	0.01	103
0.499	4.561	0.100	0.11	12
1.248	3.801	0.082	0.33	5
2.496	2.534	0.062	0.98	3

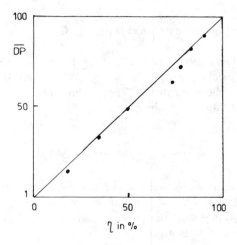

FIGURE 1.   Dependence of the degree of polymerization on the yield of the polymerization of anhydrous lauryl lactam at 280°C initiated by lauric acid ($[I]_o/[M]_o = 0.01$). The straight line corresponds to Equation 18, with $j = 1$ and $[I]_o \gg [I]$. (Adapted from Elias, H. G., *Macromolecules. Synthesis and Materials,* Vol. 2, Plenum Press, New York, chap. 18.)

If initiator fragments are incorporated intò the chain, then the degree of polymerization is given as the number of monomer units per molecule plus the initiator unit at any given yield, $\eta < 100$. If equilibrium has not been reached, then $[M]_e$ and $[I]_c$ must be replaced by the instantaneous concentrations $[M]$ and $[I]$, respectively.

$$(\overline{DP})_\eta = \overline{X}_n + 1 \tag{16}$$

where

$$\eta = ([M]_o - [M])/[M]_o \tag{17}$$

Combining Equations 15 to 17, one obtains the instaneous degree of polymerization:

$$(\overline{DP})_\eta = 1 + j\,\frac{M_o}{[I]_o - I} \tag{18}$$

Thus, a plot of $(\overline{DP})_n$ against η gives a straight line with ordinate intercept DP $= 1$ and a slope of $j\,M_o/([I]_o - [I])$ independent of yield. This is shown in Figure 1 for the polymerization of lauryl lactam by lauric acid as initiator. Even if the initiator concentration in this case is low ($[I]_o/[M]_o = 0.01$), high molecular polymers are not formed until $\eta \geq 50\%$, due to the step-wise character of the polymerization process. Therefore, high initiator concentrations and low yields favor — in the absence of transfer reactions — formation of oligomers as main polymeric species.

If new monomer is added to such a polymerization without termination and transfer reactons after achievement of equilibrium, both the yield and the molecular weight increase further. If, however, transfer reactions occur, the yield increases but the molecular weight does not necessarily increase.[1]

Under ideal conditions, the number of growing macromolecules, N_p, coincides with the amount of initiator, $[I]_o$, and the molecular mass of the polymer can be presented as:

$$\overline{M}_n = \frac{g_m}{[I]_o} \tag{19}$$

where g_m is the mass of monomer polymerized. In reality, some impurities are always present in the polymerization system and they can play the role of accidental initiators or deactivators, yielding functionless chains. In this case, the number-average molecular mass can be given as follows:

$$\overline{M}_n = \frac{g_m}{N_p} = \frac{g_m}{[I]_o + \Sigma N_i} \tag{20}$$

Here, ΣN_i is the cumulative amount (in moles) of "active" impurities in the system, introduced deliberately or not (i.e., initiator, transfer agents, or deactivators).

Transformation of Equation 20 yields the balance Equation 21:

$$N_p = \frac{g_m}{\overline{M}_n} = [I]_o + \Sigma N_i \tag{21}$$

which then leads to the expression for a number-average functionality in ionic polymerization:

$$\overline{f}_n = \frac{[I]_o}{N_p} = 1 - \frac{\Sigma N_i}{N_p} = 1 - \frac{\Sigma N_i}{g_m} \cdot \overline{M}_n = 1 - a\overline{M}_n \tag{22}$$

where $1/a = \Sigma N_i$ has a meaning of limiting the molecular weight, M_{lim}, which can be reached in the given system. On this basis, it can be predicted that the higher the molecular weight expected, the more difficult it is to attain the quantitative functionalization, unless it is not the result of the enchainment mechanism.

II. CATIONIC OLIGOMERIZATION

Cationic oligomerization is the conversion of monomeric molecules into low molecular polymeric ones (oligomers) via a mechanism involving stepwise growth of a carbenium (R_3C^+), carboxonium ($R^+O{=}C{=}$), oxonium ($R_3{}^+O$), sulfonium (R_3S^+), immonium ($R_2{}^+N{=}$), or phosphonium ($R_2{}^+P{=}$) ion. Both unsaturated and cyclic monomers are susceptible to oligomerization by such intermediates, e.g.:

$$nCH_2{=}CHY \xrightarrow{K^+} \text{www}C^+ \rightarrow -(-CH_2{-}CHY{-})_n{-} \tag{23}$$

and

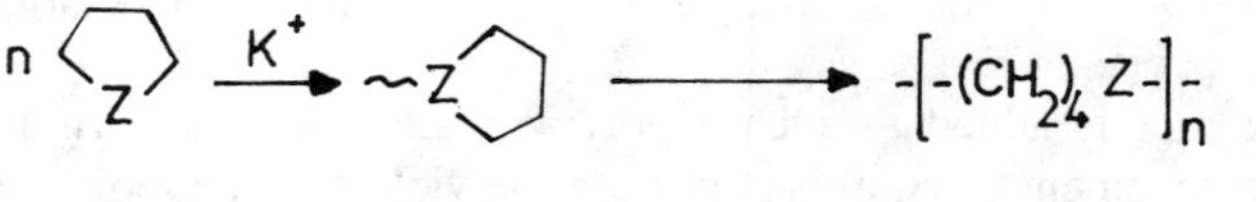

$$n \left\langle \underset{Z}{\bigcirc} \right\rangle \xrightarrow{K^+} \sim Z \bigcirc \longrightarrow -\!\left[-(CH_2)_4 Z-\right]_n^- \tag{24}$$

where K^+ represents the cationic fragment of the initiator and Y is a heteroatom.

With olefinic monomers, the substituent R is often electron donating and formally resides on the α-carbon atom, e.g.:

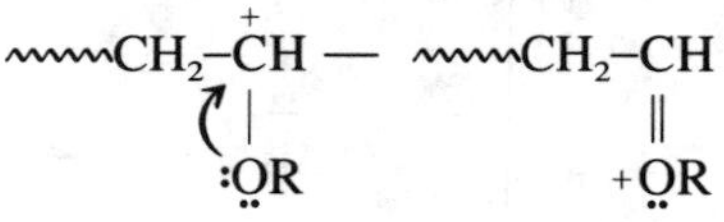

$$\sim\!\!\sim\!\!\sim CH_2-\overset{+}{C}H \ — \ \sim\!\!\sim\!\!\sim CH_2-CH \tag{25}$$
$$\qquad\quad :\underset{\cdot\cdot}{O}R \qquad\qquad\qquad +\underset{\cdot\cdot}{O}R$$

In geminally disubstituted olefins, e.g., $CH_2{=}CRR'$, the cation is easily stabilized through the inductive effects of alkyl substituents:

$$H^+ \ + \ CH_2{=}CRR' \ — \ CH_3 \rightarrow \quad \begin{array}{c} R \\ \downarrow \\ C \\ \uparrow \\ R' \end{array} + \tag{26}$$

As far as cyclic monomers are concerned, the propagating intermediate is also stabilized, since most of the positive charge resides on the basic heteroatom, e.g.,

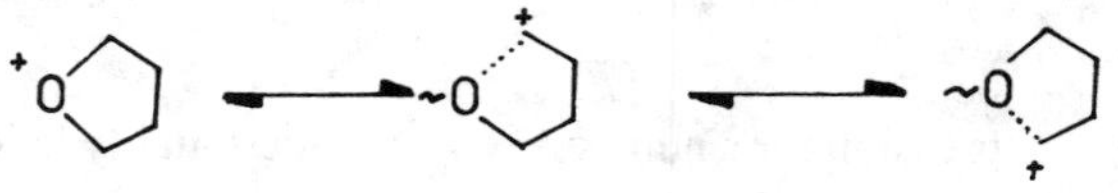

$$\overset{+}{O}\bigcirc \ \longleftrightarrow \ \sim\!\!O\overset{+}{\bigcirc} \ \longleftrightarrow \ \sim\!\!O\bigcirc_{+} \tag{27}$$

Except in very special circumstances, e.g., polymerization induced by γ-radiation, electroneutrality is maintained in cationic and anionic polymerizations by the presence of negatively and positively charged counterions or gegenions, respectively. This species has no analogue in free radical polymerization and its physical relationship with the growing ion is of vital importance in the interpretation of kinetic data.

All cationic oligomerizations (polymerizations) must first involve the generation of a positively charged species and an accompanying counterion, viz.:

$$AB \rightarrow A^+B^- \tag{28}$$

In the case of chemical initiation, as opposed to initiation by ionizing radiation *(vide infra)*, this process can be carried out *in situ* in the polymerization mixture (e.g., using iodine or Lewis acids) or, in special cases (referring especially to cyclic monomers), it may be undertaken externally by producing a stable or transient salt, such as a preformed carbonium, oxonium, or acylium ion.

Metal salts of strong acids with anions derived from either Brønsted or Lewis acids, or from both, are also active promoters of cationic polymerizations[4] and excellent catalysts for electrophilic reactions.[5] Diazonium, diaryliodonium, triarylsulfonium, and triarylselenonium salts very actively initiate cationic polymerization of virtually all known classes of cationically polymerizable monomers in the presence of light, while they possess extraordinary latency in its absence.[6,7]

Protonic acids also fall within the definition of preformed initiators.[2] Since, on irradiation with ionizing radiation (e.g., γ-rays from a ^{60}Co source), an electron is presumably ejected from a suitable liquid monomer with the generation of a radical cation which can then propagate (a process which constitutes the *initiation reaction* of a cationic encatenation, all preinitiation phenomena being absent):[8,9]

$$CH_2=CHR \xrightarrow{k_i} \overset{\cdot}{C}H_2-\overset{+}{C}HR + \bar{e} \qquad (29)$$
$$\uparrow$$
$$\gamma\text{-ray}$$

some generalizations were introduced into the cationic reaction mechanism:[2]

$$A^+B^- + monomer \xrightarrow{k_i} monomer^+X^- \qquad (30)$$

Conceptually, any chemical that can be cationated and in turn is able to attack another molecule of the original chemical is a cationically polymerizable monomer, i.e., a *cationic monomer*.[10] Cationic polymerizability necessarily implies that the cationated species will remain sufficiently reactive and will sustain propagation. For example, in vinylamine, cationic attack occurs readily but propagation does not proceed because of the high stability of the species formed:

$$CH_2=CH \xrightarrow{H^+} CH_2-\overset{+}{C}H \rightleftharpoons CH_2-CH \rightleftharpoons CH_2=CH \qquad (31)$$

Sometimes, even if the monomer can readily be cationated, the resulting carbenium ions are severely hindered and are unable to propagate because of steric compression in the transition state:

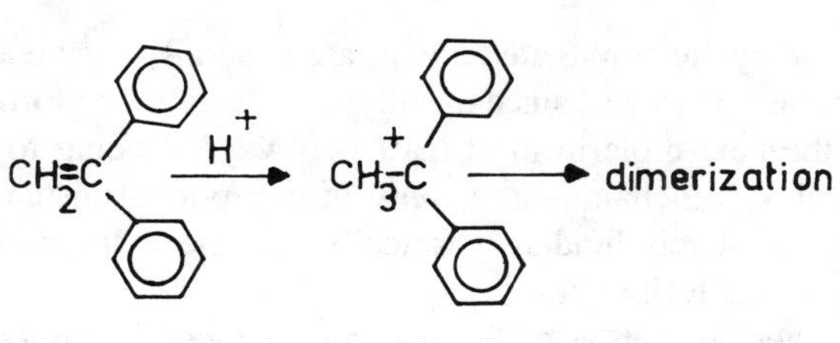

$$CH_2=C \xrightarrow{H^+} CH_3-\overset{+}{C} \longrightarrow \text{dimerization} \qquad (32)$$

and

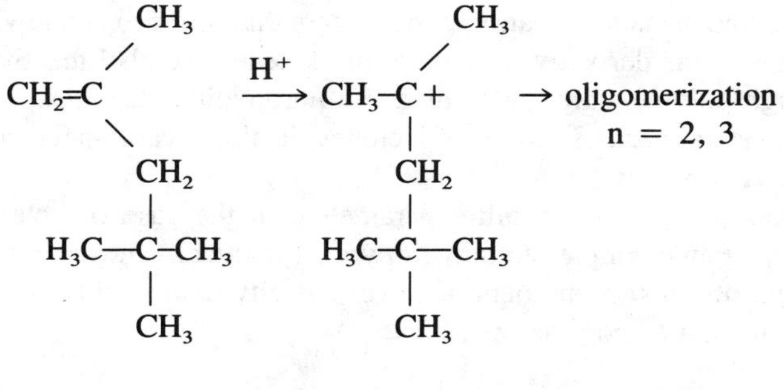

$$\longrightarrow \text{oligomerization} \qquad (33)$$
$$n = 2, 3$$

Similarly, 1,2 disubstitution also reduces cationic vinyl polymerizability. However, if a mechanism is available by which the sterically unfavorable carbenium ion can be avoided, propagation may proceed to high polymers. Carbenium ions are inherently unstable species and are prone to rearrange to a more stable structure. For example, while encatenation of methylene cyclohexane results only in oligomers (viscous oils) since the tertiary cation is sterically unable to propagate:

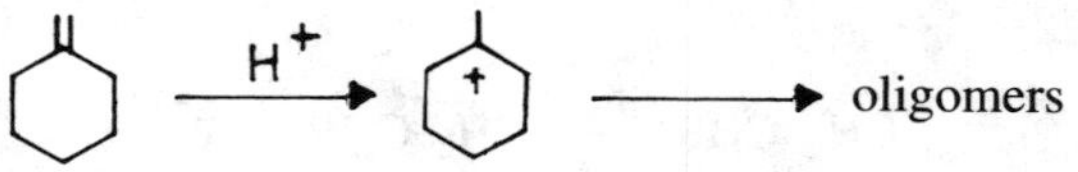

$$\text{(structure)} \xrightarrow{H^+} \text{(structure)} \longrightarrow \text{oligomers} \qquad (34)$$

that of α-pinene succeeds in forming solid polymers, since the initial attack on the exo-methylene group giving rise to a "buried" cyclic carbenium ion is followed by a rearrangement to a sterically favorable propagating cation:[10]

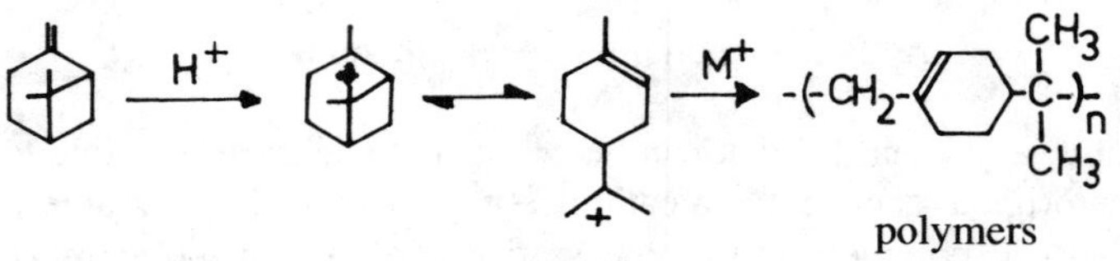

$$\text{(structures)} \qquad (35)$$

polymers

The classical example of a controlled isomerization polymerization (i.e., by hydride migration to form an energetically more favorable tertiary ion, the true propagating site) is the case of 3-methyl-1-butene:[11,12]

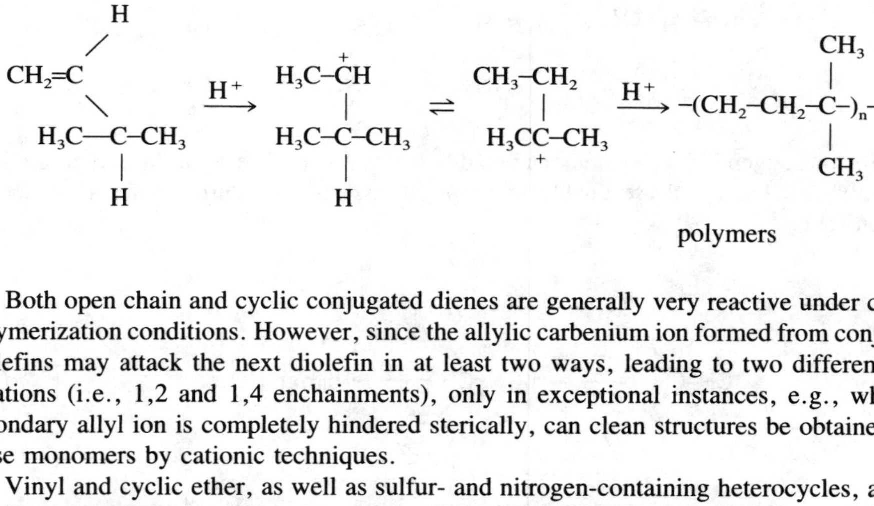

$$\text{(structures)} \qquad (36)$$

polymers

Both open chain and cyclic conjugated dienes are generally very reactive under cationic polymerization conditions. However, since the allylic carbenium ion formed from conjugated diolefins may attack the next diolefin in at least two ways, leading to two different encatenations (i.e., 1,2 and 1,4 enchainments), only in exceptional instances, e.g., when the secondary allyl ion is completely hindered sterically, can clean structures be obtained from these monomers by cationic techniques.

Vinyl and cyclic ether, as well as sulfur- and nitrogen-containing heterocycles, are also readily polymerized by cations due to the high stability of the carbenium ions, resonance stabilized by the unshared pair of electrons on the neighboring heteroatom (Equation 3).

Cyclic formals and acetals are cationic monomers that also propagate via oxonium ions. Aliphatic aldehydes, their derivatives, and simple ketenes are also reactive under cationic conditions, but high polymers are not formed from aromatic aldehydes.

Among cyclic esters, various groups of lactones, lactides, carbonates, and oxalates have been polymerized cationically.

The most important polymerizability parameters in the case of heterocycles are ring strain and basicity. For example, tetrahydropyran (practically no ring strain) and alkyl-substituted tetrahydrofurans (monomers of lower basicity than readily polymerizable tetrahydrofuran) are reluctant to polymerize ($\Delta E \geq 0$).

It now has become more and more evident that cationic polymerization of olefin (vinyl) and heterocyclic monomers may lead to polymers containing more or less considerable amounts of linear and cyclic monomers. In many cases, the oligomers are the main reaction products. If the polymers are the desired end products, oligomer formation must be regarded as an undesired side reaction. Therefore, it is technologically useful to know by which mechanism these oligomers are formed in order to be able to conduct the process under such conditions that this side reaction is minimized. On the other hand, it may be equally interesting to influence the reaction in such a manner that oligomers become the main reaction products. This is especially the case when telechelic or macrocyclic oligomers are formed.

A. CATIONIC OLIGOMERIZATION OF HYDROCARBONS
1. Olefins and Dienes

In the history of cationic polymerization, the year 1931 marks the patented discovery of Otto and Müller-Cunradi[13] that, at low temperature (below $-10°C$), isobutylene can be polymerized by the use of BF_3 or its complexes to high molecular weight solids. This landmark patent provided the basis for the entire polyisobutylene industry, an industry of great contemporary significance.[10] The highest molecular weight polymer was obtained later by the same authors[14] by carrying out the polymerization of purified isobutylene at very low temperature ($-100°C$), i.e., the practical recognition at that time that low temperatures "freeze out" chain transfer in cationic processes.

It is well known that the activation energy of cationic polymerizations is lower than 15 kcal/mol and, in some cases, even negative.[15] this is why they can proceed even at extremely low temperatures.[16] The high polymerization rate, e.g., that observed for isobutylene in CH_2Cl_2 with $AlCl_3$ coinitiated with ethyl chloride, cannot be reduced by lowering the temperature even to $-180°C$.[17]

As a result of a large number of fundamental studies, many elements of carbocationic polymerizations have been elucidated and thus became controllable, i.e., only three elementary events should be taken into consideration for cationic macromolecular engineering:[18,19]

1. *Ion generation (priming)*

$$RX + MtX_n \longrightarrow R^+MtX_{n+1}^- \tag{37}$$

2. *Cationation*

$$R^+ + M \longrightarrow RM^+ \tag{38}$$

and propagation

$$RM^+ \xrightarrow{nM} R(M)_nM^+ \tag{39}$$

3. *Termination*

$$R(M)_nM^+MtX_{n+1}^- \rightarrow R(M)_{n+1}X + MtX_n \tag{40}$$

where RX = certain halogenated organic species that can be small molecules or polymers (oligomers), MtX = certain metal halides, i.e., Friedel-Crafts acids, and M = monomer. Due to the similarity between cationation and propagation, these two processes are viewed as a single step. Catiogens (H_2O, RX) are the initiators and Lewis acids (Et_2AlCl, BCl_3) are the coinitiators. Termination is the complete and irreversible destruction of the kinetic chain and should be carefully differentiated from chain transfer.[20]

By a most versatile method for the control of chain transfer to monomer in carbocationic enchainments, it has been possible to obtain liquid telechelic prepolymers.[21] These terminally functional oligomers can be converted to high polymers by a variety of techniques, e.g., reaction injection molding (RIM), and thus may be processed by most efficient low-shear equipment.

Oligomeric and telechelic polyisobutylenes are of great interest for the manufacture of many specialty polymers where the unique combination of physical-mechanical properties of polyisobutylenes (e.g., outstanding environmental stability, low-temperature properties, barrier characteristics, high strength and modulus, etc.) is of value.[22,23] For example, polyurethanes made with polyisobutylene-diol soft segments exhibit unexpectedly good high-temperature characteristics and hydrolytic stability *(vide infra),* ω-monofunctional oligomers have fine applications as oil additives, and ω,ω'-difunctional oligomers are used as solid rocket propellant binders or to prepare rubbery materials by cross-linking reactions through the end functions.

A most important requirement of telechelic prepolymers is their perfect terminal functionality, i.e., the number-average terminal functionality $\bar{F}_n$ of linear telechelic must be exactly 2.0, or that of three-arm star telechelics 3.0. Since the high polymers are prepared from telechelic prepolymers by various stoichiometric end-linking techniques, if the $\bar{F}_n$ is not a well-defined whole number, efficient end linking is impossible to achieve. Further, to obtain the best physical properties by means of end linking, the functional groups must be at the chain ends only and not randomly distributed along the chain. End linking is extremely sensitive to accurate end-group stoichiometry;[24] for example, the degree of extension attainable with $\bar{F}_n$ = 1.80 or 1.90 is only 10 or 20, respectively. High-quality networks can only be prepared by the use of prepolymers having $\bar{F}_n$ as an integer (2, 3 or higher whole numbers).[21]

Synthesis of perfectly telechelic (i.e., with $\bar{F}_n$ = 2.0 or 3.0) prepolymers became possible with the cationic *inifer technique.*[19,21,25] In this technique, chain transfer to special chain transfer agents is forced to occur and if the rate of this step, $R_{tr,I}$, can be increased over that of chain transfer to monomer, $R_{tr,M'}$, the latter may be entirely avoided. The nature of polymer end groups can be controlled by the use of certain chain-transfer agents.

The principle of this method is that the mechanisms of initiation by organic halides and that of chain transfer to organic halides are essentially identical.[26,27] It was found that by the use of specific organic halides, both initiation and chain transfer to the initiator could be accomplished (hence the term *inifer*). In the simplest case, the following set of reactions is involved:[28]

$$RCl + BCl_3 \rightleftharpoons R^+ B\bar{Cl}_4 \tag{41}$$

$$R^+ + \,C{=}C \longrightarrow R{-}\overset{|}{C}{-}\overset{|}{C}{}^+ \tag{42}$$

$$R{-}\overset{|}{\underset{|}{C}}{-}\overset{|}{\underset{|}{C}}{}^+ \!+ nC{=}C \longrightarrow R(\overset{|}{\underset{|}{C}}{-}\overset{|}{\underset{|}{C}})_n \overset{|}{\underset{|}{C}}{-}\overset{|}{\underset{|}{C}}{}^+ \tag{43}$$

$$R{-}(\overset{|}{\underset{|}{C}}{-}\overset{|}{\underset{|}{C}}){-}\overset{|}{\underset{|}{C}}{-}\overset{|}{\underset{|}{C}}{}^+ + RCl \longrightarrow R(\overset{|}{\underset{|}{C}}{-}\overset{|}{\underset{|}{C}}){-}\overset{|}{\underset{|}{C}}{-}\overset{|}{\underset{|}{C}}Cl + R^+ \tag{44}$$

$$R{-}(\overset{|}{\underset{|}{C}}\overset{|}{\underset{|}{C}})_n {-}\overset{|}{\underset{|}{C}}\overset{|}{\underset{|}{C}}{}^+ B\bar{Cl}_4 \longrightarrow R{-}(\overset{|}{\underset{|}{C}}\overset{|}{\underset{|}{C}})_n \overset{|}{\underset{|}{C}}{-}Cl + BCl_3 \tag{45}$$

TABLE 2
Relation Between Inifers, Number-Average Functionality, and Molecular Weight Polydispersity[29]

No.	Inifer	Product	$\overline{F}_n$	$\overline{M}_w/\overline{M}_n$
1	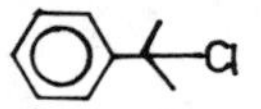	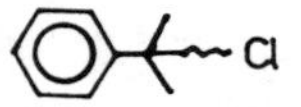	1.0	2.0
2	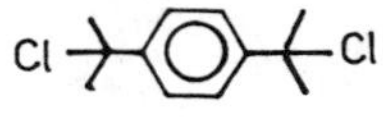	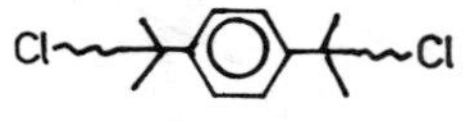	2.0	1.5
3	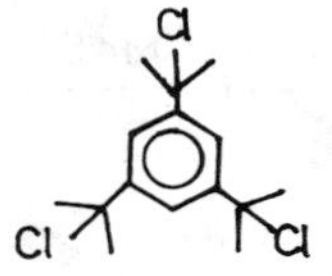	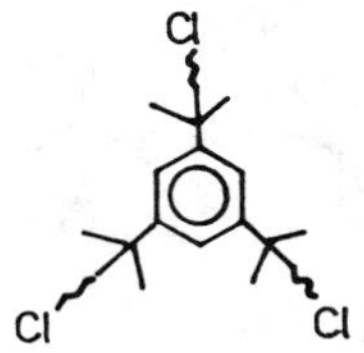	3.0	1.33

In this scheme, the inifer RX is a monofunctional halide, e.g., cumyl chloride, and the monomer $\diagup$C═C$\diagdown$ is isobutylene. The product is α-phenyl-ω-*tert*-chloroisobutylene. Inifers can be mono-, di-, tri-, or multifunctional halides (i.e., minifers, binifers, trinifers, etc.),[29] as shown in Table 2.

However, the earliest report of the deliberate exploitation of a similar principle for the synthesis of telechelics from 4-methyl-pentene-1 was from Ver Strate and Baldwin.[30,31] The bulk of the polymerization initiated by $AlBr_3$ at $-80°C$ has been directed toward obtaining a product with an $\overline{M}_n$ in the range of 1500 to 3500 with a number-average functionality near 2 and a weight-average functionality not greater than 2.

Quantitative derivatization of asymmetric and symmetric end-reactive prepolymers obtained by the inifer technique gave rise to a whole new family of unique materials,[19] representative members of which are presented below.

Chlorine telechelic polyisobutylenes (PIBs) has been used as initiators in conjunction with alkylaluminum coinitiators or certain silver salts ($AgSbF_6$, $AgBF_4$) for the preparation of two- and three-arm stars and radial block copolymers containing the PBI soft segment and a variety of other segments, e.g., polystyrene, poly(α-methylstyrene), and poly(tetrahydrofuran):[27,32]

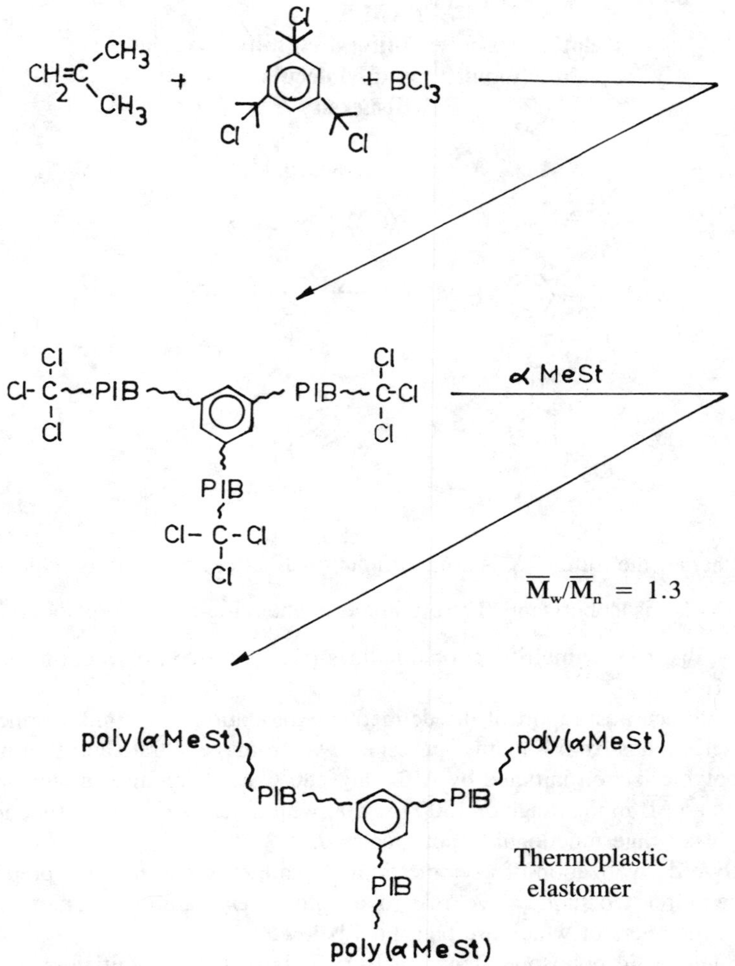

$$CH_2=C\begin{smallmatrix}CH_3\\CH_3\end{smallmatrix} \quad + \quad [\text{ring with 3 }CCl_3] \quad + \quad BCl_3 \longrightarrow$$

$$Cl-\underset{\underset{Cl}{|}}{\overset{\overset{Cl}{|}}{C}}\sim PIB\sim\!\!\!\!\!\!\bigcirc\!\!\!\!\!\!\sim PIB\sim\underset{\underset{Cl}{|}}{\overset{\overset{Cl}{|}}{C}}-Cl \qquad \xrightarrow{\quad \alpha \text{ MeSt} \quad}$$

$$\begin{array}{c}PIB\\\wr\\Cl-\underset{\underset{Cl}{|}}{C}-Cl\end{array}$$

(46)

$$\overline{M}_w/\overline{M}_n = 1.3$$

poly(αMeSt)   poly(αMeSt)

$$PIB\sim\!\!\!\bigcirc\!\!\!\sim PIB$$

(47)

PIB

Thermoplastic
elastomer

poly(αMeSt)

Phenol telechelic polyisobutylenes were prepared by quantitative Friedel-Crafts alkylation of phenol by the linear or three-arm star *tert*-chlorine telechelic PIB.[33,34]

New polycarbonates or polysulfones and epoxy resins with a built-in elastomer segment were then synthesized by reacting these phenol-telechelic oligomers with their corresponding reagents, i.e., phosgene (Equation 48) and epichlorohydrin (Equation 49).[19]

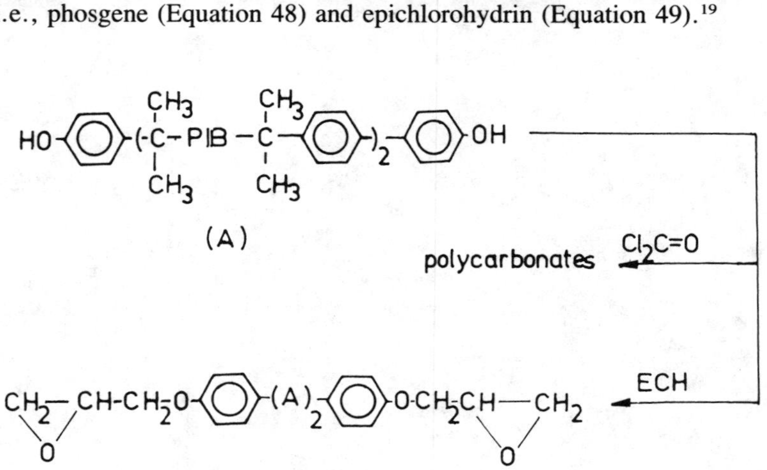

$$HO-\bigcirc\!\!-(\!\!\begin{smallmatrix}CH_3\\|\\C\\|\\CH_3\end{smallmatrix}\!\!-PIB-\begin{smallmatrix}CH_3\\|\\C\\|\\CH_3\end{smallmatrix}\!\!-\bigcirc)_2\!\!-\bigcirc\!\!-OH$$

(48)

(A)

polycarbonates $\xleftarrow{Cl_2C=O}$

$$CH_2\!\!-\!\!CH\text{-}CH_2O-\bigcirc\!\!-(A)_2\!\!-\bigcirc\!\!-O\text{-}CH_2CH\!\!-\!\!CH_2 \xleftarrow{\text{ECH}}$$

(49)

Quantitative dehydrochlorination of *tert*-chlorotelechelic polyisobutylenes by the use of strong nucleophiles ($tBuO^-$) gave rise to isopropylidene-telechelic polyisobutylenes, i.e., $[CH_2=C(CH_3)CH_2]_n PIB$, where n = 1.0, 2.0, or 3.0.[35] These intermediates are valuable for the further synthesis of many other end-reactive products. For example, oxidation with *m*-chloroperbenzoic acid yields the corresponding epoxides which, in turn, have been further converted to telechelic aldehydes, etc.:[36]

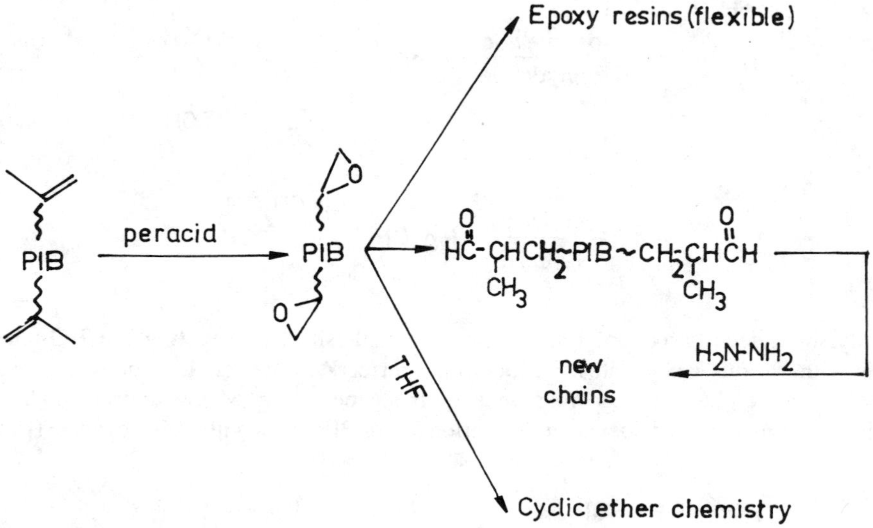

Epoxy resins (flexible)

peracid

PIB

new chains

$H_2N\text{-}NH_2$

Cyclic ether chemistry   (50)

Sulfonation by acetyl sulfate (Equation 51) or on "ene" reaction with maleic anhydride (MA) (Equation 52) gave new ionomers exhibiting a strong thermoplastic elastomer character:[37,38]

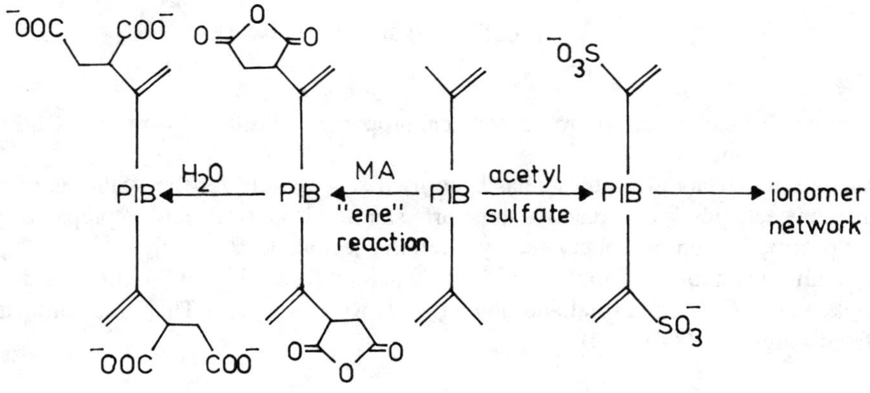

(52)   (51)

Hydroboration followed by oxidation of olefin-telechelic polyisobutylenes yielded hydroxyl-telechelic prepolymers which, in turn, have been used in a great variety of additional transformations.[39-41] A few examples are shown in Equations 53 to 55:

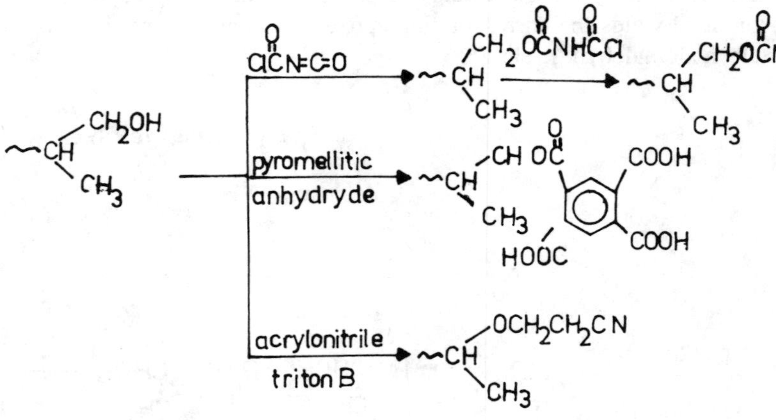

$$\text{(53)}$$

$$\text{(54)}$$

$$\text{(55)}$$

Polyisobutylene-based polyurethanes were synthesized by reacting the HO-telechelic liquids with various isocyanates.[42] Model or "perfect" networks, i.e., networks in which the molecular weights between cross-links are the same, $\overline{M}_c = \overline{M}_n$, were prepared by cross-linking PIB diols with a triisocyanate (Equation 56) or PIB triols with a diisocyanate (Equation 57):

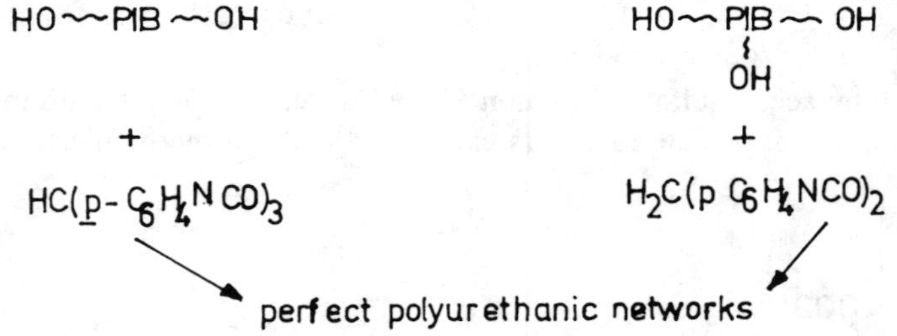

HO～PIB～OH

+

HC(p-C$_6$H$_4$NCO)$_3$

(56)

HO～PIB～OH
|
OH

+

H$_2$C(p C$_6$H$_4$NCO)$_2$

perfect polyurethanic networks

(57)

These networks exhibit satisfactory mechanical properties, excellent hydrolytic stability, and very low gas permeability.[43]

An improved Gabriel synthesis has been used successfully for the synthesis of amino-terminated telechelics.[44] End-reactive three-arm star polyisobutylenes were prepared by continuous polymerization of isobutylene by the inifer technique.[45]

The inifer technique as proposed and elucidated by Kennedy and Smith[46] was discussed by Nuyken et al.[47-54] and several additional details were provided. Thus, the elimination of HCl from inifer 1[49,50] (Table 2)

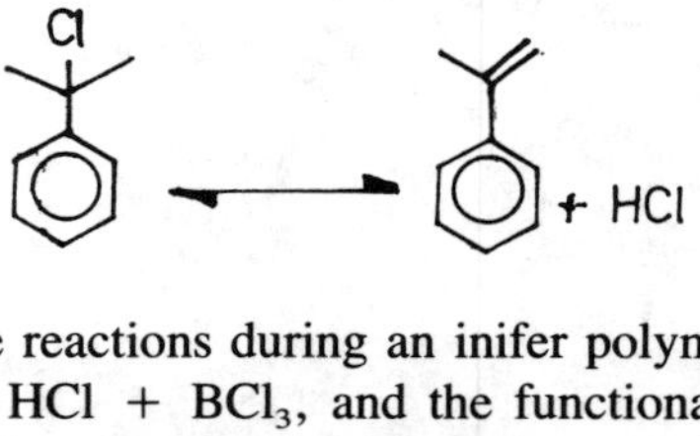

$$\text{(58)}$$

could lead to significant side reactions during an inifer polymerization, i.e., the polymerization could be initiated by HCl + BCl$_3$, and the functionality of the products obtained would be less than ideal. Only when a *proton trap* (e.g., 2,6-di-*t*-butyl pyridine) is added to the system can the molar mass and functionality of the products be controlled.[49,51] Binifers and trinifers (compounds 2 and 3, Table 2) are considerably more stable than minifer 1 with respect to a loss of HCl. Therefore, the polymerizaton of isobutylene using 2 + BCl_3 is practically ideal; even for low molecular masses, Relation 19 holds, i.e., $\overline{DP}_n = [M]_o/[2]_o$.[47,54] An analysis of the Born-Haber cycles[55] for the binary ionogenic equilibrium

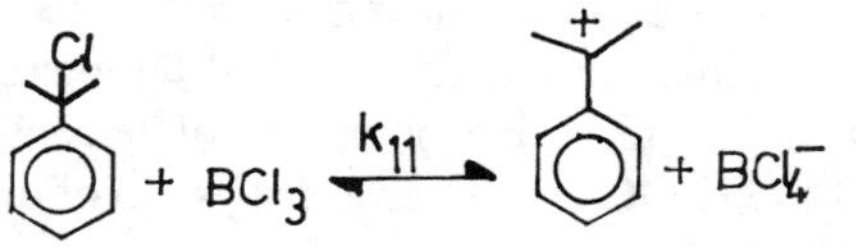

$$\text{(structure)} + BCl_3 \xrightleftharpoons{k_{11}} \text{(structure)}^+ + BCl_4^- \qquad (59)$$

has proved very useful for estimating reactivity differences and specific reaction behavior for the systems 1 + BCl_3, 2 + BCl_3, and 3 + BCl_3: $k_{11}(1 + BCl_3)$, $k_{11}(2 + BCl_3)$, $k_{11}(3 + BCl_3)$.[47] It explained the experimental results that, using 3 + BCl_3, the polymer molecules essentially all contain three polymer chains attached to a central aromatic ring, despite trinifer remaining at the end of polymerization.[56]

Furthermore, compounds 1 to 3 also undergo chemical transformation in the presence of BCl_3,[54,57,58] e.g., formation of the indane structure shown in Equation 60:

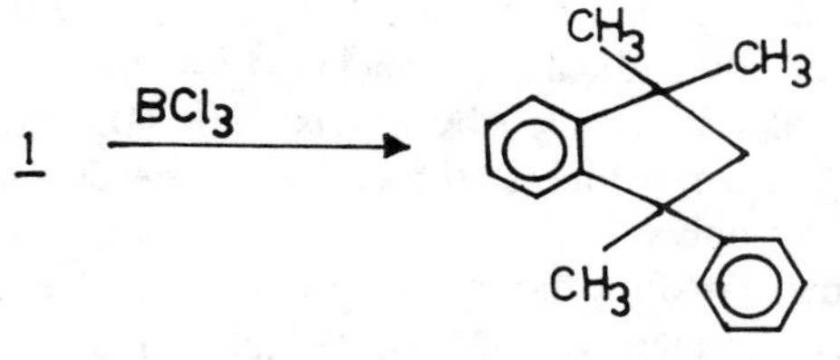

$$1 \xrightarrow{BCl_3} \text{(structure)} \qquad (60)$$

and polymers:

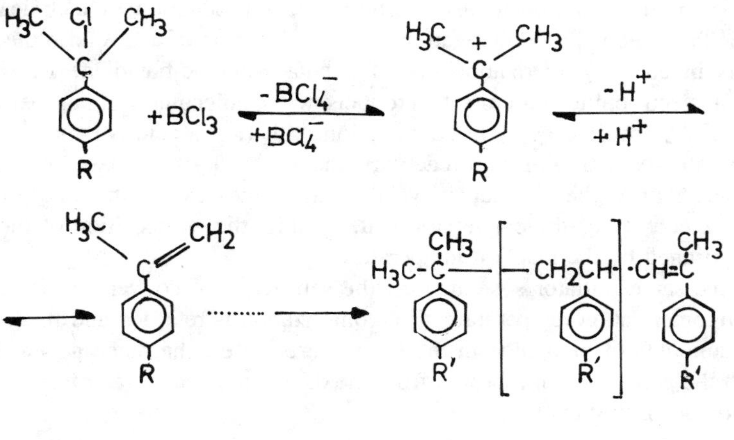

$$\text{(reaction scheme)}$$

where

$$R = -\underset{CH_3}{\overset{CH_3}{C}} - Cl \ , \quad R' = -\underset{CH_3}{\overset{CH_3}{C}} - Cl \quad or \quad -\overset{CH_2}{\underset{CH_3}{C}} \qquad (61)$$

α-Chloro and α,ω-dichlorooligoisobutylene ($\overline{M}_n$ in the range of 1000 to 3000) have been prepared using the $BCl_3/H_2O/CH_2Cl_2$ system proposed by Tessier et al.[59] and Kennedy et al.[60,61] and the inifer technique,[46] respectively, by monitoring the experimental conditions, i.e., the nature of the solvent, the $[H_2O]/[BCl_3]$ ratio, the isobutylene concentration, $[M]$, and the temperature effects on low molecular masses.

A structural H^1 NMR (nuclear magnetic resonance) analysis of dehydrochlorinated products[62] showed two kinds of end groups: 70% were separated from the inifer [*p*-di(2-chloro-2-isopropenyl)benzene] and 30% were directly linked to it.

Cationic telomerization of isobutylene using $AlCl_3$/telogene or BF_3/telogene, where the telogene is CCl_4 or CCl_3Br, gave head-to-tail regular structure compounds with molecular weights higher than those obtained by redox or radical telomerization ($\overline{DP}_n = 1$ to 60).[63] Initiators such as $CH_3CO^+SbF_6^-$ allow the obtaining of telechelic oligoisobutylenes with esteric groups,[64] and the $BF_3/C_2H_5OC_2H_5/CH_3COOH$ system of difunctional telomers having $-OH$ and $-COOH$ end groups.[65]

Olefins are oligomerized by BF_3 complexes for the production of synthetic lubricating oil base stocks.[65-67] The process can involve significant (14 to 22%) methyl migration at the monomer stage during oligomerization of 1-butene and *cis*-2-butene: the similarity of the isomers produced suggests basically the same intermediate, i.e., the *sec*-butyl carbenium ion.

C_{10} Fischer-Tropsch products were also suitable feeds for cationic oligomerization with BF_3 as catalyst, and the alcohols contained in the raw feeds were used as promotors.[68] If a high promotor-BF_3 complex concentration was combined with free BF_3, reinitiation of the oligomerization could occur, leading to products of very high viscosity.

The molecular mass of oily and waxy products ($\overline{M}_n = 200$ to 700) obtained by cationic polymerization of (cyclo)olefins initiated by Lewis acids (BF_3, $AlCl_3$, $AlBr_3$, $ZnCl_4$) and the structure of lower oligomers (dimers, trimers, etc.) are dependent mainly on reaction temperature and type of cocatalyst.[69-73]

The number of isomers, mainly those due to hydride and methanide shifts, increases with increasing oligomerization temperature, while higher molecular masses are obtained at lower temperatures.

A promising method for separating the components of a commercial technical C_4-hydrocarbon mixture, simultaneously permitting the production of technically important products, is the selective oligomerization or polymerization of a C_4 olefin out of a mixture, e.g., the isobutene oligomerization out of a butane-butene-butadiene fraction under the influence of electrophilic initiators to products with molecular weights between 300 and 1200 g/mol.[74-80] Brönsted acids in combination with Lewis acids or metal salts (sulfates) form an initiator system with high reactivity and selectivity[74,77] where, as a rule, products are originated with higher molecular weights, as in the case of the sole use of Brönsted acids. The selectivity of the conversion and, by this, the composition of the oligomers is widely determined by the reaction temperature.

Using the same initiator system, i.e., the selectivity of conversion decreases with increasing oligomerization temperature,[74] oligomerization is retarded and finally inhibited by increasing amounts of butadiene in the C_4 mixture,[10] i.e., the diene acts as a terminating agent, with the generation of a proton from the diisobutene cation and formation of a stable allylic cation according to:[74]

$$\text{\textasciitilde\textasciitilde\textasciitilde}H_2C-\underset{\underset{CH_3}{|}}{\overset{\overset{CH_3}{|}}{C}}-CH_2-\overset{+}{\underset{\underset{CH_3}{|}}{C}}-CH_3 + H_2C=CH-CH=CH_2 \rightarrow$$

$$\text{\textasciitilde\textasciitilde\textasciitilde}H_2C-\underset{\underset{CH_3}{|}}{\overset{\overset{CH_3}{|}}{C}}-CH_2-\underset{\underset{CH_3}{|}}{C}=CH_2 + H_3C-\overset{+}{C}H-CH=CH_2 \qquad (62)$$

TABLE 3
Cooligomerization of Isobutene and Piperylene Initiated by
BF_3 in CH_2Cl_2[81]

Diene (mol %)	T (K)	$\overline{M}_n$	$-CH=C(CH_3)_2$ (mol^{-1})	$-CH=CH-$ (mol^{-1})	$\overline{M}_o$	$\overline{M}_r$	f(OH) mol^{-1}
0	233	7,000	—	—	—	—	—
10	233	1,100	1.00	0.4	1,500	—	—
10	213	2,500	0.72	1.5	2,900	2,800	1.8
0	198	80,000	—	—	—	—	—
10	198	3,500	0.00	1.8	3,400	3,200	2.0
15	198	2,900	0.00	3.0	1,400	1,300	2.0

The effective transfer expressed by Equation 62 was exploited by Razzonk et al.[81] to obtain ω,ω'-difunctional oligoisobutylenes by cationic "copolymerization" of isobutylene (IB) with 1,3-pentadiene (piperylene, Pi) initiated by BF_3 in CH_2Cl_2 ($\overline{M}_n$), followed by oxidation with ruthenium tetroxide ($\overline{M}_o$) and reduction with $LiAlH_4$ ($\overline{M}_r$). The results are shown in Table 3. The average −OH functionality of the reduced oligomers (determined by PMR measurement of the −CH_2OH end groups) was 2.0 when the synthesis was carried out at 198 K ($-75°C$) and 1.8 for higher temperatures (213K, $-60°C$), since the effect of diene as transfer agent is progressively more important as the temperature is lowered, and becomes predominant around 200 K, as witnessed by the virtual absence of the typical polyisobutene terminal unsaturation. Moreover, the diene comonomer begins to be incorporated along the chain at higher temperatures.

Recently, synthesis of α,ω-di-*t*-chloropolyisobutylenes[82] and three-arm star polyisobutylene[83] has been accomplished by electron donor-stabilized living polymerization using aliphatic and aromatic tertiary esters as inhibitors in conjuncton with BCl_3 coinitiator in various solvents in the -20 to $-70°C$ range.[84-88]

Polymerization was visualized to occur by a mechanism akin to group transfer[85] (*vide infra*) and to involve a polyisobutylene chain carrying living monoacetate growing centers:[86,66]

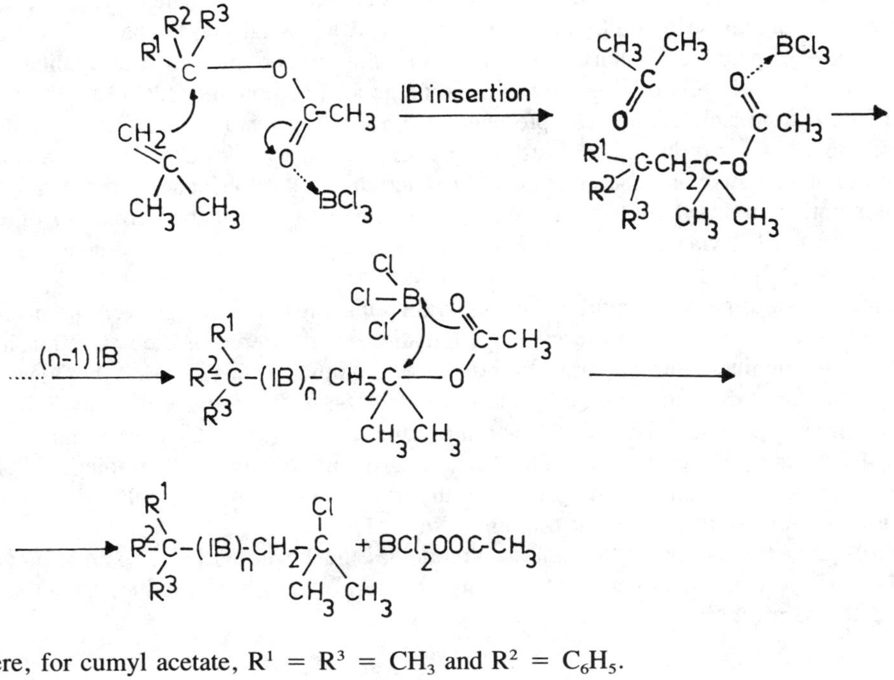

(63)

where, for cumyl acetate, $R^1 = R^3 = CH_3$ and $R^2 = C_6H_5$.

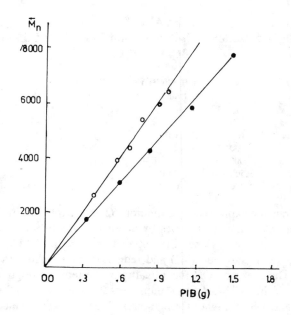

FIGURE 2.   Living polymerization of isobutylene in CH_3Cl at $-30°C$ by ($\circ$) 2,4,4,6-tetramethyl-heptane-2,6-diacetate/BCl and ($\bullet$) cumyl acetate/BCl_3. (From Faust, R., Nagy, A., and Kennedy, J. P., *J. Macromol. Sci. Chem.*, A24, 595, 1987. With permission.)

The living nature of the polymerization was demonstrated by a plot of the linear $\overline{M}_n$ vs. amount of polyisobutylene formed, starting at the origin (Figure 2).

Formation of the undesirable indanyl structures shown in Equation 60 can be suppressed by decreasing the temperature and the polarity of the polymerization medium (i.e., by using $CH_3Cl/n\text{-}C_6H_{14}$ mixtures).[82]

In the cationic polymerization of conjugated dienes, the allylic carbenium ion formed (see Equation 62) may attack the next diolefin in at least two ways, leading by two different enchainments (i.e., 1,2 and 1,4) to a mixture of ill-defined products. Therefore, the cationic oligomers of butadiene and alkylbutadienes produced by metal halides have been treated more or less as petroleum resins.[89] Presently, the most important industrial application of conjugated dienes in this field is the use of isoprene as a comonomer (2%) in butyl rubber manufacture from isobutylene. The presence of 0.3% anhydrous ("water-free") aluminum chloride in CH_3Cl solution catalyzes the copolymerization at $-90°C$.[90] Trace amounts of cyclic oligomers, such as 1-isopropenyl-2,2,4,4-tetramethylcyclohexane, were identified in the preparation of butyl rubber, and the mechanism involved in the formation of these products proceeded via initiation by protonated isoprene and cyclization after isobutylene addition.[91]

Unsaturated oligomers useful as vegetable oil substitutes in coatings were prepared by oligomerization of conjugated dienes, e.g., butadiene, piperylene, or diene fraction in the presence of aluminum or titanium chloride-based complex catalysts, i.e., homogeneous $EtAlCl_2$ and heterogeneous $TiCl_4$ or silica-supported Na or K carbonate-doped $AlCl_3$.[92-94]

Polymerization with $EtAlCl_2$ at low temperatures (-20 to $20°C$) gave relatively low molecular weight products ($\overline{M}_n \geq 1600$) containing both linear and cyclic moieties, in high yields containing low amounts of gel, while the heterogeneous $AlCl_3$ catalyzed, at $100°C$, the formation of exclusively linear oligomers with $\overline{M}_n \sim 500$.

To improve the stability of the cationic oligodienes and the properties of coatings prepared from them, the catalyst is deactivated with a mixture of propylene oxide, CH_3OH, and chlorinated paraffin.[92]

The cationic oligomerization of 2-ethyl-1,3-butadiene (2EBD) by a superacid (CF_3SO_3H) and a superacid derivative (CH_3COClO_4) accompanied monomer isomerization to 3-methyl-1,3-pentadiene (3MPD) before propagation to yield oligomers of the isomerized monomer as main products according to Equation 64, while in the presence of a metal halide catalyst (BF_3OEt_2), 2EBD reacted without isomerization (Equation 65) and yielded oligomers rich in 1,4 enchainment that were different from those produced by the foregoing superacid catalysts:[95,96]

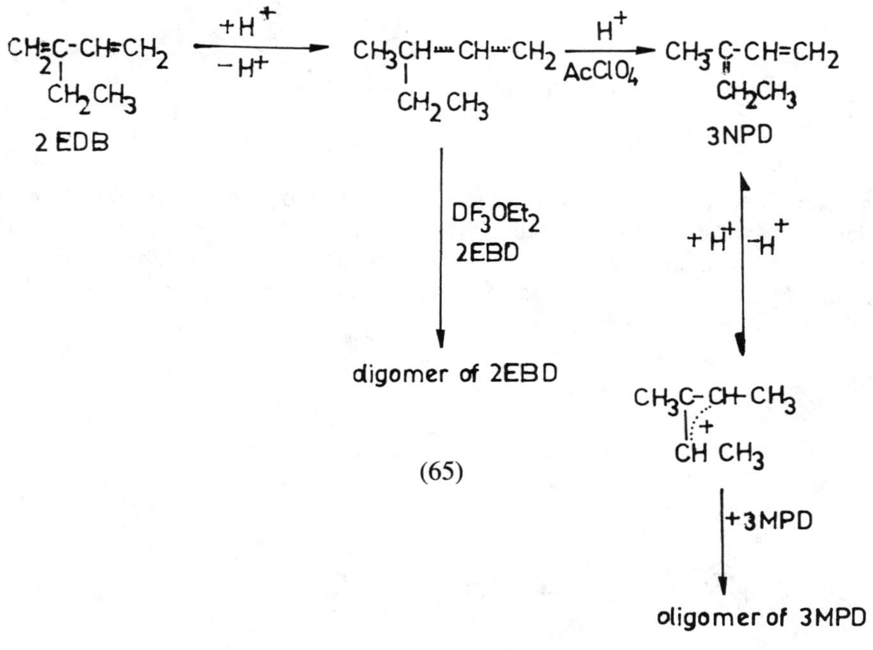

(65)

(64)

2. Vinyl, Cycloalkene, and Cyclodiene Monomers

Oligomerization of vinyl monomers such as styrene and related ring-substituted derivatives by conventional acidic catalysts (e.g., H_2SO_4) is poorly selective and usually gives a mixture of linear and cyclic dimers and higher oligomers.[97-99] Oligomerization of styrene using CF_3COOH gave products ($\overline{DP} = 1$ to 6) with O_2C-CF_3 termination, while normal oligomers were formed in the presence of CF_3-SO_3H.[100] A linear unsaturated dimer of styrene (1,3-diphenyl-1-butene) was obtained in high yield by using acetyl perchlorate ($AcClO_4$) or trifluoromethanesulfonic acid under suitable conditions.[101,102] A linear unsaturated styrene trimer, 1,3,5-triphenyl-1-hexene, could be produced as a main component when styrene was reacted with trichloroacetyl chloride or CF_3-SO_3H thermally (50 to 70°C).[103,104] On the other hand, ring-substituted methylstyrenes (p-,m-, and o-methylstyrenes) in conjunction with $AcClO_4$ or CF_3SO_3H as catalysts gave their linear unsaturated dimers (n = O, Equation 66) in high yield in benzene at the same temperatures (50 to 70°C).[105] The rate of the reaction by $AcClO_4$ decreased in the order o-methylstyrene $>$ m-methylstyrene $>$ p-methylstyrene, while with CF_3-SO_3H as catalyst, the monomer reactivity increased in the same order. Oligomer with a cyclic terminal structure (Equation 67) increased in the products at higher temperature:

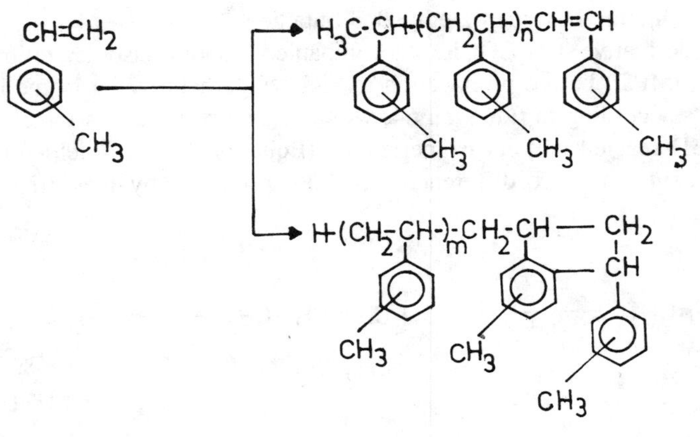

(66)

(67)

The same oxo catalysts ($AcClO_4$ and CF_3-SO_3H) promoted formation of the linear oligomers ($\overline{DP}$ = 2 to 5) *p*- and *m*-divinylbenzene in nonpolar solvents above room temperature and at low monomer concentration.[106] The polymers were soluble in aromatic and halogenated hydrocarbons and consisted of *trans* olefinic and phenyl groups in the main chain with vinyl groups at both ends:

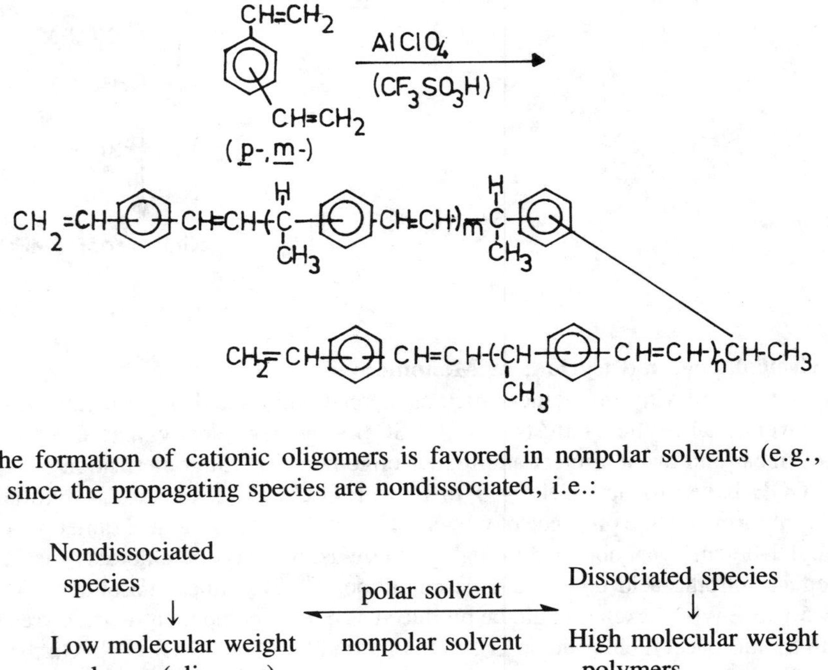

(68)

The formation of cationic oligomers is favored in nonpolar solvents (e.g., benzene, CCl_4) since the propagating species are nondissociated, i.e.:

Nondissociated
species
↓
Low molecular weight
polymer (oligomer)

← polar solvent →
nonpolar solvent

Dissociated species
↓
High molecular weight
polymers

(69)

The equilibrium of Equation 69 is reflected by the bimodal distribution of the molecular weight obtained at the cationic polymerization of styrene by acetyl perchlorate, various protonic acids (CF_3SO_3H, etc.), and iodine.[107]

Cationic oligomerization of styrene by a solid acid, i.e., the poly(styrene sulfonic acid) resin Amberlyt 15 (A-15), and by the corresponding soluble catalyst, i.e., *p*-toluenesulfonic acid, TSA, were compared with respect to their reactivity and molecular weight distribution (MWD) of products.[108] The apparent activaton energy for A-15 (20.1 kJ/mol) was close to that for TSA (15.5 kJ/mol), indicating a similar oligomerization pathway for the two catalysts, although the activity of the former was ten times larger. The striking feature of the oligomerization by A-15 disclosed in this study is the virtual absence of solvent effects, even in the polar 1,2 dichloroethane. With soluble TSA, in sharp contrast, the reaction was accelerated, with an increase of solvent polarity (i.e., with the number of dissociated species), as generally observed in cationic polymerization.

In solid catalysts, the degree of ionic dissociation of the sulfonic acid groups may no longer be influenced by solvent polarity because most of them are surrounded by the polymer molecules so as to be free from solvation. Styrene oligomers consisting mainly of trimer and tetramer were produced with cation exchange catalyst A-15 at high temperature (70 to 120°C) at low monomer concentraton, whereas a linear dimer was obtained in 85% concentration with TSA.

In acid-catalyzed homogeneous oligomerization as well as polymerization, four factors are known to control the MWD of products:[102,108,109] (1) the nature of the counterion, (2) monomer concentration, (3) solvent polarity, and (4) reaction temperature. In heterogeneous oligomerization, it was also revealed that the production of higher oligomers ($\overline{DP}$ 6) can be depressed by decreasing the monomer concentration and increasing the reaction temperature.[108]

High-boiling hydrocarbon oils, useful for heat-transfer media, solvents, and electrical insulating oils, are obtained by the reaction of alkylbenzenes with styrenes, α-methylstyrene, α-olefins, indene, coumarone, acenaphthalene, etc. in the presence of solid acid catalysts, producing mixtures of styrene-derived oligomers (from dimer to tetramer) and styrene-derived alkyl benzene adducts.[110,111]

Oligomerization of α-methyl styrene in the presence of solid sulfonic acid-type cation exchangers at 30 to 50°C gave an oligomer composition containing 0.8% saturated dimer, 74.1% unsaturated dimer, 22.1% trimer, 2.5% tetramer, and 0.4% pentamer.[112]

In the presence of chlorinated acid initiators, i.e., $ClCH_2COOH$, $Cl_2CHCOOH$, and Cl_3COOH, at 0 to 20°C, monomeric telomers accompanied by a counterion were obtained in the first two cases.[113] These adducts were liable to revert to the monomeric carbocation, which led to the dimeric carbocation through the addition of α-methylstyrene. All systems involved depolymerization and isomerization reactions of the oligomers. The decrease of the nucleophilicity of the counterion promoted the interconversion between dimer and trimer and the isomerization of 4-methyl-2,4-diphenyl-1-pentene to 4-methyl-2,4-diphenyl-2-pentene, followed by 1,3,3-trimethyl-1-phenylindane formation.

The dimer and trimer obtained by the cationic oligomerization of p-chlorostyrene were trans-3-di(p-chlorophenyl)-1-butene and a distereomeric pair of trans-1,3,5-tri(p-chlorophenyl)-1-hexenes, while those obtained by the pyrolysis of poly(chlorostyrene) were 2,4-di(p-chlorophenyl)-1-butene and 2,4,6-tri-(p-chlorophenyl)-1-hexene.[114]

Very electron-rich N-containing vinyl monomers, such as N-vinylcarbazol, NVK, or p-(dimethylamino)styrene, DAS, undergo kinetic cyclobutane formation with an electrophilic olefin without a leaving group, e.g., methyl-β,β-dicyanoacrylate, and one with a weak β-leaving group, e.g., tetracyanoethylene, while olefins having a strong β-leaving group, i.e., β,β-dicyanovinylchloride, readily initiate cationic polymerization of NVK and oligomerization of DAS. If an excess of donor DAS is used, all electrophilic olefins initiate cationic oligomerization of DAS, as conventional Brønsted initiators do.[115]

An unexpected pathway of the cationic oligomerization of 1-vinyl-4,5,6,7-tetrahydroindole, VTI, in the presence of various Brønsted and Lewis acids was reported by Trofimov et al.[116] All reported examples of the cationic polymerization of N-vinylpyrrole and related monomers under the influence of acidic compounds are considered to proceed mainly via the N-vinyl group or to afford charge-transfer complexes.[117] Only in the BF$_3$·OEt$_2$-induced polymerization of N-vinylpyrroles was chain termination via an intramolecular attack by the propagating cation at a neighboring pyrrole ring detected.[118] VTI undergoes an alternative oligomerization involving both the double bond and the α-position of the pyrrole ring to form the dimer 1-vinyl-2-[1-(4,5,6,7-te-trahydro-1-indolyl)ethyl]-4,5,6,7-tetrahydroindole in a yield up to 68% and with oligomers ($\overline{M}_n$ = 1000 to 2000) of the same backbone:[119]

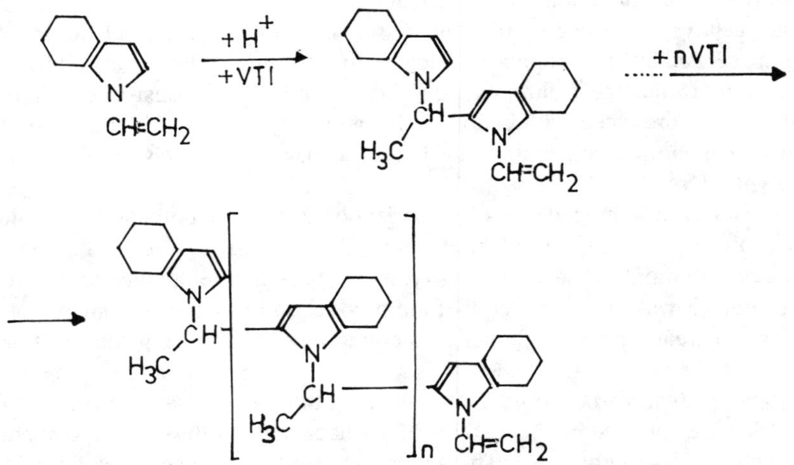

$$(70)$$

Low molecular weight oligomers, mainly trimers, were obtained by irradiation of α-methylstyrene with ^{60}Co γ-rays (dose rate, 5.10^5 rad/h) for 1 hour in dilute hydrocarbon solution at 0°C. The oligomers are produced under rigorously dry conditions and their $\overline{DP}$ decreases with decreasing $[M]_o$. Since the oligomers contain alkyl groups derived from the solvent molecules, their formation is explained in terms of chain transfer to the solvent.[120]

The combination of vinyl ether monomers and vinyl ether-terminated urethane (VEU) oligomers along with onium salt catalysts provides a versatile new radiation-curable coating systems. These coatings may be cured at high speeds using either UV or electron beam (EB) irradiation to produce coatings with a desirable set of physical and functional properties.[121-125]

One interesting feature of the production of VEU oligomers is the way that it may be tied to the initial acetylene reaction for synthesis of vinyl ethers. Hydroxyvinyl ethers were chosen as intermediates in the synthesis of vinyl ether-functionalized oligomers since they are easily obtained as byproducts of the synthesis of divinyl ether (DVE) monomers:[124]

$$HO-R-OH + 2HC \equiv CH \xrightarrow[\text{pressure}]{KOH, 200°C} (CH_2=CHO)_2R + HO-R-O-CH=CH_2 \quad (71)$$

If the vinyl ethers cannot withstand the hot basic conditions of the vinylation reaction, the exchange or transvinylation reaction can be used:

$$CH_2=CH + R'OH \xrightarrow{Hg^{2+}} CH_2=CH + ROH \quad (72)$$
$$\quad | \qquad\qquad\qquad\qquad | $$
$$OR \qquad\qquad\qquad\qquad OR'$$

Certain vinyl ethers have been produced by dehydrochlorinaton reactions, both commercially and in laboratory preparations. Reaction of 2.2'-dichlorodiethyl ether with sodium hydroxide is commercially used to manufacture 2-chloroethyl vinyl ether:

$$ClCH_2CH_2OCH_2CH_2Cl \xrightarrow{NaOH} CH_2=CHOCH_2CH_2Cl + NaCl + H_2O \quad (73)$$

New synthetic routes include vinylation of alcohols:

$$ROH + CH_2=CH_2 + {}^1/_2O_2 \xrightarrow{PdCl_2 \cdot CuCl_2 \cdot HCl} ROCH-CH_2 + H_2O \qquad (74)$$

or vinyl exchange with vinyl acetate:

$$ROH + CH_2=CHOCCH_3 \xrightarrow{PdCl_2 \cdot Na_2WO_4} ROCH=CH_2 + CH_3COH \qquad (75)$$

Alcohols add to isocyanate to form carbamates, the most important and well-studied of isocyanate reactions. The ether urethane is commonly used by analogy with ethyl carbamate:

$$RNCO + R'OH \rightarrow RNH-\overset{\overset{\displaystyle O}{\|}}{C}-OR' \qquad (76)$$

urethane

It may actually be desirable to use the mixture of the acetylene reaction products of Equation 71 in the synthesis of VEU oligomer.[126] The diol serves to chain extend the resin. Each hydroxyl group on the diol reacts with an isocyanate group, forming a higher molecular weight material. The total concentration of hydroxy groups from the diol and the hydroxy vinyl ether can be determined and then combined with an equivalent concentration of isocyanate:

$$HO-R-OH + HO-R-O-CH=CH_2 + OCN-R'-NCO + DVE \rightarrow$$

$$H_2C=CH-O-R-O-\underset{O}{\overset{O}{\underset{\|}{C}}}-NH-R'-NH-\underset{O}{\overset{O}{\underset{\|}{C}}}-O-R-O-\underset{O}{\overset{O}{\underset{\|}{C}}}-NH-R'-NH-\underset{O}{\overset{O}{\underset{\|}{C}}}-O-R-O-CH=CH_2 \qquad (77)$$

The DVE does not have free hydroxy groups; therefore it will not react with the diisocyanate. The divinyl ether serves as a diluent which lowers the viscosity of the mixture during oligomer synthesis. Additonal DVE may be added after oligomer synthesis is complete in order to adjust the viscosity of the formulation. When the mixture is finally cured, the divinyl ether copolymerizes with the VEU oligomer, effectively increasing the cross-link density.[124] Many VEU oligomers were synthesized using triethyleneglycol divinylether (TEGDVE) as a reaction solvent. Multifunctional VEU oligomers may be prepared by reacting hydroxyvinylethers with polyisocyanate or by chain extending diisocyanate monomers with multifunctional alcohol, e.g., the product of the reaction of TEGMVE and *p,p'*-diisocyanate diphenyl methane (MDI) with trimethylpropane (TMP) in a 3:3:1 ratio:

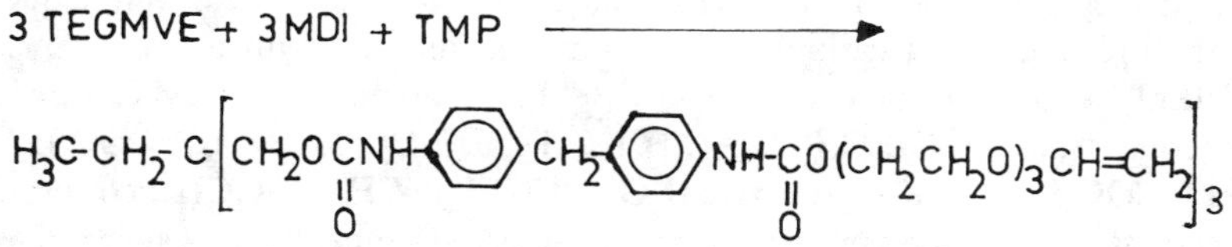

$$3\,TEGMVE + 3\,MDI + TMP \longrightarrow$$

$$H_3C-CH_2-C\left[CH_2O\underset{O}{\overset{}{\underset{\|}{C}}}NH-\bigcirc-CH_2-\bigcirc-NH-\underset{O}{\overset{}{\underset{\|}{C}}}O(CH_2CH_2O)_3CH=CH_2\right]_3 \qquad (78)$$

Recently, it has been shown that telechelics and macromonomers can be prepared by living polymerization of vinyl ethers.[127-129] To accomplish living cationic polymerization, the growing cation is stabilized by a strong nucleophilic counterion or by an externally added

TABLE 4
Synthesis of Poly(vinyl ether) Macromonomers E by Living
Cationic Polymerization[129]

X	$H_2C=CH-OR$	$\overline{M}_n$	$\overline{M}_w/\overline{M}_n$	$\overline{F}_n$
$H_2C=C-CO-$ $\quad\mid\ \ \parallel$ $H_3C\ \ O$	$H_2C=CH-OC_2H_5$	532 3772 7372	1.15 1.05 1.08	0.99 1.05 0.99
$H_2C=C-C-O$ $\quad\mid\ \ \parallel$ $H_3C\ \ O$	$H_2C=CH-CH_2CH_2OH$	622	1.08	1.03
$H_2C-CH-CH_2-$ $\ \ \diagdown\ \ \diagup$ $\qquad O$	$H_2C=CH-O-C_2H_5$	640 1720	1.05 1.07	1.04 1.05

base.[128-132] The first approach used the HI/I$_2$ initiating system that generates the iodine and allows clean syntheses of terminally functionalized oligomers:[129,130]

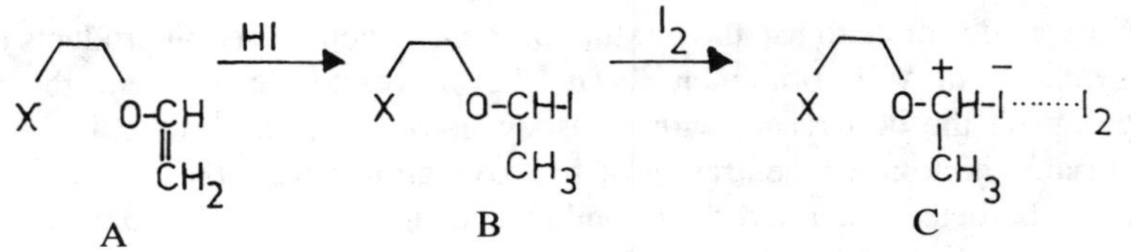

$$(79)$$

$$C + nH_2C=CH-OR \rightarrow X \text{\large\sim\sim\sim} O-CH-(-CH_2-CH-)_n-I \cdots I_2 \qquad (80)$$
$$\qquad\qquad\qquad\qquad\qquad\qquad\quad \mid\qquad\qquad \mid$$
$$\qquad\qquad\qquad\qquad\qquad\qquad\quad CH_3\qquad\quad OR$$

D

$$D + CH_3OH \rightarrow X\text{\large\sim\sim\sim}O-CH-(-CH_2-CH-)_n-OCH_3 \qquad (81)$$
$$\qquad\qquad\qquad\qquad\qquad\qquad \mid\qquad\qquad \mid$$
$$\qquad\qquad\qquad\qquad\qquad\qquad CH_3\qquad\quad OR$$

E

The first step is a quantitative addition of HI to a vinyl ether A that carries a pendant functional group X. The C–I group of the resultant adduct B (initiator) is then activated by iodine and thereby commences the living polymerization of vinyl ethers CH$_2$=CH–OR. All macromonomers E formed have controlled molecular weights ($\overline{DP}_n$ = [A]$_o$/[B]$_o$), narrow MWDs ($\overline{M}_w/\overline{M}_n \sim 1.1$), and exactly one terminal group per chain (F$_n$ = 1.0 $\mp$ 0.05). Some examples are given in Table 4.

A new trimer and tetramer of vinyl ethers with controlled repeat unit sequences were prepared through the living cationic polymerization initiated with a hydrogen iodide/zinc iodide (HI/ZnI$_2$) system in toluene at $-40°C$.[132] For example, the tetramer consisted of the following sequences, H–CH$_2$CH(O–n–C$_4$H$_9$)–CH$_2$CH–[–OCH$_2$CH$_2$CH(COOC$_2$H$_5$)$_2$–CH$_2$CH–(OCH$_2$CH$_2$OCOC$_6$H$_5$)–CH$_2$CH$_2$–OCH$_2$CH$_2$OCOC(CH$_3$) =CH$_2$–]$_4$–OCH$_3$, corresponding to a new methacrylate-type monomer. The synthesis involved sequential and successive reactions of the corresponding four vinyl, i.e. n-butylvinyl ether, ethyl 2-(vinyloxy) ethyl malonate, 2-(benzoyloxy) ethyl vinyl ether, and 2(vinyloxy)ethyl methacrylate (each equimolecular to hydrogen iodide), with the HI/ZnI$_2$-generated living oligomeric growing species, starting from the quantitative addition of hydrogen iodide to *n*-butyl vinyl ether and subsequent activation of the resulting adduct with ZnI_2.

3. Small-Ring Paraffins

It has long been known that the chemical properties of cyclopropane are more similar to those of the lower olefins, ethylene and propene, than to those of higher molecular, such as cyclopentane, of the cycloparaffin series. Determination of certain physical properties of cyclopropane derivatives have supported the view that the ring has some double bond character.[133-135]

Cyclopropane has a polymerizable structure because it possesses a high electron density at the center of the ring. These carbon-carbon bonds are formed by the incomplete overlapping of atomic orbitals sp with a strong, marked p-character, which gives the ring partial unsaturation taken from the tricentric structure. This characteristic allowed Pinazzi et al.[136] to explain the polymerizability, assuming a π-complex between the active site and cyclopropane compound. Cyclobutane is similar to cyclopropane from a structural viewpoint. Actually, cyclobutane is made more reactive toward cationic catalysts in the presence of appropriate substituents.

Cationic ring-opening polymerization of cyclic paraffins and related compounds was reviewed by Kennedy.[137] Some data are discussed below.

Cyclopropane was polymerized in heptane in the presence of an $AlBr_3/HBr$ system to give products of low molecular weight (>700). The degree of polymerization increased from $\overline{DP} \sim 6$ at $+25°C$ to ~ 16 at $-78°C$ and the oligomers appeared to be mainly long chain-alkanes with a terminal double bond:[138]

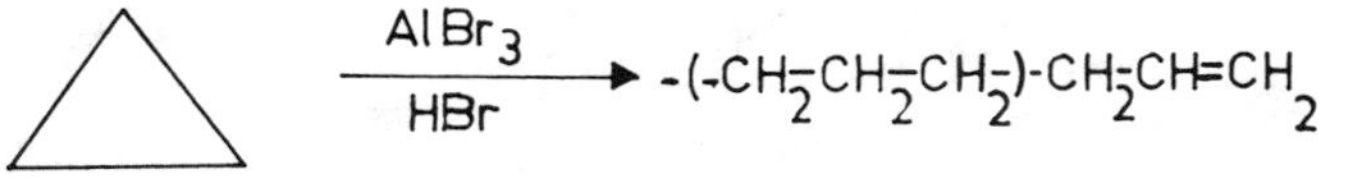

$$\triangle \quad \xrightarrow[\text{HBr}]{\text{AlBr}_3} \quad -(-CH_2-CH_2-CH_2-)-CH_2-CH=CH_2 \qquad (82)$$

Isopropylcyclopropane was polymerized with $AlBr_3$ in the range of -10 to $-50°C$ to liquid oligomers ($\overline{DP} \sim 3$ to 4).[139] Spectroscopic data indicated the following repeat unit:

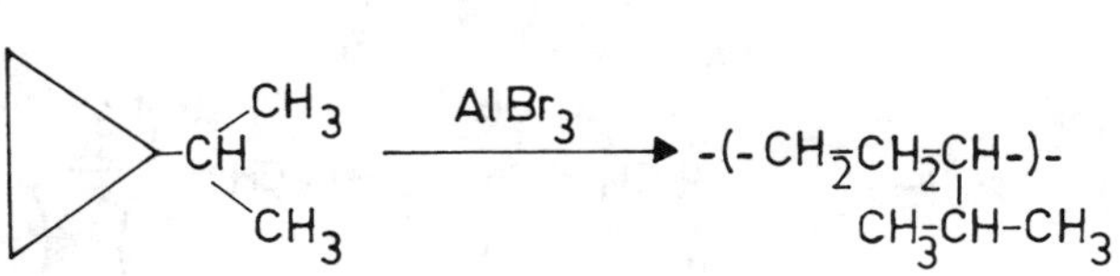

$$\triangleright CH \begin{array}{c} CH_3 \\ CH_3 \end{array} \quad \xrightarrow{\text{AlBr}_3} \quad -(-CH_2-CH_2-CH-)- \atop CH_3-CH-CH_3 \qquad (83)$$

which suggested[137] that the polymerization did not proceed by the conventional carbenium ion mechanism, i.e., the implication of the rearrangement propagation:

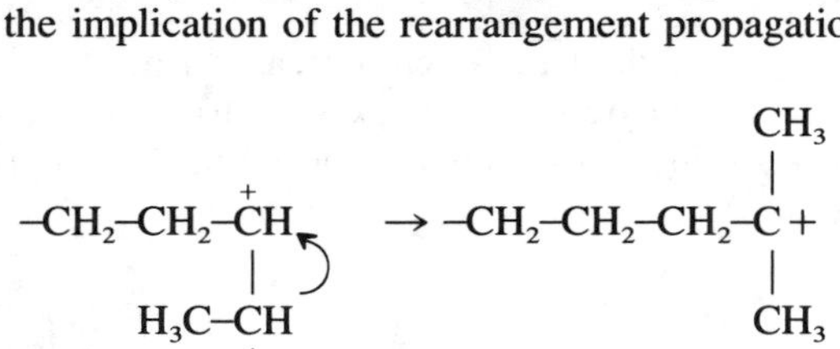

$$-CH_2-CH_2-\overset{+}{C}H \atop \underset{CH_3}{\overset{|}{H_3C-CH}} \quad \rightarrow \quad -CH_2-CH_2-CH_2-\overset{\overset{\displaystyle CH_3}{|}}{\underset{\underset{\displaystyle CH_3}{|}}{C}}+ \qquad (84)$$

1,1-Dimethylcyclopropane was polymerized in bulk or in chlorinated hydrocarbons to low molecular weight (~ 1000) products in the presence of Lewis acids.[140,141] Effective polymerization to a viscous liquid oligomer ($\overline{DP} \sim 5$) also occurred in *n*-hexane. The structure of the oligomers was identical to that of cationic poly(3-methyl-1-butene):[140]

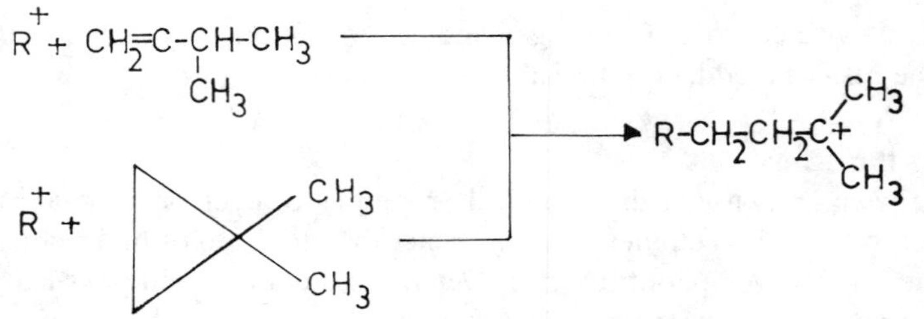

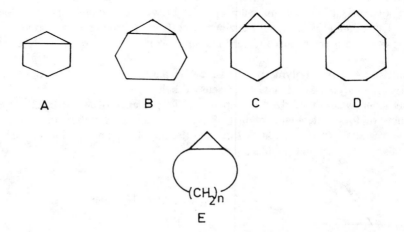

The bicyclo[n.1.0]alkanes shown below, (A to E), where n = 3 to 6 and 10, i.e., bicyclo[3.1.0]hexane, bicyclo[4.1.0]heptane, bicyclo[5.1.0]octane, bicyclo[6.1.0]nonane, and bicyclo[10.1.0]tridecane, respectively,

behaved like olefins toward Lewis acids.[136,142] The polymers obtained had a "bag" configuration with methyl groups attached to the rings:

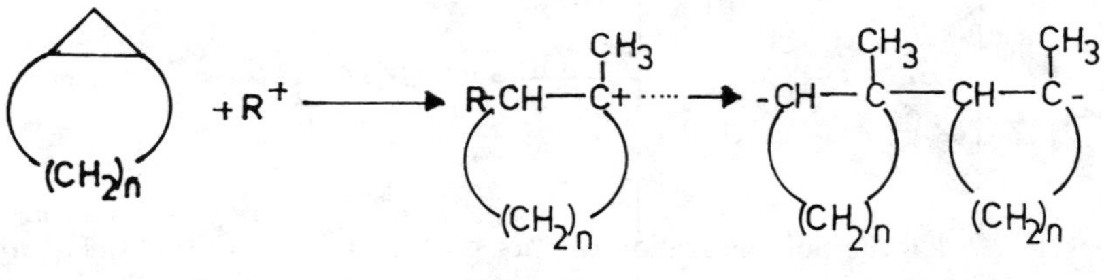

The size of the rings affected the $\overline{DP}$, which decreased from 11 to 2 when the n of the bicyclic [n.1.0] alkanes increased from n = 0 to n = 10. It was probably not the only cause, since the deprotonation in the α-position occurred rapidly to give unsaturated chain ends:

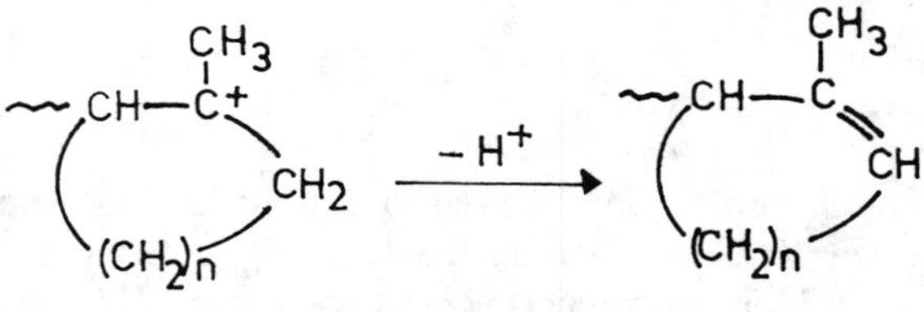

The cationic polymerization of spiro (2-n)alkanes F to K shown below, with n = 2, 4, 5, 6, 7, and 11, i.e., spiropentane, 2,4-spiroheptane, 2,5-spirooctane, 2,6-spirononane, 2,7-spirodecane, and 2,11-spirotetradecane, respectively, proceeded by ring opening of cyclopropane.[136,143]

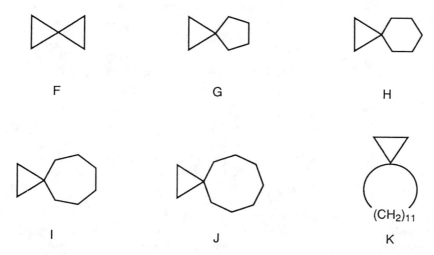

The oligomers formed had a structure, according to their monomer units, consisting of the large original ring and one methyl in a side chain:

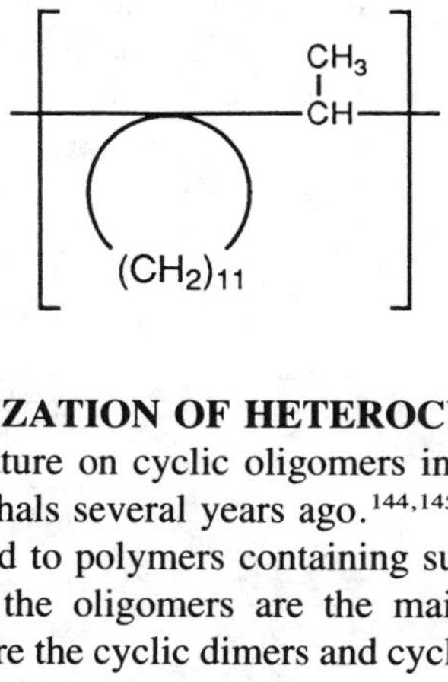

B. CATIONIC OLIGOMERIZATION OF HETEROCYCLES

A survey of the basic literature on cyclic oligomers in the cationic oligomerization of heterocycles was made by Goethals several years ago.[144,145] The cationic polymerization of heterocyclic monomers may lead to polymers containing substantial, yet variable, amounts of oligomers. In many cases, the oligomers are the main reaction products. The most frequently observed oligomers are the cyclic dimers and cyclic tetramers, but other structures are also possible. From studies with oxiranes, thiiranes, and some aziridines, it follows that the oligomers are produced by degradation of first-formed polymer and not directly from monomer.

1. Cyclic Ethers and Acetals

The best-known example of an oligomer-forming monomer is ethylene oxide. When this monomer is polymerized by borontrifluoride in 1,2-dichloromethane, the polymerization proceeds only to a molecular weight of about 700 ($\overline{DP}$ = 28), although monomer continues to disappear more or less indefinitely, but this is due almost entirely to the formation not of new polymer, but of 1,4-dioxane.[146] It is believed that the 1,4-dioxane (i.e., the cyclic dimer of ethylene oxide) is formed by degradation of the polymer chain via oxonium salts. This was deduced from the observation that, if high molecular weight polyglycols are introduced to the reaction mixture, their molecular weight is reduced until the average molecular weight is once again 700. With sufficient oxonium salt (e.g., triethyloxonium tetrafluoroborate), the polymers ($\overline{M}_n$ = 1500 to 20,000) were completely degraded and the only reaction product was 1,4-dioxane.[147]

In the polymerization of heterocyclic monomers, two reaction pathways are possible for the ionization reaction, i.e., internal ionization (Equation 88) and external ionization (Equation 89):[148]

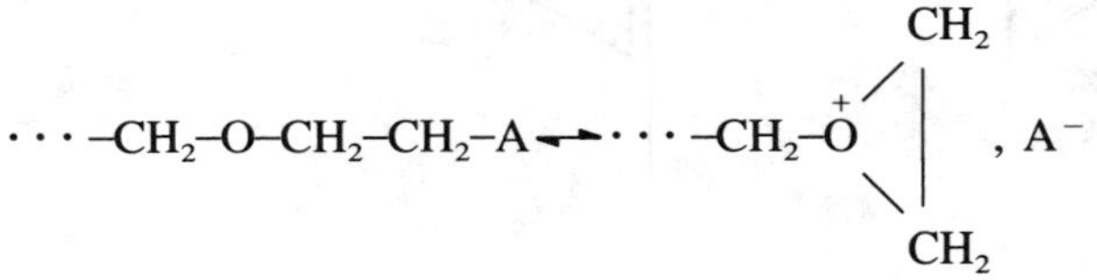

$$\cdots -CH_2-O-CH_2-CH_2-A \longleftrightarrow \cdots -CH_2-\overset{+}{O} \Big\langle \quad, \ A^- \qquad (88)$$

$$\cdots -CH_2-O-CH_2-CH_2-A \ + \ O\Big\langle \quad \rightleftharpoons \cdots -CH_2-O-CH_2-CH_2-\overset{+}{O}\Big\langle \qquad (89)$$

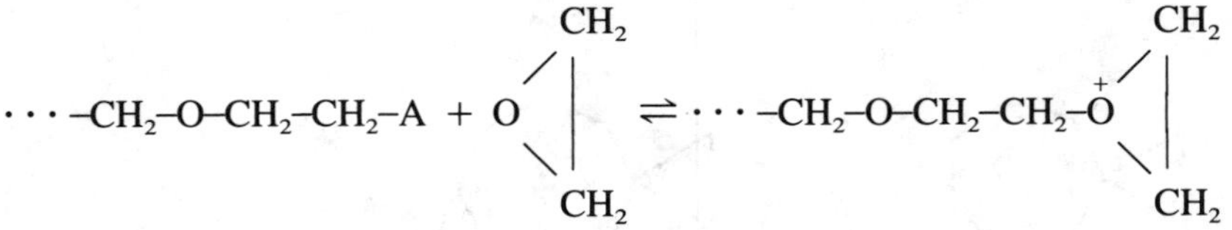

The contribution of these two mechanisms, operating simultaneously, may be estimated on the basis of the dependence of the ion/ester ratio on conversion. Ethylene oxide is converted to 1,4-dioxane in yields of up to 96% when the monomer is treated with a catalytic amount of a superacid.[149] 1,4-Dioxonium and other trialkyloxonium ions were observed by ^{1}H-NMR spectroscopy in the polymerization of oxiranes initiated with $(C_6H_5)_3C^+$.[150] The accepted mechanism of dimerization involves therefore an internal ionization (Equation 90), followed by backbiting of the polymer chain (Equation 91) on the tertiary oxonium ion-chain end and expulsion of a 1,4-dioxane unit (Equation 92):

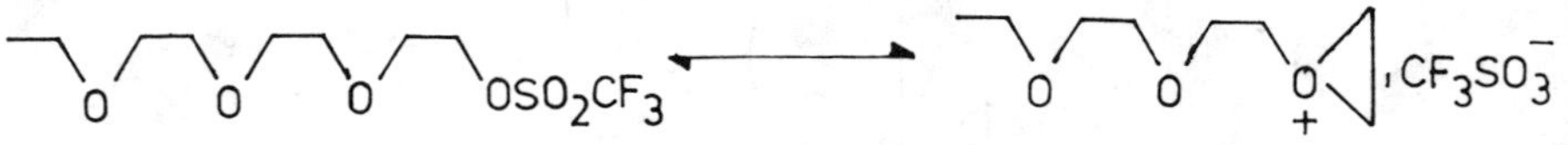

$$ \cdots OSO_2CF_3 \longleftarrow \cdots \ , CF_3SO_3^- \qquad (90)$$

$$ \cdots \ , CF_3SO_3^- \longleftarrow \cdots + O \quad O, CF_3SO_3^- \qquad (91)$$

$$ \cdots + O \quad O, CF_3SO_3^- \longrightarrow \cdots \ OSO_2CF_3 + O \quad O \qquad (92)$$

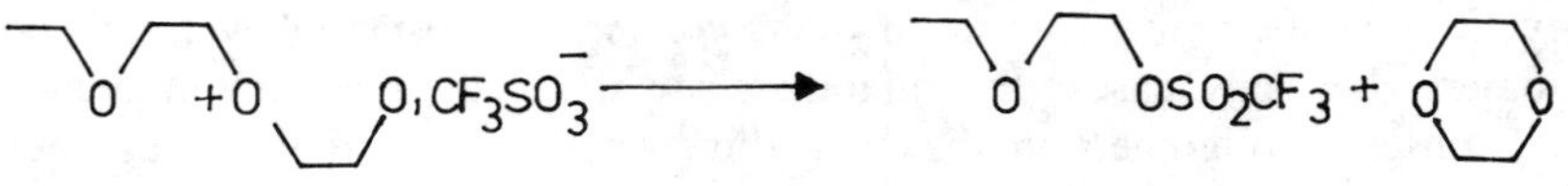

When treated with fluorine-containing catalysts (SbF_5, HBF_4), ethylene oxide produces a mixture of cyclic oligomers of $\overline{DP}$ = 3, 4, 5, 6, 7, 8, 9, (7:4:6:6:2:1), unaccompanied by open-chain oligomers and polymers.[151-153] The production of cyclic tetramers, pentamers, and hexamers as well as the linear oligomers is increased by the additon of alkali metal cation as templates.

Trans-2,5-diphenyl-1,4-dioxane has been isolated during the cationic polymerization of styrene oxide.[154-156] Interestingly, the maximal yield of dimer was 10% when Friedel-Crafts catalysts (i.e., Lewis acids) were used, but the same catalysts gave no oligomers with ethylene oxide or propylene oxide.[156]

Both cyclic and linear oligomers were formed when the polymerization was initiated by nitronium tetrafluoroborate in nitromethane and dichloromethane at 5 to 20°C.[157] Cyclic dimer and trimer in CH_3NO_2 were 2-benzyl-4-phenyl-1,3-dioxolane and 1,3,5-tribenzyltrioxane, respectively. In CH_2Cl_2, 2,5-diphenyldioxane was isolated. In CH_3NO_2, mainly isomerized structures with acetal linkages were produced, while in CH_2Cl_2, isomerization was not produced. By NMR and IR spectra, the presence of carbonyl and hydroxyl end groups in the linear oligomers was shown. These were indications that oligomers formed both directly from the monomer and by degradation of the polymer.

A number of other epoxides have been found to produce cyclic oligomers. Propylene oxide, when treated with triethyloxonium tetrafluoroborate or boron trifluoride or with trimethylaluminum, produces a mixture of cyclic oligomers, the most important one being the tetramer.[158-160] Because of the occurrence of head-to-tail and head-to-head structures and *cis-trans* isomerism, the cyclic tetramer can occur as 23 geometrical isomers; 22 were identified and characterized.[161]

The structure of cationic isopropyloxirane oligomers determined from ^{13}C-NMR spectra are consistent, however, with the nonrearrangement of polymeric structural units. This is related to the nature of the bridged propagating oxonium ion intermediate in the cationic enchainment.[162]

The cationic ring-opening polymerization of oxiranes in conjunction with a glycol or water as a modifier produces hydroxyl-terminated liquid polymers.[163-167] Lower molecular weight liquid polymers ($\overline{M}_n < 1200$) are free from cyclic oligomers; higher molecular weight polymers show a bimodal MWD and contain 5 to 20% cyclic oligomers.[166] Neither macromonomers nor telechelics can be synthesized from oxiranes by normal cationic ring-opening methods because cyclic oligomers are formed preferentially by backbiting of the polymer chain on the tertiary oxonium ion from the chain end,[168] as exemplified above in the case of ethylene oxide (Equation 91).

Conditions have been found, however, that suppress the formation of cyclic oligomers in the cation ring-opening polymerization of cyclic ethers.[148,169-173] The mechanism for which Penczek et al.[148] coined the name *activated monomer mechanism*, AMM, consists of the step-by-step addition of the protonated (activated) monomer, AM, or monomer activated (in the ground state) in another way, to the growing macromolecule, fitted with a nucleophilic end group (e.g., hydroxyl end group):[170]

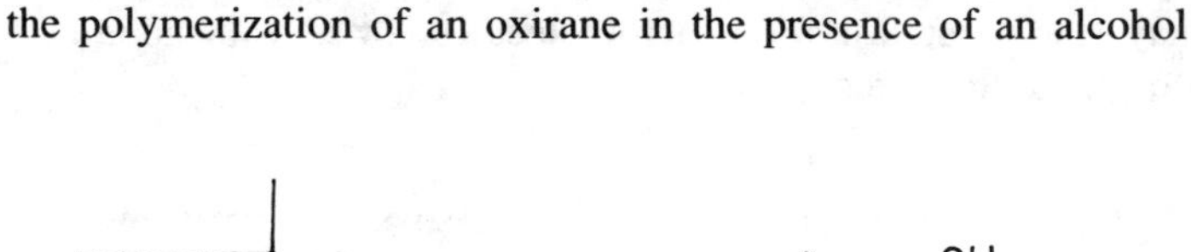

$$\cdots-CH_2CH_2OH + \underset{\substack{\diagdown\diagup\\O+\\|\\H\\ \text{AM}}}{CH_2{-}CH_2} \longrightarrow \cdots-CH_2CH_2O\text{-}CH_2CH_2OH \qquad (93)$$

The liberated protons may activate by protonation the next monomer molecule. Considering also the *active chain end* (ACE) mechanism, the following initiation reactions can take place at the polymerization of an oxirane in the presence of an alcohol and protonic acid:[171]

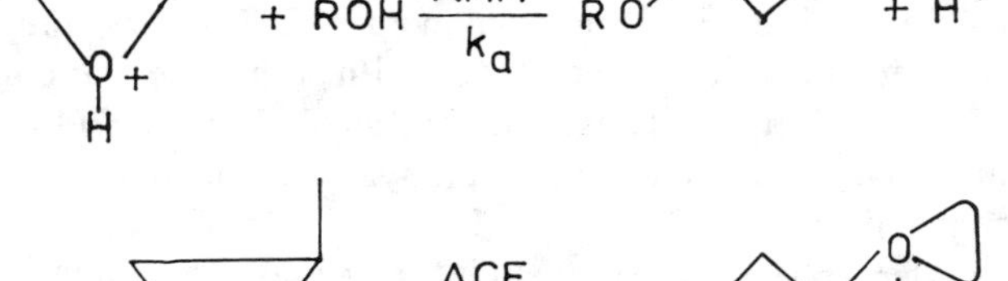

$$\underset{\substack{O+\\|\\H}}{\triangle} + ROH \xrightarrow{\underset{k_a}{AMM}} RO\diagup\diagdown\diagup^{OH} + H^+ \qquad (94)$$

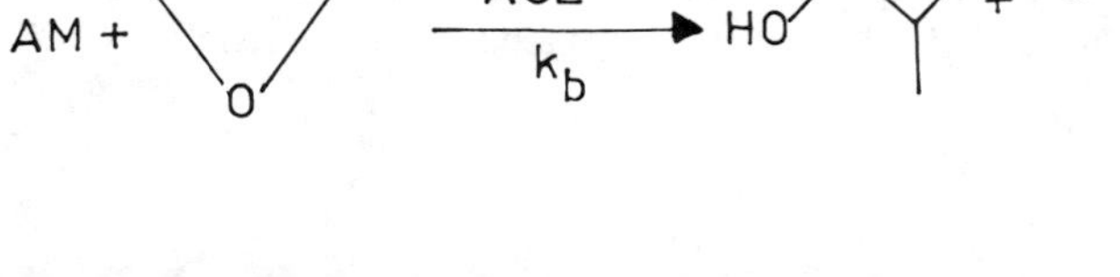

$$AM + \underset{O}{\triangle} \xrightarrow{\underset{k_b}{ACE}} HO\diagup\diagdown\diagup\diagdown_{O+}\triangle \qquad (95)$$

Hydroxyl-containing compounds present in the system compete with monomers for reaction with protonated (activated) monomer. If the concentration of hydroxyl groups is low or the difference in nucleophilicities is not large enough, the contribution of Reaction 94 may be negligible and the reaction proceeds by the conventional oxonium ion mechanism. If, however, the nucleophilicity of the terminal group is high and its concentration is large, Reaction 95 may be the prevailing or even practically exclusive route of polymer formation. It follows that in order to enhance the proportion of AMM propagation (i.e., simultaneously suppress the ACE propagation), one has to increase the rate of the AMM process. The contribution of AMM (C_{AM}) is proportional to $k_a[AM][ROH]$ and inversely proportional to $k_b[AM][M]$. Therefore

$$C_{AM} = k_a[ROH]/k_b[M] \tag{96}$$

This relation derived from the initial process is also true for the further stages of polymerization. Thus, in order to have high C_{AM}, one has to increase [ROH] and decrease [M]. However, every molecule of initiator (ROH) gives one chain; therefore, the $[M]_o/[ROH]_o$ ratio is equal to $\overline{DP}_n$. In the rate expression (Equation 96), [M] denotes the instantaneous concentration of monomer. The slow introduction of a monomer into the polymerization mixture allows keeping the concentration low enough and the total amount of monomer introduced provides the required $\overline{DP}_n$. For example, in the polymerization of propylene oxide by ACE, there is over 40% cyclic tetramer formation, whereas for products with $\overline{M}_n$ up to several thousands, the tetramer content can be decreased below 1% when slow monomer addition is applied ($[M]_{inst} \sim 10^{-2}$ mol/l).[172] In the polymerization of ethylene oxide, the overall rate of polymerization (expressed as d[M]/dt) increases with monomer conversion and the MWD largely deviates from the Poisson-type distribution.[173]

The behavior of the system was explained by assuming that besides the terminal hydroxyl groups, the ethereal bonds within a chain also participate in propagation, with their importance increasing with the polymer chain length (i.e., total monomer consumption):

$$(97)$$

The tertiary oxonium ion created in this way can react further in a number of ways, intra- or intermolecularly, upsetting the living character of the polymerization. This is particularly important when macromonomers are prepared. The intermolecular reaction (scrambling) may lead to undesirable macromolecules, fitted with none and two reactive groups instead of just one, as required:[171]

$$(98)$$

A typical α-epichlorohydrin (ECH) macromonomer (EPCM) was prepared as follows: ECH was polymerized in CH_2Cl_2 at 25°C with ethylene glycol methacrylate, $CH_2=C(CH_3)COOCH_2CH_2OH$ (EGM), as initiating alcohol and $BF_3 \cdot O(C_2H_5)_2$ as cationic initiator. In a similar way, macromonomers of propylene oxide and some copolymers (copolyethers) were prepared.

Taking hexamethylene glycol instead of EGM as the corresponding initiator and HBF_4 as catalyst, α,ω–OH-ended telechelics, propylene oxide telechelics (POT), and epichloro-

TABLE 5
Macromonomers and α,ω-OH-Ended Telechelics
Obtained by Activated Monomer Mechanism[171]

Product	$\overline{M}_n$ (calcd)	$\overline{M}_n$ (VPO)	$\overline{M}_n$ (^{1}H NMR)	Cyclic tetramer (mol %)
EPCM	780	750	735	0.04
ECMT	600	570	670	—
	705	510	720	—
	1,410	1,390	1,425	—
POT	2,100	2,230	2,200	—
	2,745	2,735	2,625	0.1

hydrine telechelics (ECHT) were obtained from propylene oxide and ECH, respectively. In the case of POT, 60% of the secondary and 40% of the primary −OH groups were found by ^{1}H-NMR spectroscopy after reacting −OH groups with $(CF_3CO)_2O$. Some example are given in Table 5.

Copolymerization of oxiranes with THF at the conditions enhancing the AMM participation leads to telechelic copolymers. The mechanism of copolymerization leading to α,ω−OH-ended telechelics in a range up to several thousands combines simultaneous propagation according to the ACE mechanism (mostly THF) and AMM (exclusively oxiranes), since THF does not homopolymerize in the presence of −OH-containing compounds. The cyclic oligoether ($\overline{M}_n = 270$) content of cationic THF-epoxide copolymers ($\overline{M}_n = 1250$) obtained in the absence of AMM amounted to up to 16%.[174]

The synthetic methods used to make functional derivatives of poly(ethylene oxide) and their applications have been extensively reviewed.[175] Some examples of recent developments are as follows. Radically curable oligomers ($\overline{M}_n = 850$) which give polymers with excellent weather resistance, flexibility, and impact resistance useful for coatings, adhesives, etc. are manufactured by AMM cationic ring-opening copolymerization of substituted oxiranes of the type (R, R′ = H, CH₃):

$$CH_2=CRCCOCH_2-CR' \quad CH_2 \qquad (99)$$

and cyclic ethers in the presence of monohydric alcohols and BF_3OEt_2.[176-178]

Cationic telomerization of ethylene glycoldiglycidylether with ethylene glycol and $CH_3C-O-O-CH_2OH$ (telogen) in the presence of BF_3Et_2O gave peroxytelomer by AMM with a functionality of 2.0, suitable as a cross-linking agent for unsaturated oligoesters.[179]

Oligoether diols were prepared by cationic polymerization, at 50 to 100°C, of cyclic ethers with three- to seven-membered rings in the presence of a diol and Brønsted acids.[167]

The oligomerization of oxetane (trimethylene oxide) initiated by sulfuric acid in methylene chloride in the presence of 1,3-propanediol (trimethylene glycol) led to the formation of α-hydro-ω-hydroxyoligo(oxytrimethylene)s.[167,180] Polymers from dimethylterephthalate and dimer, and copolymers from 1,3-propanediol, dimer, and dimethyltetrephthalate, were synthesized and characterized. It was concluded that, depending on the chemical structure of the polyesters formed, the increase in flexibility, caused by the presence of the oxytrimethylene units in the poly(oxytrimethyleneoxyterephthaloyl) chain, decreases the temperatures of glass transition ($T_g = -5$ to 45°C), cold crystallization, and melting ($T_m = 85$ to 225°C), compared to PET.[180]

Cycloaliphatic diepoxy chain-stopped linear siloxane monomer and oligomers, when combined with low concentrations of soluble iodonium photocatalysts, display extremely fast ultraviolet cure response. The surface and bulk properties of these cured materials are easily modified and enhanced by co-cure with mono- and polyfunctional alcohols.[181,182]

Four-, five-, six-, and seven-membered cyclic ethers, shown below, form both linear and cyclic oligomers when treated with typical cationic initiators.

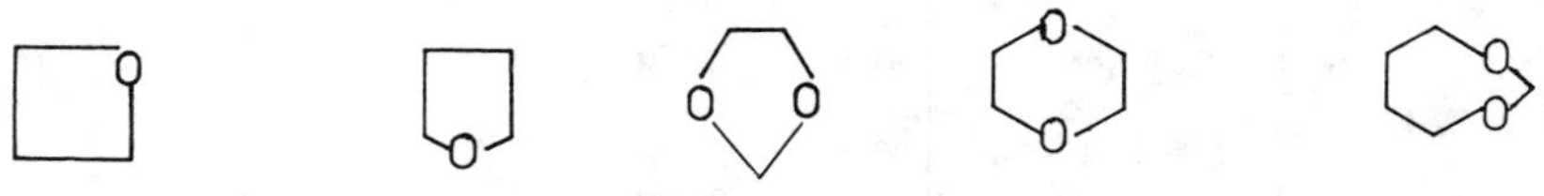

Oxetane	Tetrahydrofuran	1,3-Dioxolane	1,3-Dioxane	1,3-Dioxepane

The nature and proportion of oligomers are determined by reaction conditions, i.e., initiating system, temperature, and concentration. Oxetane, for example, gives 4% tetramer (16-membered rings) at $-80°C$, but at $50°C$, 66% of the reaction product (yield, 65%) was tetramer.[183,184] Cyclic trimer has been isolated at the polymerization of 3,3-bischloromethyl-oxetane with triethylaluminum.[185] It was found that due to excessive transfer to monomer, only oligomers were obtained from 2,2-dimethyloxycyclobutane. These oligomers have a chain structure corresponding to alternating isobutylene and oxyethylene groups and contain at one end an isopropenyl group and at the other a primary hydroxyl group. After regio-selective hydroboration, ω,ω'-dihydroxy functional oligomers (F = 2.0) have been obtained:[186]

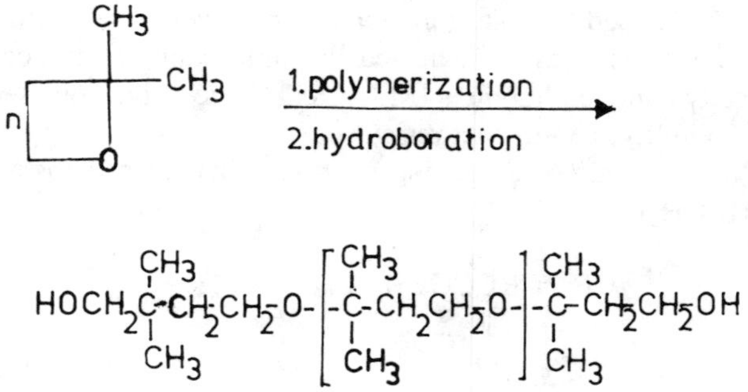

$$HOCH_2\underset{CH_3}{\overset{CH_3}{C}}CH_2CH_2O-\left[-\underset{CH_3}{\overset{CH_3}{C}}-CH_2CH_2O-\right]-\underset{CH_3}{\overset{CH_3}{C}}-CH_2CH_2OH \qquad (100)$$

The nature of oligomers in tetrahydrofuran (THF) polymerization systems is very much dependent on the kind of initiator. In polymerization systems initiated with ethyl trifluoromethanesulfonate, linear and cyclic oligomers were found by gas chromatography/chemical ionization mass spectrometry, while in systems initiated with trifluoromethanesulfonic acid (triflic acid), only cyclic oligomers were formed under similar polymerization conditions.[187,188] Other cationic systems such as trimethyloxonium borate ($Me_3O^+ BF_4^-$) and methyltriflate (CF_3SO_3Me) also led to a mixture of linear dimethoxyethers (2, 3, 4, . . . , 8) and macrocyclic oligomers (3, 4, . . . , 9), i.e., crown ethers: 15-crown-3(trimer), 20-crown-4(tetramer), etc., the largest ring detectable being 45-crown-9(nonamer). The predominance of 20-crown-4 at short polymerization times may be due to a tail-biting cyclization:

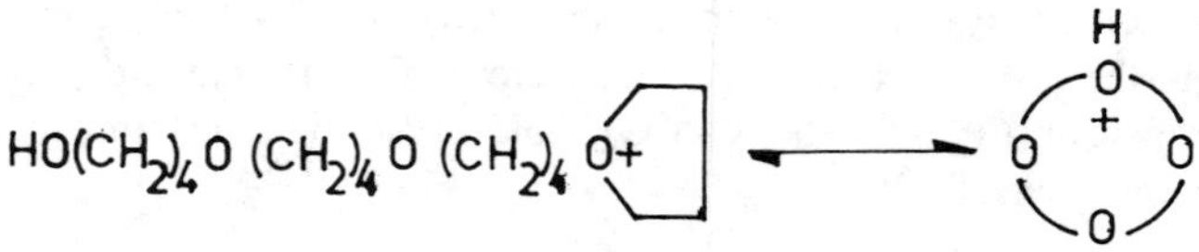

$$HO(CH_2)_4O\,(CH_2)_4\,O\,(CH_2)_4\,O+ \qquad \longleftarrow \qquad (101)$$

and higher macrocyclic oligomers are probably formed by backbiting.

The molecular weight of poly(tetrahydrofuran) could be increased by using a mixture of fuming sulfuric acid and aromatic hydrocarbons or aromatic sulfonic acids.[189,190] When benzene was used, $\overline{M}_n$ increased from 739 (no benzene added) to 3425 (0.05 M C_6H_6).[189] The explanation may imply control of the SO_3 content by the formation of benzene pyrosulfuric acid *in situ:*[190]

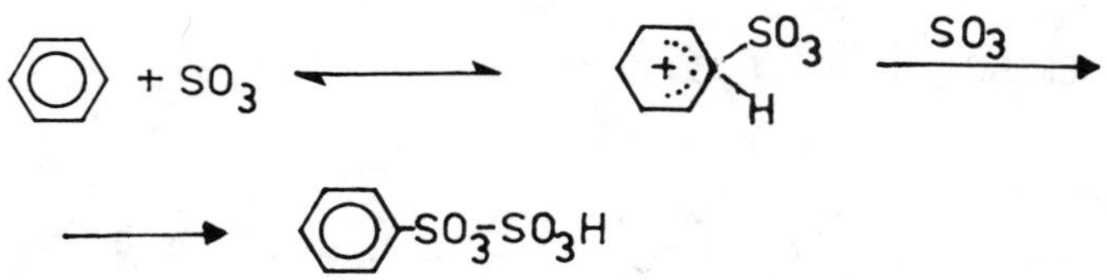

$$\text{(102)}$$

The cationic polymerization of THF is living in the sense that the concentration of oxonium ions remains constant under certain conditions.[191,192] The active species are highly reactive and can therefore be transformed into interesting groups by reaction with suitable nucleophiles. However, since the chain end is equally as reactive or even more reactive than the monomer, control of $\overline{M}_w$ and F is complicated by the fact that chain coupling may occur. Monofunctional low $\overline{M}_w$ polymers, usually $\overline{M}_n < 10,000$, obtained by end capping of living macromolecules with monofunctional initiators (e.g., anhydrides), should be used for the preparation of bifunctional telechelics (Equations 103 and 104):[193,194]

$$\text{(103)}$$

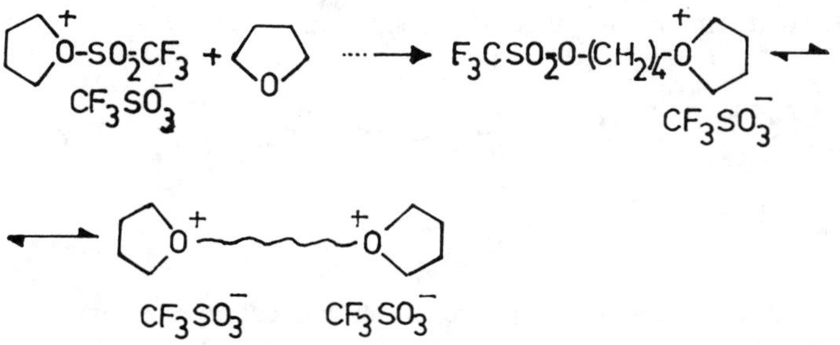

$$\text{(104)}$$

Successful control of the molecular mass and functionality has only been achieved at low initiator concentrations due to low solubility of the bis-oxonium salts.[194] Due to the reversibility of the polymerization process of THF ($T_c = 85°C$) and the high reactivity of the oxonium functions, the bifunctional living polymers had to be end-capped in order to be handled as telechelics. This new functionalization is easily achieved if a nucleophile is introduced in the solution of the living polymer.[195]

Various anionic derivatives have been used,[196-198] new active species being formed by the addition of cyclic amines:[199-201]

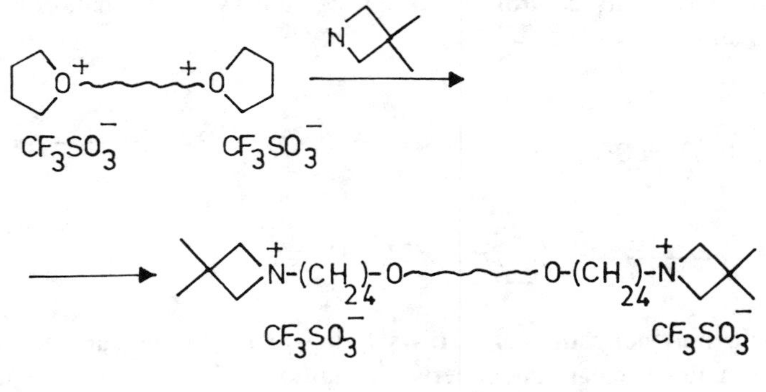

(105)

or thiolane:

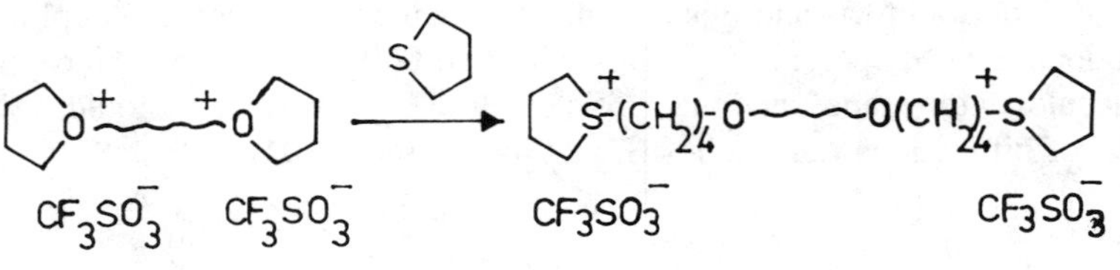

(106)

Addition of thiolane[202] to the reaction mixture leads to immediate termination, with formation of the corresponding α,ω-dithiolonium-terminated poly(tetrahydrofuran) (Equation 106). The thiolonium end groups are inert toward water, but do react with carboxylate anions at slightly elevated temperature under dry conditions.

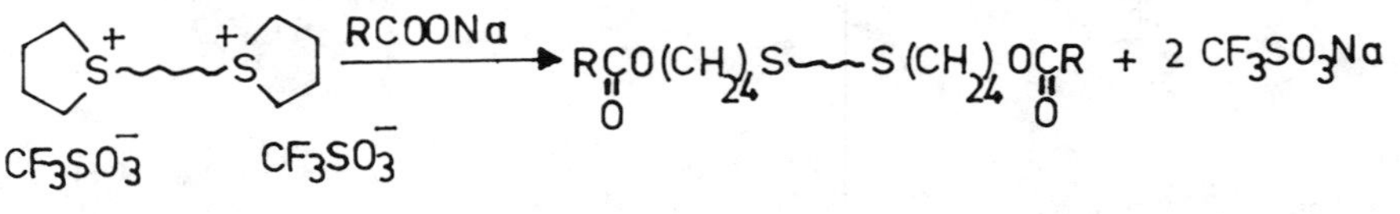

(107)

This reaction has been used to prepare block copolymers with carboxyterminated polymers (Equation 107) such as polystyrene and polybutadiene. With multifunctional carboxylates such as pyromellitic acid, poly(tetrahydrofuran) networks are formed. An important feature of this reaction is that it does not occur in solution as long as a polar solvent is present (e.g., 2% water). Casting of a solution of a prepolymer ($\overline{M}_n$ = 8000) followed by curing at 60°C for 2 h led to elastomeric films. A typical stress strain curve of such a cross-linked poly(tetrahydrofuran) film at 22°C had an inflexion point at 3 MPa/250% and 500% elongation, for a stress of 10 MPa.[202]

The less reactive, weakly electrophilic azetidinium ions undergo ring-opening reactions with charged nucleophiles such as di-, tri-, and tetrafunctional carboxylates.[203,204] However, block copolymers are formed when more reactive cyclic amines such as aziridines *(vide infra)* are added to the living poly(tetrahydrofuran).[199,200]

When polymerizaton of tetrahydrofuran is initiated by protons in the presence of anhydrides as chain-transfer agents, the obtained oligomers have ester-terminated groups.[205,206] The corresponding α,ω-dihydroxy telechelics are then formed by hydrolysis. End-capping reagents and the resulting end groups used in the synthesis of poly(tetrahydrofuran) have been recently reviewed.[207]

TABLE 6
Equilibrium Concentration of Cyclic Oligomers of Macrocyclic Formals[221]

Monomer	T °C	[M]₀ mol/l	Equilibrium conc (10^2 mol/l)									
			x = 1	x = 2	x = 3	x = 4	x = 5	x = 6	x = 7	x = 8	x = 9	x > 9
	+30	0.394	7.2	5.8	3.3	2.4	1.4	1.1	0.8	0.6	0.5	15.80
TOC	0	0.118	3.6	3.2	2.0	1.1	0.5	0.3	0.2	—	—	0.95
	0	0.387	3.7	5.4	4.1	2.5	1.7	1.3	1.0	0.8	—	17.91
	−30	0.399	1.3	4.4	3.7	2.2	1.4	1.1	0.9	0.7	0.6	23.63
	+30	0.182	2.1	4.5	2.2	1.3	0.8	0.6	0.5	—	—	6.10
11-CF-4	0	0.196	1.2	3.6	1.8	1.3	1.0	0.7	0.7	0.6	—	9.31
	0	0.327	1.4	3.4	2.1	1.6	1.1	—	—	—	—	24.16
	−20	0.195	0.7	3.2	1.6	1.1	0.9	0.8	0.6	—	—	10.70
	+30	0.172	2.8	3.0	1.7	1.0	0.6	0.5	0.4	—	—	6.73
	0	0.151	2.2	2.2	1.5	1.0	0.7	0.5	0.4	—	—	8.35
	0	0.297	2.1	2.0	1.8	1.1	0.7	0.6	0.5	—	—	20.87
	−30	0.185	1.0	1.4	1.8	1.0	0.7	0.5	0.4	—	—	11.63

In the case of 1,3-dioxolane, small amounts of only cyclic oligomers going from dimer to nonamer have been isolated.[208] The formation of 1,3,5-trioxepane under the influence of boron trifluoride etherate[209] was taken as an indication of the existence of a backbiting process in the polymerization of 1,3-dioxolane.[145] Experimental data and a theoretical approach have been presented,[210,211] indicating that the backbiting mechanism seems to be the most probable interpretation of ring formation in the cationic polymerization of this monomer.

The occurrence of a backbiting reaction is the reason for the formation of cyclic oligomers in the cationic polymerization of trioxane.[212-214] During the reacton, an eight-membered ring, tetroxane, and minor amounts of a ten-membered ring, pentoxyne, are usually formed.

Appreciable quantities of dimer were produced under the influence of perchloric acid from 1,3-dioxane[145,215] and 1,3-dioxepane.[215,216] Since in the latter case $\overline{DP}_n$ increased with conversion until a maximum value was attained and then decreased drastically at higher conversions, this phenomenon was attributed to a possible degradation of polymer chains.

In the polymerization of the seven-membered cyclic ether, i.e., 1,3-dioxepane, both intra- and intermolecular ionizations have to be considered.[148] Moreover, in the presence of very reactive initiators such as the trimethylsilylesters of strong acids,[217] the polymerization of this oxacyclic monomer proceeds via ion pairs or even free ions ($\overline{M}_n$ = 700 to 5000).[218]

1,3-Dioxacycloalkanes, e.g., 1,3-dioxacyclooctane, were found to polymerize in two stages, forming cyclic oligomers in the first stage and mainly high polymers in the second stage with boron trifluoride ether complex as an initiator.[219,220] Some of the formed cyclic oligomers, i.e., macrocyclic formals having ether oxygens along with acetal oxygens, were isolated and polymerized. Thus, macrocyclic formals, such as 1,3,6-trioxacyclooctane (TOC), 1,3,6,9-tetraoxacycloundecane (11-CF-4), and 1,3,6,9,12-pentaoxacycloheptadecane (17-CF-6), were polymerized in dichloromethane by using boron trifluoride ethyl etherate as a catalyst.[221] The polymerization was accompanied by formation of cyclic oligomers with monomer units n ≤ 9. The equilibrium concentration of these oligomers was dependent on the reaction temperature. As expected from the Jacobson-Stockmayer theory,[222] the concentration of each product at equilibrium at a given temperature (0°C) was influenced exclusively by the concentration of total monomer present. Some results are given in Table 6.[221]

Although the enthalpy changes for polymerization, ΔH_p, were observed to be finite even for the 34-membered ring (i.e., dimer of 17-CF-6 ΔH_p = 3.18 kJ/mol), those for the cyclic oligomers having rings with more than 40 members were virtually equal to 0. The polymerizability of these macrocyclic oligomers was therefore controlled only by the entropy term. The molar cyclization equilibrium constant decreased in proportion to the -2.5 power of the ring size x,[221] in accordance with the Jacobson-Stockmayer theory, when ΔH_p is negligibly small.

Dioxabicyclo- derivatives form preferentially cyclic dimers and low molecular weight polymers when cationic catalysts are used. For example, bis-7, 9-dioxabicyclo-4,3,0-nonane was transformed quantitatively into *cis-anti-cis* dimer (Equation 108) with phosphorous pentafluoride, whereas the *trans*-isomer formed the *trans-anti-trans* dimer (Equation 109).[223] The dimerization mechanism

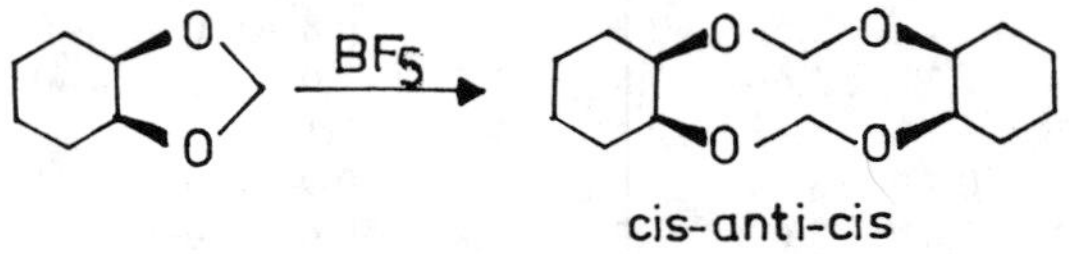

$$\text{(108)}$$

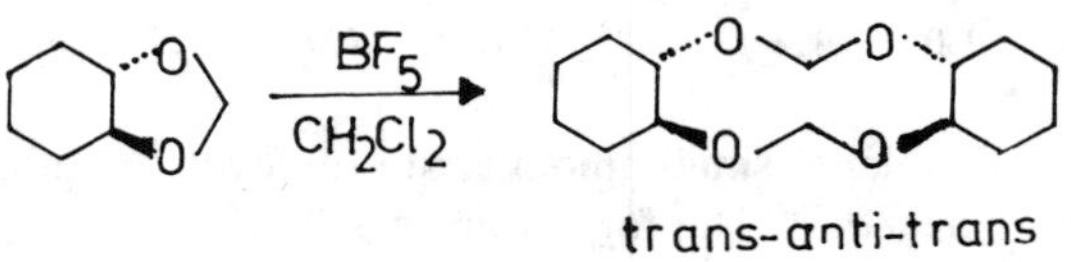

cis-anti-cis

$$\text{(109)}$$

trans-anti-trans

resembled the ring-expansion mechanism proposed by Plesch and Westerman for the polymerization of cyclic acetals:[215]

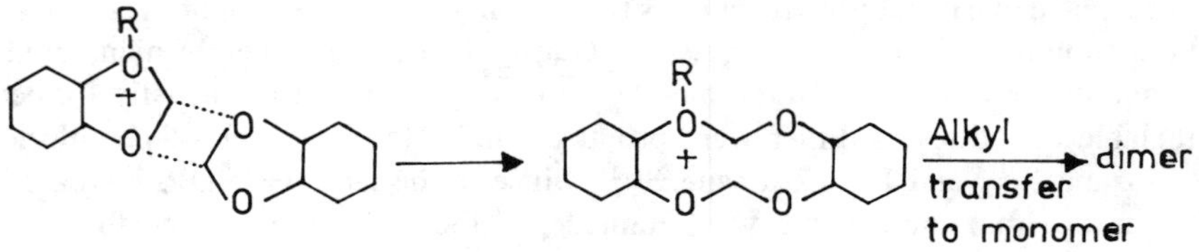

$$\text{(110)}$$

Similarly, the axial isomers of 4-bromo-6,8-dioxabicyclo 3.2.1-octane-7-one showed a tendency to cyclodimerize in the presence of PF_5 or $SbCl_5$, particularly at higher temperatures, whereas the equatorial isomer was much less reactive.[224]

2. Lactones

In contrast to the five- and six-membered rings, all four-, seven-, and eight-membered lactone rings are polymerizable.[225] A perfectly controllable oligomerization of ε-caprolactone was observed when the polymerization was initiated with a trialkyloxonium salt in combination with an alcohol.[226,227] This process resembles the activated monomer polymerization mechanism of oxiranes.[170-173]

A hydroxo mechanism was proposed for the cationic oligomerization (20 to 60°C) in CH_2Cl_2 of ε-caprolactone initiated by *p*-toluene sulfonic acid in the presence of ethylene glycol in which chain growth occurred at monoalkyloxonium cation-active centers.[228] The oligomerization rate sharply increased on increasing the diol concentration in the reaction mixture and no polymerization was observed in the absence of glycol. As expected, the molecular weights of the obtained oligomers decreased on increasing the diol concentration, but were independent of acid concentration. The rate of the bulk oligomerization (30 to 60°C) of ε-caprolactone with diols such as ethylene glycol, hexanediol, butanediol, and hydroquinone in the presence of the onium complex catalysts ($(C_6H_5)_4PFeCl_4$ and $(C_4H_9)_4NFeCl_4$ was two to three orders higher than in the absence of catalysts.[229] The molecular weight of the obtained oligomers was controlled exclusively by the chain transfer to the diol in reactions carried out at $[\text{diol}]_o \gg [\text{catalyst}]_o$. The oligomerization mechanism involved catalyst activation of intermediate monomer-diol complexes by binding with the lactone ring to form a new ternary reactive complex. Chain growth occurred via two complexing processes: (1) monomolecular decompositon of the monomer-diol-catalyst complex and (2) bimolecular reaction of the complex with lactone. The results indicated that the contribution of the monomolecular reaction with respect to the bimolecular reaction decreased with increasing temperature.

The acid number of oligocaprolactone diols is decreased by using perfluorinated sulfonic acids such as perfluorocyclohexane sulfonic acid.[230]

3. Thiiranes

Both the three-membered thiirane rings and the four-membered thietane rings can be polymerized cationically, while the five-membered thiolane ring cannot be polymerized at all.[231]

All substituted ethylene sulfides (thiiranes) seem to behave in a similar way in cationic conditions: a rapid polymerization followed by degradation to cyclic oligomers.[145,232,233] For example, poly(propylene sulfide) obtained in a two-stage polymerization with triethyl oxonium tetrafluoroborate slowly degraded to form a mixture of low molecular weight polymer, a small amount of cyclic pentamer, and about 50% cyclic tetramer.[232,233] If high molecular weight poly(propylene sulfide) obtained by an anionic or a coordinative polymerization was treated with a catalytic amount of oxonium salt, the polymer degraded rapidly to again form a mixture of cyclic pentamer and tetramer in a ratio of approximately 1:10.[144]

Trans-2,3-butylene sulfide polymerized quantitatively with $(C_2H_5)_3O^+BF_4^-$ (Equation 111), but the polymer rapidly degraded to form equimolecular amounts of 3,4,6,7-tetramethyl-1,2,5-trithiacycloheptane and *trans*-butene.[234]

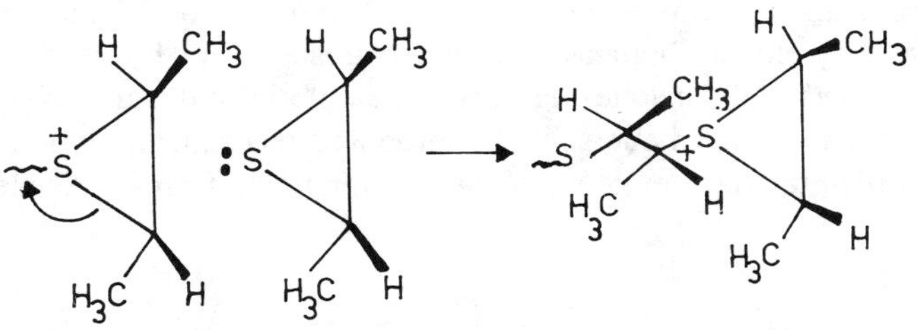

$$(111)$$

The polymer of *cis*-2,3-butylene sulfide formed a mixture of tetramer, trithiacycloheptane derivative, and *cis*-butene.[234] *Trans*-2,3-butylene sulfide has a three configuration (RR or SS). During the S_N2 attack of the sulfide function of a monomer molecule (propagation step, Equation 111) on the α-carbon of a *three*-membered ring sulfonium salt, one of the two asymmetric carbons will undergo an inversion of configuration leading to a polymer in which the units have *erithro* configurations (RS, nSR, and meso form). Since in the transition state of elimination reactions the two leaving groups generally must occur in an anti- position to each other, only *trans*-butene and no *cis*-butene should be formed (Equation 112):

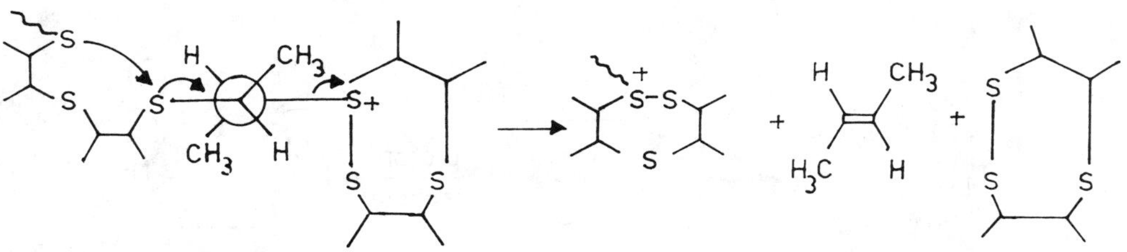

$$(112)$$

The exclusive formation of *cis*-butene from cationic *cis*-2,3-butylene sulfide polymers can be explained in a similar way.[144]

With triethyloxonium tetrafluoroborate as initiator, a rapid and quantitative polymerization of styrene sulfide was observed, followed by a slow degradation of the polymer to a mixture of *cis*- and *trans*-2,5-diphenyl-1,4-thiane and *cis*- and *trans*-2,6-diphenyl-1,4-dithiane.[234,235] Since BF_4^- counterion is not capable of forming a covalent bond, a backbiting reaction with sulfonium ion was assumed as the plausible mechanism for the dimer formation.

Because of these shortcomings, i.e., the rapid chain transfer to polymer and degradation of the active polymer to cyclic products, the cationic oligomerization has not yet been reported as a method for synthesizing telechelics from thiiranes.

4. Cyclic Amines

It seems that cyclic amines can be polymerized only cationically and the polymerizations have a living character.[236-240] It appears that, in all cases, the presence of substituents retards the termination more than it does the propagation. Consequently, higher molecular weight polymers are obtained when substituted monomers are used.[237] Whether an aziridine (ethylene imine) forms a polymer, an oligomer, or a mixture of the two reacton products depends on small changes in monomer structure and on the initiator and reaction conditions.

The polymerization of *N*-alkylaziridines is generally characterized by a fast propagation and a fast termination reaction which consists of a nucleophilic attack of a polymer amino function on the active species, the aziridinium ion. Therefore, these polymerizations generally stop at limited conversions, producing low molecular weight polymers. *N-tert*-butylaziridine is an exception to the rule which is ascribed to the "steric deactivation" of the polymer amino functions of the bulky *tert*-butyl group.[240] If the aziridine ring carries an additional substituent on one of the carbon atoms, the polymerization behavior changes dramatically (Table 7).

A number of 1-alkyl-aziridines were converted into the corresponding dimer, 1,4-di-alkylpiperazines, in yields as high as 95% when treated with hydrohalic acids in polar solvents (acetone, methylethylcetone). However, perchloric acid and *p*-toluenesulfonic acid resulted in the formation of polymers.[241] The proposed mechanism involves the formation of the dimer in which the halogen atom of the initiating acid forms a covalent bond with a carbon atom:

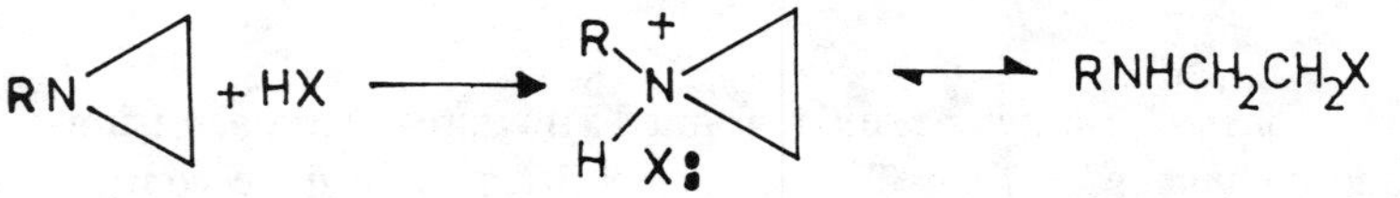

$$\text{(113)}$$

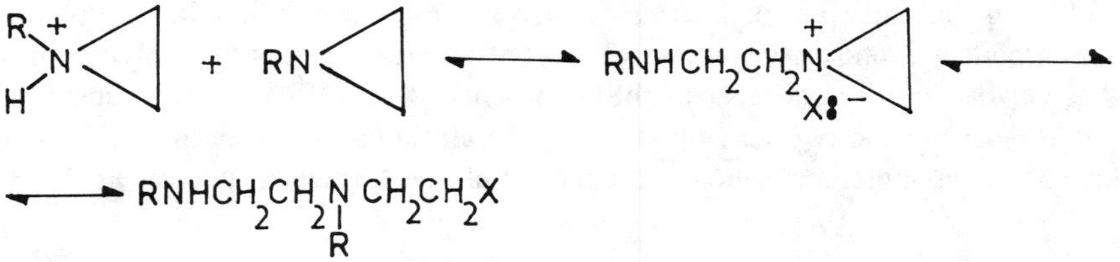

$$\text{(114)}$$

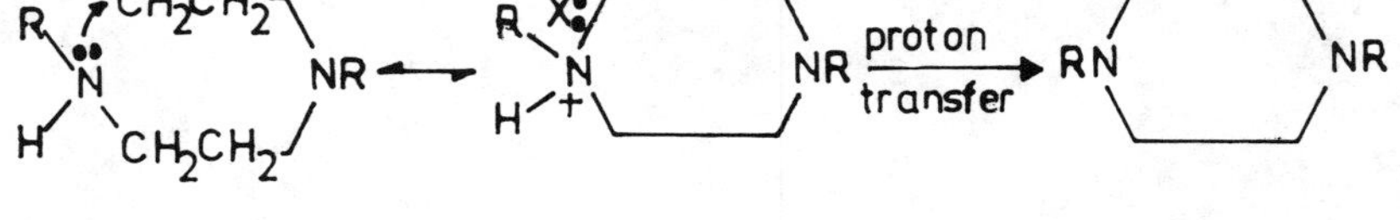

$$\text{(115)}$$

TABLE 7
Polymerizability of Aziridine Monomers[240]

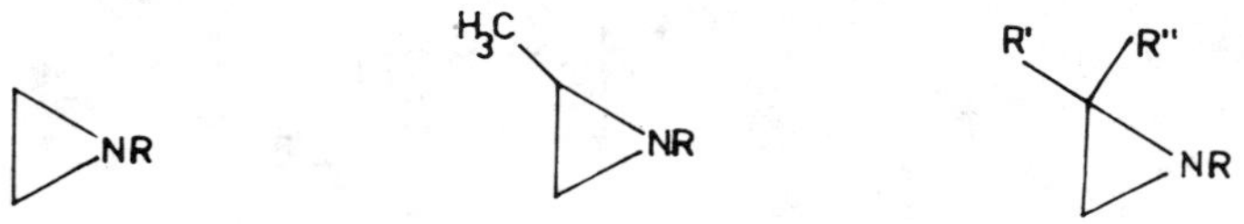

Fast polymerization	Slow polymerization	No polymerization

R: i–C_3H_7	R: –$CH_2C_6H_5$	R',R''H,CH_3H,C_6H_5 CH_3,CH_3
t–C_4H_9	–$CH_2CH_2C_6H_5$	R t–C_4H_9 t –C_4H_9 –$CH_2C_6H_5$
–$CH_2C_6H_5$	–CH_2CH_2CN	
–$CH_2CH_2C_6H_5$		
–CH_2CH_2CN		

When alkyl halides or methyl *p*-toluenesulfonate were used as the initiator, the oligo-merization stopped at the *N*-alkylpiperazinium salt stage.[241]

A mixture of polymers and the corresponding piperazine derivatives were formed when 1-substituted aziridines were treated with triethyl oxonium tetrafluoroborate in CH_2Cl_2. Since BF_4^- ion is not able to form a covalent bond, degradation of polymer involving quaternary ammonium salts (Equation 116) is the most probable mechanism:[145]

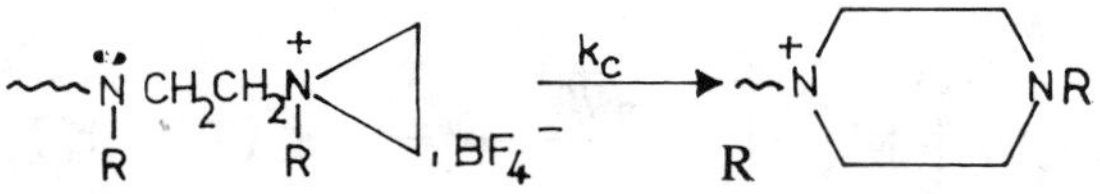

$$(116)$$

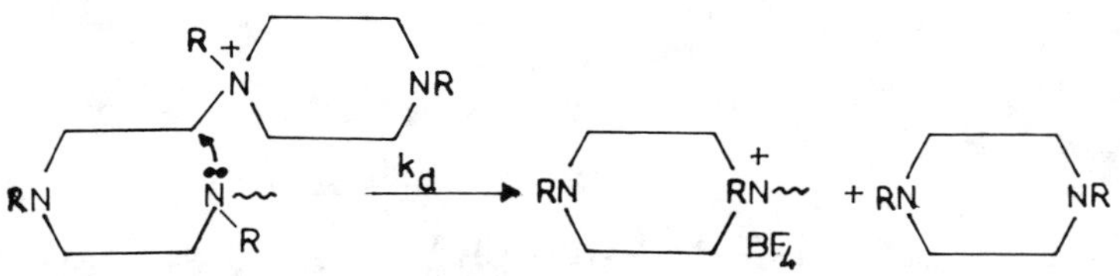

$$(117)$$

Formation of the first piperazinium salt may be regarded as a termination reaction for the propagation (Equation 117). Dimer formation through a backside attack of the penultimate nitrogen was also considered the most probable mechanism for the formation of dimers in the systems 1-(2-triethylsilyl)aziridine/$AlCl_3$[242] and 1-methylaziridine/$HClO_4$[243] as well as 1-ethyl-, 1-allyl-, and 1-(2-hydroxyethyl)aziridine-formed dimers.[145] However, 1-(2-hydrox-yethyl)aziridine did not form the corresponding piperazine,[239] whereas 1-(2-cyano-ethyl)aziridine formed a mixture of polymer and tetramer.[244] 1-Benzylaziridine was reported to form a cyclic tetramer in nearly quantitative yields with *p*-toluensulfonic acid.[245] 1-Isopropyl-1-*tert*-butyl- and 1-phenyl-aziridine did not form oligomers when treated with oxonium ions.[145]

Azetidine (trimethylene imine) resembles aziridine in polymerizability.[236] Conidine, a cyclic aziridine, and 7-ethyl-, 8-methyl-, and 6,8-dimethyl conidine have been polymerized in bulk with boron fluoride etherate:[246]

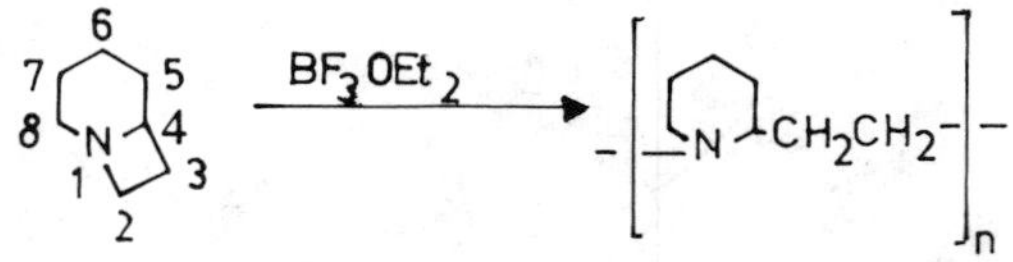

$$\text{(118)}$$

Isotactic crystalline polymers melting at 94°C were obtained by cationic polymerization (BF$_3$·OEt$_2$) of *d*- and *1*-2(2-hydroxyethyl)piperidine, whereas the polymer (oligomer) of the *dl* monomer was soft and tacky at room temperature.[247]

The high living character of the cationic ring-opening polymerization of aziridines has been used to synthesize a number of poly(1-*tert*-butylaziridine)s with functional end groups[239] and AB and ABA block copolymers.[200] The latter ones were produced by the two classical methods: (1) initiation from a living polymer and (2) coupling reaction of the active end group with nucleophile end groups of another prepolymer. These methods provided the possibility of producing a variety of block copolymers with polyamine segments which can be transformed into hydrophilic polyelectrolyte chains by quaternization with proton acids or with strong alkylating agents.

An early example of the synthesis of block copolymers with 1-*tert*-butylaziridine (TBA) segments by the "blocking form" method has been described using "dormant" cationic polystyrene (prepared by perchloric acid in methylene chloride at $-78°C$) as the macro-initiator.[248] The perchloric ester end groups, unable to induce the cationic vinyl polymerization, were able to initiate the much stronger nucleophilic aziridine:

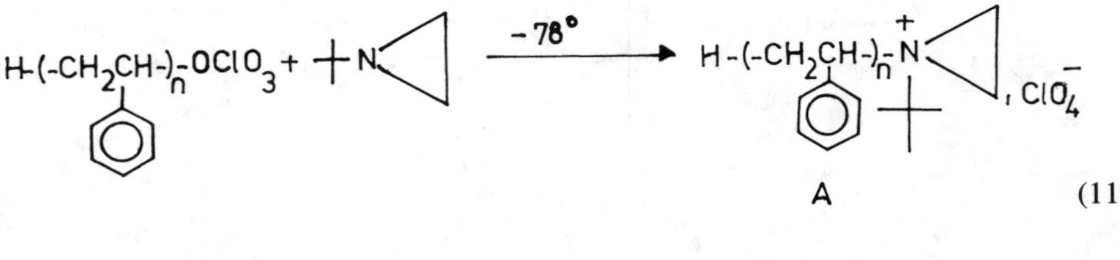

$$\text{A} \qquad \text{(119)}$$

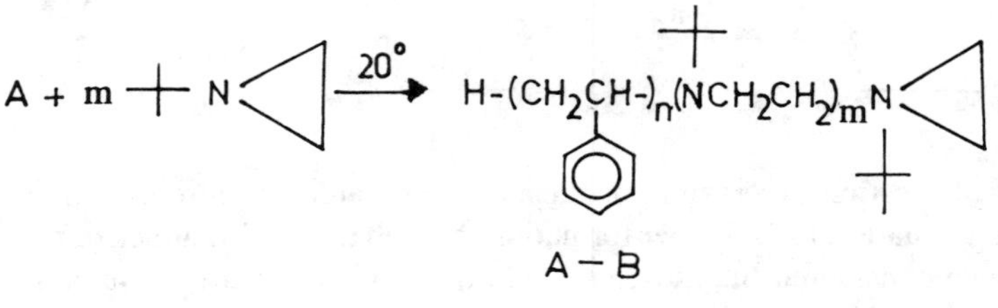

$$\text{A} - \text{B} \qquad \text{(120)}$$

Since oxonium ions are strong electrophiles, a block copolymer was formed when TBA was added to living polytetrahydrofuran:[200]

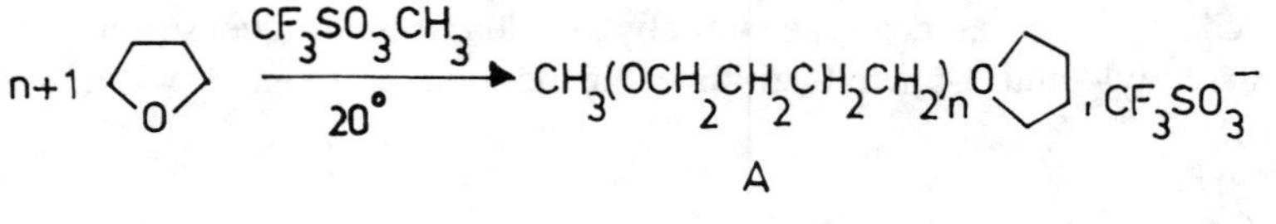

$$\text{(121)}$$

A

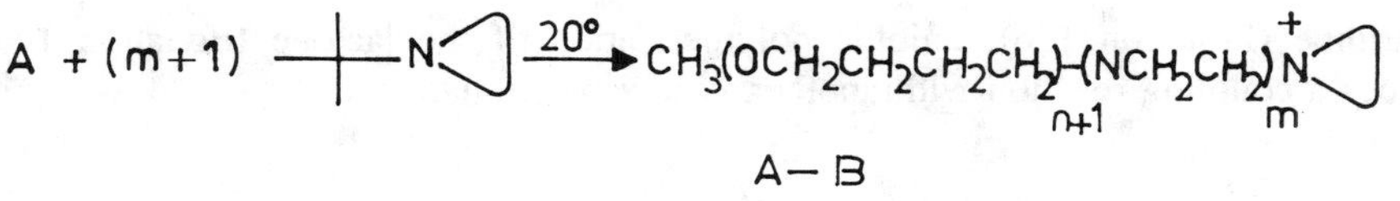

$$A + (m+1) \quad \underset{N}{\overset{20°}{\longrightarrow}} \quad CH_3(OCH_2CH_2CH_2CH_2)_{n+1}(NCH_2CH_2)_m \overset{+}{N} \quad (122)$$

$$A-B$$

The block lengths of segments A and B can be varied over a broad range by appropriate choice of the initiator concentration, reaction time for the THF oligomerization, and amount of TBA added. A representative block copolymer had a polyether segment of molecular weight 3000 and polyamine segment of $\overline{M}_n = 3000$ with $\overline{M}_w/\overline{M}_n = 1.2$.[200] Addition of TBA to bifunctional living poly(THF) gave the corresponding BAB block copolymers (Equation 105). With triflic anhydride as an initiator, poly(THF) blocks with low polydispersity ($\overline{M}_w/\overline{M}_n < 1.3$) are obtained if the $[M]_o/[I]_o$ ratio is higher than 500.[194]

Reaction of equivalent amounts of telechelic carboxy-terminated polybutadiene (CTB, $\overline{M}_n = 4800$, $F = 2.1$) and living poly(TBA) led to the ABA type of block copolymers with poly(TBA) segments of $\overline{M}_n = 3000$:

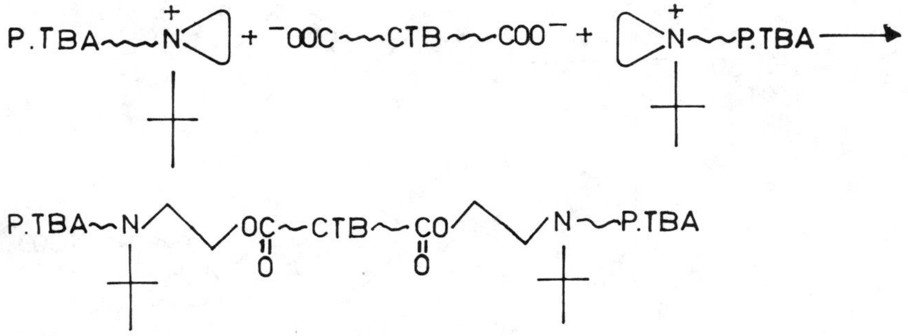

$$P.TBA\sim\overset{+}{N} + {}^-OOC\sim CTB\sim COO^- + \overset{+}{N}\sim P.TBA \longrightarrow$$

$$P.TBA\sim N\diagdown OC\sim CTB\sim CO\diagup N\sim P.TBA \qquad (123)$$
$$\underset{O}{\overset{\parallel}{}} \qquad \underset{O}{\overset{\parallel}{}}$$

The aziridinium end group of living poly(TBA) is readily opened by primary amino functions[239] and, consequently, amino-terminated ABA block copolymers were formed by reacting living poly(TBA) with telechelic amino-terminated poly(ethylene oxide) (Equation 124).[200] The latter was prepared from hydroxy-terminated polyether $\overline{M}_n = 7000$ via the tosylate and Gabriel synthesis with phthalimide.[249]

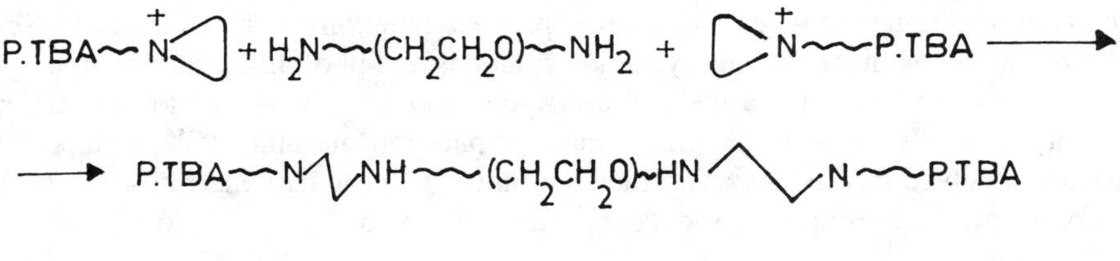

$$P.TBA\sim\overset{+}{N} + H_2N\sim(CH_2CH_2O)\sim NH_2 + \overset{+}{N}\sim P.TBA \longrightarrow$$

$$\longrightarrow P.TBA\sim N\diagdown NH\sim(CH_2CH_2O)\sim HN\diagup N\sim P.TBA \qquad (124)$$

The block copolymers are water soluble if the polyether block is larger than the sum of the two P·TBA blocks, but all are soluble in aqueous acid.

If a tertiary amine (e.g., an N-substituted aziridine) is reacted with a lactone, a zwitterion containing an ammonium ion and a carboxylate ion is formed. The zwitterion formed from N-alkyl aziridine and β-propiolactone (Equation 125) initiates a cationic polymerization of the aziridine (Equation 126), anionic polymerization of the lactone (Equation 127), or undergoes a coupling reaction with another zwitterion (Equation 128):

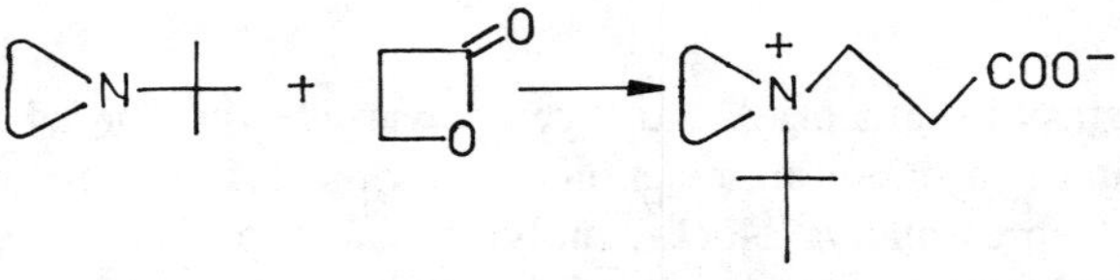

$$\tag{125}$$

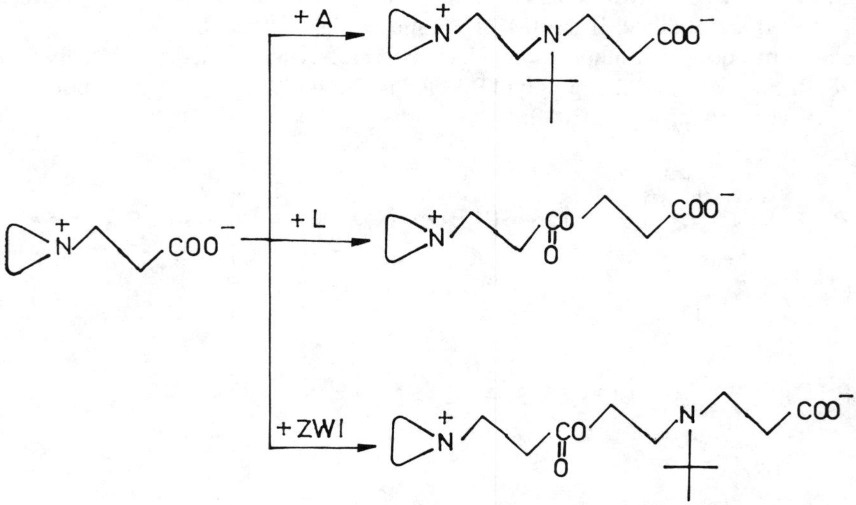

$$\tag{126}$$

$$\tag{127}$$

$$\tag{128}$$

The last reaction is the reason for the formation of cyclic oligomers (up to 50%) with $\overline{M}_w$ = 342 to 468 (determined by mass spectrometry) and the general formula $A_n L_m$ with n = 2, 3, 4 and m = 1, 2, the most abundant being the four-unit cooligomer AAAL and the penta-unit AAAAL oligomer.[240]

5. Cyclic Phosphorous Monomers

The ring-opening polymerization of cyclic phosphorous monomers is a versatile method for producing functional polymers having phosphorous functions in the main chain. Several trivalent phosphorous monomers of cyclic phosphinites and phosphites have been polymerized with a cationic initiator via the Arbuzov-type reaction, giving rise to low molecular weight polymers ($\overline{M}_w \sim 10^3$) of a phosphinate or phosphonate unit.[250-256] A side reaction was usually involved to produce an isomerized unit.[250,251] The cationic ring-opening polymerization of cyclic phosphinates (deoxophosphonates) A to C shown below

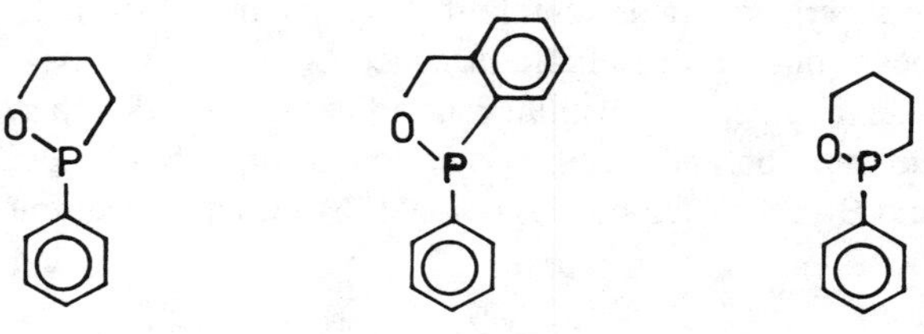

A   B   C

yielded polymers made exclusively of a phosphine oxide repeating unit.[252-256] It was found that the polymerization of A initiated by methyltriflate, $CH_3OSO_2CF_3$ (MeOTf), involved a stable cyclic phosphonium species and produced a powdery material of $\overline{M}_w = 3900$.[252,253]

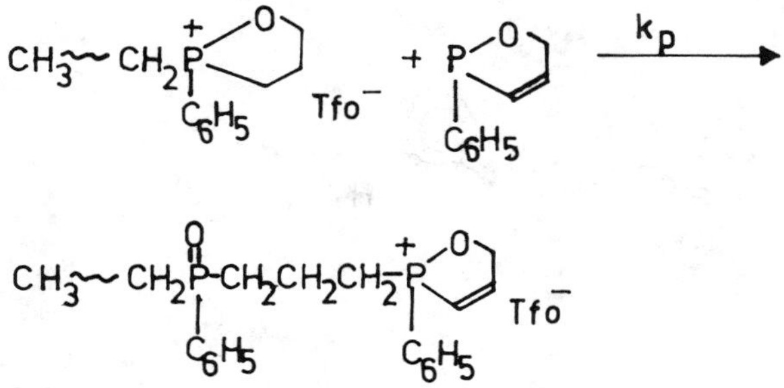

$$CH_3 \sim CH_2\overset{+}{P}\underset{C_6H_5}{\overset{O}{|}} \; Tfo^- \;+\; \overset{O}{\underset{C_6H_5}{P}} \xrightarrow{k_p}$$

$$CH_3 \sim CH_2\overset{O}{\underset{C_6H_5}{P}}-CH_2CH_2CH_2\overset{+}{\underset{C_6H_5}{P}}\overset{O}{} \; Tfo^- \tag{129}$$

The propagation as described by Equation 129 is an S_N2 reaction of the Arbuzov type between a P(IV) cation and a P(III) dipole to produce a P(V) phosphine oxide unit.

An intramolecular Arbuzov-type reaction seemed to be involved in the alkyl halide-initiated systems (e.g., CH_3I) since the cyclic phosphonium species was detected:[256]

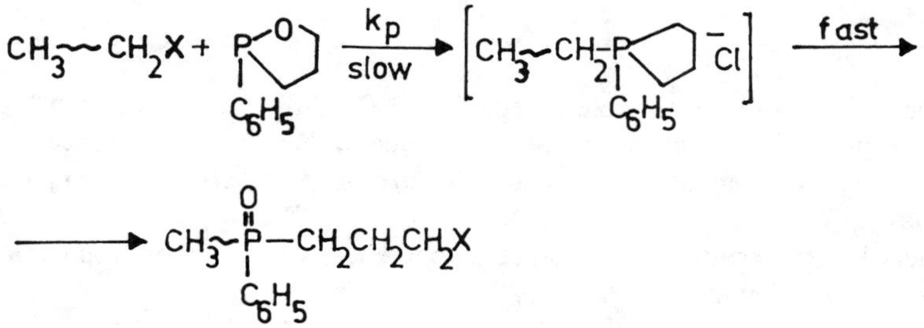

$$CH_3 \sim CH_2X + \overset{O}{\underset{C_6H_5}{P}} \xrightarrow[slow]{k_p} \left[CH_3 \sim CH_2\overset{+}{\underset{C_6H_5}{P}} \; Cl^- \right] \xrightarrow{fast}$$

$$\longrightarrow CH_3 \sim \overset{O}{\underset{C_6H_5}{P}}-CH_2CH_2CH_2X \tag{130}$$

Typical cationic initiators of the MeOTf and $Et_3O^+BF_4$ type reacted with B to produce stable cyclic phosphonium species which did not propagate

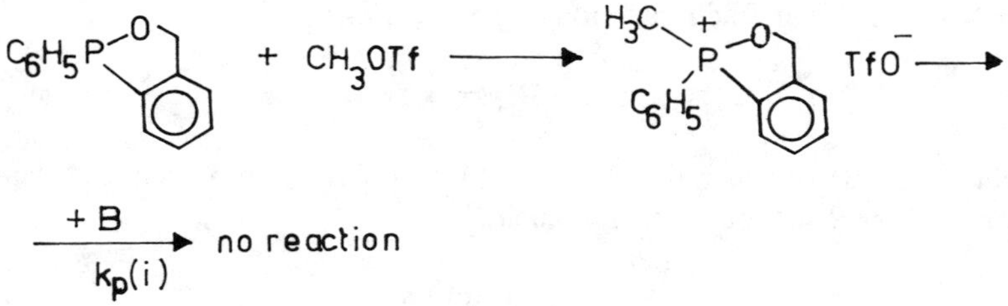

$$C_6H_5P\overset{O}{\bigcirc} + CH_3OTf \longrightarrow \underset{C_6H_5}{\overset{H_3C}{\underset{}{P}}}\overset{+}{\overset{O}{\bigcirc}} \; TfO^- \longrightarrow$$

$$\xrightarrow[k_p(i)]{+B} \text{no reaction} \tag{131}$$

On the other hand, CH_3I induced the polymerization of B:

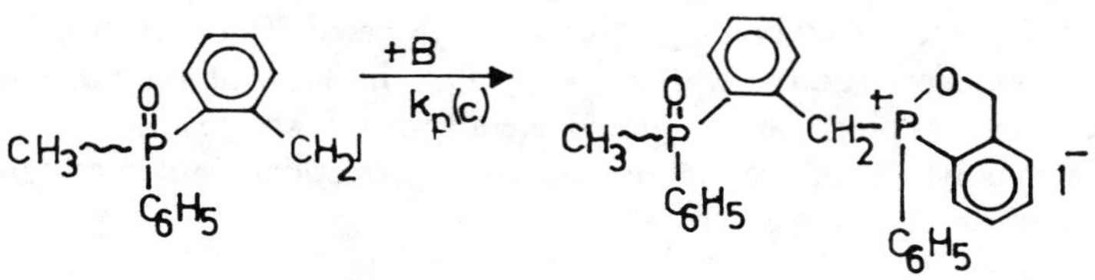

$$CH_3 \sim \overset{O}{\underset{C_6H_5}{P}}\overset{\bigcirc}{}CH_2I \xrightarrow[k_p(c)]{+B} CH_3 \sim \overset{O}{\underset{C_6H_5}{P}}\overset{\bigcirc}{}CH_2\overset{+}{\underset{C_6H_5}{P}}\overset{O}{\bigcirc} \; I^- \tag{132}$$

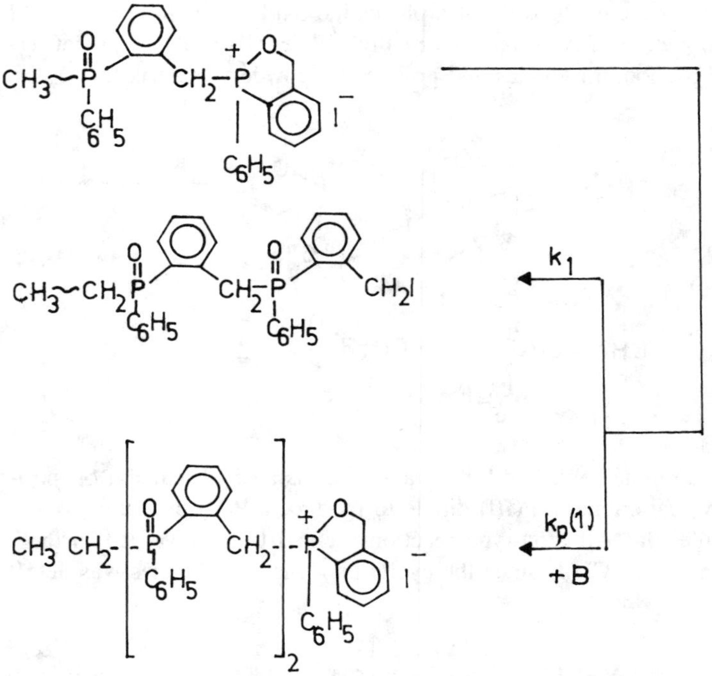

Cyclic phosphonium species were detected by ^{31}P NMR analysis only in highly polar solvents like acetonitrile and not in a less polar toluene due to the difference of the intramolecular rate (k_1) to open the ring via the Arbuzov-type reaction, i.e., k_1 (toluene) > k_1(acetonitrile).

Preliminary kinetic results revealed that the apparent rate constant of propagation, k_p(ap), was larger in toluene. Taking

$$k_p(ap) = k_p(c) \cdot X_c + k_{pi}(X_i) \qquad (135)$$

where k_p(c) and k_p(i) denote the propagation rate constants due to covalent species (Equation 132) and ionic species (Equation 134), respectively, X_c is the molar fraction of covalent species, and X_i is the molar fraction of ionic species, i.e.,

$$X_c + X_i = 1 \qquad (136)$$

In toluene, $X_i = 0$ and therefore $k_p(ap) = k_p(c)$. Taking into account Equation 131, and assuming $k_p(i) = 0$ for ionic species, then

$$k_p(ap) = k_p(c) \cdot x_c \qquad (137)$$

It was found[256] that k_p(ap)(toluene) > k_p(ap)(acetonitrile). Considering that the larger X_c in toluene ($X_c = 1$) compared with X_c in acetonitrile is more than enough to govern k_p(ap) value, notwithstanding the countervailing k_p(c) value in acetonitrile which is higher than that in toluene, the polymerization of B is one of the first instances in ionic polymerizations where ionic species do not propagate, while covalent (neutral) species do.

Monomer C also polymerized via S_N2 ring opening by a cationic initiator to give poly(*p*-phenyl)tetramethylenephosphine oxide.[255]

Macroporous-type cross-linked halomethylated polystyrene was employed instead of CH_3I to initiate the polymerization of A:[255,256]

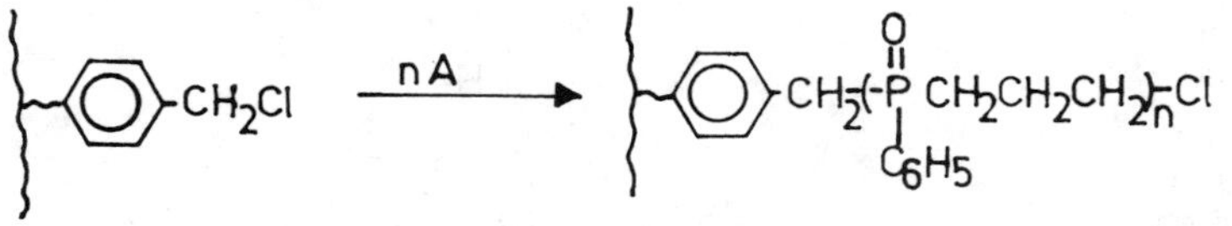

$$(138)$$

The product of grafted oligomer (n = 4 to 10) was a white bead-like resin which showed efficient chelating properties toward heavy metal ions such as UO_2^{2+}, Th^{4+}, and Pd^{2+}.[256]

Polydialkylphosphates related to biopolymers (nucleic and teichoic acids) were prepared by ionic ring-opening polymerization of appropriate monomers.[257-267]

Although the five- and six-membered cyclic esters of phosphoric acid, i.e., 2-alkyl-2-oxo-1,3,2-dioxaphospholane (m = 2) and 2-alkyl-2-oxo-1,3,2-dioxaphosphorinane (m = 3):

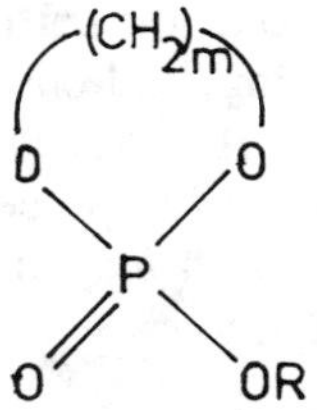

polymerize readily, such as with anionic or cationic initiation, the polymerization degrees are low because of an extensive chain transfer to monomer which competes successfully with chain propagation.[258,262]

Cationic polymerization in bulk of, e.g., 2-methyl-2-oxo-1,3,2-dioxaphosphorinane[259] induced with an initiator-bearing stable anion (like AsF_6^-, PF_6^-, or $CF_3SO_3^-$) proceeded rapidly above 100°C ($\overline{M}_n$ < 3000). At a sufficiently high concentration of initiator, polymerization proceeded to its monomer-polymer equilibrium while, at a lower concentration of initiator, it stopped before the equilibrium was reached. Polymerization in ethylene chloride was first order with respect to monomer. Considering the initiation through a cationation at the 2-oxo-group:

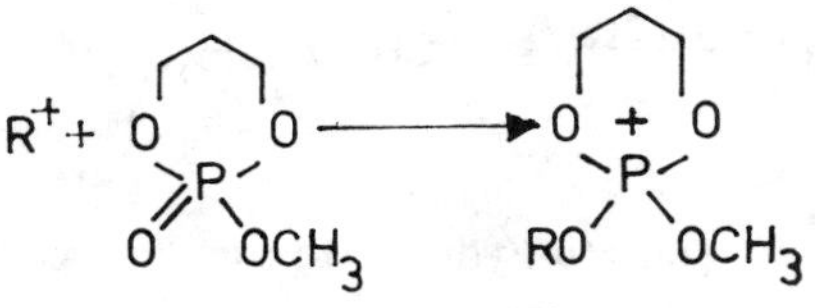

$$(139)$$

(anion not shown)

the first step of chain growth was formulated in the following way:

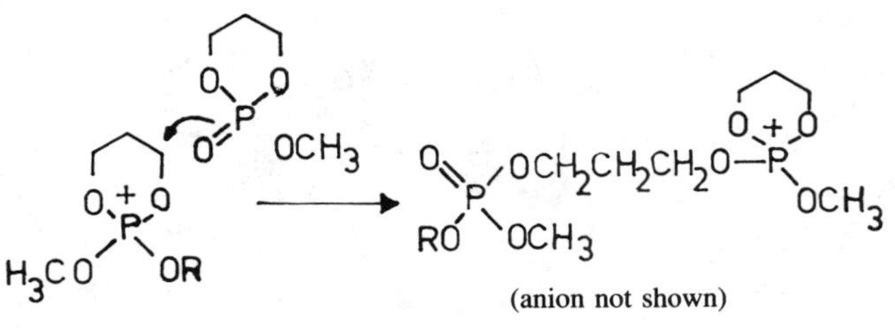

$$(140)$$

(anion not shown)

The competition between chain growth and chain transfer (i.e., oligomer formation) was represented as A below (where O=P is a part of the incoming molecule of monomer):

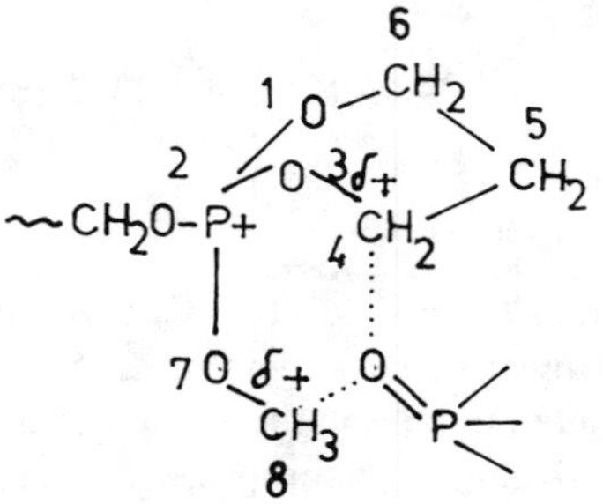

(141)

(anion not shown)

A

In the growing tetraalkyloxonium ions, the partial positive charge is localized on carbon atoms C_4 and C_6, being a part of the ring, and on the exocyclic C_8 atom. Therefore, when an O_3-C_4 bond (or equivalent O_1-C_6 bond) is broken, chain propagation occurs; if, however, the O_7-C_8 bond is broken, then chain transfer proceeds.

Chain transfer leaves an oligomer with a cyclic end group and leads to reinitiation by forming a new tetraalkoxyphosphonium ion:

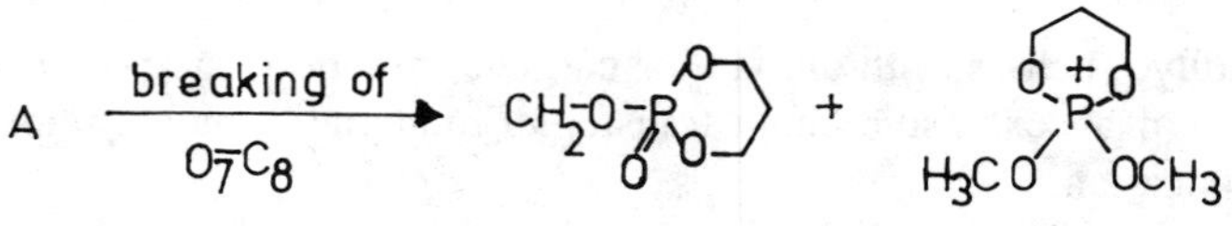

(142)

(anion not shown)

Chain transfer has no kinetic effect because the reinitiation is fast. If the exocyclic ester group was absent (e.g., 2-hydro-2-oxo-1,3,2-dioxophosphorinane), the chain transfer did not occur and the product of polymerization of the corresponding acid was a high molecular weight polyphosphonate:[258,262]

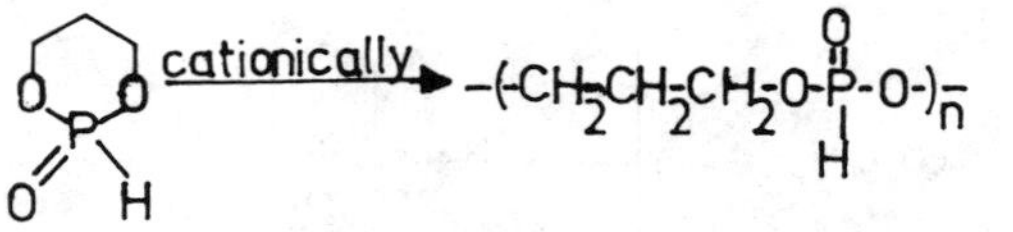

(143)

poly(p-hydro-1,3-propylene phosphonate)

A series of oligomers of phosphine sulfides were prepared by cationic ring-opening polymerization of deoxothiolphostones:[256,263]

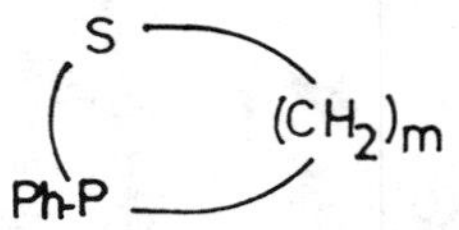

Thus, 2-phenyl-1,2-thiaphospholone (m = 3) was polymerized thermally or by using a cationic initiator (CH_3I, $C_6H_5CH_2BR$, CH_3OTf) to give $[P(S)Ph(CH_2)_3]_n$, while 2-phenyl-1,2-thiaphosphorinone (m = 4) could be polymerized only cationically to give $[P(S)Ph(CH_2)_4]_n$. The polymerizations involved a C–S bond cleavage via an Arbuzov-type reaction (Equation 129) and the $\overline{DP}_n$ was ≤ 25.

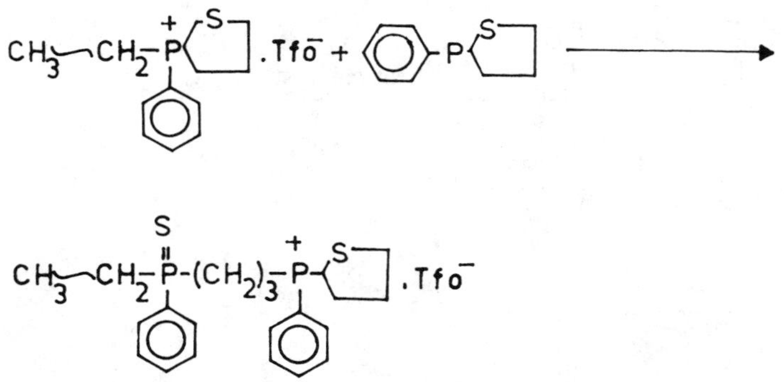

$$\tag{144}$$

The oligomer chelated Hg and Pd quantitatively over a wide pH range, Cu was adsorbed more from weakly basic solutions, and uranyl ion and UO_2^{2+} adsorbtion were unnoticeable.[263]

6. Cyclic Siloxanes

Siloxane oligomers are generally synthesized via anionic or cationic ring-opening polymerization of cyclic siloxane monomers such as octamethylcyclotetrasiloxane, D_4. Molecular weight is regulated through the incorporation of controlled amounts of monofunctional endblockers into the system.[264-266] As a result of the nature of the configurations of siloxane chains coupled with the similar reactivity of siloxane bonds in linear as compared to cyclic species, the anionic or cationic catalysts attack both the rings and chains during polymerization. These so-called *redistribution* or *equilibration* polymerizations involve reactions such as those listed in Equations 145 to 148, which occur through the process:[266, 267]

$$-D_x- + D_4 \rightarrow -D_{(x+4)} \tag{145}$$

$$-D_x- + MM \rightarrow MD_xM \tag{146}$$

$$MD_xM + MM \rightarrow MD_{(x-5)}M + MD_5M \tag{147}$$

$$MD_xM + MD_yM \rightarrow MD_{(x+w)} + MD_{(y-w)}M \tag{148}$$

In siloxane nomenclature, M denotes a monofunctional siloxane unit, whereas D refers to a difunctional siloxane unit; D_4 therefore represents the cyclic siloxane tetramer, while MM is the linear dimer. D nomenclature is normally associated with dimethylsiloxy units.

At thermodynamic equilibrium, these reactions result in a Gaussian distribution of molecular weights among the chain molecules together with an approximately decreasing distribution of ring species as ring size increases.

The use of strong acids such as sulfuric acid and its derivatives, alkyl- and arylsulfuric acid, for the synthesis of silicone oils are described in the literature.[264,266-271] Obtaining a well-defined hydroxy-terminated polysiloxane has not been straightforward since, during the acid- or base-catalyzed equilibrations, there are various side reactions involving the terminal hydroxyl groups.[270,271] Recently, the synthesis of –OH-terminated systems has been successfully demonstrated by cationic ring-opening and equilibration polymerization of oc-tamethylcyclotetrasiloxane catalyzed by superacids.[272-275]

The synthesis of functionally terminated siloxane oligomers via ring-opening equilibration polymerization for obtaining block and grafted copolymers[267-284] is illustrated in the following reaction scheme:[285]

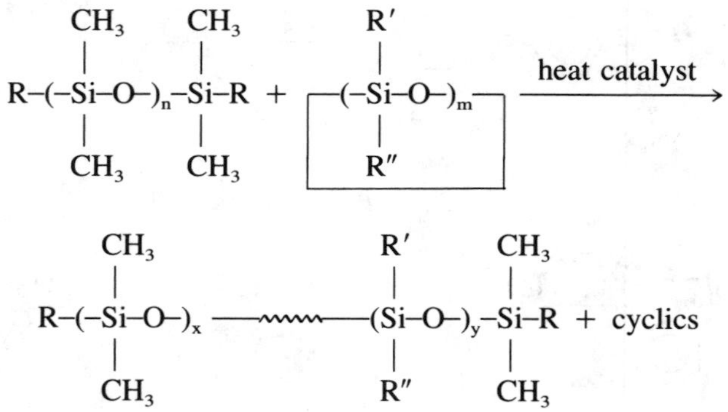

$$
\begin{array}{c}
\underset{\substack{\displaystyle | \\ CH_3}}{\overset{\substack{CH_3 \\ \displaystyle |}}{}} \quad \underset{\substack{\displaystyle | \\ CH_3}}{\overset{\substack{CH_3 \\ \displaystyle |}}{}} \qquad \underset{\substack{\displaystyle | \\ R''}}{\overset{\substack{R' \\ \displaystyle |}}{}}
\end{array}
$$

R$-$($-$Si$-$O$-$)$_n$$-Si-$R + $\overline{|\!-\!(-Si-O-)_m\!-\!|}$ $\xrightarrow{\text{heat catalyst}}$

$$
\begin{array}{c}
CH_3 \qquad\qquad R' \quad CH_3 \\
| \qquad\qquad\quad | \qquad | \\
R\text{–}(–Si–O–)_x \text{–}\!\sim\!\sim\!\sim\!\sim\text{–}(Si–O–)_y\text{–}Si\text{–}R \; + \; \text{cyclics} \\
| \qquad\qquad\quad | \qquad | \\
CH_3 \qquad\qquad R'' \quad CH_3
\end{array}
\qquad (149)
$$

R = $-$(CH$_2$)$_3$$-NH_2$, n = 1 　　　R′ = R″ = CH$_3$, m = 4
R = $-$N(CH$_3$)$_2$, n = 10 　　　R′ = R″ = C$_6$H$_5$, m = 4
x and y random 　　　R′ = $-$CH$_3$, R″ = $-$CH$_2$CH$_2$CF$_3$, m = 3

In the catonic polymerization of cyclosiloxanes initiated by acids (e.g., CF$_3$SO$_3$H), various cyclics are formed, but the mechanism of their formation is not fully understood.[286] Silicenium ions[287] and primary or tertiary[288,289] oxonium ions have been postulated as active growing centers.

When allowed to react for a very long time, a mixture of D$_4$ and TfOH reaches an equilibrium in which four distinct families of siloxanes are present (Figure 3): (1) small cyclic oligomers D$_n$ (mainly D$_4$ to D$_o$, with proportions decreasing in this order), (2) higher oligomers ($\overline{M}_n \sim 700$ to 5000), (3) a low molecular weight polymer (LP, $\overline{M}_n \sim 2.10^4$, $\overline{DP}_n \sim 300$ D units), and (4) a high molecular weight polymer (HP, $\overline{M}_n \sim 10^5$, $\overline{DP}_n \sim 1300$ D units).

For the polymerization of D$_4$, D$_5$, and D$_6$, there is general agreement that small cyclics are formed by backbiting reactions that give all possible cyclics in decreasing concentration when their size increases.[290]

For D$_3$, the rate of polymerization of which is much higher, there is a preferential formation of cyclics D$_{3x}$ with $[D_6] > [D_9] > [D_{12}]$, etc., which was explained either by a competitive end-to-end ring closure[291] or by a ring-expansion mechanism.

For all monomers, a first-order rate for monomer consumption has generally been observed, indicating a constant concentration of active centers. This stationary state could result from a continuous reinitiation reaction by the acid formed through the end-to-end ring-closure reaction, e.g., the case of D$_3$,[291] or from another type of reaction that later start a new kinetic chain, e.g., the case of D$_4$.[290]

The increase in molecular weight with conversion had been assigned to heterocondensation (e.g., between ester and silanol group) for both D$_3$ and D$_4$.

Some recent data[290] indicate that the cationic ring-opening polymerization of cyclic silanes has a living character. The assumption of the existence of two populations of growing macromolecules was made in order to explain the formation of both a living polymer and the cyclics.

III. ANIONIC OLIGOMERIZATION

It is generally accepted[292] that the anionic mechanism applies to those chain-addition polymerizations in which the growing chain end has a negative charge (real or formal), and that these are initiated by bases of various strengths. The classes of derivatives susceptible to anionic enchainment are of three main types: (1) monomers based on the carbon-carbon double bond (e.g., diienes), (2) heterocyclic monomers (e.g., cyclic oxides, cyclic sulfides), and (3) monomers based on the carbon-heteroatom multiple bond (e.g., aldehydes and nitriles). Accordingly, these different classes of monomers require initiators of differing basicity.

Anionic polymerization reactions involving the carbon-carbon double bond can be initiated principally by the following types of basic initiators: (1) alkali metals, (2) aromatic complexes of alkali metals, and (3) organoalkali compounds. Heterocyclic and heterounsaturated monomers can, however, be polymerized by weaker bases such as hydroxides and alkoxides.

A. OLIGOMERIZATION OF UNSATURATED HYDROCARBONS
1. Ethylene

The recent interest in the synthesis of polyethylene (PE) oligomers is based on the use of macromolecular compounds to immobilize homogeneous catalysts. The PE oligomers with $\overline{M}_w \geq 1000$ are soluble at 90 to 100°C in a variety of solvents, whereas at ambient temperature, they are insoluble and can be easily separated from the reaction medium. Based on this solubility feature, a novel concept for polymer-supported reagents and catalysts using functionalized oligomers of polyethylene has been developed: the oligomers behave either as catalysts supported on soluble polymers at high temperatures or as catalysts supported on insoluble polymers at low temperatures.[293-301]

Anionic oligomerization of ethylene is easily achieved in hexane with the homogeneous system n-BuLi/N,N,N',N'-tetramethylethylenediamine (TMDA)[302-305] or $tert$-butyllithium/ TMDA catalyst.[306] The living species obtained can be functionalized by adding a variety of reagents.

It has been shown that n-BuLi-TMDA and n-Bu(CH$_2$CH$_2$)$_n$Li species have the same reactivity and that the RLi-TMDA complex in a 1:1 stoichiometry is the active species.[303,306] The following kinetic equation has been established:[303]

$$-\frac{d[C_2H_4]}{dt} = k_p K_D^{1/2} [n\text{-BuLi-TMDA}]^{1/2} [C_2H_4] \tag{150}$$

where K_D is the dissociation constant of the reaction

$$(RLi\text{-}TMDA]_2 \rightleftharpoons 2RLi\text{-}TMDA \tag{151}$$

Equation 150 reflects the intervention of associated species $[n\text{-BuLi-TMDA}]_2$ as well as the influence of the concentration of the complexing agent on the kinetics of oligomerization.

The nucleophilic growing chains of anionic ethylene oligomers can be functionalized as shown in the following examples. Functionalized PE oligomers can be entrapped quantitatively in PE powders by coprecipitation with high-density polyethylene (HDPE), i.e., entrapment functionalization.[294] As an example, Equations 152 to 155 show the synthesis of copper(II)-carboxylated PE oligomers, introduction of copper(II) onto these oligomers, and subsequent entrapment in virgin HDPE.[301]

$$nH_2C{=}CH_2 \xrightarrow[\substack{\text{TMDA} \\ \text{hexane}}]{n\text{-BuLi}} n\text{-BuCH}_2CH_2{-}(-CH_2CH_2-)_{n-2}CH_2CH_2Li \qquad (152)$$
$$\text{PE-Li}$$

$$\text{PE-Li} \xrightarrow[\substack{2.\ H_3O^+}]{1.\ CO_2,\ -78°C} n\text{-BuCH}_2CH_2{-}(CH_2CH_2)_{n-2}CH_2CH_2COOH \qquad (153)$$
$$\text{PE-COOH}$$

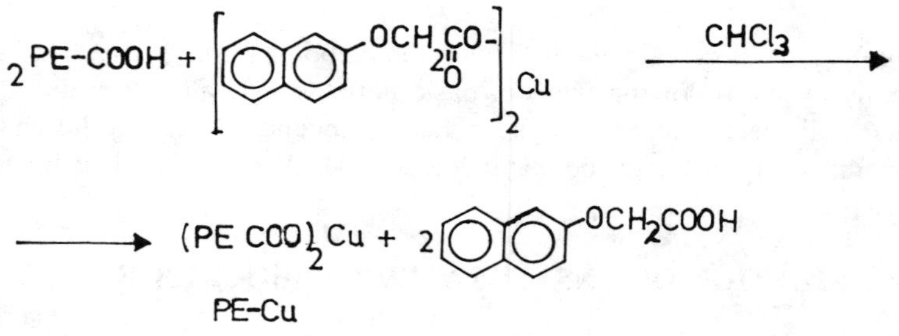

$$(154)$$

$$\text{PE-Cu}$$

$$\text{PE-Cu} + \text{HDPE} \xrightarrow[\substack{3.\ \text{coprecipitation}}]{1.\ 100°C,\ C_6H_5{-}CH_3} \begin{array}{l}\text{PE oligomer entrapped} \\ \text{in HDPE matrix}\end{array} \qquad (155)$$

The product of Reaction 155 was allowed to react with a solution of dithizone (Equation 156). A gradual decrease in the dithizone concentration in the UV-visible spectra (λ_{max} = 620 and 450 nm) and a concomitant appearance of a cupric dithizonate peak (λ_{max} = 545 nm) occured over 6 h at 25°C.

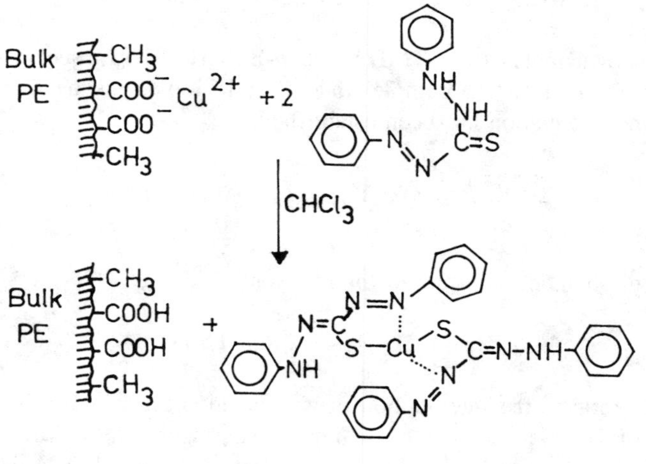

$$(156)$$

The reaction was first order in PE-Cu and the results suggested that Cu(II) ions in powder were at the polyethylene-solution surface.

The synthesis of diphenylphosphinated ethylene oligomers used as homogeneous catalysts for the synthesis of alkylchlorides from alcohols is shown in Equation 157.[296]

$$\text{PE-Li} + \text{ClP}(C_6H_5)_2 \rightarrow n\text{-BuCH}_2CH_2(CH_2CH_2)_nCH_2CH_2P(C_6H_5)_2 \qquad (157)$$

$$\text{PE-P}(C_6H_5)_2$$

Cyclooligomerization[297,298] and stereospecific polymerization of butadiene[299] were achieved with PE-entrapped Ni(0) catalyst and PE-bound neodium carboxylate, respectively. Another example of a polymerization-supported phase-transfer catalyst for both liquid-liquid and solid-liquid phase-transfer reactions is the ethylene oligomer containing crown ether chain ends:[300]

$$PE-COOH + SO_2Cl \rightarrow n\text{-}BuCH_2CH_2-(-CH_2CH_2-)_n-CH_2CH_2COCl \quad (158)$$

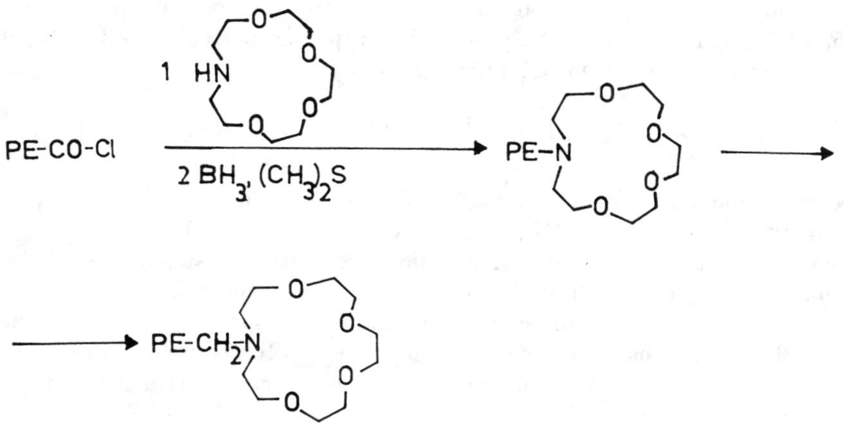

$$PE-COCl$$

$$(159)$$

A homogeneous catalyst of rhodium ligated by PE-P(C$_6$H$_5$)$_2$ oligomers reduced the double bonds of a substrate that was oxidized at the same time by an insoluble poly(vinyl pyridine)-bound chromium(VI) oxidant.[299]

Another method developed for the preparation of these types of ethylene oligomers with functionalized chain ends, which did not start from CH$_2$=CH$_2$, involved the anionic oligo-merization of butadiene *(vide infra)*, functionalization of the growing nucleophilic chains with various electrophiles, and reduction of the unsaturated carbon-carbon double bonds of the polybutadiene chains.[299]

Ethylene was polymerized in parafinic hydrocarbon solutions to low molecular weight polyethylene wax having terminal olefinic unsaturation by tertiary amine-chelated alkyl-lithium compounds.[306,307] Other olefins with available allylic hydrogen were metallized in this position. The C$_4$ olefins or aromatic hydrocarbons in conjunction with ethylene were found to undergo a telomerization reaction involving the former as telogen and ethylene as taxogen. The reaction mechanism was suggested to involve transmetallation from aliphatic carbon to a benzylic or allylic position of the telogen and addition of a carbon-lithium bond to ethylene. The assistance provided by the amine to both reactions was attributed to the formation of a coordination complex.[308]

2. Vinyl and Diene Monomers

Of the organoalkali coumpounds which can initiate polymerization of vinyl (and other) monomers, the most versatile are the organolithium type, since these are soluble in both polar and nonpolar solvents (the higher organoalkali compounds generally require ethers as solvents). These initiators operate by a direct anionic (nucleophilic) attack rather than by an electron-transfer mechanism. The stoichiometry is also very simple since each initiator molecule presumably generates a polymer chain, i.e.,

$$MW = Wt \text{ of monomer (in g)/mole of initiator} \quad (160)$$

In those cases, e.g., styrene and the dienes, where there is an absence of any noticeable termination or transfer reactions, the kinetics of both initiation and propagation steps have invariably been found to be *first order* with respect to the monomer concentration, irrespective of solvent type, temperature, or other variable. However, both of these steps have been found to show various *fractional-order* dependencies on the initiator concentration, depending on the solvent used and even on the monomer concentration.[292]

The alkyllithium-initiated anionic polymerization of vinyl and diene monomers can often be performed without the incursion of spontaneous termination or chain-transfer reactions.[309] The nonterminating (''living'') nature of these reactions has provided methods for the synthesis of polymers with predictable molecular weights and narrow molecular weight distribution.[310] If all chains have equal probability of growth, i.e., the initiation step is either similar in velocity or faster than the propagation step, the molecular weight distribution assumes the Poisson form recognized initially by Flory:[311]

$$\overline{DP}_w/\overline{DP}_n = 1 + (\overline{DP}_n - 1)/\overline{DP}_n^2 \sim 1 + 1/\overline{DP}_n \tag{161}$$

This type of distribution ''narrows'' as the chain length ($\overline{DP}_n$) increases, being, in principle, already remarkably narrow when $\overline{DP}_n = 100$ (i.e., $\overline{DP}_w/\overline{DP}_n = 1.01$).

In considering the effect of ''chain termination'' processes, a distinction should be made between the following possibilities: (1) partial destruction of initiator by impurities — the actual $\overline{M}_w$ is higher than predicted, but the molecular weight distribution is not affected, (2) termination of growing chains by impurities, monomer, solvent, or bound rearrangement — $\overline{M}_w$ is not affected, but the distribution is broadened, and (3) termination of growing chains by transfer reaction.

It should be noted in this connection that spontaneous termination can occur under certain conditions in organoalkali polymerizations by decomposition of the growing chain end into an alkali hydride (e.g., LiH or NaH) and a terminal double bond, the hydride apparently being incapable of initiating new chains.[292]

The organosodium polymerization of butadiene in a toluene-THF mixture was found to exhibit so much chain transfer to solvent that only oligomers of ten units or less were obtained.[312-314] Organolithium polymerization of butadiene in the presence of toluene has been shown to produce low molecular weight polymers of any desired molecular weight or chain structure. Diamines and potassium *tert*-butoxide are used to promote the transfer reaction and to control the microstructure.[315-318]

Such chain transfer processes, being random in nature, not only drastically reduce molecular weight, but also broaden the distribution, which tends to approach the ''most probable'' type, i.e., where $\overline{DP}_w/\overline{DP}_n = 2$.[319] The occurrence of transfer to monomer has been postulated in the case of *n*-butyllithium polymerization of 9-vinylanthracene in THF, only oligomers having 4 to 12 units being obtained even at 100% conversion.[320,321]

Gatzke[322] showed the presence of chain transfer between polystyryllithium and toluene at 60°C and derived the following relation between the molecular weight and the transfer constant:

$$DP_n = [M]_o X/[-StLi] - C_{RH}[RH]\ln[1 - X] \tag{162}$$

where $[M]_o$ is the initial monomer concentration, X the degree of conversion, [−StLi] the chain-end concentration (initiator concentration), [RH] the toluene concentration, and C_{RH} the transfer constant, i.e.,

$$C_{RH} = \frac{k_{tr}}{k_p} \tag{163}$$

Equation 162 is presumably applicable to any case of chain transfer in a nontermination polymerization. In the above case, the value of C_{RH} for toluene was found to be 5.10^{-6}, indicating a small extent of transfer.

Rapid-injection NMR (RINMR) analysis, a conceptually simple technique for deserving short-lived intermediates,[323,324] suggested that the rate of initiation (k_i) was considerably slower than the rate of propagation (k_p) in the early stages of the polymerization of styrene with n-BuLi in toluene, i.e., $k_p/k_i = 28$.[324] With this ratio, the theory predicts that monodisperse polystyrene would be obtained when the molecular weight reached 150,000 amu. However, the commercial polymerization of styrene with n-BuLi can produce polystyrene with $\overline{M}_w = 1500$ amu and MWD < 1.1.[324]

^{1}H and ^{7}Li RINMR spectroscopy of the reaction of styrene in toluene-d_8 with n-BuLi in THF-d_8 at $-80°$C evidenced both the tetramer/dimer equilibrium aggregation state of n-BuLi and the styryl and polystyryl (DP > 8) oligomers.[324]

The "living" anionic polymerization of styrene by the homogeneous system sodium naphthalene discovered by Swarc et al.[325] has a rapid initiation step through an electron transfer from the naphthalene to the styrene, as follows:

$$\left[\bigcirc\bigcirc \right]^{-} Na^{+} + \underset{\bigcirc}{\overset{CH=CH_2}{\Big|}} \longrightarrow \bigcirc\bigcirc + \left[\underset{\bigcirc}{\overset{CH=CH_2}{\Big|}} \right]^{-} Na^{+} \qquad (164)$$

The electron transfer process is not governed solely by electron affinity considerations, as illustrated by the equilibrium shown in Equation 164, but is also a function of the equilibrium concentration of the particular radical anion, which may be strongly affected by the possibility of side reactions, e.g., coupling of radical anions to stable dianions. The latter is apparently favored in the case of styrene:[292]

$$2\left[\underset{\bigcirc}{\overset{CH=CH_2}{\Big|}} \right]^{-} Na^{+} \longrightarrow Na^{+}\overset{-}{:}\underset{\bigcirc}{\overset{CH}{\Big|}} CH_2 CH_2 \underset{\bigcirc}{\overset{CH}{\Big|}} {:}^{-} Na^{+} \qquad (165)$$

The dianion formed is capable of propagating a chain of styrene units. Both Reactions 164 and 165 occur quite rapidly, so that the initiation step is completed long before chain growth occurs to any extent. Under these conditions, the molecular weights can be predicted from simple stoichiometric considerations, i.e.,

$$MW = \text{Wt of monomer (in g)}/0.5 \text{ mol of initiator} \qquad (166)$$

The MWD would, on theoretical grounds, be of the Poisson type. The same type of electron-transfer initiation presumably occurs with the alkali metal as well, but is obscured by their heterogeneous character.

The GPC distributions of α-methylstyrene polymers obtained in THF and potassium as initiator were found to be multimodal in character, yielding the four components D, A, B, and C with an average degree of polymerization of about 4, 16, 250, and 1000, respectively.[326] Some results are shown in Figure 3. It has been speculated that these multimodal distributions are a result of dead (D) and dormant (A) polymers in combination with polymers due to different ion pairs (B and C). Bimodal distributions of anionic polystyrene prepared in 3-ethyl tetrahydrofuran have also been attributed to the presence of different ion pairs.[327]

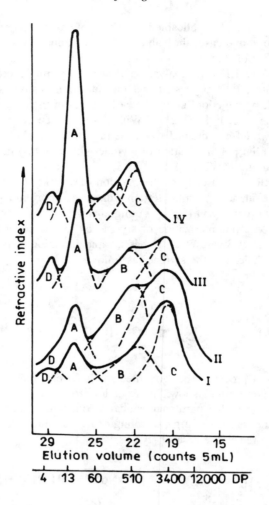

FIGURE 3.   GPC distributions of anionic poly(α-methylstyrene) samples synthesized with K in THF with different concentrations of living ends, [LE] mol/l. (I) 0.007; (II) 0.008; (III) 0.009; (IV) 0.017. [LE] $= 2w/M_v V_e$, where w is the weight of the polymer (in grams) present at equilibrium in the volume V_e (in l). (From Leonard, J. and Malhorta, S. L., *J. Macromol. Sci. Chem.*, A10, 1279, 1976. With permission.)

Furthermore, under identical conditions of $[M]_o$ and concentration of living ends [LE], the GPC distributions of poly(α-methylstyrene) prepared in cyclohexane, *p*-dioxane, and THF were the same, in spite of their different dielectric constant.[328-330] Under identical conditions of $[M]_o$ but with different [LE], the effect of excessive [LE] on the GPC distributions of the polymers prepared in cyclohexane was not limited to the component D + A, as was the case when THF or *p*-dioxane were the solvents, but also on component C, which increased its contribution to the polymer at equilibrium.[330]

The structure of dicarbanionic oligomers as well as of the corresponding protonated species formed in the polymerization of α-methylstyrene and styrene, respectively, in THF initiated by potassium, determined by MS and UV analyses, pointed to the following mechanism of enchainment:[309,331,332]

Initiation

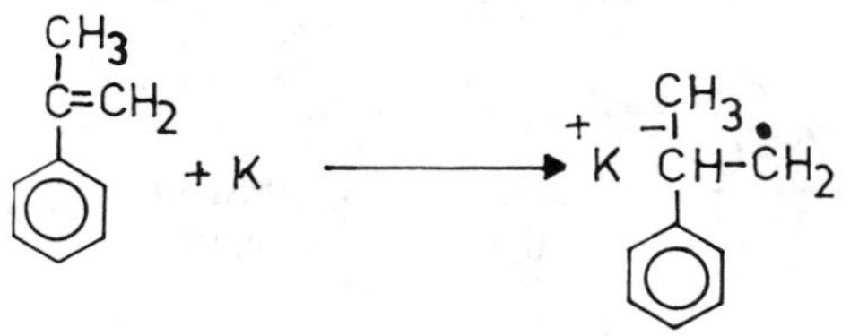

$$
\begin{array}{c}
\underset{\displaystyle\bigcirc}{\overset{\displaystyle CH_3}{\underset{|}{C=CH_2}}} + K \longrightarrow K^+ \ \ \underset{\displaystyle\bigcirc}{\overset{CH_3}{\underset{|}{CH}}}\text{-}\dot{C}H_2
\end{array} \tag{167}
$$

Radical dimerization

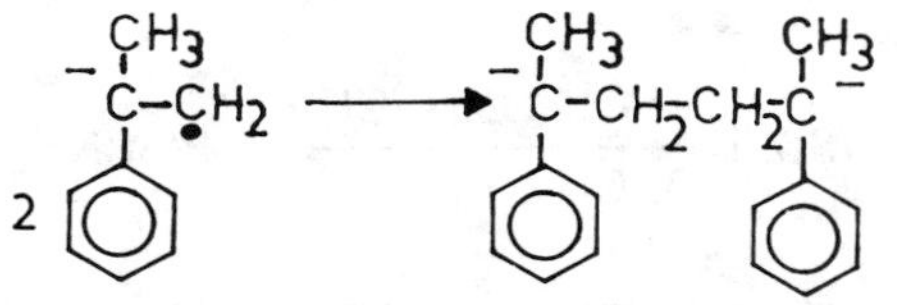

$$
2\ \underset{\displaystyle\bigcirc}{\overset{CH_3}{\underset{|}{C}}}\text{-}\overset{\bullet}{C}H_2^- \longrightarrow \ ^-\underset{\displaystyle\bigcirc}{\overset{CH_3}{\underset{|}{C}}}\text{-}CH_2CH_2\underset{\displaystyle\bigcirc}{\overset{CH_3}{\underset{|}{C}}}{}^- \tag{168}
$$

Propagation

$$
^-\underset{\displaystyle\bigcirc}{\overset{CH_3}{\underset{|}{C}}}\text{-}CH_2CH_2\underset{\displaystyle\bigcirc}{\overset{CH_3}{\underset{|}{C}}}{}^- + \underset{\displaystyle\bigcirc}{\overset{CH_3}{\underset{|}{C}}}H_2\!=\!CH \longrightarrow \ ^-\underset{\displaystyle\bigcirc}{\overset{CH_3}{\underset{/}{C}}}\text{-}CH_2CH_2\underset{\displaystyle\bigcirc}{\overset{CH_3}{\underset{|}{C}}}\text{-}CH_2\underset{\displaystyle\bigcirc}{\overset{CH_3}{\underset{|}{CH}}}{}^- \tag{169}
$$

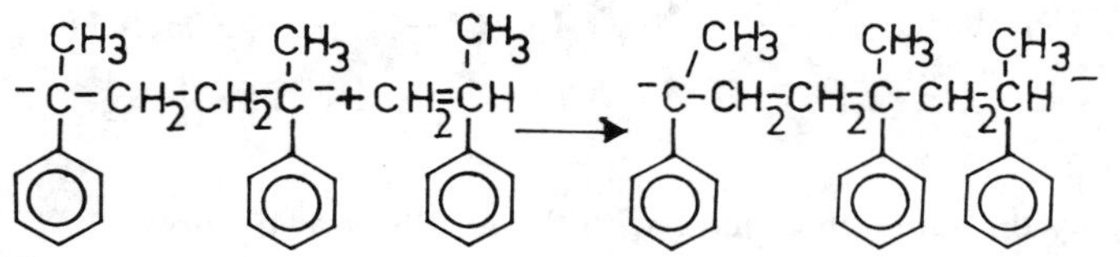

However, because the propagation shown by Equation 169 is very slow, transfer to the solvent and monomer and transformation of the reversible species into irreversible ones occur before the polymerization can reach the equilibrium position predicted by thermodynamics.[333]

The stable carbanionic chain ends generated by anionic oligomerizations, in principle, can be converted into a diverse array of functional end groups. For example, the carbonation of poly(styryl)lithium, P(St)Li, synthesized by polymerization of styrene with alkyllithium,[310] in a 75/25 mixture (by volume) of benzene and THF with gaseous carbon dioxide occurred quantitatively to produce the corresponding carboxylic acid chain ends:[334]

$$
\begin{array}{c}
\text{P(St)Li} \\
\text{in } C_6H_6/\text{THF}
\end{array}
\xrightarrow[\text{2. H}^+]{\text{1. CO}_2\text{(g)}}
\text{P(St)COOH} \tag{170}
$$

If the carboxylation of polystyryllithium ($\overline{M}_n = 4200$) was carried out in solid state, in which the chain ends are associated, the corresponding ketone (90% yield) was formed:[335]

$$
\begin{array}{c}
\text{P(St)Li} \\
\text{freeze dried}
\end{array}
\xrightarrow{\text{CO}_2}
\text{P(St)COOLi/P(St)Li} \rightarrow
[\text{P(St)}]_2\overset{\displaystyle \overset{OLi}{|}}{\underset{\displaystyle \underset{OLi}{|}}{C}}
\xrightarrow{\text{H}_2\text{O}}
\text{P(St)}-\overset{\displaystyle}{\underset{\displaystyle \underset{O}{\parallel}}{C}}-\text{P(St)} \tag{171}
$$

No carboxylic acid functionality was detected and the remaining 10% of the product corresponded to polystyrene homopolymer.

Carbonation of P(St)Li in benzene produced only a 60% yield of carboxylic acid, which was contaminated with significant amounts of the corresponding ketone dimer and tertiary alcohol trimer:[336]

$$P(St)Li \xrightarrow[\text{2. H}^+]{\text{1. CO}_2(g)} P(St)COOH + [P(St)]_2CO + [P(St)]_3COH \qquad (172)$$
$$\text{in C}_6\text{H}_6 \qquad\qquad 60\% \qquad\quad 28\% \qquad\quad 12\%$$

Amination of polymeric organolithium compounds was achieved using the reagent generated from methoxyamine and methyllithium.[335,337,338] Thus, P(St)Li ($\overline{M}_n = 2000$) has been aminated with two equivalents of this aminating agent to produce a 92% yield of aminated polystyrene.[338]

$$P(St)Li \xrightarrow[\text{2. CH}_3\text{OH}]{\text{1. CH}_3\text{ONH}_2/\text{CH}_3\text{Li}} P(St)NH_2 \qquad (173)$$

The α,ω-dilithium polystyrene, prepared from styrene and lithium naphthalene in benzene/THF mixture

$$2nCH_2{=}CH \xrightarrow{\text{Li}^+(\text{C}_{10}\text{H}_8)^-} Li{-}(CH{-}CH_2{-})_n{-}({-}CH_2{-}CH{-})_m{-}Li \qquad (174)$$
$$\quad\ \ |\qquad\qquad\qquad\qquad\qquad\quad |\qquad\qquad\qquad\quad |$$
$$\quad\ \ C_6H_5\qquad\qquad\qquad\qquad\quad\ C_6H_5\qquad\qquad\quad C_6H_5$$

was aminated by the same procedure, with the results shown below:[335,338]

$$[-P(St)Li]_2 \xrightarrow[\substack{\text{1. }-78°C \\ \text{2. }-15°C}]{\text{CH}_3\text{ONH}_2/\text{CH}_3\text{Li}} [-P(St)NH_2]_2 + P(St)NH_2 + P(St)H \qquad (175)$$
$$\qquad\qquad\qquad\qquad\qquad\qquad\qquad\quad 80\% \qquad\qquad 8\% \qquad\quad 12\%$$

Concerning the synthesis of telechelic oligomers, a one-step process has been developed implying the simultaneous presence of initiator, monomer, and deactivating agent.[339] Three reactions are involved: (1) initiation of vinyl monomers by an alkali metal in a polar solvent, (2) anionic polymerization of the monomer, and (3) functionalization of the carbanionic species.[340]

Several types of telechelic polymers were prepared by this technique: ω-hydroperoxide polymers, diphosphonated polymers, and polysulfide polymers.[341-345]

The hydroperoxide polystyrene was synthesized by reactions of the living polystyrene ($\overline{M}_n \geq 8500$) in solution or in the solid state with oxygen. When oxygen reacts with a carbanion, the most likely reaction is the electron transfer from the carbanion to the oxygen:[341-343,345]

$$R^- + O_2 \rightarrow R^{\bullet} + O_2^- \qquad (176)$$

The hydroperoxide yield depends on the mode of reaction (direct or inverse oxidation), living end concentration, solvent, temperature, and carbanion structure. The general scheme of the anion oxidation is shown below:

$$P(St)^- + O_2 \rightarrow P(St)^{\bullet} + O_2^{\overline{\bullet}} \qquad (177)$$

$$P(St)^{\bullet} + O_2 \rightarrow P(St)O{-}O^{\bullet} \qquad (178)$$

$$P(St)^- + P(St)O{-}O^{\bullet} \rightarrow P(St)^{\bullet} + P(St)O{-}O^- \qquad (179)$$

To prevent diffusion phenomena which could lead to several side reactions, such as coupling reactions

$$P(St)^{\cdot} + P(St)^{\cdot} \rightarrow [P(St)-]_2 \tag{180}$$

or alcoholate formation

$$P(St)O-O^- + P(St)^- \rightarrow 2P(St)O^- \tag{181}$$

the reaction of oxygen on the living polymers was carried out in the solid state.[345]

Diphosphonated telechelic oligomers ($\overline{M}_n$ 500) of styrene and α-methyl styrene were prepared by deactivating the propagating anionic chain with $POCl_3$.[345] In this case, the initiation step (Equation 182) took place in the heterogeneous phase, whereas propagation (Equation 183) and the deactivation step (Equation 184) occurred in the heterogeneous phase:

$$2M^{\cdot} + 2Li \xrightarrow{k_i} 2(M^-Li^+) \rightarrow {}^-M-M^-, 2Li \tag{182}$$

$${}^-M-M^-, 2Li^+ + M \xrightarrow{k_p} {}^-M-M-M^-, 2Li, \text{ etc.} \tag{183}$$

$$\text{wwwM}^-, Li^+ + POCl_3 \xrightarrow{k_t} M-(PO)Cl_2 + LiCl \tag{184}$$

where M is the styrene of α-methylstyrene monomer unit.

The deactivation reaction of the carbanions with $POCl_3$ competed with the propagation. For a $[POCl_3]/[M]$ ratio $\ll 3$, a mixture of functionalized dimers and trimers was obtained. The dichlorophosphonic chain ends of oligomers can be changed to monochlorophosphonic ester or to phosphonic diester by adjusting the acidity of the medium.

The synthesis of polysulfide polymers was carried out with an alkali metal (Na) in THF in the presence of elemental sulfur S_8,[344,345] the latter being partially soluble in the solvent (THF) at room temperature. The reaction proceeded by a nucleophilic attack of the cyclic S_8 to give polysulfide polymer:

$$P^-M_n^- + PS_8 \xrightarrow{k_t} -[-M_n-S_x-]_p- + PS_{8-x}^{2-} \tag{185}$$

where M can be a diene, styrene, or methyl methacrylate (MMA).

In the case of styrene, when the ratio $K = [Na]/[S_8]$ was low, i.e., $K = 2$ or 3, the sulfur rank (x) corresponded to polysulfide-polystyrene, i.e. $x > 1$. For $k = 8$, the polymer obtained was a monosulfide ($x = 1$). In this case, the organic chain length corresponded on the average to a styrene trimer, while the molecular weight of the polysulfide polymer was high ($\overline{M}_n = 32,000$), i.e., for a repeating unit

$$-[-(CH_2-CH-)_3-S-]-$$
$$|$$
$$C_6H_5$$

the degree of polymerization was $\overline{DP} = 93$.

Diorganomagnesium compounds initiate the anionic polymerization of vinyl and diene monomers in hexamethylphosphoric triamide (HMPA), but the chain transfers and the instability of the magnesium-carbon bond determine the formation of low oligomers exclusively.[346-348] The oligomer of α-methylstyrene, initiated by magnesiacyclohexane, was treated with dimethyldichlorosilane in order to obtain a stable cyclic oligomer:[347]

$$\text{(186)}$$

The yield of cyclic oligomer ($\overline{DP}$ = 3 to 5) was so low that cyclic compounds could not be discriminated from the linear oligomer. However, the oligo-α-methylstyrylanion was active enough to initiate the anionic polymerization of styrene ($\overline{DP}$ = 60 to 130) in HMPA.

Trans-stilbene, 1,1-diphenylethylene, and α-methylstyrene were allowed to react with dibenzylmagnesium to form their oligomers in HMPA.[348] One and two molecules of stilbene and 1,1-diphenylethylene were incorporated into the magnesium-carbon bond, and the carbanions obtained in HMPA were stable, analogous to the anionic living polymer having an alkali cation as gegenion in ethers. The low molecular weight ($\overline{DP}$ = 2 to 4) products which were formed in the reaction between α-methylstyrene and magnesium catalyst were found to have no magnesium-carbon bond. It was considered that the cleavage of the propagating chain occurred gradually after the rapid propagation had proceeded to consume monomer.

2-Methyl-2-butene-1,4-diylmagnesium exhibited a high catalytic activity in the oligomerization of isoprene, while dialkylmagnesium compounds showed relatively low catalytic activity.[349] The degree of oligomerization paralleled the initial molar ratio of isoprene to 2-methylbutenediylmagnesium until the ratio reached 15. The oligomers involved the 3,4-polyisoprene unit exclusively.

The polymerization and oligomerization of isoprene with the use of lithium alkylamides have been extensively studied.[350-354] The polymerization process is markedly affected by the presence of additives such as free amine. The reaction of lithium *N*-alkylbutylamide with isoprene in the presence of *N*-alkylbenzylamine produced oligomers of molecular weight ranging from 200 to 700.[354] The molecular weight was dependent on the bulkiness of N-substituted alkyl groups, e.g., lithium *N*-methylbenzylamide gave an oligomer with $\overline{MW}$ = 360 and *N-t*-butylbenzylamide yielded an oligomer having a molecular weight of 560.

The amino-ended oligoisoprenes with MW ~ 10^3 are interesting for they can introduce soft segments into polymeric materials through reaction of the amino end group (such as quaternization). Lithium diisopropylamide initiated the oligomerization of isoprene in cyclohexane at 80°C to form products with a tertiary amino group having $\overline{M}_n$ ~ 1000, provided the reaction was run at an initial amine/amide mole ratio of 1:2.[351,352] The microstructure of the oligomers obtained was 37% *cis*-1,4, 53% *trans*-1,4, and 10% 3,4 enchainment.

Selective formation of secondary amino-ended oligoisoprenes ($\overline{M}_n$ ~ 800) was found to be effective when lithium monoalkylamide initiated the polymerization of isoprene in the presence of monoalkylamine ([isoprene]$_o$/[amine]/[amide] = 60/1/8.[253] The oligoisoprenes contained 50% *cis*-1,4, 33% *trans*-1,4, and 17% 3,4 enchainments. The oligomerization was considered to proceed via Equations 187 to 192.

$$RNHLi + Ip \rightarrow RNH(Ip)Li \tag{187}$$

$$RNH(Ip)Li + (n - 1)Ip \rightarrow RNH-(I_p)_nLi \tag{188}$$

$$RNH-(Ip)_nLi + RNH_2 \rightarrow RNH-(I_p)_nH + RNHLi \tag{189}$$

$$RNH-(Ip)_n-H + RNHLi \rightarrow \underset{\underset{Li}{|}}{RN}-(I_p)_nH + RNH_2 \tag{190}$$

$$\underset{\underset{Li}{|}}{RN}-(I_p)_n-H + mI_p \rightarrow RN \begin{matrix} \nearrow (I_p)_n-H \\ \searrow (I_p)_m-Li \end{matrix} \tag{191}$$

$$RN \begin{matrix} \nearrow (Ip)_n-H \\ \searrow (Ip)_m-Li \end{matrix} + RNH_2 \xrightarrow{-RNHLi} RN \begin{matrix} \nearrow (Ip)_n-H \\ \searrow (Ip)_m-H \end{matrix} \tag{192}$$

where Ip is isoprene monomer or an enchained isoprene unit. For selective formation of secondary amino-ended isoprene oligomers in Equation 189, the reaction described by Equation 191 should be suppressed.

Oligomerization of 1,3-butadiene by n-BuLi as initiator in toluene in the presence of dipiperidinoethane (DPE) gave a highly 1,2 telomer. The propagation rate with respect to initiator concentration was first order for [DPE]/[Li] $\geq$ 1, but when [DPE]/[Li] $<$ 1, the reaction rate was independent of initiator concentration.[354] The reaction of butadiene with Et-Li also occurred as 1,4- and 1,2-addition to the exclusively 1,2-addition.[355] The composition of end groups of a product obtained at the start of oligomerization differed significantly from that of a product obtained at the stage of chain termination.[355,356] Terminated vinyl groups were formed mainly from monolithium adducts by the addition of EtOH.

Isoprene oligomerization in toluene catalyzed by lithiated diethylene triamine in the presence of diethyleneamine gave a product which showed a unimodal GPC peak at M_{GPC} = 500 (polystyrene standard).[357] According to ^{1}H NMR and GS-MS analysis, the reaction product was a 1:4 adduct of diethylenetriamine and isoprene formed by 1,4-addition, i.e., 2,14-dimethyl-5,8-(or 5,11)-bis(3-methyl-2-butenyl)-5,8,11-triaza-2,13-pentadecadiene.

1,3-Butadiene was reacted with dixanthenyl barium as anionic initiator in diglyme or $CH_3OCH_2CH_2OCH_3/C_6H_6$ mixture and the oligomer ($\overline{M}_n$ = 1850 to 3000) had 50% trans-1,4 units and 25% vinyl enchainment.[358] However, the activity of alkaline earth metal (Mg, Ba) organometallic compounds as anionic catalysts for vinyl and diene oligomerization is much lower than that of alkaline metals and their alkyl or vinyl derivatives.

A vast array of new synthetic procedures and polymer structures have resulted from utilization of these well-defined anionic vinyl diene oligomers[359-365] for grafting, copolymerization, and linking reactions,[366-368] as well as for initiation of the polymerization of other monomers ("mechanism switching").[369] For example, a SKDSR rubber was prepared by solution polymerization of butadiene in the presence of oligomeric poly(α-methyl)-styrenedi-Na salt as initiator stabilized with (iso-Bu)$_3$Al.[370] The plasticity and physicomechanical (good) properties of the rubber were monitored by varying the concentration of the oligomeric initiator and the reaction temperature.

Polymerization of butadiene in toluene initiated by oligomeric dilithiopolyisoprene, termination of the reaction with gaseous CO_2, and additon of HCl gave a carboxy-terminated polybutadiene (MW = 2500) containing 3.53% carboxylic groups, corresponding to a functionality of 1.93.[371]

Anionic oligomerization of cyclodienes,[372] triene, tetraene,[373] and aromatic[374-376] or polar acetylenes[377] was easily achieved, in particular with alkyllithium catalysts or alkaline metals (K, Na). For example, phenyl acetylene was oligomerized with n-BuLi to a conjugated product (MW = 2500),[374] while poly(diphenylacetylene) (MW = 610 to 1830) and poly(diphenylbutadiene) (MW = 1300 to 2000) were prepared by polymerization of the respective monomers in the presence of K or Na suspension, naphthalene lithium, naphthalene sodium, or n-BuLi.[375] Paramagnetic, conjugated polymeric oligomers (MW $\sim 10^3$) were obtained by polymerization of dicyanoacetylene with the same types of catalysts.[377] The materials were soluble in both water and organic solvents. Linear, highly unsaturated products with MW = 1000 to 7000 were prepared by anionic polymerization with n-BuLi of 1,3,5-hexatriene and 1,3,5,7-octatetraene.[373] The postpolymerization cyclization and cross-linking reactions were inhibited by addition of *tert*-butylpyrocatechol in the system.

Polymerization of 1,3-cyclohexadiene in the presence of an n-BuLi/THF complex was characterized by a rapid initiation which determined the formation in high yields (88 to 100%) of polymers with narrow MWI. The molecular weight (≥ 2000) could be monitored by changing the BuLi/THF ratio and reaction temperature.

3. Polar Vinyl Monomers

Vinyl monomers with heteroatom-containing substituents, such as acrylates, vinyl sulfoxide, vinyl ketones, vinylpyridines, etc., fall into the class of *polar monomers*. The group of polar vinyl monomers which have been most extensively investigated are the acrylates, specifically, *methyl methacrylate*.[292] In this case, it has been shown[378,379] that organometallic compounds, besides initiating polymerization through the vinyl group, also react substantially with the carbonyl function, e.g.

$$RLi + H_2C=\overset{\overset{\displaystyle CH_3}{|}}{\underset{\underset{\displaystyle O=C-OCH_3}{|}}{C}} \longrightarrow H_2C=\overset{\overset{\displaystyle CH_3}{|}}{\underset{\underset{\underset{\displaystyle R}{|}}{LiO-C-OCH_3}}{C}} \longrightarrow H_2C=\overset{\overset{\displaystyle CH_3}{|}}{\underset{\underset{\underset{\displaystyle R}{|}}{C=O}}{C}} + CH_3OLi \qquad (193)$$

Methanol is generated upon hydrolysis.

Other important potential complications include[380] (1) activation of protons in the α-position to the carbonyl group, leading to chain transfer. (2) Due to the bidentate character of the active centers (Equation 194), they may attack the monomer not only by the carbanion (1,2 addition, Equation 195), but also by the enolate oxygen (1,4 addition, Equation 196):

$$\left[\overset{\overset{\displaystyle CH_3}{|}}{\underset{\underset{\displaystyle R-C=O}{|}}{\sim C^-}} \longleftrightarrow \overset{\overset{\displaystyle CH_3}{|}}{\underset{\underset{\displaystyle R}{|}}{\sim C}} \overset{\delta-}{\cdots\cdots} C \overset{\delta-}{\cdots\cdots} O \longleftrightarrow \overset{\overset{\displaystyle CH_3}{|}}{\underset{\underset{\displaystyle R}{|}}{\sim C}} = C - O^- \right] \qquad (194)$$

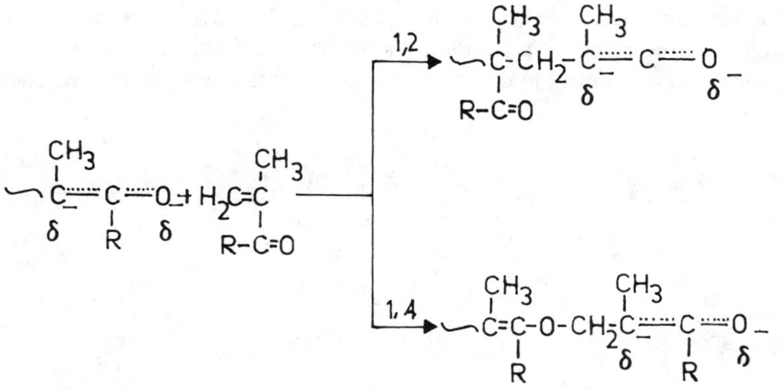

$$\text{(195)}$$

$$\text{(196)}$$

(3) The carbonyl group may coordinate with the counterion of the living chain end *(intramolecular solvation)* or may lead to association of ion pairs and (4) the chain end may interact with electron donors. Most of these side reactions were reported by Hatada et al.[381,382] upon the polymerization of MMA-d_8 with n-C_4H_9MgCl in toluene at $-78°C$. It was concluded that n-C_4H_9MgCl attacks both vinyl and carbonyl double bonds of MMA in the initiating process to form the initiating species (Equation 197) and butyl pentadeuteroisopropylketone (Equation 197) and butyl pentadeuteroisopropylketone (Equation 198):

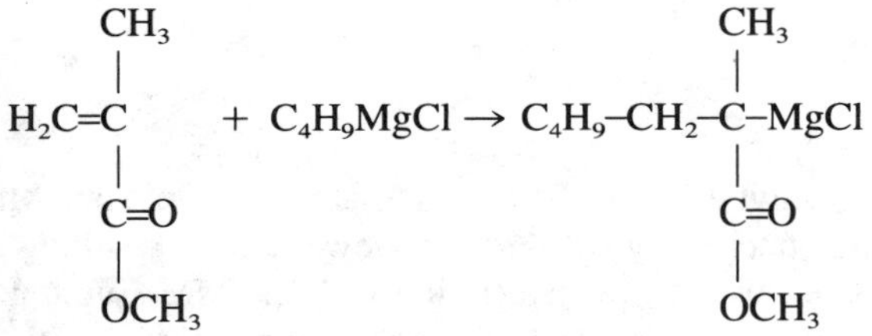

$$\text{(197)}$$

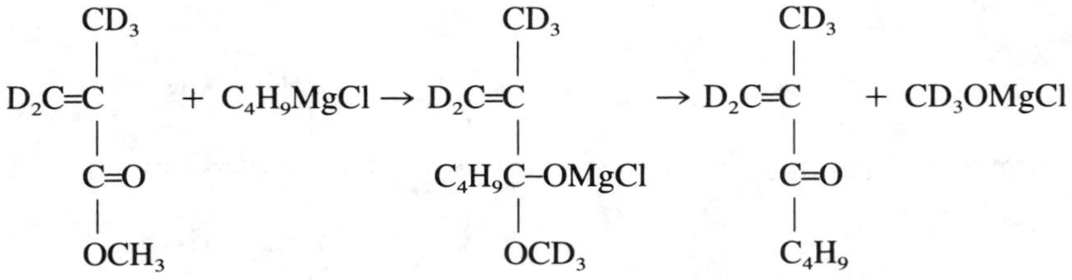

$$\text{(198)}$$

Mass spectra of the fractionated lower molecular weight oligomers also indicated that there were two types of oligomers, which were ended with MMA (I) and ketene (II) units, respectively:

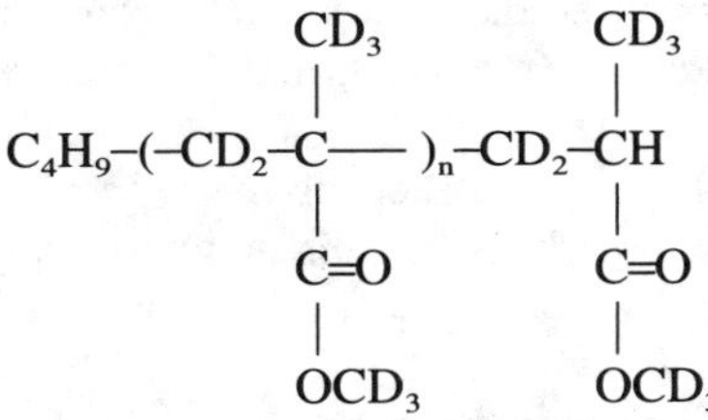

I

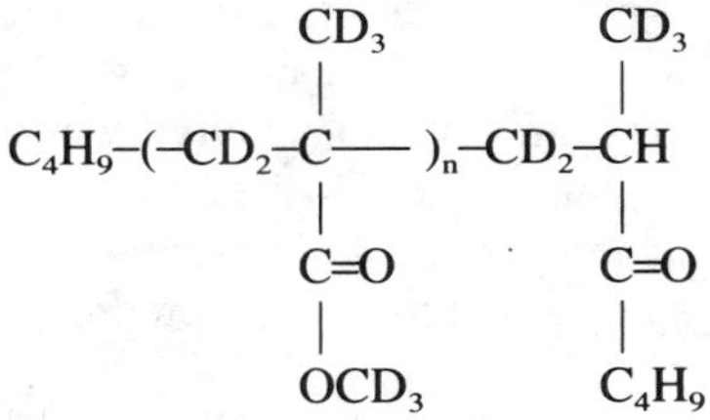

II

The following three structures of the undeuterated oligomers were noted: structure III formed through intramolecular cyclization and structures IV and V formed by addition of the propagating chain end onto the carbonyl groups of ketene and MMA, respectively.

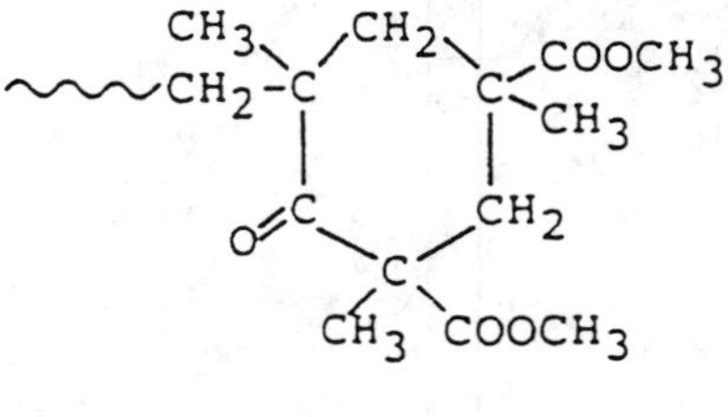

$$
\begin{array}{c}
CH_3 \quad\quad CH_2 \\
\sim\!\!\sim\!\!CH_2-C \quad\quad C-COOCH_3 \\
CH_3 \\
O=C \quad\quad CH_2 \\
C \\
CH_3 \quad COOCH_3
\end{array}
$$

III

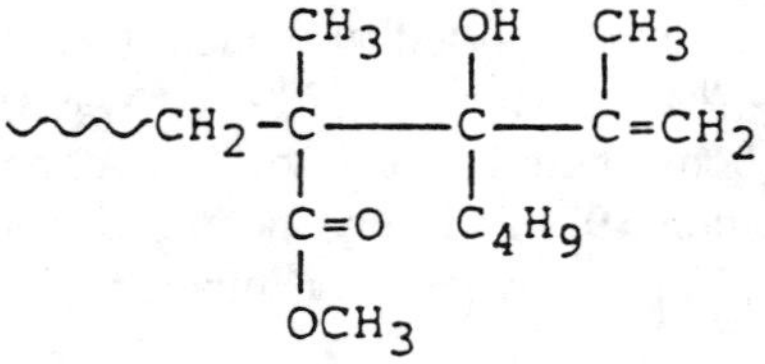

$$
\begin{array}{c}
CH_3 \quad OH \quad CH_3 \\
\sim\!\!\sim\!\!CH_2-C\!\!-\!\!-\!\!-\!\!C-C=CH_2 \\
C=O \quad C_4H_9 \\
OCH_3
\end{array}
$$

IV

$$
\begin{array}{c}
CH_3 \quad CH_3 \\
\sim\!\!\sim\!\!CH_2-C\!\!-\!\!-\!\!-\!\!C-C=CH_2 \\
C=O \quad O \\
OCH_3
\end{array}
$$

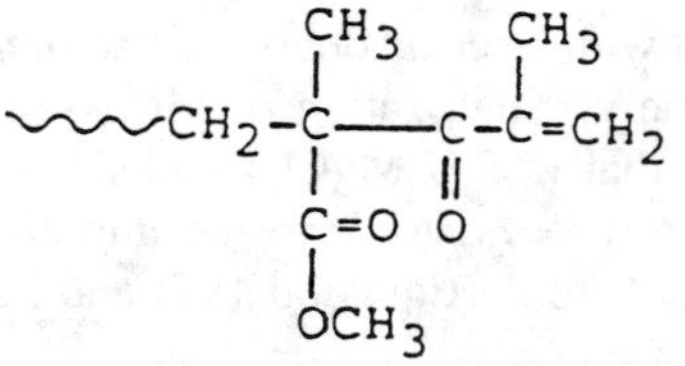

V

Since butyl isopropenyl ketone, BIPK, is more reactive than MMA, a part of the propagating species are attacked by BIPK.[382,384] However, the resulting species ending with the BIPK unit are less reactive than those ending with the MMA unit and remain unreacted during the polymerization (Equation 199). This is why most of the BIPK units are located at the chain end.

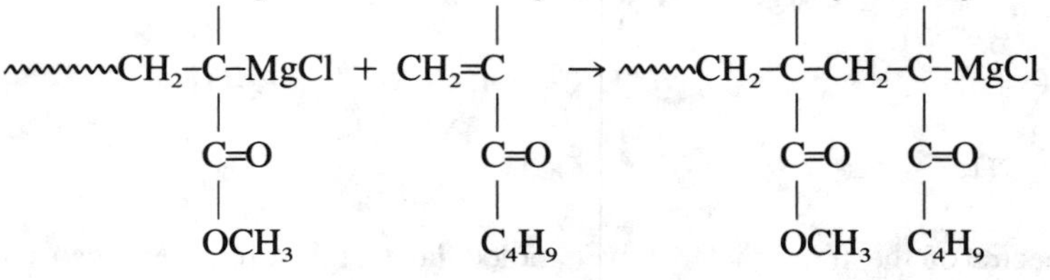

$$
\begin{array}{ccc}
CH_3 & CH_3 & CH_3 \quad CH_3 \\
| & | & | \quad\quad | \\
\sim\!\!\sim\!\!\sim\!\!CH_2-C-MgCl + CH_2=C & \rightarrow & \sim\!\!\sim\!\!\sim\!\!CH_2-C-CH_2-C-MgCl \\
| & | & | \quad\quad | \\
C=O & C=O & C=O \quad C=O \\
| & | & | \quad\quad | \\
OCH_3 & C_4H_9 & OCH_3 \quad C_4H_9
\end{array}
\qquad (199)
$$

Thus, participation of BIPK in the polymerization made the MWD broad ($\overline{M}_w/\overline{M}_n = 3.2$). CH_3OMgCl formed in Equation 198 coordinated with the propagating chain end to produce species of different activity and stereoregularity. This caused the multiple active species and stereoblend polymer formation, e.g., I:H:S = 31:18:51, where I, H, and S represent (%) iso-, hetero-, and syndiotacticity, respectively.

Of the initiator used, the sum total of the butyl groups in the polymer and oligomer chains amounted to 58%, 18% remained unreacted (and perhaps coordinated with the propagating species affecting the stereoregularity), and 8.8% was found in the product of side reactions.

The competition between these reactions and initiation of the polymerization of acrylic monomers varies with the type of organometallic compound, solvent, and temperature. Thus, in the system MMA/n-butyllithium in toluene at $-30°C$, about 60% of the initiator disappears by side reactions, while 1,1-diphenylhexyllithium (i.e., the adduct of n-BuLi and 1,1-diphenylethylene) undergoes none of these side reactions with MMA. In the reaction of acrylonitrile in toluene at $-70°C$ with n-BuLi, no more than 2.7% of the initiator is consumed during polymerization.[284] About half the initiator is used up during the formation of acrylonitrile oligomers with molecular weights of 170 to 330. The nitrile groups of methacrylonitrile units participate in both intra- and intermolecular interactions with the counterion.[385] In the anionic oligomerization of acrylonitrile with lithium *tert*-butoxide, it has been found that the monomer formed complexes with the active centers in a 1:1 and/or 2:1 stoichiometry.[386] Oligomer formation was found to be the main cause of low initiation efficiency in the polymerization of acrylonitrile initiated by metal alkyls in hydrocarbon media.[387]

Cyclization by the Ziegler-Thorpe reaction is the main side reaction in the propagation step of the anionic polymerization of acrylonitrile and methacrylonitrile.[388,389] The living trimer P_3^* is able to participate simultaneously in two competitive reactions, i.e., propagation and cyclization. The following relation was deduced for the anionic polymerization of acrylic monomers:[390]

$$\frac{[P_3^c]}{[P_4^*]} = \frac{k_c}{k_p} \cdot \frac{1}{M_o} \tag{200}$$

where P_3^c and P_4^* represent the concentrations of cyclic oligomer and linear tetramer, respectively, and k_c and k_p are the corresponding rate constants for cyclization and propagation. This relation shows that the linear tetramer P_4^* is predominantly formed at a higher initial monomer concentration $[M]_o$. Moreover, heterocyclics are formed in the anionic polymerization of methacrylonitrile through the polymerization of nitrile groups. The conjugated $-(-C=N-)_n-$ system is capable of accepting electrons from anions in which anion and dianion radicals are obtained. The lack of this side reaction in the anionic polymerization of acrylonitrile can be explained by the presence of an acidic H-atom at the carbon atom adjacent to the nitrile group which promotes the metallation reaction under splitting of an H-atom.

The problem of side reactions to the methacrylate carbonyl has been well documented in the literature.[387] Methods have been developed to avoid these deleterious side reactions, namely, hindered and less basic initiators (e.g., diphenylhexyllithium), polar solvents, and low temperatures. (For example, allowing the polymerization to warm before termination often leads to side reactions such as ketone formation and subsequent broadening of the polydispersity).

The sensitive problem of acrylate monomer purification, in particular the elimination of alcoholic, terminating impurities, has recently been solved by treating the monomers with trialkyl aluminum reagents.[391-393] The method allowed for both the synthesis of narrow-distribution poly(alkyl methacrylate)s of controlled and predictable molecular weights and for the utilization of a wide variety of methacrylate monomers.[391,392]

The anionic polymerization of alkylacrylates cannot be controlled by conventional anionic methods. It was proposed to minimize the relative importance of secondary transfer and termination reactions (usually ascribed to the presence of the unhindered carbonyl group and the α-acidic hydrogen atom) by creating enough steric hindrance around the propagating ion pair through the use of electronically well-balanced ligands sich as LiCl.[394,395] It appeared that μ-type hindered lignads were quite efficient in controlling a perfectly "living", quantitatively fast polymerization of hindered acrylates by dilithium (ar)alkyls:[380,395-397] linear dependence of $\overline{M}_n$ on conversion, $\overline{DP}$ equal to the monomer over catalyst ratio, narrow MWD, and complete resumption of additional monomer polymerization with the same characteristics.

TABLE 8
Oligomerization of MMA with Grignard
Reagent RMgBr in Toluene at $-78°C$[382]

		Tacticity			
R	M_n	I	M	S	$\overline{M}_w/\overline{M}_n$
n-C_4H_9	7420	20.5	15.2	64.3	11.2
iso-C_4H_9	5540	92.5	5.4	2.1	2.49
sec-C_4H_9	4930	95.5	4.5	0.0	1.29
t-C_4H_9	5010	97.4	2.6	0.0	1.18

^{7}Li NMR spectroscopy confirmed the formation of a LiR:LiCl complex controlling the propagation process, and remaining efficient up to brine temperature in THF.[395]

In the oligomerization of methyl methacrylate initiated by methyl α-lithioisobutyrate or the corresponding sodium compound, the rate constants of partial reactions considerably depended on $\overline{DP}$.[390,398,399] The rate constant of initiation was higher than that of the subsequent propagation steps, k_p, by about two orders of magnitude, thus providing conditions for the formation of a polymer with a narrow MWD. Cyclization of living oligomers led to inactive ketoesters. The rate constant of cyclization, k_c, strongly decreased with $\overline{DP}$, therefore being less important in the later stages of polymerization.

Addition of alkali *tert*-butoxide had a marked effect in decreasing both k_p and k_c.[399] Since the latter dropped ten times more than the propagation rate, the addition of butoxide resulted in a higher limiting conversion, narrower MWD, and an enhanced livingness of the active centers.[399-402]

In the oligomerization of *tert*-butyl acrylate initiated by *tert*-butyl α-lithioisobutyrate, t-BuLiiB (molar ratio, 2:1) in THF, more than 95 mol% of the products consisted of linear oligomers, the rest being ketoesters and other derivatives. A pronounced dependence of the propagation rate constants on $\overline{DP}$ was observed. The kinetics and MWD were strongly dependent on the nature of the additions. In the presence of lithium *tert*-butoxide, t-BuOLi ([t-BuOLi]:[t-BuLiiB] = 3:1), the dimer was the main product besides high oligomers and polymers ($\overline{M}_w/\overline{M}_n \sim 3.5$), but the content of the side products was drastically decreased. When LiCl was added instead of butoxide (LiCl:t-BuLiiB = 1:3), the MWD of oligomers decreased to 1.15.[399]

Both the tacticity and the $\overline{M}_w/\overline{M}_n$ ratio of anionic acrylic polymers are very much dependent on the initiating system. Thus, it was found that t-C_4H_9MgBr in toluene caused no side reaction and formed highly isotactic PMMA with a narrow MWD.[403-405]

Upon the polymerization of MMA with Grignard reagents (RMgBr) in toluene, the isotacticity of the polymer increased greatly and the MWD became narrower as the alkyl group of the initiator became bulkier.[381,382] The bulky alkyl group prevents the side reactions shown for $-C_4H_9$MgCl (Equations 197 and 198) and gives a living character to the polymerization process. This leads to the formation of highly isotactic PMMA with a narrow MWD (Table 8).[381,382] Polymerization of MMA by t-C_4H_9MgBr in THF at $-78°C$ gave a syndiotactic polymer (85% triad).[406]

Highly syndiotactic PMMAS with narrow MWD were also obtained using the initiating system t-C_4H_9Li/AlR$_3$, generally at an Al/Li ratio $\gg 3$.[407,408] A stereoblend-type polymer with a bimodal MWD was obtained at an Al/Li ratio of 1.5. The high molecular weight fraction was found to be isotactic and the low molecular weight fraction syndiotactic. The triad syndiotacticity increased with decreasing polymerization temperatures, and that of PMMA prepared at $-93°C$ reached 96%.[407]

Polymerization of other alkylmethacrylates (except t-Bu methacrylate) with this initiating system also gave highly syndiotactic and narrow MWD polymers (Table 9).

TABLE 9
Polymerization of $H_2C=C(CH_3)-COOR$
with $t\text{-}C_4H_9Li/Al(C_2H_5)_3$ in Toluene at
$-78°C$[407]

R	$\overline{M}_n$	Tacticity (%)			$\overline{M}_w/\overline{M}_n$
		mm	mr	rr	
CH_3	5170	0	10	90	1.21
C_2H_5	4810	3	7	90	1.10
$i\text{-}C_3H_7$	6340	2	9	84	1.31
$n\text{-}C_4H_9$	7230	2	7	91	1.20
$i\text{-}C_4H_9$	7130	3	8	89	1.15
$t\text{-}C_4H_9$	7140	10	33	57	1.64

Highly isotactic PMMA macromer was prepared from the living PMMA formed with $t\text{-}C_4H_9MgBr$ in toluene at $-78°C$ by endcapping with p-bromomethylstyrene (p-BrSt) in the presence of HMPA.

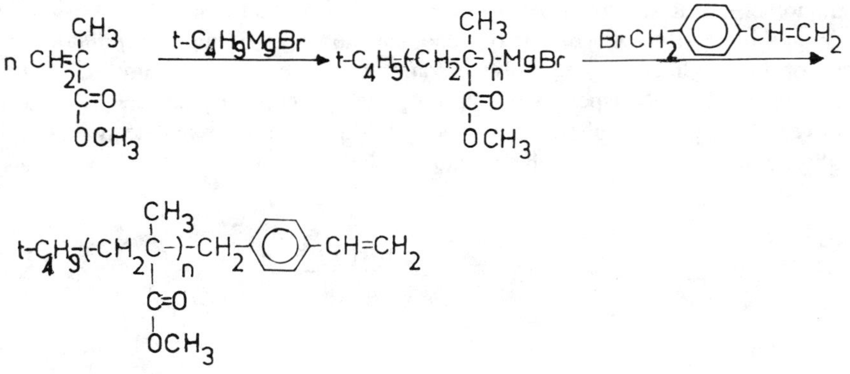

$$ (201) $$

Highly syndiotactic PMMA macromer ($\overline{M}_n \sim 2500$) with the same chemical structure was obtained from the living PMMA prepared with $t\text{-}C_4H_9Li/Al(C_2H_5)_3$ in toluene at $-78°C$ in the presence of TMEDA. Syndiotactic PMMA macromers were also prepared from the living PMMA obtained by polymerization of 1,1-diphenylhexyllithium (DPhHLi), 3,3-dimethyl-1,1-diphenylbutyllithium, and fluorenyllithium in THF at $-78°C$.[407]

Macromers of desired molecular weight and functionality were reported[409] by functional initiation of the methacrylate chain via the hydroxy functional initiator 1-(-p-hydroxyphenyl-)-1-phenylethylene. The initially formed phenoxide anion was treated with additional $s\text{-}C_4H_9Li$ to form the dianion:

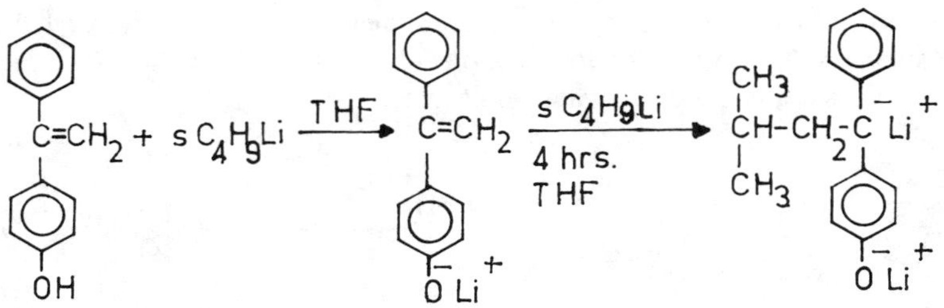

$$ (202) $$

The macromers obtained are perhaps mostly syndiotactic.

TABLE 10
Pyridine-Mediated Anionic Polymerization of Methacrylate Monomers

Monomer	Initiating system	T (°C)	Yield (%)	$\overline{M}_n$	$\overline{M}_w/\overline{M}_n$	Ref.
MMA	DPHLi/Pyridine	−78	100	1,000	1.11	412
MMA	*s*-C_4H_9/Pyridine	−45	100	4,700	1.02	413
MMA	*s*-C_4H_9/Pyridine	0	50	10,000	1.60	413
MMA	*s*-C_4H_9/Pyridine	25	7	300	2	413
MTFMA	Pyridine	RT	91	13,423	1.91	413
MTFMA	Pyridine/H_2O(5:1)	RT	66	6,370	1.60	414
MTFMA	Pyridine/H_2O(1:1)	RT	76	1,840	1.41	414
MTFMA	Pyridine/H_2O(1:1)	RT	41	1,020	1.29	414

Pure-isotactic and pure-syndiotactic MMA dimers to octamers isolated by HPLC[410] from the oligomer mixtures prepared in toluene with *t*-C_4H_9MgBr and *t*-$C_4H_9Li/AlEt_3$, respectively, were fully investigated by ^1H and ^{13}C NMR spectroscopy.[408] A lower segmental mobility of the syndiotactic oligomers was suggested since their ^1H- and ^{13}C-T_1S were smaller in every monomeric unit than those of the corresponding isotactic oligoMMAs.

Using pyridine (Pyr) as solvent or cosolvent and alkyllithium as initiator, the living character of the alkylmethacrylate polymerization can be controlled from −78°C to as high as −20°C.[411-413] The polydispersity of oligomers and polymers is extremely low (Table 10). It is believed[413] that pyridine dissociates the alkyllithium aggregates and decreases the basicity of the alkyllithium initiator through formation of a "σ complex":

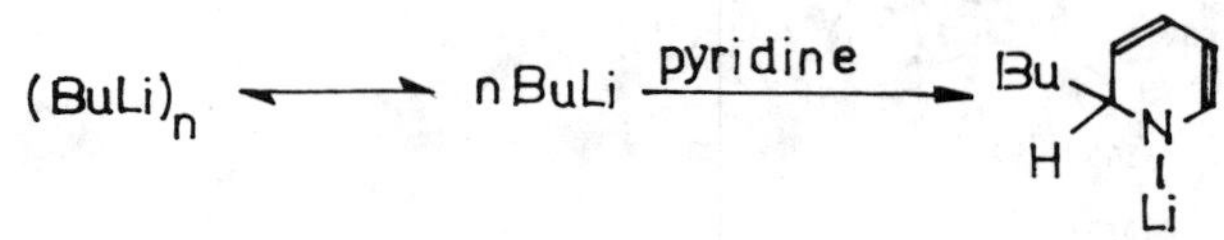

(BuLi)n
nBuLi
pyridine
Bu
H
N
Li

(203)

The "σ complex" is maintained during the polymerization so that the side reactions (attack of the ester by carbanion) may be avoided:

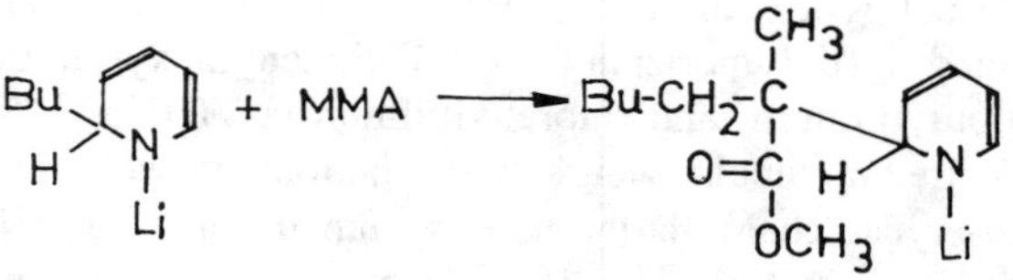

Bu
H
N
Li
+ MMA
Bu-CH2-C
CH3
O=C
OCH3
H
N
Li

(204)

Pyridine itself is capable of initiating anionic polymerization of α-(trifluoromethyl)acrylate, MTFMA, while typical organometallic initiators for MMA were not as effective as pyridine for the polymerization of this highly polarized monomer.[414] Addition of water to the MTFMA/pyridine system resulted in a reduction of molecular weight (Table 10). The initiation step proposed in this case involved the water molecule:

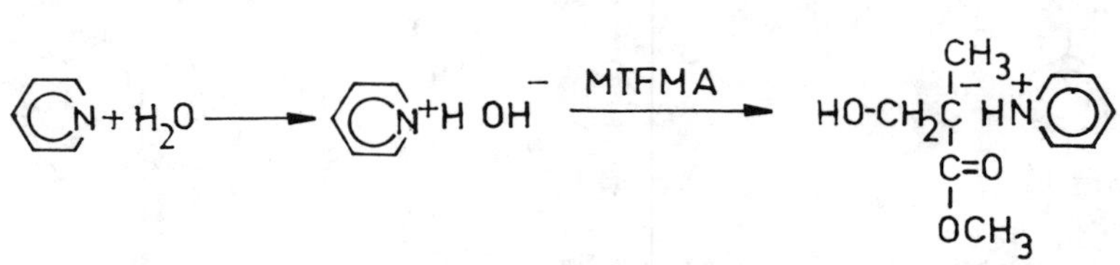

N + H2O
N+H OH−
MTFMA
HO-CH2C
CH3
HN
C=O
OCH3

(205)

TABLE 11
Tacticity (Triad) of Anionic
PMMA Prepared in the Presence
and Absence of Pyridine with
Alkyllithium Initiators

Solvent	T (°C)	Tacticity			Ref.
		I	M	S	
THF	−78	7	36	57	415
THF	−60	10	39	51	413
Pyr.	−60	7	33	60	411
Pyr.	−60	6	26	68	413
Pyr.	−40	8	32	60	413
Pyr.	0	9	35	56	413

while the most likely mechanism in the absence of water was supposed to involve one electron transfer from pyridine to MTFMA through the charge transfer to form a pyridine radical cation:[414]

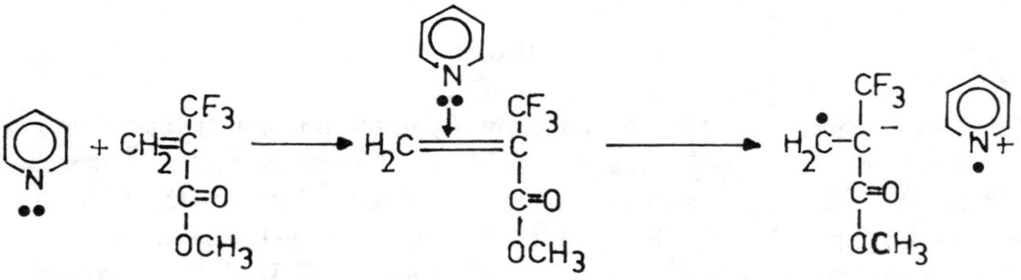

$$(206)$$

The addition of water to the organometallic/pyridine/MMA system led only to deactivated oligomers:[413]

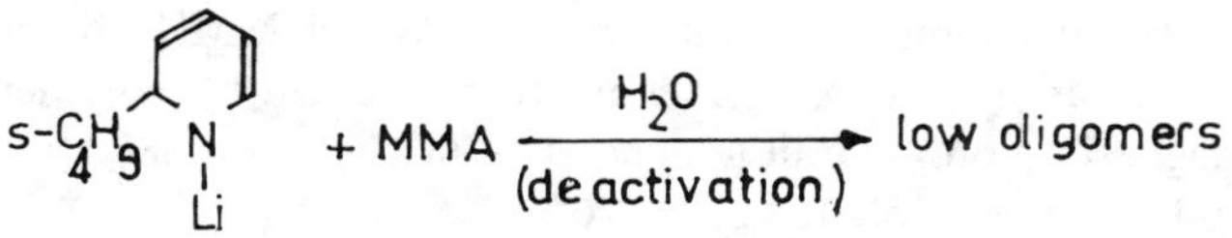

$$(207)$$

The syndiotactic triad content of PMMA prepared in pyridine as solvent was higher than that prepared in THF, but the products were not highly syndiotactic. Moreover, the tacticity was not very dependent on the reaction temperature (Table 11).[415]

Maleic anhydride was also oligomerized anionically with pyridine as initiator.[416] The oligomer conversion increased with increasing concentration of initiator. The region ^{1}H and ^{13}C NMR peaks of the material were consistent with the presence of pyridine, succinic anhydride units, and (probably) a substituted cyclobutane ring.

Optically active, isotactic oligo- and polymethacrylates with high optical rotation ($[\alpha]_{578}$ = 70 to 530°) were reported[417] to be obtainable by asymmetric polymerization of MMA and several other methacrylates with t-BuOK/chiral crown ether or n-BuLi/chiral diamine complexes in toluene at −78°C. The optical activity has been ascribed to the helical structure of the oligomers and polymers. The slow decrease of activity in solution at room temperature (RT) has been attributed to uncoiling (mutarotation) of the helix. One-handed helical methacrylate polymers were also prepared by asymmetric, anionic polymerization of bulky methacrylates initiated by chiral systems,[418-424] i.e., triphenylmethyl methacrylate (M^1), diphenyl-

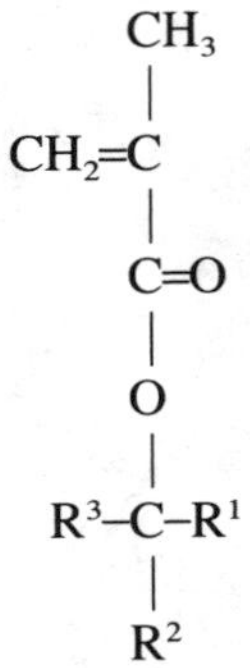

	R¹	R²	R³
M¹	Ph	Ph	Ph
M²	Ph	Ph	2-Pyr
M³	Ph	2-Pyr	o-Tol
M⁴	Ph	2-Pyr	m-Tol

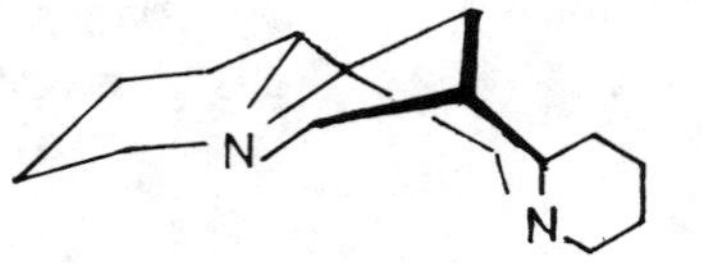

CH₃
|
CH₂=C
|
C=O
|
O
|
R³–C–R¹
|
R²

(−)-Sp

	R
(+)–DDB	–N(CH₃)₂
(+)–DHB	–N(CH₂)₆

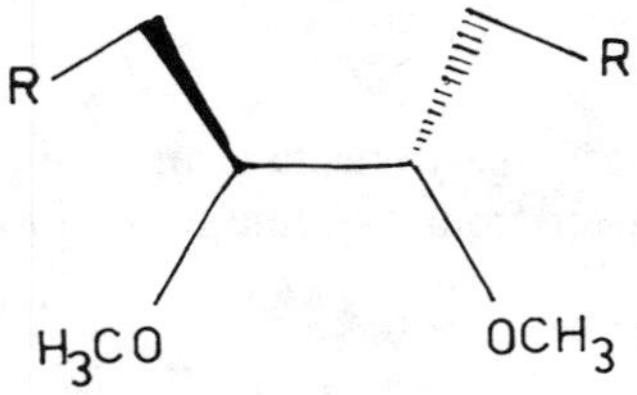

SCHEME 1

2-pyridylmethylmethacrylate (M²), phenyl-2-pyridyl-o-tolylmethyl methacrylate (M³), and phenyl-2-pyridyl-m-tolylmethyl methacrylate (M⁴) as monomers, and (−) sparteine-9-fluorenyllithium (Sp-FlLi), (−) sparteine-9-methyl-9-fluorenyllithium (Sp-9MeFlLi), (−)-sparteine-n-BuLi (Sp-n-BuLi), (+)- and (−)-2,3-dimethoxy-1,4-bis(dimethylamino)-butane-N,N′-diphenylethylenediamine monolithium amide (DDB-DPEDE-Li) complexes, and (+)-(S,S)-2,3-dimethoxy-1,4-bis(1-perhydroazepinyl)butane-9-fluorenyllithium (DHP-FlLi) and DMO-DPEDA-Li complexes as initiators (Scheme 1).

However, optically active polymers wer not obtained from less bulky methacrylates having ester groups such as MMA, diphenylmethyl-methacrylate, and 1,1-diphenylethyl-methacrylate.[423] The enantiomers of both 5-mer and 8-mer MMH oligomers (resolved by GPC in hexane from an isotactic oligomer mixture obtained by polymerization of MMA with naphthyl magnesium bromide in toluene at −78°C, $\overline{M}_n$ = 3800 and MWD = 1.65) showed very small rotation (i.e., $[\alpha]_{365}$ = 26° and 23°, respectively), indicating that the methyl ester is too small to sustain a helical structure.

Uniform, optically MMA oligomers with $\overline{DP}$ = 1, 2, 3, and 4 were obtained by the asymmetric anionic polymerization of triphenylmethyl methacrylate *(vide infra)*, followed by the substitution of a trityl group for methyl and subsequent fractionation by GPC and separation into diastereomers and enantiomers by HPLC or GC.[424] Packed-column super-critical fluid chromatography has also been found to be suitable for the rapid and detailed analysis of the isotactic and syndiotactic oligomers ($\overline{DP}$ = 2 to 20) of MMA prepared by stereoregular living polymerizations with t-C₄H₉MgBr and t-C₄H₉Li/Al(C₂H₅)₃ complexes, respectively, in toluene at −78°C.[425]

M¹ formed a highly isotactic one-handed helical polymer ($[\alpha]_D$ = 360°) by asymmetric polymerization with Sp-FlLi,[419-421] Sp-9MeFlLi,[422] Sp-DPhHLi,[424] DDB-FlLi, and DDB-DPEDA-Li[421] complexes in toluene at −78°C. Detailed analysis of oligomers by GPC, chiral HPLC, and NMR indicated that various stereoisomers of dimers (2-mer), trimers, and tetramers were formed in the early stages of the polymerization and only one or two of the stereoisomers propagated to the higher isotactic oligomers. A stable one-handed helical structure seemed to be formed at 6 to 9 mer.

Only DHB-FlLi or DHB-DPEDA-Li complexes afforded poly(M^2) with a high one-handed helicity in good yield. When the optically active monomer ($+M^3$) was polymerized with ($+$)- and ($-$)-DDB complexes at different temperatures, large positive or negative rotations were noted, i.e., $[\alpha]_D^{-78}$ about $+400°$ and $[\alpha]_D^{+25}$ about $-320°$, which indicated that the helicity of the chains is governed by the chirality not of the ligands, but of the ester groups. Enantioasymmetric (stereoselective or asymmetric-selective) polymerization of racemic M^3 was possible. However, the helicity of M^4 polymer chains was little influenced by the chirality of the ester group, but, rather, was controlled by the chirality of the ligands, as observed for M^1 and M^2.[421]

Detailed structure and stereochemical information on the anionic oligomerization of a variety of polar vinyl monomers of the type

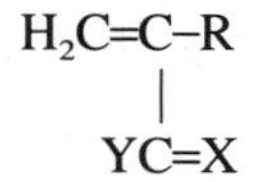

$$H_2C=C-R$$
$$|$$
$$YC=X$$

where R is H or alkyl and X, Y is O, N, or C have been provided mostly by Hogen-Esch et al.[426-450] These oligomer studies have dealt with acrylates,[426] vinylpyridines,[427-439] vinyl-sulfoxides,[440,442-446] and vinyl ketones.[450] The anionic polymerization of some monomers of this type proceeded through relatively stable intermediates that were studied by spectroscopic methods (^1H and ^{13}NMR). For instance, the carbanions that initiated the oligomerization of methyl methacrylates were taken as models for the growing chain end, and upon termination of the growing chain ends by methylation, symmetrical oligomers (n = 1 to 6, n $\sim$ 15) were formed:[426]

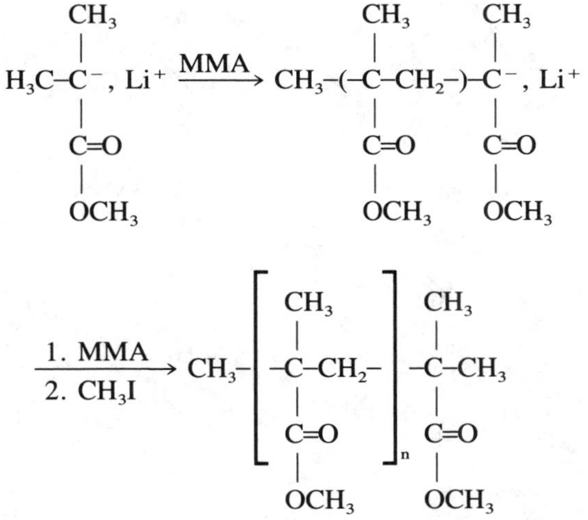

$$\text{(208)}$$

These symmetrical oligomers were excellent models for the elucidation of the stereo-chemistry of the PMMA chain ends (formed by methylation of the living polymer using $^{13}CH_3I$) in that, by comparing the stereochemistry of the main chain and chain end, the consistency can be tested with a Bernoullian or Markoffian statistical process.[427,428]

The absence of polymer stereoregularity observed in the anionic polymerization of 2-vinylpyridines in THF in the presence of Li as counterion (the occurrence of cation coordination notwithstanding)[432] has been plausibly attributed to the existence of (E)- and (Z)-geometric isomers in the oligomeric carbanion intermediate in a E/Z ratio kinetically determined to be close to one.[429] The highly delocalized carbanion has been shown to be sp^2 hybridized and has a large energy barrier to rotation about the C$_2$-C$_\alpha$ bond.[433] Therefore, the possibility of *E:Z* isomerism exists in the active center of polymerization

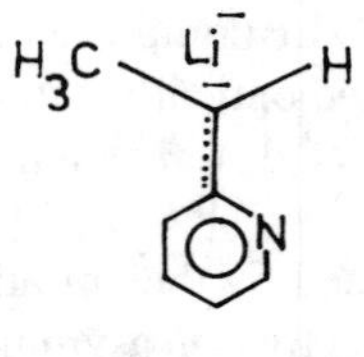

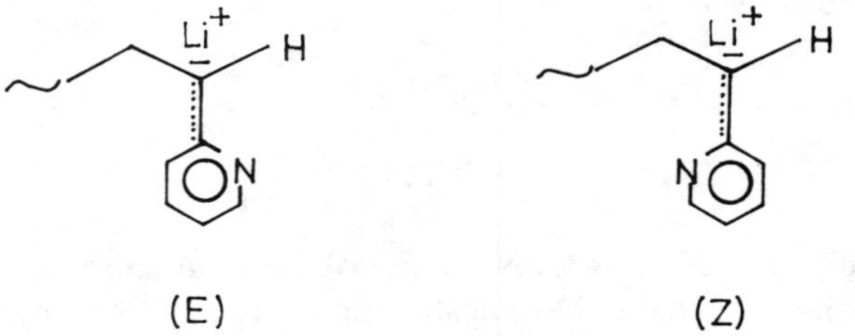

(E)   (Z)

the occurrence of which was probably a direct consequence of the lack of selectivity in monomer presentation *(S-cis or S-trans)* during the transition state:

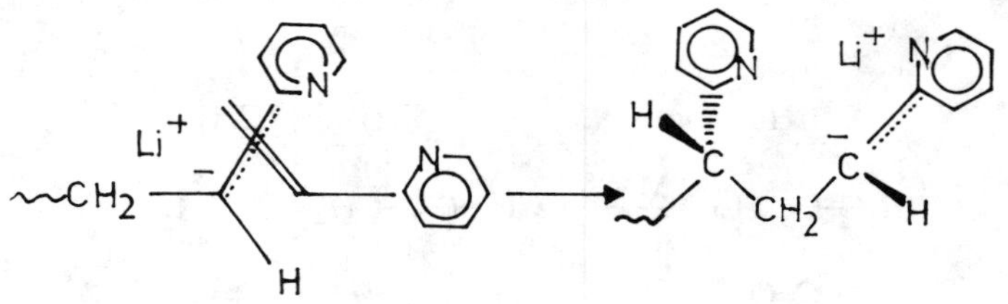

S-cis addition   *(Z)-isomer*   (209)

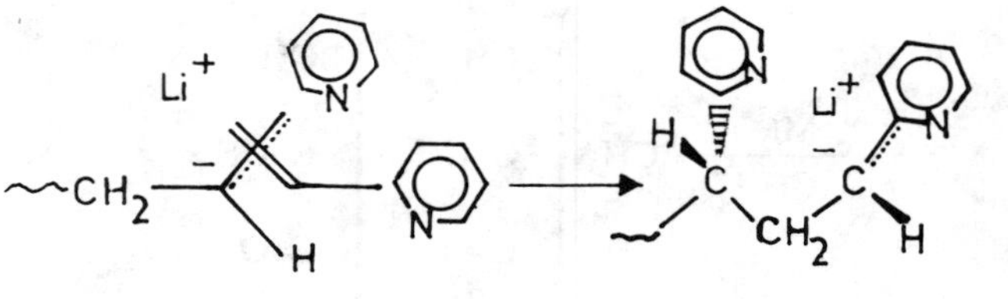

S-trans addition   *(E)-isomer*   (210)

^{13}C- and ^{1}H-NMR data of oligomeric anions of the type:

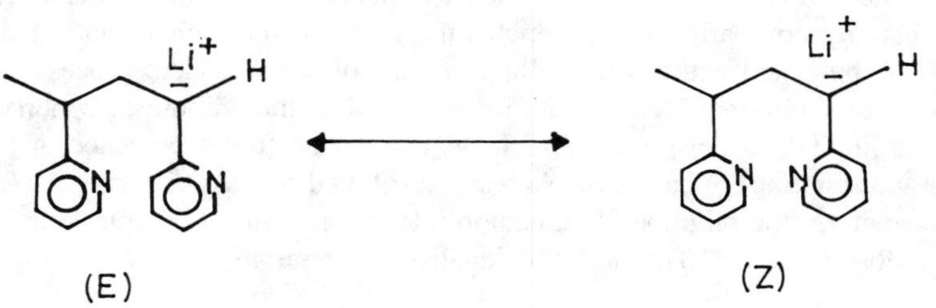

(E)   (Z)   (211)

TABLE 12
Lewis Acid-Mediated Anionic Polymerization of 2-Vinylpyridine Initiated by t-C_4H_9Li in Toluene at $-78°C$[434]

$\dfrac{[AlEt_3]}{[t\text{-}C_4H_9Li]}$	$\overline{M}_n$	$\overline{M}_w/\overline{M}_n$	Fraction of triads		
			mm	mr	rr
0.0	3.000	1.7	0.50	0.37	0.13
0.7	3.250	1.4	0.10	0.45	0.45
2.5	4.300	2.1	0.03	0.40	0.57
6.0	5.300	2.7	0.01	0.33	0.66
9.0	7.200	1.9	0.01	0.31	0.69

TABLE 13
Triad Tacticity of Poly(2-Vinylpyridine) and Poly(3-Methyl-2-Pyridine)

Polymer	Type of initiation	T (°C)	Fraction of triads			Ref.
			mm	mr	rr	
P3M2VP	Anionic(Li)	-71	0.85	0.14	0.01	435
P2VP	Anionic(Li)	-78	0.63	0.30	0.07	434
P2VP	Anionic (Li + AlEt$_3$)	-78	0.00	0.31	0.69	434
P3M2VP	Radicalic	60	0.43	0.36	0.21	435

and

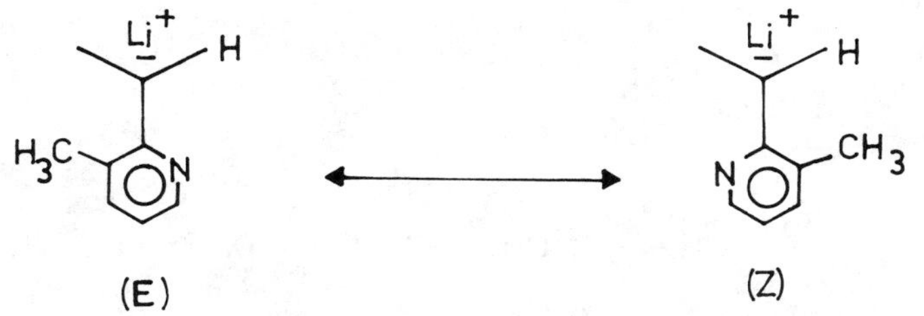

$$(212)$$

have demonstrated the existence of stable E- and Z-isomers in the polymerization of both 2-vinylpyridine (Equation 211) and 3-methyl-2-vinylpyridine (Equation 212), which interconvert slowly on the NMR and laboratory time scale.[429,433]

The formation of approximately equal fractions of E- and Z-isomers was considered to be caused by the failure of the 2-vinylpyridine monomer to compete effectively with THF that is of comparable coordinating power, but that was present at much higher concentration (i.e., as solvent). However, in the presence of Lewis acids (AlEt$_3$), t-C_4H_9Li readily initiated polymerization of 2-vinylpyridine in toluene at $-78°C$ and yielded syndiotactic products.[434] Increasing the $Al(C_2H_5)/t$-C_4H_9Li ratio from 1 to 9 raised the syndiotactic triad content from 0.50 to 0.69 (Table 12). The explanation lay in the complexation of the nitrogen with the Lewis acid, which favors the formation of E-isomers. In this regard, it is interesting to note that the stereochemistry of polymerization of MMA in toluene initiated by alkyllithiums was changed from isotactic to syndiotactic upon addition of triethylaluminum.[407]

The greater stereoregularity (i.e., an isotactic placement) of poly(3-methyl-2-vinylpyridine), P3M2VP (shown in Table 13), compared to poly(2-vinylpyridine), P2VP, was explained by the interaction between the 3-methyl groups with previous pyridine rings favoring a helical conformation.[435] In comparison with the radical polymerization of the more bulky M[1], which gave helical isotactic polymers,[436] the radical polymerization of 3-methyl-2-vinylpyridine was not stereoselective (Table 13).

Macrocyclic oligomers and polymers ($\overline{M}_n$ = 4000) of 2-vinylpyridine were prepared by the end-to-end linking of difunctional living P2VP using 1,4-bis(bromo-ethyl)benzene as the coupling agent, followed by fractional precipitation.[437-439] The polymers were found to have a narrow MWD ($\overline{M}_w/\overline{M}_n$ = 1.15). Preparation of two-ended living P2VP was carried out at $-78°C$ in THF, by vapor-phase addition of 2VP monomer onto 1,4-dithio-1,1,4,4-tetraphenylbutane.[437] The T_g of linear P2VP increased with increasing molecular weight, as expected, while the T_g of cyclic P2VP decreased with increasing molecular weight.[438] Blends of cyclic and linear oligo-2VP were perfectly compatible, as shown by the linear dependence of T_g vs. composition.[439]

The anionic polymerization of *vinyl sulfoxide* is of considerable interest because the pronounced dipole moment and cation coordinating ability of the sulfoxy group may predispose these monomers toward stereoregular and stereoselective polymerization. Due to the sulfur chirality, the stereochemistry of polymerization is considerably more complicated when compared to that of achiral vinyl monomer.[440] Unlike vinyl sulfones,[441] vinyl sulfoxides are rather reluctant to undergo vinyl addition with radical and anionic initiators.[442] However, vinyl phenyl sulfoxide, VPhSO, was successfully converted to poly(vinyl phenyl sulfoxide) by methyllithium in THF at $-78°C$.[440] The stereochemical analysis of dimerization (Equation 213) indicated that the vinyl addition reaction was highly stereoselective as only the R configuration formed at the α-sulfinyl carbon adjacent to sulfur in a R configuration.

$$CH_3-\overset{|}{\underset{|}{C}}-H, Li^+ \xrightarrow[\substack{THF \\ -78°C}]{nVPhSO} CH_3-(-\overset{*}{\underset{|}{C}}H-CH_2)_n-\overset{|}{\underset{|}{C}}-H, Li^+$$

$$\begin{array}{ccc} & *S{\rightarrow}O & \\ & | & \\ & C_6H_5 & \end{array} \qquad \begin{array}{cc} *S{\rightarrow}O & *S{\rightarrow}O \\ | & | \\ C_6H_5 & C_6H_5 \end{array}$$

aR

$$\xrightarrow[\substack{THF \\ -78°C}]{Ex} CH_3-(-\overset{*}{\underset{|}{C}}H-CH_2-)_n-\overset{*}{\underset{|}{C}}H-E + LiX \qquad (213)$$

$$\begin{array}{cc} *S{\rightarrow}O & *S{\rightarrow}O \\ | & | \\ C_6H_5 & C_6H_5 \end{array}$$

where Ex is CH_3I or CH_3OH.

The addition reaction (i.e., dimerization, trimerization, etc.) was also highly stereoselective, with a 12-to-1 preference for incorporation of S monomer to anion aR (over incorporation of R monomer to aR).[443-445] The highly stereoselective polymerization did not generate what is normally considered a stereoregular polymer when racemic monomer was used. A highly stereoregular polymer was produced from optically active monomer.[445,446]

Due to the facile stabilization of the carbanionic active center, by the conjunctive effect of the carbonyl group, *vinyl ketones* can be easily polymerized by anionic catalysts.[447] By varying the mole ratio of monomer to initiator, yields of specific oligomers can be optimized. However, lithium alkyls, Grignard reagents, and organoaluminum compounds are known to react with α,ω-unsaturated ketones by either carbonyl, conjugate, or vinyl addition, e.g.:[448]

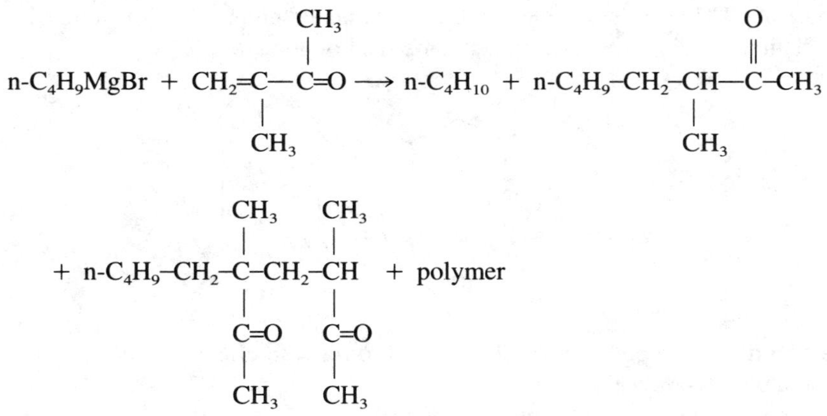

$$n\text{-}C_4H_9MgBr + CH_2=C-C=O \longrightarrow n\text{-}C_4H_{10} + n\text{-}C_4H_9-CH_2-CH-C-CH_3$$

(with CH_3 above and CH_3 below the central carbon of the reactant; product has O double bond on carbonyl and CH_3 below)

$$+ \; n\text{-}C_4H_9-CH_2-C-CH_2-CH \; + \; polymer \tag{214}$$

(with CH_3 and CH_3 above; $C=O$, $C=O$ and CH_3, CH_3 below)

Since the electron density on the β-carbon of isopropenyl *t*-butyl ketone, IPTBK, is much higher than that of other vinyl ketones (due to the steric hindrance which allows a twisted conformation with a limited conjugation between vinyl and carbonyl groups), this monomer does not undergo radical homopolymerization and the anionic polymerization is a slow equilibrium process with a low ceiling temperature ($<0°C$).[449] As shown in Scheme 2, an initiating nucleophile RM (e.g., alkyllithium, aryllithium, Grignard reagents) could add to either the C=C or C=O of IPTBK:

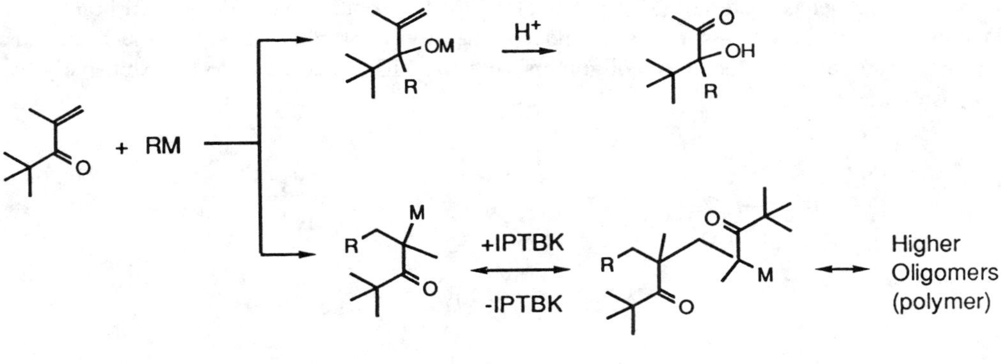

SCHEME 2.

Therefore, for elucidation of the stereochemistry of the oligomerization of vinyl ketones, the monomer *tert*-butyl vinyl ketone, *t*-BVK, chosen since it has no acidic hydrogen alpha to the carbonyl, and the *t*-butyl might be expected to protect the carbonyl from nucleophilic attack.[450] The model initiator was the lithioenolate of *t*-butyl ethyl ketone, *t*-BEK, a bidentate nucleophile:

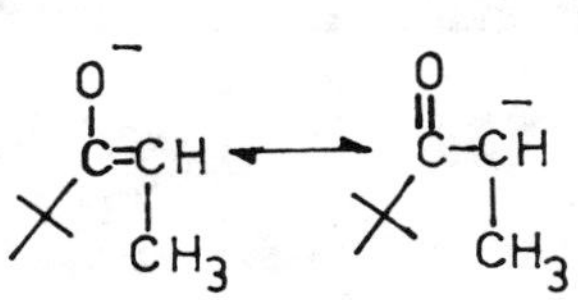

$$\tag{215}$$

Choice of CH$_3$I as a "soft" electrophile to terminate propagation maximized the extent of C-alkylation, leading to the desired symmetrical oligomers:

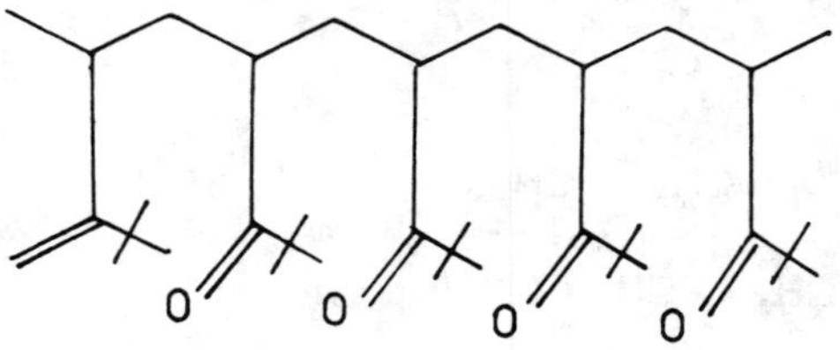

The "hard" electrophile Me$_3$SiCl was used to trap the oligomeric geometrical isomers of the enolate by O-alkylation.

From the two possible conformations of *t*-BVK

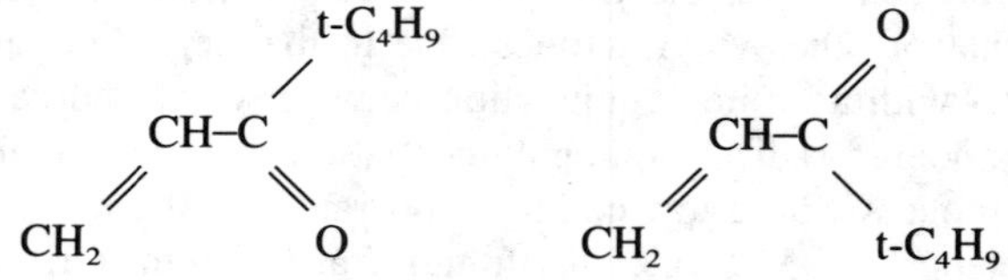

s-*cis* s-*trans*

the s-*cis* rotamer is preferred (92%). The attack of the enolate to the β-vinyl carbon would therefore result in a *(Z)*-isomer and, indeed, the silyl enol ether of the "dimer"-enolate, trapped by quenching the active oligomers with Me$_3$SiCl, was found to be exclusively the *(Z)*-isomer:

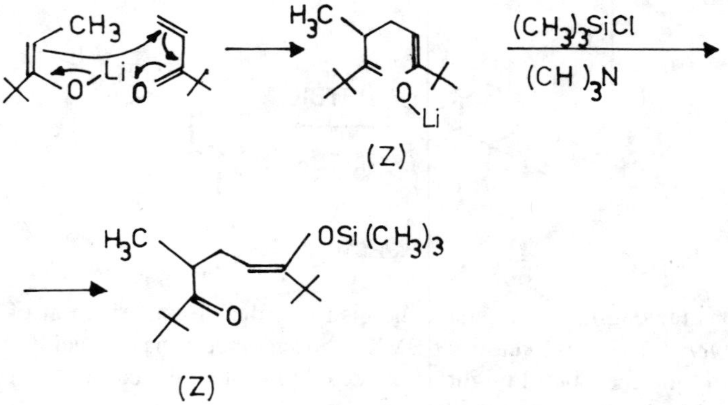

(216)

In THF, the *meso-* and racemic dimers were found by NMR in a m/n ratio of 1.6. Of the four possible stereoisomers of trimers, only the heterotactic is unsymmetrical:

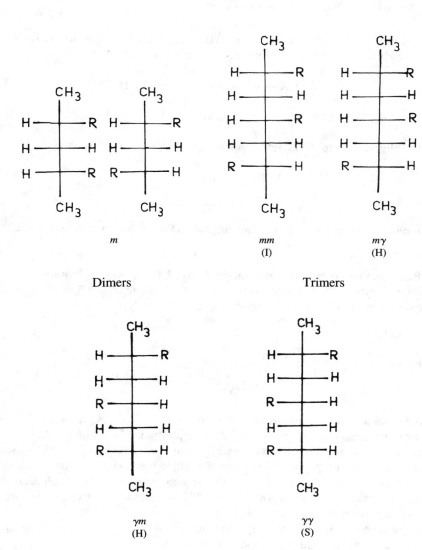

Dimers

m

Trimers

mm
(I)

mγ
(H)

γm
(H)

γγ
(S)

Trimers

where R is

$$-C \overset{\displaystyle O}{\underset{\displaystyle t\text{-}C_4H_9}{\big\|}}$$

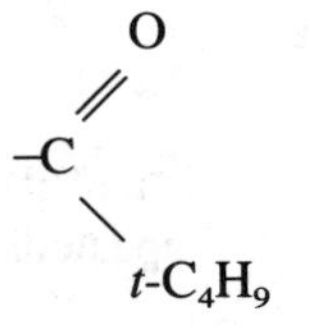

For the oligomerization done in THF, the reactions I:H:S = 35:57:7 were typical values. The isotactic isomer mmm of the tetramer was one of two (of six possible) diastereomers which dominated the oligomerization.

Anionic oligomerization with strongly basic initiators (*n*-butyllithium or sodium *tert*-butoxide) of *unsaturated amides* such as acrylamide, maleamide, or mesaconamides occurs by hydrogen transfer polymerization through a step-growth mechanism.[451-452]

An α-peptide residue was considered to arise from the addition of the $-CO-NH-^{(-1)}$ anion to the carbon atom of the monomer carrying the amide group:

$$R^- + CH{=}CHX \longrightarrow \underset{\substack{|\\CONH_2}}{R-CH-}\underset{\substack{|\\CONH_2}}{\overline{C}HX} \longrightarrow \underset{\substack{|\\CONH^-}}{R-CH-CH_2X} \qquad (217)$$

A cyclic oligomer (2 to 3%) corresponding to a cyclic trimer of β-alanine

$$\boxed{-[-NH-(CH_2)_2-CO-]_3-}$$

has been found in the earliest stages of the polymerization of acrylamide with sodium *tert*-butoxide.[452]

4. Miscellaneous Oligomers and Oligomerizations of Polar Vinyl Monomers

Alkyl methacrylates[453,454] and methacrylonitrile[455-457] were each polymerized in an alcohol-alcoxide solution to give mixtures of oligomers of the type

$$\underset{\substack{|\\X}}{RO-(CH_2-\overset{\overset{\textstyle CH_3}{|}}{C}-)_n-H,} \qquad \text{where} \qquad X = \underset{\substack{\|\\O}}{-C-OCH_3,} \qquad \underset{\substack{\|\\O}}{-C-O(s-C_4H_9),} \qquad -CN$$

Terminally unsaturated dimers and trimers of acrylonitrile[458,459] and acrylic esters[460,461] were also prepared in alcohols using various alkoxides or phosphines as basic catalyst.

Gas-phase anionic oligomerization of methyl acrylate initiated by F_3C^-, $NC-CH_2^-$, and alkyl anions was shown to involve a 1,4 Michael addition reaction to the vinyl monomer.[462] Termination occurred when the trimeric anion underwent competitive intramolecular proton transfer and Dieckmann cyclization with loss of CH_3OH, with the relative amounts of these two reactions depending on the initiation anion.

The preparation of oligomers by polymerization has the disadvantage that molecules of different sizes are obtained and separation is very tedious.

A general synthesis method was developed[463] for the preparation of terminally unsaturated oligomers (DP = 2, 3, 4, 5, . . .) of the structure:

$$\underset{\substack{|\\X}}{CH_2{=}C-}\underset{\substack{|\\X}}{(-CH_2-CH-)_n}\underset{\substack{|\\Y}}{-CH_2-CH-R} \qquad (218)$$

where X = Y, X $\neq$ Y, X,Y = COOR,CN, and R = H, CH$_3$.

A set of these steps was carried out repeatedly to introduce one moiety of a vinyl monomer, as follows:

1. Base-catalyzed addition of an acetic acid ester derivative to an activated olefin

$$X-CH_2-COOR + \underset{\substack{|\\Y}}{CH_2{=}C-R} \xrightarrow{B^-} \underset{\substack{|\\X}}{H-\overset{\overset{\textstyle COOR}{|}}{C}-CH_2-}\underset{\substack{|\\Y}}{CH-R} \qquad (219)$$

A

2. Hydrolysis of one ester group at C_1

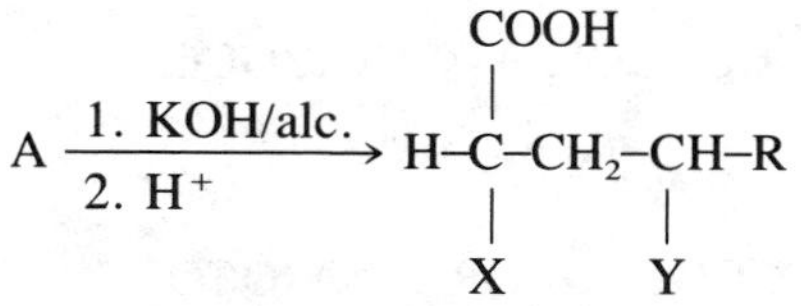

$$A \xrightarrow[\text{2. H}^+]{\text{1. KOH/alc.}} \overset{\displaystyle \text{COOH}}{\underset{\displaystyle \underset{X}{|}\ \ \ \underset{Y}{|}}{\text{H–C–CH}_2\text{–CH–R}}} \tag{220}$$

$$B$$

3. A Mannich reaction of the carboxylic acid B to introduce the terminal double bond

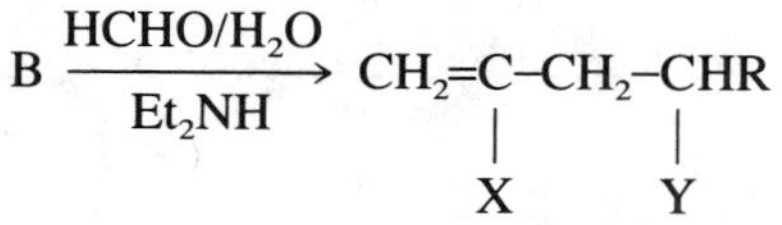

$$B \xrightarrow[\text{Et}_2\text{NH}]{\text{HCHO/H}_2\text{O}} \underset{\underset{X}{|}\ \ \ \underset{Y}{|}}{\text{CH}_2\text{=C–CH}_2\text{–CHR}} \tag{221}$$

Dimer C

In order to obtain the corresponding trimer D, the dimer C is subjected to the same three steps:

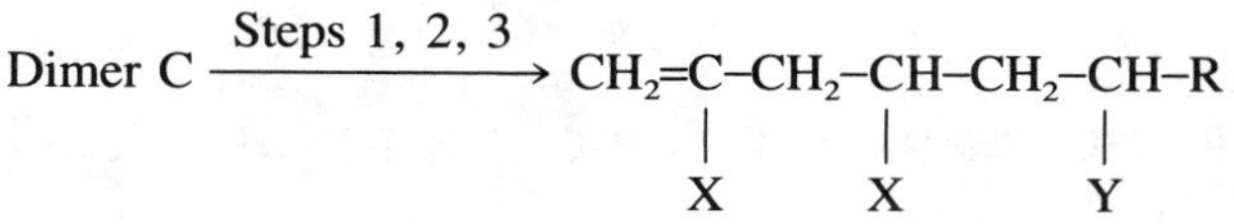

$$\text{Dimer C} \xrightarrow{\text{Steps 1, 2, 3}} \underset{\underset{X}{|}\ \ \ \underset{X}{|}\ \ \ \underset{Y}{|}}{\text{CH}_2\text{=C–CH}_2\text{–CH–CH}_2\text{–CH–R}}$$

Trimer D

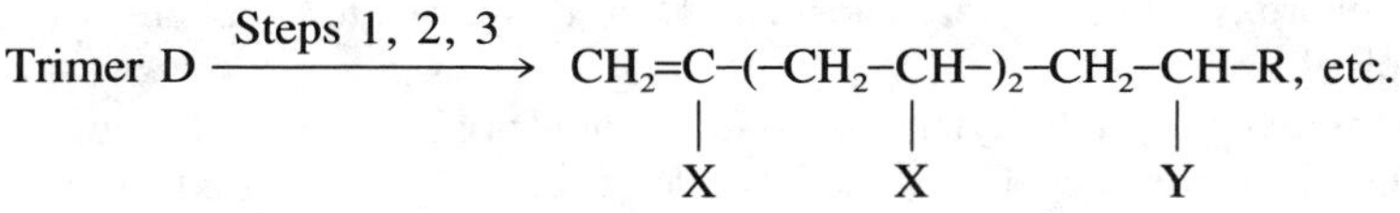

$$\text{Trimer D} \xrightarrow{\text{Steps 1, 2, 3}} \underset{\underset{X}{|}\ \ \ \underset{X}{|}\ \ \ \underset{Y}{|}}{\text{CH}_2\text{=C–(–CH}_2\text{–CH–)}_2\text{–CH}_2\text{–CH–R, etc.}} \tag{222}$$

Tetramer

Chemical transformations through the functional groups might be used to prepare different oligomers, e.g., acrylic *n*-mer→ acrylic acid *n*-mer→ acrylamide *n*-mer, etc.

Last but not at least, *acrylated* oligomers are an especially important commercial class of oligomers. Acrylated oligomers can be divided into six main classes: oligoester acrylates, oligocarbonate acrylates, oligourethane acrylates, oligoalkyleneoxyphosphinate acrylates, oligooxyalkylene acrylates, and oligosiloxane acrylates. Their synthesis is schematized below.

Oligoester acrylates[464-466]

$$nR(COY)_2 + (n + 1)R'(OH)_2 + 2CH_2\text{=C(X)COY}' \xrightarrow{-H_2O(HCl, ROH)}$$

$$CH_2\text{=C(X)COOR'O[OCRCOOR'O]}_n\text{OCC(X)=CH}_2 \tag{223}$$

where R and R′ are alkyl or aryl radicals, X = –H, –CH$_3$, –CN, –halogen, etc., Y = –OH, –OR, or –Cl, and Y′ = –OH, –OR, –OROH, or –Cl.

Oligocarbonate acrylates[467]

$$2CH_2{=}C(X)COOROH \ + \ ClCOR'OCCl \ \xrightarrow[\ -\ 2(R)_3N \cdot HCl\ ]{2(R)_3N}$$
$$\overset{\|}{O} \qquad \overset{\|}{O}$$

$$CH_2{=}C(X)COOROCOR'OCOROCOC(X){=}CH_2 \qquad (224)$$
$$\qquad\quad \overset{\|}{O} \qquad\ \overset{\|}{O}$$

where R and R' are alkyl or aryl radicals, and X $= -H, -CH_3,$ or $-$halogen.

Oligourethane acrylates are formed by the reaction of a diisocyanate with a polyol and a hydroxyalkyl acrylate

$$CH_2{=}CH{-}ROH \ + \ OCN{-}D{-}NCO \rightarrow CH_2{=}CH{-}R{-}O{-}C{-}NH{-}D{-}NCO \qquad (225)$$
$$\overset{\|}{O}$$

$$2CH_2{=}CH{-}R{-}O{-}C{-}NH{-}D{-}NCO \ + \ HO{-}P{-}OH \rightarrow$$
$$\overset{\|}{O}$$

$$CH_2{=}CH{-}R{-}O{-}C{-}NH{-}D{-}NH{-}C{-}O{-}P{-}O{-}C{-}NH{-}D{-}NH{-}C{-}O{-}R{-}CH{=}CH_2 \qquad (226)$$
$$\qquad\quad \overset{\|}{O} \qquad\qquad \overset{\|}{O} \quad\ \overset{\|}{O} \qquad\qquad \overset{\|}{O}$$

where R $=$ hydroxy alkyl acrylate backbone D $=$ diisocyanate backbone (aliphatic, aryl), and P $=$ polyol backbone.

Oligomer based on poly(ethylene oxide)diols could not be prepared because the attempted syntheses invariably resulted in gelation.[468] However, oligomer based on ethylene oxide-capped poly(propylene oxide)diols could be made.

Oligooxyalkylene acrylates are usually prepared by cationic telomerization of oxyranes or other alkylene oxides (THF, dioxolane) in the presence of acrylic derivatives as chain-transfer agents, e.g., the oligomerization of tetrahydrofurane in the presence of methacrylic anhydride:[469]

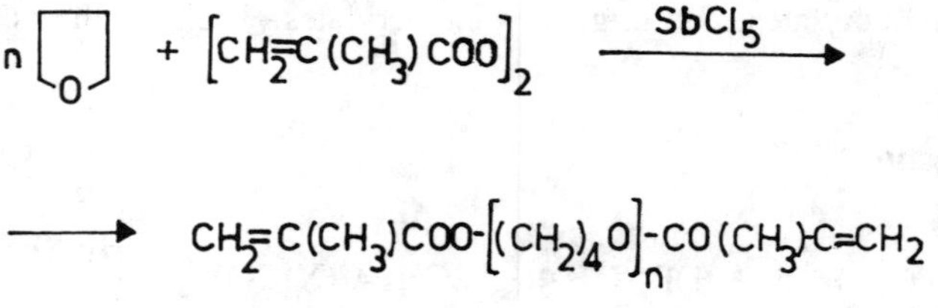

$$n \ \underset{O}{\boxed{}} \ + \ \left[CH_2{=}C(CH_3)COO\right]_2 \ \xrightarrow{\ SbCl_5\ }$$

$$\longrightarrow \ CH_2{=}C(CH_3)COO{-}\left[(CH_2)_4O\right]_n{-}CO(CH_3)C{=}CH_2 \qquad (227)$$

Oligoalkylene oxyphosphinate acrylates are synthesized by the oligomerization of haloanhydrides of phosphinic acids with oxyalkyl(meth)acrylates:

$$RPOCl_2 + 2CH_2=C(X)COOCH_2CH_2OH \xrightarrow[-2HCl]{C_5H_5N}$$

$$\underset{\overset{\|}{O}}{CH_2=C(X)COCH_2CH_2OP}\underset{\overset{\|}{O}}{(R)OCH_2CH_2OC}\underset{\overset{\|}{O}}{(X)C=CH_2} \qquad (228)$$

where X = −H or −CH$_3$, and R = alkyl.

Oligosiloxane acrylates are prepared by oligocondensation of α,ω-dichlorooligosiloxanes and oxyalkyl(meth)acrylates:[470,471]

$$2CH_2=C(CH_3)COO(CH_2)_2OH + Cl-(-SiR_2-O-)_{n-1}-SiR_2-Cl \xrightarrow[+2HCl]{R_3N}$$

$$\underset{\overset{\|}{O}}{CH_2=C(CH_3)CO(CH_2)_2O-(-SiR_2-O-)_n-O(CH_2)_2OC}\underset{\overset{\|}{O}}{(CH_3)C=CH_2} \qquad (229)$$

where R = alkyl, aryl.

Other macromonomers of methyl methacrylate,[472-475] 2-vinylpyridine,[476,477] and 4-vinylpyridine[478] were anionically synthesized by the termination technique.[479]

B. ANIONIC OLIGOMERIZATION OF HETEROCYCLIC MONOMERS
1. Specific Considerations
The large majority of anionic ring-opening polymerizations belong to the nucleophilic substitution reactions. A typical example is the anionic polymerization of oxiranes:

$$\text{wwwCH}_2CH_2O^- + CH_2 \underset{\diagdown\diagup}{\overset{\underset{CH_2}{}}{\rule{2cm}{0.4pt}}} O \rightarrow -[-CH_2CH_2\bar{O} \cdots CH_2 \cdots \bar{C}H_2] \qquad (230)$$

The propagation reaction differs considerably in its basic features from the anionic polymerization of unsaturated monomers such as dienes or styrene derivatives. Several similarities were observed, however, with the polymerization of polar vinyl monomers such as (meth)acrylates or vinyl ketones.[480]

For some heterocyclic monomers, the unique chemical structure of the growing species follows unequivocally from the monomer structure: the bond involving the heteroatom is the most labile one because of the electronegative character of the heteroatom (e.g., oxygen or sulfur), so that the active chain end is headed by that same heteroatom. Isomeric structures have to be taken into account when the bidentate character of the active species becomes possible, e.g., the polymerization of thietane, where the carbanion, but not the thiolate anion, was proposed for certain conditions of polymerization:[481]

$$RLi + \underset{\overset{|}{CH_2-CH_2}}{CH_2-S} \longrightarrow R-S-CH_2-CH_2-CH_2-Li^+ \qquad (231)$$

The fact that the growing chain is *carbanionic* in nature is confirmed by the fact that it can initiate a block copolymerization of styrene.[482] At the same time, lithium thiolates were found to be ineffective in initiating the polymerization of these thietanes.[481]

Unsymmetrically substituted monomers can provide active species by α- or β-ring scission, as in the polymerization of substituted oxiranes:

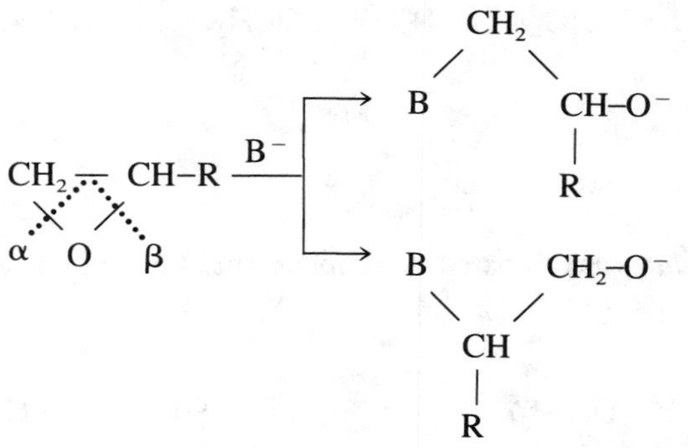

$$
CH_2 = CH-R \xrightarrow{B^-}
\begin{cases}
\begin{array}{c} CH_2 \\ B \diagup \quad \diagdown CH-O^- \\ \big| \\ R \end{array} \quad (232) \\[2em]
\begin{array}{c} B \diagdown \quad CH_2-O^- \\ CH \\ \big| \\ R \end{array} \quad (233)
\end{cases}
$$

For example, noncatalytic oligomerization of phenylglycidyl ether by interaction with phenol proceeds via ordinary (α) and anomalous (β) ring opening with formation of the secondary and primary alcoholic groups, respectively, in the 4:3 ratio, regardless of the reaction conditions or extent.[483]

Lactones and dimers of α-hydroxycarboxylic acids can polymerize by O-acyl or O-alkyl bond scission, giving the corresponding carboxylic or alcoholate anions as growing species. Transition from one kind of species, formed in the initiation, to the other kind, actually propagating, has also been observed.[484] These structures usually cannot distinguished directly by spectrophotometric methods applied to the polymerizing mixtures. Penczek et al.[485] developed a method based on anion capping with $Cl(O)P(OC_6H_5)_2$, followed by determination of the structure of the parent anion in ^{31}P and ^{1}H NMR spectra, and comparing chemical shifts with the independently studied model compounds. Some examples based on equation 234 are given in Table 14.

$$\text{wwwX}^-, M^+ + Cl(O)P(OC_6H_5)_2 \rightarrow \text{wwwX-}\underset{\underset{O}{\parallel}}{P}(OC_6H_5)_2 + MCl \qquad (234)$$

where X is the heteroatom of the growing oligomer and M^+ the metallic counterion.

Another method is based on the formation of different structures in the reaction of oligomeric carboxylate and alcoholate anions with 2,4,6-trinitroanisole.[483] Carboxylate anions react with 2,4,6-trinitroanisole, giving picrate anions (Equation 235), while alcoholate anions participate in the reversible reaction (Equation 236), leading to the formation of a Meisenheimer complex:[486]

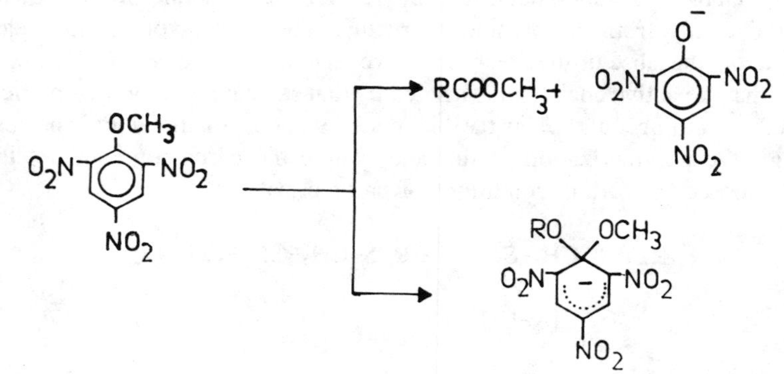

$$ \rightarrow RCOOCH_3 + \quad (235) $$

$$ (236) $$

TABLE 14

Active Species in Anionic Ring Opening Polymerization Determined from ^{31}P Chemical Shifts of Phosphate Groups of Capped Oligomers[485]

Type	Monomer	Product of trapping	Growing species
1. Oxiranes	 (oxirane)	$\cdots-CH_2OP(OC_6H_5)_2$ $\parallel$ O	$-CH_2O^-$
	(phenyloxirane, C_6H_5)	O $\parallel$ $\cdots-CH_2CHOP(OC_6H_5)_2$ $\mid$ C_6H_5	$-CH_2CHO^-$ $\mid$ C_6H_5
		$\cdots-CHCH_2OP(OC_6H_5)_2$ $\mid$ $\quad\parallel$ C_6H_5 $\quad O$	$-CHCH_2O^-$ $\mid$ C_6H_5
2. Lactones	(β-lactone)	$\cdots-CH_2COP(OC_6H_5)_2$ $\parallel$ $\parallel$ O O pyrophosphate	O $\diagup$ $-CH_2C\;\;(-)$ $\diagdown$ O
	(δ-valerolactone)	$\cdots-(CH_2)_4OP(OC_6H_5)_2$ $\parallel$ O	$-CH_2O^-$
	(ε-caprolactone)	$\cdots-(CH_2)_5OP(OC_6H_5)_2$ $\parallel$ O	$-CH_2O^-$
3. Cyclosiloxanes	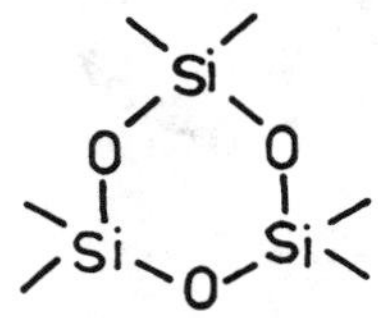 (cyclosiloxane)	O $\parallel$ $-Si-OP(OC_6H_5)_2$ $\downarrow$ pyrophosphate	$\mid$ $-Si-O^-$ $\mid$
4. Cyclic sulfides	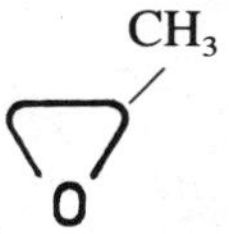 (methyl thiirane, CH_3)	$\cdots-CH_2CHSP(OC_6H_5)_2$ $\mid$ $\quad\parallel$ CH_3 $\quad O$	$-CH_2CHS^-$ $\mid$ CH_3

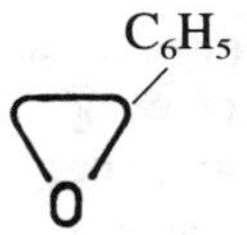

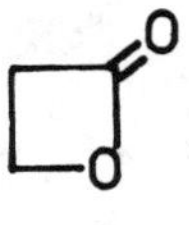

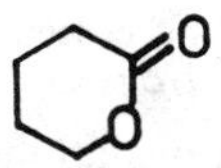

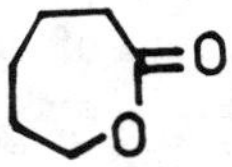

Since the picrate anions absorb at $\lambda = 380$ nm, whereas the Meisenheimer complex absorbs at $\lambda = 420$ and 495 nm, the differences in the UV spectra of polymerizing mixtures with added 2,4,6-trinitroanisole can be used to discriminate between carboxylate and alcoholate active species.

In all of the systems described in Table 14, active species are located at the end of the chain, and these processes were classified as "active chain end" polymerizations, in contrast to the "activated monomer (AM)" polymerizations, where a monomer molecule bears an electrical charge in the actual propagation step.[486] The latter mechanism is operative in the anionic polymerization of lactams[487-489] and is considered to be the main mechanism of the anionic polymerization of N-carboxy-anhydrides (NCA) under certain conditions.[490] A kinetic analysis[491] showed that the maximum degree of polymerization obtainable in polymerizations of NCA proceeding by the AM mechanism is given by $\sqrt{k_p/k_i}$. However, it was argued that since $\overline{DP}$ deduced from kinetic data, *assuming an AM mechanism,* are too low to account for the actual molecular weights obtained in NCA polymerizations, the AM mechanisms of NCA and lactam polymerization are not similar.

The strongly localized negative charge on heteroatoms and high nucleophilicity of anions facilitate aggregation. The tendency for aggregation is also affected by the size of the counterions. For example, in the polymerization of ε-caprolactone,[492] replacing Na^+ with a larger K^+, and complexing the potassium counterions with cryptands in the anionic polymerization of D_3 (see Section III.B.6) increases an effective cation radius[493] and disrupts aggregation, giving systems in which only ion pairs are present.

Comparison of the selective reactivities of ions and ion pairs (k_P^*/k_P^-) shows that in the polymerization of propylene sulfide and β-lactone, in contrast to the general behavior of the vinyl systems, the reactivity of ion pairs may be similar to or higher than that of ions. This is due to the different contributions of the desolvation energy of macroion pairs and macroions.[486]

Heterocyclic monomers and the corresponding polymers are able to solvate counterions, due to the nucleophilicity of heteroatoms. Unsaturated polar monomers and their polymers, e.g., MMA and 2-VP polymers, behave similarly.

It was established that polyethylene oxide (PEO) growing chains solvate counterions[494,495] and that this ability depends on the polymer $\overline{DP}$. In the solvation shell of metal cation, six oxygen atoms are involved. Solvation of cations with PEO has a cooperative character, i.e., the first coordination (k_1) is a process with a loss of translational degrees of freedom, while the further coordination with oxygen atoms ($k_2, k_3 \ldots k_6$) is accompanied only by a reduction of vibrational and rotational degrees of freedom of the polymer segments. Thus, the loss of entropy is the highest in the first step, i.e., $k_1 \gg k_2 > k_3 > k_4 > k_5 > k_6$:[486]

(237)

Similar phenomena were observed in the polymerization of polar vinyl macromers containing nucleophilic nitrogen and oxygen atoms, such as vinyl pyridines[496] and acrylates.[497] It was also observed that highly polar β-propiolactone participates in solvation of the propagating carboxylate anions,[498] i.e., ΔH_p^* depends on the starting monomer concentration, irrespective of solvent polarity.[486]

Whenever active centers are able to react with polymer chains (backbiting) or end groups (endbiting), macrocycles are produced. A computer simulation indicated that when endbiting is faster than backbiting (because of higher inherent reactivity or a more favorable steric arrangement), then for strained heterocycles in the early stages of polymerization, the concentration of rings should be higher than their equilibrium concentration.[486,499] With increasing chain length of a statistical linear chain, this enhancement vanishes because the preferred conditions for chain ends gradually disappear. For polymerizations with slow backbiting and endbiting, the formation of linear polymer may be followed by its partial conversion into macrocyclics after the monomer is completely reacted. Thus, there are polymerizations with kinetic enhancement in macrocyclics as well as in linear polymer units.[486] Taking ω-caprolactone as a model compound, it was concluded from the dependence of propagation rate constants, with the participation of macrocycles, on the number of monomer units in the cyclic oligomer that the reactivity of ester groups decreases slightly with increasing ring size.[486] This effect was attributed to "conformational hindrance", resulting from the puckered conformation of rings of medium size, which makes some of the ester groups less accessible. Thus, the anionic polymerization of ϵ-caprolactone is an example of kinetic enhancement in linear polymers.

In some heterocyclic monomers, there are *endo-* and *exo*cyclic groups linked to the size of attack by the same kinds of chemical bonds. The *endo*cyclic ones may participate in chain transfer. For example, in the anionic polymerization of cyclic esters of phosphoric acid (Equation 238), the chain transfer is accompanied by a "wrong" monomer addition (Equation 239), giving a cyclic end group instead of the linear polymer unit:[258,486,500]

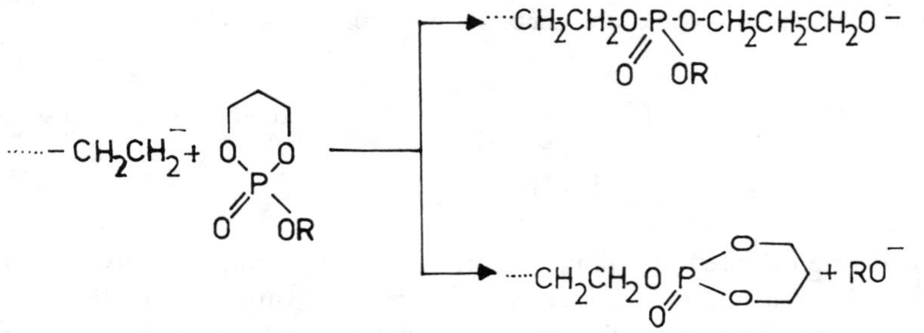

$$\cdots CH_2CH_2O\text{-}P\text{-}O\text{-}CH_2CH_2CH_2O^- \qquad (238)$$

$$\cdots CH_2CH_2O\text{-}P\Big\langle \qquad + RO^- \qquad (239)$$

In the polymerization of 2-methoxy-2-oxo-1,3,2-dioxaphosphorinane (R = $-CH_3$) initiated with $C_2H_5\text{-}O^-Na^+$ at 135°C, the proportion between proper and "wrong" monomer addition was about 4.[486] Every oligomer molecule contained one cyclic end group (Equation 240), as evidenced by a comparison of the $\overline{DP}$ and the proportion of phosphorous atoms of end groups determined by ^{31}P NMR.

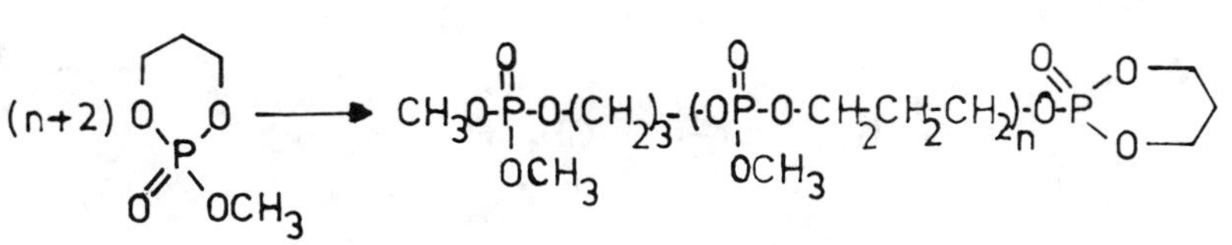

$$\overline{M}_n < 5000$$

The proper choice of *exo*cyclic group, e.g., H instead of OR, eliminated chain transfer.[501]

2. Oxiranes

The anionic polymerization of ethylene oxide is an example of the synthesis of a commercially important water-soluble macromolecule.

$$nCH_2\text{---}CH_2 \xrightarrow{RO^-} -(CH_2CH_2O\text{-})_n\text{-} \qquad (241)$$

The materials of the structure shown in Equation 241 are currently produced at a low molecular weight of a few hundred to a few thousand as well as at a very high molecular weight, into the millions.[502] The closely related poly(propylene oxide), PPO, is an important intermediate for polyurethane foams, and the production of nearly all of the automobile seats in the U.S. is based upon this technology.[503] Some of the basic chemistry for anionic oligomerization of EO and PO is discussed below.[503]

Potassium hydroxide, largely used in industry as an anionic initiators, can attack PO at the primary carbon atom, thus generating alkoxide A:

$$KOH + CH_2\underset{\underset{O}{\diagdown\diagup}}{\text{---}}CH\text{--}CH_3 \rightarrow HO\text{--}CH_2\text{--}\underset{\underset{CH_3}{|}}{CHO^-}K^+ \qquad (242)$$

A

Several different ions and ion pairs may be present under these conditions and, in precise work, one needs to address the kinetic parameters in terms of their exact ionic nature (e.g., k^+, k^-, etc.) and not simply in terms of a global rate constant.[503,504] In any event, it is possible to prepare oligomers (MW = 2000 to 4000) that are predominantly hydroxyl terminated with one primary and one secondary group.

$$A + (n-1)\, CH_2\underset{\underset{O}{\diagdown\diagup}}{\text{---}}CH\text{--}CH_3 \xrightarrow[\text{2. } H^+]{\text{1. Propagation}} HO\text{--}(\text{--}CH_2\text{--}\underset{\underset{CH_3}{|}}{CH}\text{--}O\text{--})_n\text{--}H \quad (243)$$

Since it is often desired that polyurethane foams be chemically cross-linked, average functionalities higher than 2 are required and one approach to this aim is outlined in Equations 243 and 244. The intermediate oligomeric alkoxide can undergo exchange reactions with polyhydroxy compounds such as glycerol to terminate one chain forming PPO and initiate another from the alkoxide derived from the polyhydroxy compound:

$$HO\text{--}(\text{--}CH_2CH\text{--}O\text{--})_{n-1}\text{--}\underset{\underset{CH_3}{|}}{CH_2}\text{--}\underset{\underset{CH_3}{|}}{CH}\text{--}O^-K^+ \xrightarrow{+\ glycerol}$$

A

$$HO\text{--}(\text{--}CH_2CH\text{--}O\text{--})_n\text{--}H + \underset{\underset{\underset{CH_2O^-K^+}{|}}{CHOH}}{\underset{|}{CH_2OH}}$$

PPO

$$\qquad\qquad (244)$$

$$HO\text{--}CH_2\text{--}\underset{\underset{OH}{|}}{CH}\text{--}CH_2O^-K^+ \xrightarrow{+\ m\ PO} \text{average } f = 3 \text{ chains} \quad (245)$$

The polyhydroxy compound also would function as a chain-transfer agent to regulate molecular weight as well as functionality. In order to produce an intermediate oligomer with a more reactive primary hydroxyl terminal, the chain ends bearing secondary alkoxide groups are capped with EO. However, the initiation of ethylene oxide by alkoxide A is slower than the subsequent propagation and a significant amount of EO must be incorporated if one wishes to have predominantly primary alcohol reactivity. Moreover, several side reactions are known for the anionic polymerization of PO which are both molecular weight limiting as well as detrimental to the development of hydroxyl functionality. For example, alkoxide A may attack the PO monomer by abstracting a hydrogen atom and generating the strong base B:

$$HO\text{-}(\text{-}CH_2CHO\text{-})_{n-1}\text{-}CH_2CHO^-, K^+ + PO \rightarrow PPO + K^{+\,-}CH_2\text{-}CH\text{---}CH_2 \quad (246)$$

with CH₃ substituents shown below the respective carbons (labeled A), and an epoxide ring O on structure B.

$$\text{A} \qquad\qquad\qquad\qquad \text{B}$$

This base quickly isomerizes to the allyl alkoxide (Equation 247) which is capable of further reaction with monomer (Equation 248) to produce a propylene oxide oligomer which can only bear, at best, one −OH group per molecule.

$$K^{+\,-}CH_2\text{-}CH\text{---}CH_2 \rightarrow CH_2\text{=}CH\text{-}CH_2O^{-\,+}K \quad (247)$$

$$\text{B} \qquad\qquad\qquad\qquad \text{C}$$

$$C + mCH_2\text{---}CH\text{-}CH_3 \xrightarrow[\text{2. H}^+]{\text{1. propagation}} CH_2\text{=}CH\text{-}CH_2\text{-}O\text{-}(\text{-}CH_2CHO\text{-})_m\text{-}H \quad (248)$$

with CH₃ substituent shown below.

These reactions can be a serious shortcoming, particularly for the preparation of linear oligopropylene oxide polyols. In recent years, important advances have been made in the coordination polymerization of propylene oxide (*vide infra*) which overcome this problem.

The use of an unsaturated anionic initiator — such as potassium *p*-vinylbenzoxide — is possible for the ring-opening polymerization of oxirane to obtain polyester macromers:[505]

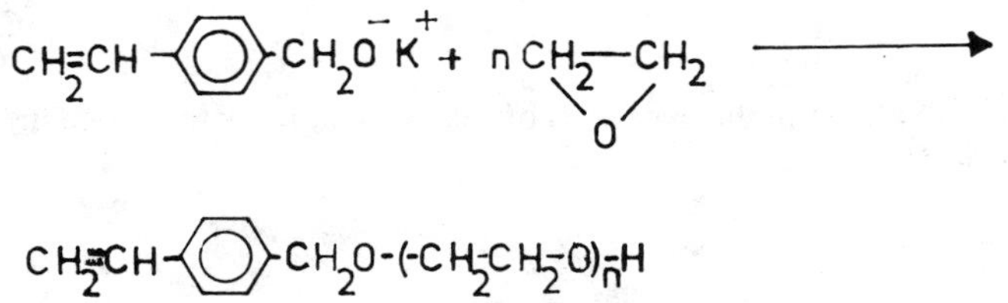

$$CH_2\text{=}CH\text{-}\bigcirc\text{-}CH_2\bar{O}\,K^+ + n\,CH_2\text{---}CH_2 \longrightarrow$$

$$CH_2\text{=}CH\text{-}\bigcirc\text{-}CH_2O\text{-}(\text{-}CH_2CH_2O)_n\text{-}H \quad (249)$$

Although initiation is generally heterogeneous, the polymers exhibit the molecular weight expected and a low polydispersity. In this case, the styrene-type unsaturated at the chain end cannot become involved in the process, as the propagation sites are oxanions. Similarly, ω-oxazolinyl PEO macromonomer samples have been made by anionic initiator with *p*-oxazolinyl-phenoxide:[506]

$$\underset{O}{\overset{N}{\diagup}}C\text{-}\bigcirc\text{-}\bar{O}\,Li^+ + n\,CH_2\text{-}CH_2 \longrightarrow \underset{O}{\overset{N}{\diagup}}C\text{-}\bigcirc\text{-}O\text{-}(CH_2CH_2O)_n\text{-}H$$

$$(250)$$

The oxazoline cycle is insensitive to anionic attacks, but it polymerizes cationically quite effectively.

A direct one-step preparation of macromers in the polymerization of EO with trimethylamine (TMA) as initiator gives polymers with up to one double bond per macromolecule through a mechanism similar to the Hofmann method of olefin synthesis:[507]

$$Me_3N + EO \longrightarrow Me_3\overset{+}{N}-CH_2CH_2O^- \longrightarrow H_2C=CHO^-Me_3\overset{+}{N}H$$

zwitterion   ion pair

$$\xrightarrow{+\ (n+1)\ EO} H_2C=CHO(CH_2CH_2O)_nCH_2CH_2O^-Me_3\overset{+}{N}H \qquad (251)$$

The molecular weights of macromers are in the range of 900 to 3000, with $\overline{M}_w/\overline{M}_n = 1.1$ to 1.3. The ion pair formed upon the decomposition of the zwitterion by a β-elimination mechanism is the real growing center. Propagation species are inclined to spontaneous deactivation in an equilibrium:

$$H_2C=CHO\text{-}\wedge\wedge\wedge\text{-}CH_2CH_2O^-Me_3\overset{+}{N}H \rightleftharpoons H_2C=CHO\text{-}\wedge\wedge\wedge\text{-}CH_2CH_2OH + Me_3N \qquad (252)$$

and TMA is able to reactivate macromolecules or start a new chain. High values of functionality, which are close to 1, indicate that the reaction between TMA and EO, followed by the decomposition of the starting group, is practically the only way to form the macromolecules. Thus, the given system corresponds to the criteria of effective functionalization. Such a type of macromonomer can be classified as a "natural" one, because the polymerizable groups are the result of the reaction mechanism. Another well-known case of "natural" macromonomer formation is the cationic polymerization of isobutene with an intensive chain transfer to monomer.[508]

Unsaturated EO monomers were synthesized using an unsaturated deactivator to endcap the living chains:[505,509]

$$\cdots\cdots -CH_2CH_2O^-K^+ + ClOC-\underset{\underset{CH_3}{|}}{C}=CH_2 \longrightarrow \cdots -CH_2CH_2OC-\underset{\underset{O\ \ CH_3}{\|\ \ |}}{C}=CH_2 \qquad (253)$$

Oxiranes are also added in the synthesis of macromonomers to reduce the nucleophilicity of living ends, e.g.:[510]

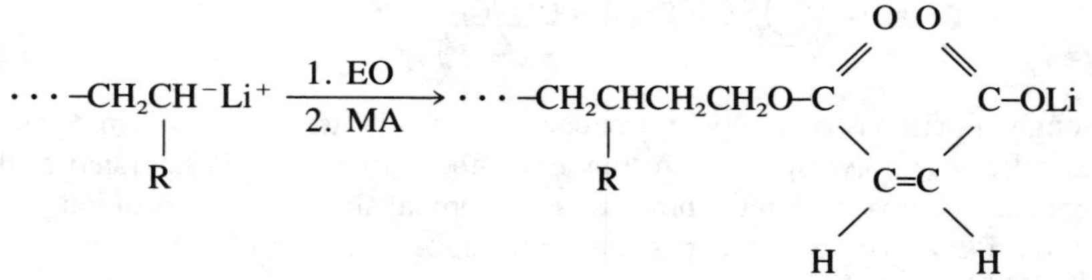

$$\cdots -CH_2\underset{\underset{R}{|}}{C}H^-Li^+ \xrightarrow[\text{2. MA}]{\text{1. EO}} \cdots -CH_2\underset{\underset{R}{|}}{C}HCH_2CH_2O-C\overset{\overset{O\ \ O}{\diagup\diagdown}}{\underset{\underset{\underset{H\ \ \ \ H}{\diagup\ \ \diagdown}}{C=C}}{}}C-OLi \qquad (254)$$

Equation 254 is used to obtain amphiphilic macromonomers possessing a central unsaturation. They are polystyrene (PS)/PEO diesters of maleic acid with PS and PEO blocks of $\overline{M}_n = 2500$ to 8000:[511]

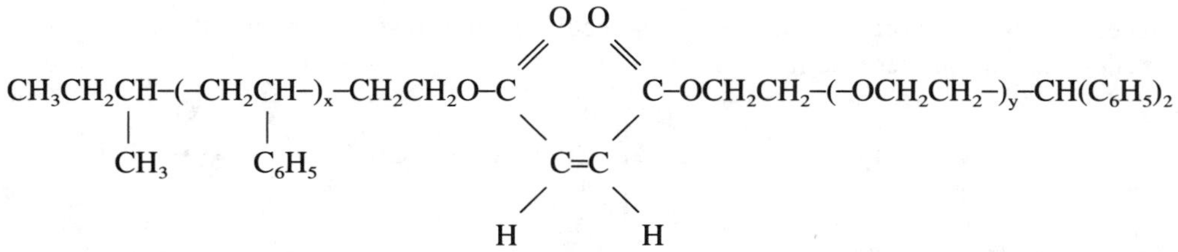

$$CH_3CH_2CH-(-CH_2CH-)_x-CH_2CH_2O-C \overset{O}{\overset{\|}{\diagdown}} \quad \overset{O}{\overset{\|}{\diagup}} C-OCH_2CH_2-(-OCH_2CH_2-)_y-CH(C_6H_5)_2$$

with side groups CH_3, C_6H_5 and central $C=C$ bearing H, H.

$$x{:}y = 0.50\text{---}0.15 \tag{255}$$

Synthesis of PEO chains fitted at one end with two hydroxyl functions was achieved by protecting the two functions by acetalization and releasing them afterwards by acid hydrolysis.[512] Such α,α-ω-trihydroxy telechelics can participate in polycondensation with diacids, diisocyanates, or dichlorosilanes, thus giving access to graft polycondensates.

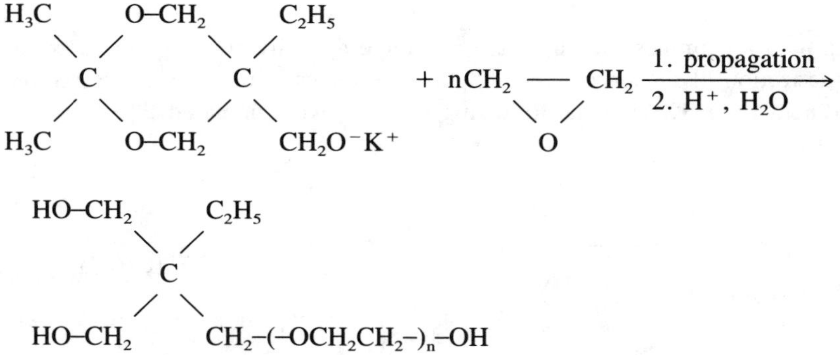

$$
\begin{array}{c}
H_3C \diagdown \diagup^{O-CH_2} \diagdown \diagup^{C_2H_5} \\
\quad C \qquad\qquad C \qquad + \; nCH_2\!\!-\!\!CH_2 \xrightarrow[\;2.\; H^+,\, H_2O\;]{\;1.\; \text{propagation}\;} \\
H_3C \diagup \diagdown_{O-CH_2} \diagup \diagdown_{CH_2O^-K^+} \qquad\qquad \overset{\diagdown\,/}{O}
\end{array}
$$

$$
\begin{array}{c}
HO-CH_2 \diagdown \diagup C_2H_5 \\
\qquad C \\
HO-CH_2 \diagup \diagdown CH_2-(-OCH_2CH_2-)_n-OH
\end{array}
\tag{256}
$$

As pointed out earlier, initiation with sodium naphthalene yields two simultaneous growing chains. Therefore, polymerization of oxirane with this kind of initiator gives α,ω-telechelics by proper end-cappings of living ends.[513,514]

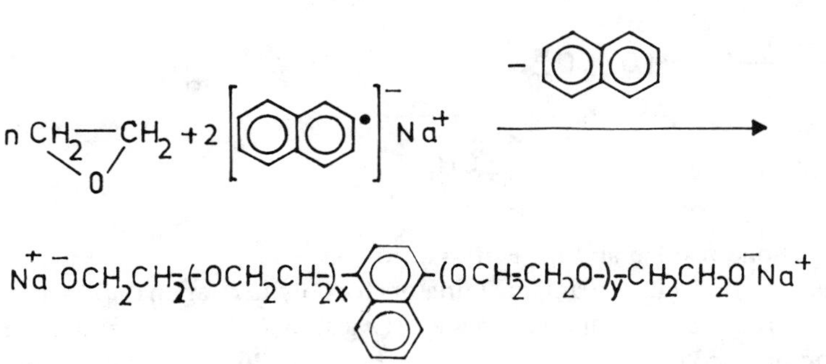

$$n\, CH_2\!\!-\!\!CH_2 + 2\left[\text{(naphthalene)}\right]^{\bullet -} Na^+ \longrightarrow$$

$$Na^+\,{}^-OCH_2CH_2(OCH_2CH_2)_{2x}\text{-(naphthalene)-}(OCH_2CH_2O)_{7}CH_2CH_2O^-\,Na^+ \tag{257}$$

3. Lactones

The anionic polymerization of ϵ-caprolactone with alkoxides results in a living ring-chain equilibrium system.[515,516] The product distribution is essentially determined by the entropy term, the lower cyclics being favored over the linear chains at higher dilution. Thus, the equilibrium state is readily changed by dilution or concentration of the living system. However, the alkoxide end group also attacks the ester functionalities of the polymer backbone, resulting in a broadening of the molecular weight distribution and the formation of cyclics (backbiting degradation):

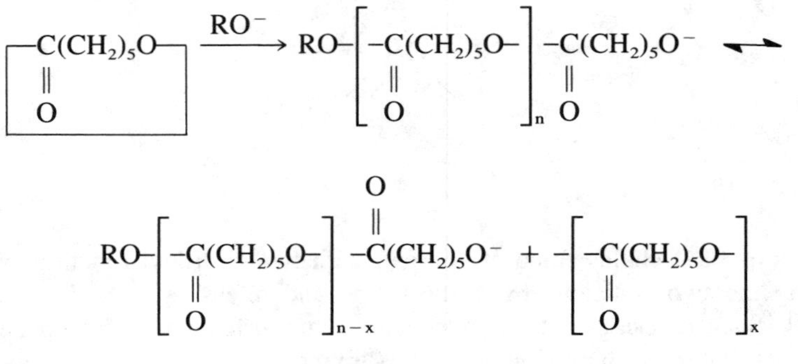

$$\begin{array}{c}\left[\begin{array}{c}-C(CH_2)_5O-\\ \parallel\\ O\end{array}\right] \xrightarrow{RO^-} RO-\left[\begin{array}{c}-C(CH_2)_5O-\\ \parallel\\ O\end{array}\right]_n \begin{array}{c}-C(CH_2)_5O^-\\ \parallel\\ O\end{array} \curvearrowright \end{array}$$

$$RO-\left[\begin{array}{c}-C(CH_2)_5O-\\ \parallel\\ O\end{array}\right]_{n-x}\begin{array}{c}O\\ \parallel\\ -C(CH_2)_5O^-\end{array} + \left[\begin{array}{c}-C(CH_2)_5O-\\ \parallel\\ O\end{array}\right]_x$$

$$x = 2, 3, 4, 5,\dots \tag{258}$$

Using hydroxy amines such as 2,2,6,6-tetramethyl-4-hydroxypiperidine as the initiator for polymerization of ϵ-caprolactone, oligomeric polyester diols ($\overline{M}_w = 2000$, OH value 56.8, and acid value 1.24) of the following structure were prepared.[516]

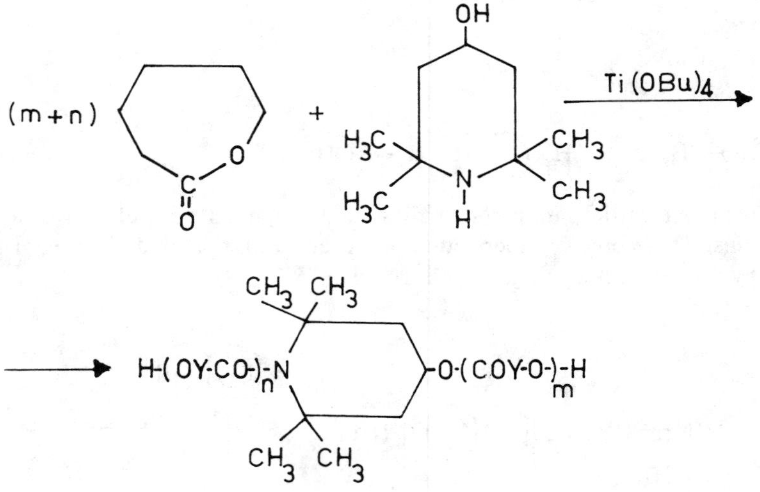

$$\tag{259}$$

where Y = lactone residue and m + n = 2 to 15.

The most important class of anionic initiators for polymerization of β-lactones are organic bases such as tertiary amines and phosphines (Equation 260).[517-519] To avoid chain transfer by abstraction of a proton from the α-position (Equation 261), α,α-disubstituted β-lactones are generally used.[520-522]

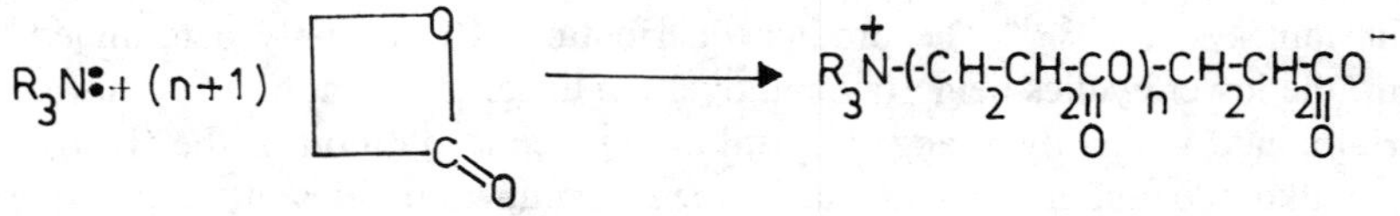

$$\tag{260}$$

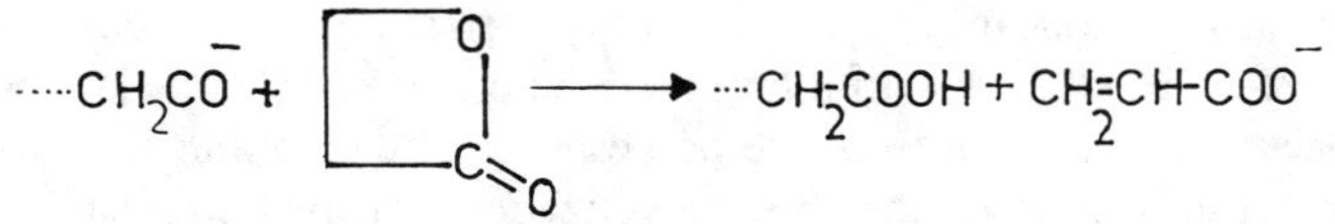

$$\cdots CH_2\overset{-}{CO} + \underset{C=O}{\overbrace{}} \longrightarrow \cdots CH_2COOH + CH_2=CH-COO^- \tag{261}$$

The formation of macrozwitterion by attack of the nucleophile (R_3N:, R_3P:) on the methylene group of pivalolactone, PVL (Equation 262), has been supported by NMR and chemical studies on propiolactone.[517-519]

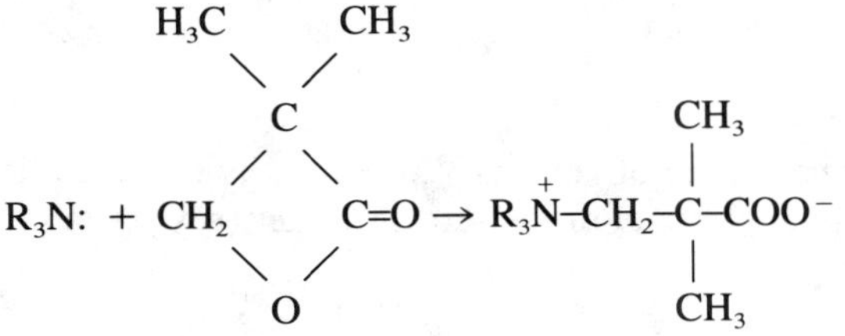

$$R_3N: + \underset{O}{\overset{H_3C \quad CH_3}{\underset{\diagdown \diagup}{\underset{CH_2}{\diagup}\underset{C}{\diagdown}\underset{\diagdown \diagup}{C=O}}}} \rightarrow R_3\overset{+}{N}-CH_2-\underset{CH_3}{\overset{CH_3}{\underset{|}{\overset{|}{C}}}}-COO^- \tag{262}$$

PVL

Pivalolactone (α,α-dimetyl-β-propiolactone) has been converted into several telechelics and used to prepare block copolymers.[520-523] A segmented telechelomer containing monodisperse PEO and poly(PLV) blocks[524] was synthesized as follows. Initiator I was reacted with EO to give the potassium salt of masked polyether II having an alkoxide ion at its end (Equation 263). This anion was modified with succinic anhydride to give a carboxylate anion which polymerized PVL, giving the masked polyoxythylene-co-pivalolactone-co-polymeric salt III (Equation 264). A carboxylate anion is preferred in the ring-opening polymerization of PVL since the alkoxide ion has led to side reactions.[525] Compound III was converted to the telechelomer (Equation 265) by treating it with acids ($\overline{M}_n$ = 2700, MWD = 1.4).

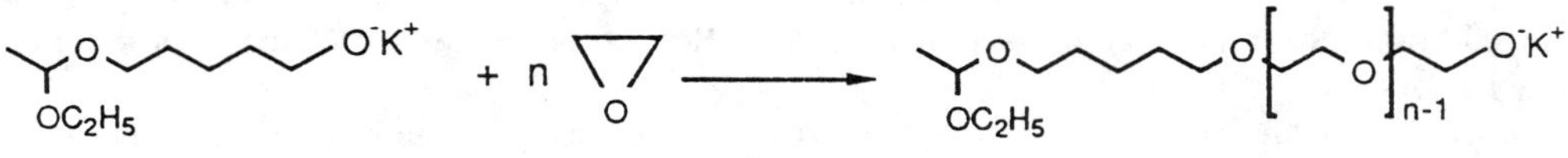

$$(263)$$

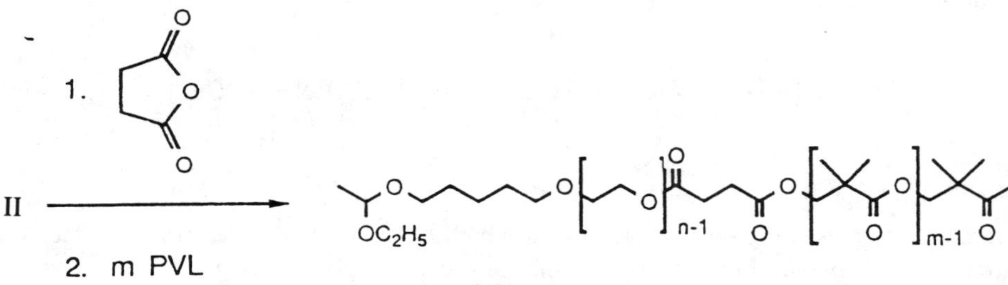

$$(264)$$

$$\text{III} + \xrightarrow{\text{3N HCl}} HO(CH_2)_5-O-(CH_2CH_2O)_n-\overset{O}{\overset{\|}{C}}(CH_2)_2\overset{O}{\overset{\|}{C}}-O-(CH_2\overset{CH_3}{\underset{\underset{CH_3}{|}}{\overset{|}{C}}}-COO-)_m-H \tag{265}$$

Telechelomer, n = 29, m = 12

The product distribution in the anionic polymerization of the six-membered ring δ-valerolactone, VL, was studied with lithium *tert*-butoxide in THF.[526] The oligomers were formed by the backbiting reaction from the polyester until the equilibrium distribution was established. The log-log plots of the molar cyclization constant against ring size for δ-valerolactone and ε-caprolactone were straight lines with a slope of -2.5, in accord with the Iacobson-Stockmayer theory. Dimer of VL and trimer of ω-caprolactone deviated from the line, perhaps due to strained unfavorable conformations, while for other oligomers, the decreased entropy of ring closure determined the stability of the oligomers. Anionic polymerization of five-membered γ-butyrolactone was studied[516] and found to be rather stable, and little oligomer was observed.

4. Lactams

The anionic polymerization of lactams[517] proceeds by nucleophilic attack of the lactam anion at the cyclic carbonyl group of the growth center (acyl-lactam).

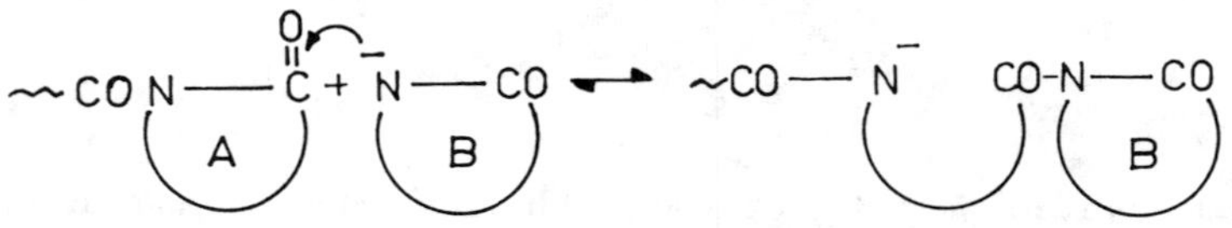

$$\sim CON\!-\!\overset{\overset{O}{\|}}{C} + \bar{N}\!-\!CO \rightleftharpoons \sim CO\!-\!\bar{N} \quad CO\!-\!N\!-\!CO \tag{266}$$

$$A \qquad B \qquad \qquad B$$

An attack at the exocyclic carbonyl group results only in an exchange of the terminal acylated lactam:

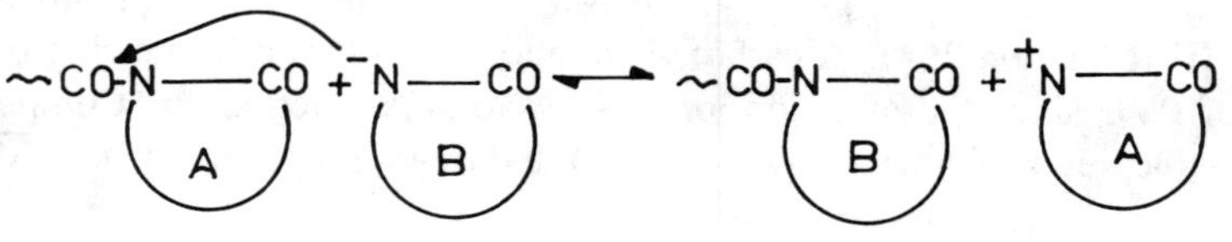

$$\sim CO\text{-}\bar{N}\!-\!CO + \bar{N}\!-\!CO \rightleftharpoons \sim CO\text{-}N\!-\!CO + {}^{+}N\!-\!CO \tag{267}$$

$$A \qquad B \qquad \qquad B \qquad A$$

The ratio of the rates of Reactions 266 and 267 depends on ring size and substitution. In addition, the reactivity of lactam anions in Reactions 266 and 267 depends on their basicity.

Chain growth, Reaction 266, involves two activated species: lactam anions (i.e., particles of increased nucleophilicity) and *N*-acylated lactam units (i.e., growth centers of increased acylating ability). Lactam anions are able to produce growth centers via Reaction 268 and, thus, can initiate an anionic polymerization.

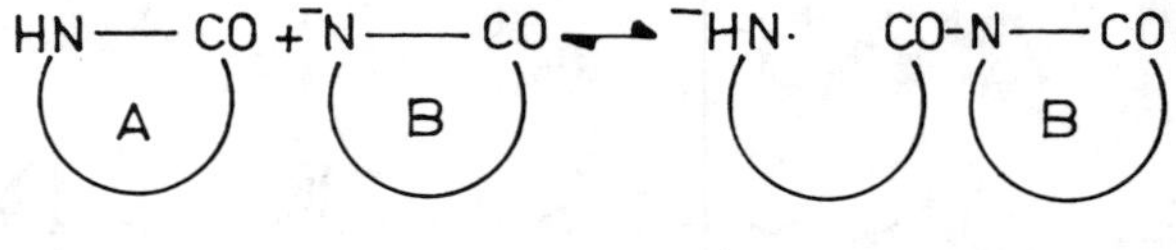

$$HN\!-\!CO + \bar{N}\!-\!CO \rightleftharpoons \bar{H}N\cdot \quad CO\text{-}N\!-\!CO \tag{268}$$

$$A \qquad B \qquad \qquad B$$

Therefore, lactamates, i.e., salts of lactams in which the anion is derived from the lactam, and strong bases capable of producing lactamates (e.g., NaOH) are designated as *initiators*. Acyl-lactams or precursors of growth centers (e.g., acyl chlorides) are termed *activators* because they significantly enhance the effect of initiators. Since the discovery of the fast anionic lactam polymerization with string bases,[518] a great number of initiators have been described in the literature — e.g., the production of initiator solution in dry monomer just by passing a solution of NaOH in molten caprolactam through a column with a molecular sieve.[517]

Cyclization reactions in which cyclic oligomers of lactams are formed is an important problem:

$$A-[-NH-(CH_2)_5-CO-]_z-B \rightleftharpoons A-[-NH-(CH_2)_5-CO-]_x-B$$

$$+ \boxed{-[-NH-(CH_2)_5-CO-]_y-} \tag{269}$$

where A and B are initiator residues and z = x + y.

Irrespective of the reaction mechanism, the same equilibrium concentrations of cyclic oligomers are achieved at a given temperature, provided the number of polymer molecules is the same. Assuming that chains of caprolactam polymers have end-to-end distances similar to those measured under θ-point conditions and taking into account the effect of temperature as well as chain length on the dimensions of the unperturbed chains, the cyclization constant k_x was related to $(r_x^2)_o$ by the equation:[519]

$$k_x = \left[\frac{3}{2}(r_x^2)_o\right]^{3/2}\left(\frac{1}{N_A x}\right) \tag{270}$$

where N_A is the Avogadro constant and x the number of monomer units in cyclics. The equilibrium concentration of x-meric cyclics in caprolactam polymers calculated from Equation 270 was higher as the experimental values for the polymer equilibrated in melting. It is possible that strong hydrogen bonding between amide groups will decrease with increasing temperature and chain expansion may occur at temperatures above the melting point of polyamides. In the solid state, only the amorphous regions participate in the polymer-cyclics equilibria. Since polymer molecules are partly incorporated into the crystalline region, the average length of the amorphous part of the polymer chain is shorter than the whole molecule. Fixation of parts of the polymer chain in crystalline regions prevents the amorphous part from adopting random-coil conformations similar to those in the molten state. Polymer chains with end groups anchored in crystalline regions cannot participate in the equilibrium shown by Equation 269. For these reasons, the experimental values were much lower than the calculated ones.[517]

Optimal conditions for the anionic polymerization of ε-caprolactam by reaction molding technology were achieved by using equal concentrations of N-acetyl caprolactam (activator) and sodium caprolactamate (initiator) at 1 to 2 mol%.[520] The formation of high cyclic oligomers (up to heptamer) decreased with increasing concentration of catalytic species. Both the overall oligomeric fraction in the product mixture and the crystallinity of the Nylon-6 formed increased with increasing annealing temperature.

5. Cyclic Sulfides

Polymerization of racemic mixtures of episulfides may lead to stereoregular polymers through a *stereoselective* process that gives either atactic or isotactic polymers.[521-523]

The polymerization of monosubstituted racemic thiiranes

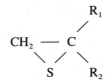

TABLE 15

Relation[a] of Racemic Thiirane Using Chiral Initiator System(s) 2,2'-Binaphtholate and the Amount of Oligomers Formed[523]

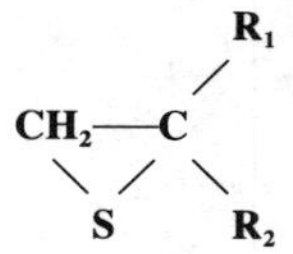

$$CH_2 \underset{S}{\overset{R_1}{\underset{\diagdown \diagup \diagdown}{C}}} R_2$$

Substituents R_1, R_2	Me,H	Et,H	iPr,H	n-Bu,H	t-Bu,H	Me,Et	$-CH_2OCH_3$
$(\alpha/\alpha_o)_{x/2}$, %	80	63	31	—	—	—	16
Oligomer (%)	10	26	43	48	—	100	—

[a] $(\alpha/\alpha_o)_{x/2}$ = optical purity of unreacted monomer at half reaction.

with zinc (s) 2,2'-binaphthalate as chiral initiator system occurs with high enantioselectivity for methyl- or ethyl-substituted monomers ($k_s/k_R \sim 15$ to 20), although magnitude of the resolution decreases with the bulkiness of the substituent group (Table 15).[523]

The high molecular weight polymer fraction and low molecular weight polymer fraction ($\overline{M}_w = 6$ to 10×10^3) appear as distinct species in GPC curves. The proportion of these products increases with the bulkiness and degree of substitution on the asymmetric carbon, as seen from Table 15. Simultaneously with the formation of low molecular weight products, a side reaction takes place during polymerization. The products formed contain an excess of sulfur compared to the normal linkage, for which y = 1 in the monomer unit:

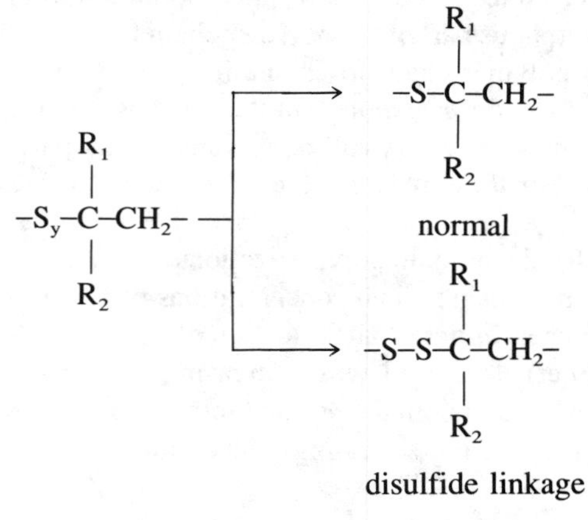

$$-S_y-\underset{R_2}{\overset{R_1}{C}}-CH_2- \quad \begin{cases} \longrightarrow & -S-\underset{R_2}{\overset{R_1}{C}}-CH_2- \quad (271) \\ & \text{normal} \\ \longrightarrow & -S-S-\underset{R_2}{\overset{R_1}{C}}-CH_2- \quad (272) \end{cases}$$

disulfide linkage

The amount of disulfide linkage varies with the degree of substitution and the bulkiness of R_1 and R_2. While for methyl thiirane practically no disulfide linkage was detected, in the case of 2-methyl, 2-ethyl thiirane, the obtained oligomers contain an exclusively disulfide linkage (y = 2). Cyclohexene sulfide polymers ($\overline{M}_n \sim 4500$) contain an excess of sulfur, but show no optical activity at D line. There are many more disulfide linkages in oligomers than in the corresponding high molecular weight polymers. Formation of the disulfide linkage is accompanied by evolution of the corresponding olefin, which is detected in the recovered monomer.

$$t\,CH_2 \underset{S}{\overset{R_1}{\underset{\diagup}{\overset{\diagup}{C}}}} R_2 \longrightarrow -(-S_y-\underset{R_2}{\overset{R_1}{\underset{|}{C}}}-CH_2-)_n- + nCH_2=\underset{R_2}{\overset{R_1}{\underset{\diagdown}{\overset{\diagup}{C}}}} \tag{273}$$

where $t = m + n$ and $m = (y - 1)n$. The amount of olefin increases with the increase of bulkiness on the asymmetric carbon.

Polymerization of methyl thiirane with sodium naphthalene and terminated with 1-chloromethyl naphthalene yields dinaphthalene-terminated telechelics.[524] This reaction was used to synthesize α,ω-mercaptopoly(propylene sulfide)s.[525,526]

6. Cyclic Siloxanes

Catalysts involving bases generally do not polymerize D_3 at moderate temperatures, except in the presence of donor solvents.[527] For example, the cyclic trimer can be polymerized in solvents such as THF with organolithiums. The initiation reaction could be represented by the simple structure shown in Equation 274, although it is recognized as possibly more complicated.[503]

$$RLi + D_3 \xrightarrow{\text{THF}} R-(-\underset{CH_3}{\overset{CH_3}{\underset{|}{Si}}}-O-)_2-\underset{CH_3}{\overset{CH_3}{\underset{|}{Si}}}-O^-Li^+$$

$$\xrightarrow{(3n-2)D_3} R-(-\underset{CH_3}{\overset{CH_3}{\underset{|}{Si}}}-O-)_{3n}-\underset{CH_3}{\overset{CH_3}{\underset{|}{Si}}}-O^-Li^+ \tag{274}$$

The cyclic trimer is known to be somewhat strained and to polymerize with a significant exothermic enthalpy. If initiation is comparable to propagation, the polymerization behaves like a living process and produces fairly narrow molecular weight distributions of predictable molecular weights.

Lithium siloxanates do not readily undergo interchange (equilibrium) processes, which are quite common for other alkali metals with siloxane systems. The polar solvent THF is also necessary to allow the reaction to proceed, and this involves generation of various ionic intermediates.[528] The common route to polysiloxanes utilizes the more readily available cyclic tetramer D_4.[503] In the presence of potassium hydroxide, the ring is attacked to generate a dianion intermediate (Equation 275) which is capable of dimerizing (Equation 276) to a new dianionic structure:

$$KOH + D_4 \rightleftharpoons HO-(-\underset{CH_3}{\overset{CH_3}{\underset{|}{Si}}}-O-)_3-\underset{CH_3}{\overset{CH_3}{\underset{|}{Si}}}-O^-K^+ \tag{275}$$

$$2\,HO-(-\underset{CH_3}{\overset{CH_3}{\underset{|}{Si}}}-O-)_4^-K^+ \rightarrow K^{+\,-}O-(-\underset{CH_3}{\overset{CH_3}{\underset{|}{Si}}}-O-)_8^-K^+ \tag{276}$$

Both monomeric and dimeric dianion structures can interact with more of the cyclic tetramer D_4 to produce a long-chain macromolecule (Equation 277). This route is often identified with the preparation of silicone oils or silicone rubber.

$$\underset{\underset{CH_3}{|}}{\overset{\overset{CH_3}{|}}{K^{+\,-}O\!-\!(\!-\!Si\!-\!O\!-\!)_x^-\,K^+}} \xrightarrow{\;D_4\;} \underset{\underset{CH_3}{|}\;\;\underset{CH_3}{|}}{\overset{\overset{CH_3}{|}\;\;\overset{CH_3}{|}}{K^{+\,-}O\!-\!(\!-\!Si\!-\!O\!-\!)_n\!-\!Si\!-\!O^-\,K^+}} \tag{277}$$

$$x = 4 \text{ or } 8$$

An important variation in this type of chemistry is the utilization of a disiloxane derivative as both chain-transfer agent and end blocker:[275,277,278]

$$\underset{\underset{CH_3}{|}}{\overset{\overset{CH_3}{|}}{K^{+\,-}O\!-\!(\!-\!Si\!-\!O\!-\!)_x^-\,K^+}} + R_3Si\!-\!O\!-\!SiR_3 \rightarrow \underset{\underset{CH_3}{|}}{\overset{\overset{CH_3}{|}}{R_3Si\!-\!(\!-\!O\!-\!Si\!-\!)_n\!-\!OSiR_3}} \tag{278}$$

The growing chain can undergo various exchange reactions. These are quite rapid in the case of the potassium cation, but rather slow in the case of lithium. One of the important features of the blocking compound is that, although it contains a silicon-oxygen-silicon bond, it also contains, usually, silicon-carbon bonds which cannot undergo the interchange reaction. As a result, one can generate oligomers, such as that obtained in Equation 279, which are effectively capped at both ends by the group present on the disiloxane reagent. In most cases it is hexamethyl disiloxane and, therefore, the end groups are all the relatively inert and stable methyl groups. However, if one of the R groups contains an effective functionality, it is possible to generate difunctional oligomers, wherein the molecular weight is controlled by the molar ratio of dianion species to blocking derivative. Such an approach has been used effectively to produce valuable oligomeric intermediates for the ultimate synthesis of block and segmented copolymer.[529-531] Some of the functional disiloxanes that have been used to prepare functionally terminated siloxane oligomers[530] are shown below.

$$\underset{\underset{CH_3}{|}\;\;\underset{CH_3}{|}}{\overset{\overset{CH_3}{|}\;\;\overset{CH_3}{|}}{H_2N\!-\!(CH_2)_3\!-\!Si\!-\!O\!-\!Si\!-\!(CH_2)_3\!-\!NH_2}} \qquad \begin{array}{l} \alpha,\omega\text{-aminopropyl-1,3-} \\ \text{tetramethyldisiloxane} \end{array}$$

$$\underset{\underset{CH_3}{|}\;\;\underset{CH_3}{|}}{\overset{\overset{CH_3}{|}\;\;\overset{CH_3}{|}}{(CH_3)_2N\!-\!(\!-\!Si\!-\!O\!-\!)_n\!-\!Si\!-\!N(CH_3)_2}} \qquad \begin{array}{l} \text{low molecular} \\ \text{weight silylamine} \end{array}$$

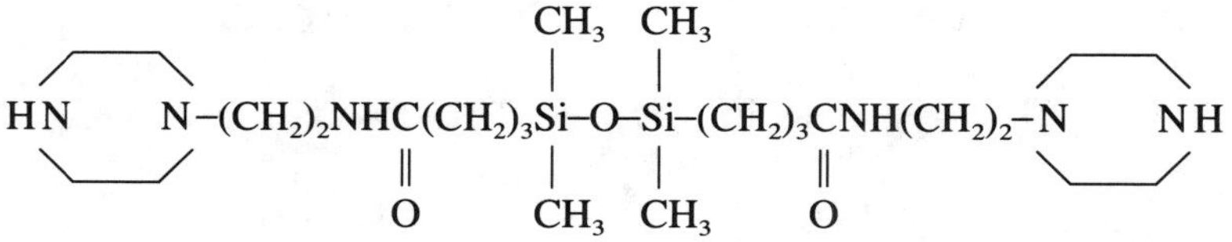

piperazine-terminated disiloxane

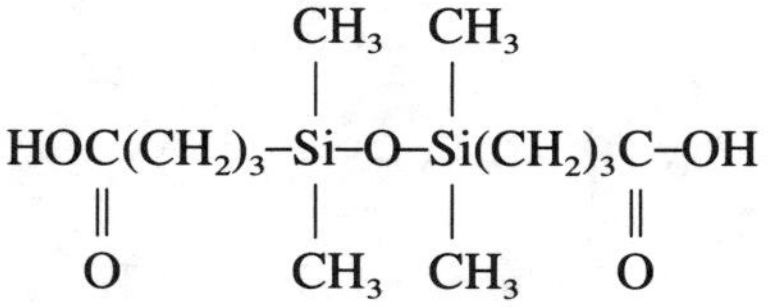

α,ω-carboxypropyl-1,3-tetramethyldisiloxane

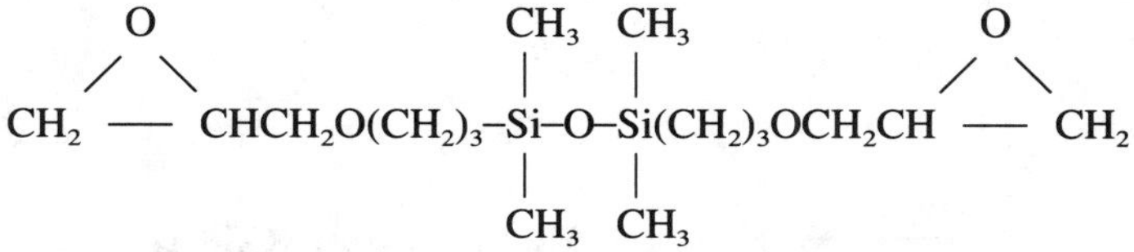

α,ω-glycidoxy-propyl-1,3-tetramethyldisiloxane

As shown earlier with the cationic oligomerization of cyclic siloxanes, the ring-opening polymerizations of these heterocycles are referred to as equilibrium reactions. It is generally convenient to use "D" to refer to a difunctional siloxane unit (i.e., dimethylsiloxane unit) and "M" to refer to a monofunctional siloxane unit (i.e., trimethylsiloxane unit).

Some redistribution reactions occurring during a siloxane equilibration are shown in Equations 145 to 148.

Since a variety of interchange reactions can take place, a quantitative conversion of the tetramer to high polymer is not achieved, and there is, at thermodynamic equilibrium, a mixture of linear and cyclic species present. The choice of catalyst depends upon the temperature of the equilibrium as well as the type of functional disiloxane that is used. Bases such as hydroxides, alcoholates, phenolates, and siloxanates of the alkali metals, quaternary ammonium and phosphonium based and the corresponding siloxanilates and fluorides, and organoalkali-metal compounds have all been found to catalyze the oligomerization of cyclic siloxanes.[277,530,532] It is believed that all catalysts generate the siloxanate anion *in situ,* and it is this species which breaks the Si-O bond in either the linear or cyclic siloxanes present. The reactivities of the disiloxane and various cyclic siloxanes differ. For example, the rate or reaction increases in the order[530] MM < MDM < MD_2M < D_4 < D_3, where MM represents the nonfunctional end blocker hexamethyldisiloxane.

Telechelic poly(dimethyl siloxane) with acetoxy, hydroxyl, or 2,3-epoxy-prepoxy functions as end groups were synthesized starting from chloro-3-propyl-dimethylsilanes or 3-(-2,3-epoxy-prepoxy)-propyl dimethylsilanes.[531] The starting materials were prepared by hydrosilylation of the corresponding allylic derivatives with chlorodimethylsilane, and subsequently condensed with α-hydro-ω-hydroxyoligo-(or poly)-dimethylsiloxanes ($\overline{DP}$ = 2 to 56) in the presence of pyridine.

Polyorganosiloxane macromers with a methacrylyl function (Equation 279) or styryl function (Equation 280) at one end were synthesized by terminating the anionic polymerization of hexamethylcyclotrisiloxane with a functionalized chlorosilane-terminating agent.[533]

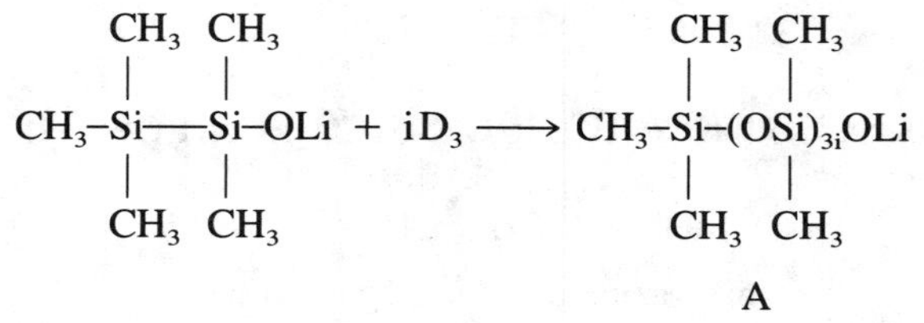

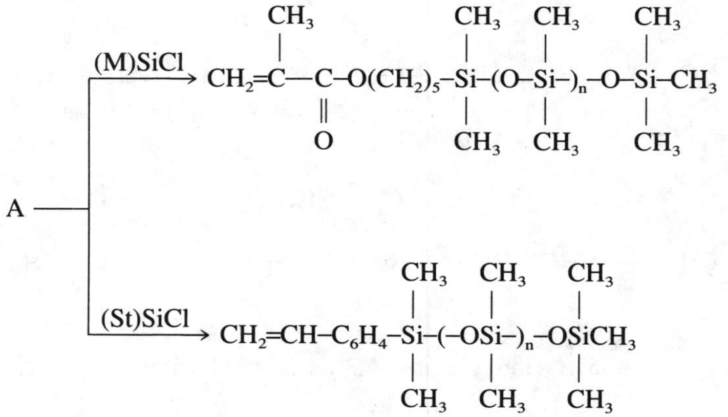

$$\underset{\substack{| \\ CH_3}}{\overset{\substack{CH_3 \\ |}}{CH_3-Si}}-\underset{\substack{| \\ CH_3}}{\overset{\substack{CH_3 \\ |}}{Si}}-OLi \;+\; iD_3 \;\longrightarrow\; \underset{\substack{| \\ CH_3}}{\overset{\substack{CH_3 \\ |}}{CH_3-Si}}-(OSi)_{3i}OLi$$

A

$$\xrightarrow{\text{(M)SiCl}} \; CH_2{=}\underset{\substack{\| \\ O}}{\overset{\substack{CH_3 \\ |}}{C}}-C-O(CH_2)_5-\underset{\substack{| \\ CH_3}}{\overset{\substack{CH_3 \\ |}}{Si}}-(O-\underset{\substack{| \\ CH_3}}{\overset{\substack{CH_3 \\ |}}{Si}}-)_n-O-\underset{\substack{| \\ CH_3}}{\overset{\substack{CH_3 \\ |}}{Si}}-CH_3 \qquad (279)$$

A

$$\xrightarrow{\text{(St)SiCl}} \; CH_2{=}CH-C_6H_4-\underset{\substack{| \\ CH_3}}{\overset{\substack{CH_3 \\ |}}{Si}}-(-O\underset{\substack{| \\ CH_3}}{\overset{\substack{CH_3 \\ |}}{Si}}-)_n-O\underset{\substack{| \\ CH_3}}{\overset{\substack{CH_3 \\ |}}{Si}}CH_3 \qquad (280)$$

The terminating agents, *p*-oligosiloxane-substituted styrenes, (St)SiCl, and methacrylate-type oligosiloxanes, (M)SiCl, were synthesized according to Schemes 3 and 4, respectively.

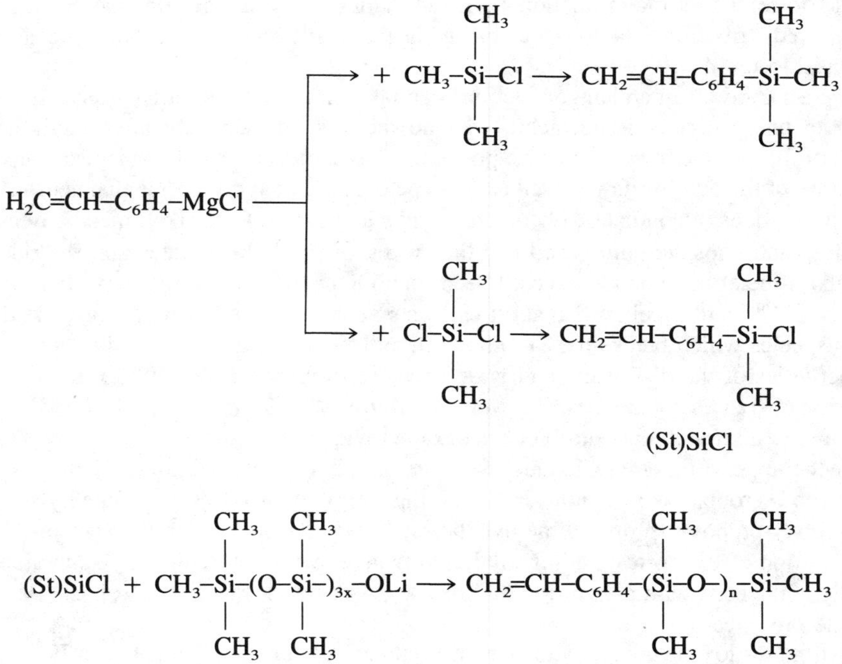

$$\longrightarrow \; + \; CH_3-\underset{\substack{| \\ CH_3}}{\overset{\substack{CH_3 \\ |}}{Si}}-Cl \;\longrightarrow\; CH_2{=}CH-C_6H_4-\underset{\substack{| \\ CH_3}}{\overset{\substack{CH_3 \\ |}}{Si}}-CH_3$$

$$H_2C{=}CH-C_6H_4-MgCl$$

$$\longrightarrow \; + \; Cl-\underset{\substack{| \\ CH_3}}{\overset{\substack{CH_3 \\ |}}{Si}}-Cl \;\longrightarrow\; CH_2{=}CH-C_6H_4-\underset{\substack{| \\ CH_3}}{\overset{\substack{CH_3 \\ |}}{Si}}-Cl$$

(St)SiCl

$$\text{(St)SiCl} \;+\; CH_3-\underset{\substack{| \\ CH_3}}{\overset{\substack{CH_3 \\ |}}{Si}}-(O-\underset{\substack{| \\ CH_3}}{\overset{\substack{CH_3 \\ |}}{Si}}-)_{3x}-OLi \;\longrightarrow\; CH_2{=}CH-C_6H_4-(\underset{\substack{| \\ CH_3}}{\overset{\substack{CH_3 \\ |}}{Si}}-O-)_n-\underset{\substack{| \\ CH_3}}{\overset{\substack{CH_3 \\ |}}{Si}}-CH_3$$

x	0	1
n	1	4

SCHEME 3. Synthetic routes to styrene type oligosiloxane monomers.

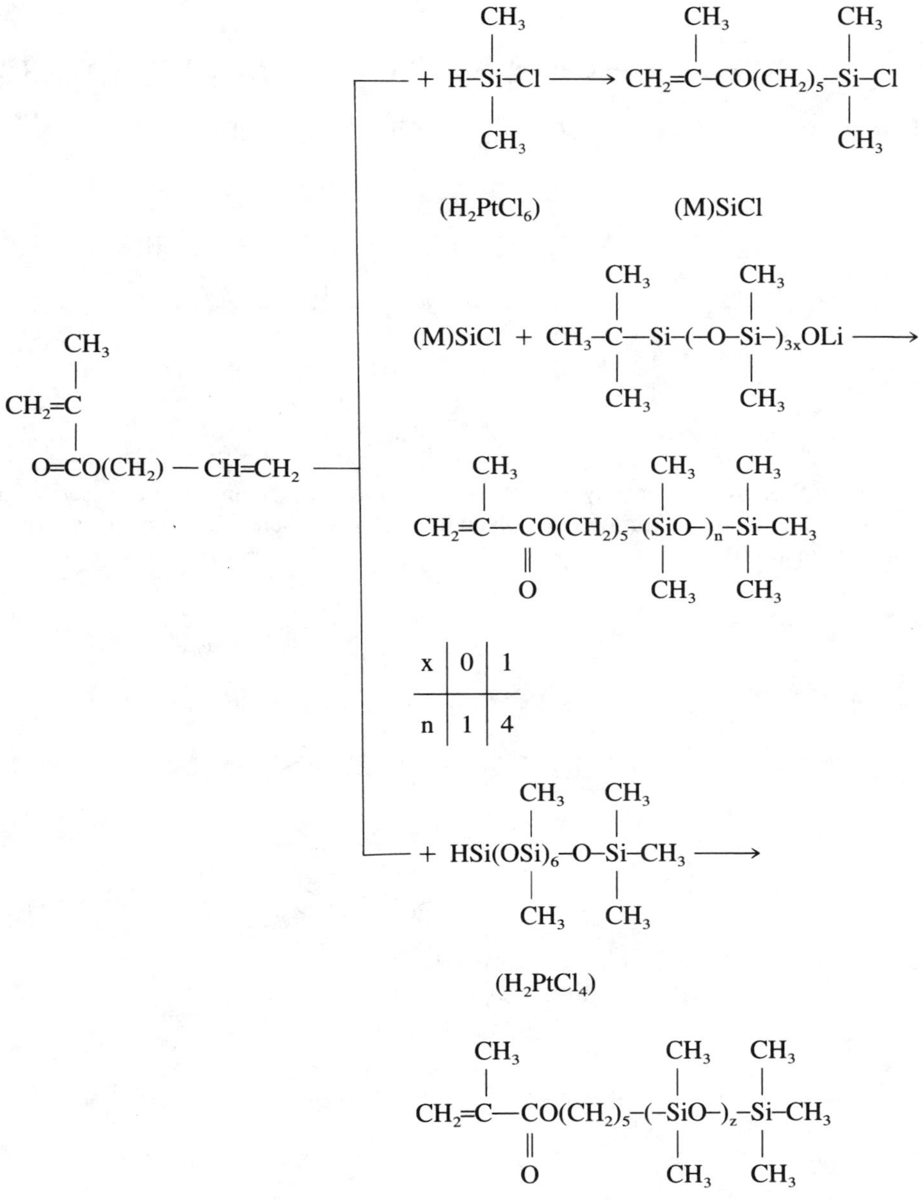

$$
\begin{array}{c}
\underset{\underset{CH_3}{|}}{H\text{–}Si\text{–}Cl} \xrightarrow{\quad} \underset{\underset{CH_3}{|}}{CH_2{=}C\text{–}CO(CH_2)_5\text{–}Si\text{–}Cl}
\end{array}
$$

(H$_2$PtCl$_6$)   (M)SiCl

$$
(M)SiCl \;+\; CH_3\text{–}\underset{\underset{CH_3}{|}}{\overset{\overset{CH_3}{|}}{C}}\text{–}Si\text{–}(\text{–}O\text{–}Si\text{–})_{3x}OLi \longrightarrow
$$

$$
CH_2{=}\underset{\underset{O}{\|}}{\overset{\overset{CH_3}{|}}{C}}\text{–}CO(CH_2)_5\text{–}(SiO\text{–})_n\text{–}Si\text{–}CH_3
$$

x	0	1
n	1	4

$$
+ \; HSi(OSi)_6\text{–}O\text{–}Si\text{–}CH_3 \longrightarrow
$$

(H$_2$PtCl$_4$)

$$
CH_2{=}C\text{–}CO(CH_2)_5\text{–}(\text{–}SiO\text{–})_z\text{–}Si\text{–}CH_3
$$

$$
CH_2{=}C \quad ,\quad O{=}CO(CH_2)\text{–}CH{=}CH_2
$$

SCHEME 4. Synthetic routes to methacrylate type oligosiloxane monomers.

The monomers obtained according to these schemes had $\overline{M}_n$ = 3000 to 9000. Fluorine-containing oligomers were also synthesized by reacting (St)SiCl with fluorinated derivatives.[533]

Different reaction routes were reported recently[534] for the synthesis of unsaturated siloxane macromers, with the ultimate aim of obtaining specialty polymeric materials.

C. ANIONIC OLIGOMERIZATION OF CARBON-HETEROATOM MULTIPLE BONDS

1. Carbon-Nitrogen Bonds

The way in which the oligomers and polymers of nitriles are formed may be depicted as follows.[535,536]

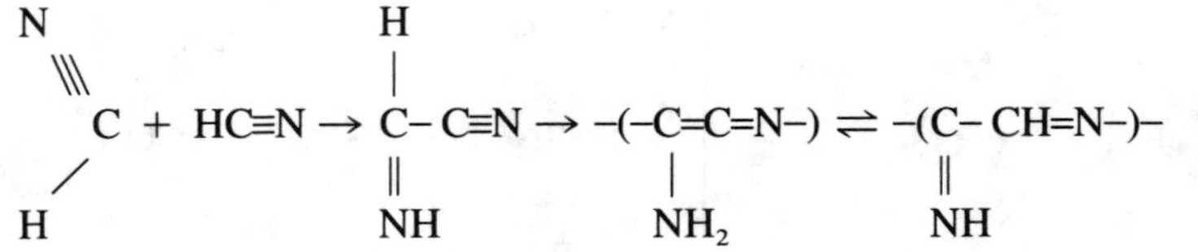

$$
\begin{array}{c}
N \\
\| \\
C \\
\diagup \\
H
\end{array}
+ \;
\begin{array}{c}
H \\
| \\
HC\!\equiv\!N \;\rightarrow\; C\!-\!C\!\equiv\!N \;\rightarrow\; -(\!-\!C\!=\!C\!=\!N\!-\!) \;\rightleftharpoons\; -(C\!-\!CH\!=\!N\!-\!)- \\
\qquad\qquad \| \qquad\quad\; | \qquad\qquad\quad \| \\
\qquad\qquad NH \qquad\; NH_2 \qquad\qquad\; NH
\end{array}
\qquad (281)
$$

and

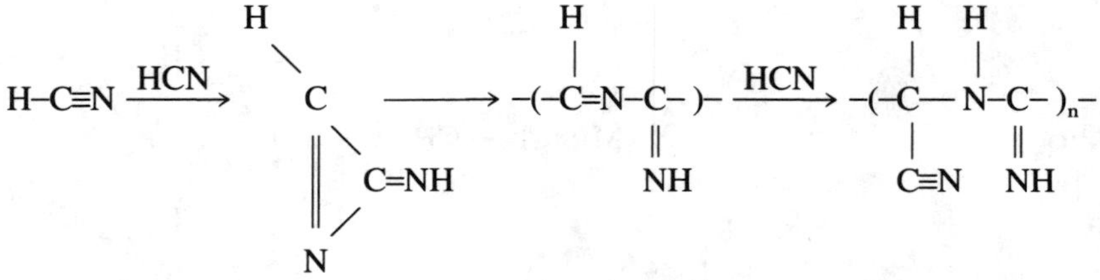

$$
H\!-\!C\!\equiv\!N \xrightarrow{\;HCN\;}
\begin{array}{c}
H \\
\diagdown \\
C \\
\| \diagdown \; C\!=\!NH \\
\| \diagup \\
N
\end{array}
\longrightarrow -(\!-\!C\!=\!N\!-\!C\!-\!)- \xrightarrow{\;HCN\;} -(\!-\!C\!-\!N\!-\!C\!-\!)_n\!-
\qquad (282)
$$

Other HCN oligomeric or polymeric structures formed in the presence of bases include the tetramer diaminomaleonitrile (A), polyaminocyanomethylene (B), and block azulmin polymers (C) and (D):

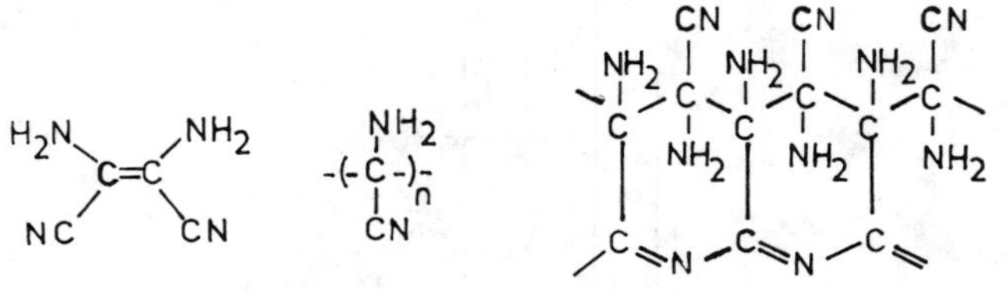

A   B   C

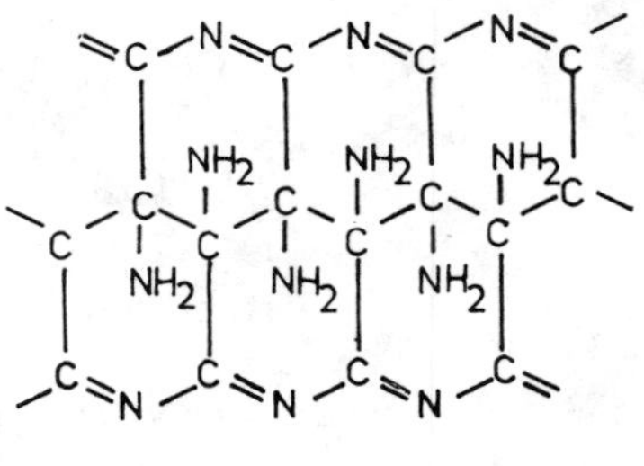

D

All these structures are believed by some authors[535-537] to have been involved in the abiotic origin of protein-like polymers (i.e., protobiopolymers).

The cyclotrimerization reaction is characteristic of multiple carbon-nitrogen bonds, e.g., the trimerization of $-N\!=\!CO$ groups when isocyanurate rings are formed:

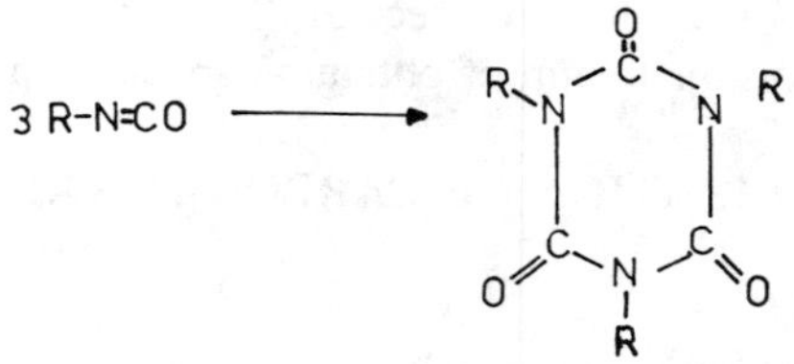

$$
3\;R\!-\!N\!=\!CO \longrightarrow
\begin{array}{c}
O \\
\| \\
R\!-\!N \diagup C \diagdown N\!-\!R \\
| \qquad\qquad | \\
O\!=\!C \diagdown \;\; \diagup C\!=\!O \\
N \\
| \\
R
\end{array}
\qquad (283)
$$

During the initial stages of cyclotrimerization of difunctional isocyanates, the formation of polyfunctional NCO-terminated oligomers of the types shown below takes place:[540]

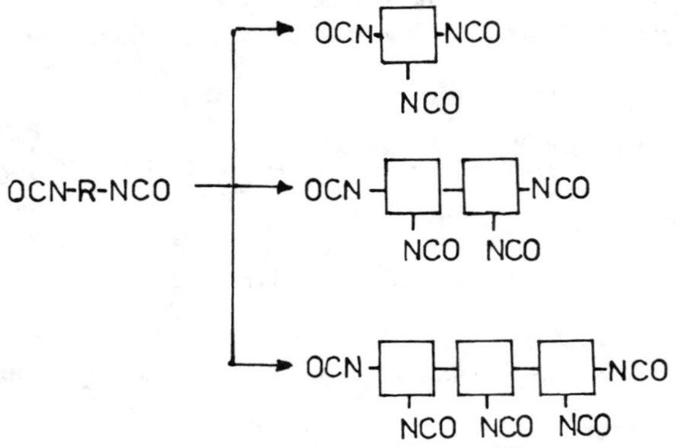

$$OCN-R-NCO \longrightarrow \qquad (284)$$
$$\qquad (285)$$
$$\qquad (286)$$

where -▢- represents the triazinic ring.

In the case of isocyanates, this reaction (Equation 283) is initiated by an anionic type of catalyst and proceeds via propagation, transfer, and termination:[538-540]

Initiation:
$$M + C \rightleftharpoons X_1 \qquad (287)$$

Propagation:
$$X_1 + M \longrightarrow X_2 \qquad (288)$$
$$X_2 + M \longrightarrow X_3 \qquad (289)$$

Transfer:
$$X_3 + M \longrightarrow X_1 + T \qquad (290)$$

Termination:
$$X_3 \longrightarrow T + C \qquad (291)$$

where X_1, X_2, and X_3 are complexes, X = isocyanate, T = trimer, and C = catalyst.

These oligomers are of considerable interest to urethane chemistry. The reactivity of isocyanates in the cyclotrimerization reaction depends on the donor-acceptor strength of substituents adjacent to the −NCO group.[541] Isocyanates with electron-withdrawing groups are more reactive than isocyanates with electron-donating groups. Therefore, aromatic isocyanates cyclotrimerize significantly faster than aliphatic or cycloaliphatic isocyanates.

The cyclotrimerization of substituted phenyl isocyanates showed that isocyanates containing electron-withdrawing substituents in the *para* position cyclotrimerize significantly faster than isocyanates with electron-donating groups. The kinetic data correlate with the Hammett equation:

$$\log\left[\frac{k_x}{k_H}\right] = \sigma\rho \qquad (292)$$

giving a ρ value of $+1.57$.[541] The large positive value indicates that the cyclotrimerization proceeds via nucleophilic attack of anion on the electrophilic carbon atom of the isocyanate group.

The nucleophilicity of the catalytic center of the catalyst plays an important role in the cyclotrimerization of isocyanates. It was found that the cyclic sulfonium zwitterions (szw) shown in Equation 293 are very active cyclotrimerization catalysts.[542,543] These derivatives are unstable and polymerize due to the nucleophilic attack of the anion $-O^-$ on the tetra-hydrothiophenium ring, followed by ring opening with consecutive polymerization:

$$\text{(293)}$$

szw; R_1, R_2 = H, R   Zwitterionic dimer(szwd)

$$\text{szwd} \xrightarrow{\ +\ \text{szw}\ } \text{cyclic dimer} + \text{cyclic trimer} + \text{linear oligomers} \qquad \text{(294)}$$

Products of polymerization (Equations 293 and 294) are low molecular weight linear and cyclic dimers and trimers.[544,545] Since cyclic sulfurium zwitterions in stable from usually contain two molecules of water, under reaction conditions (concentration, temperature) they may act as anionic initiators and permit control of cyclotrimerization of isocyanates at a desired conversion in a one-step procedure.[540] The rate constant k_{cat} for cyclotrimerization increased with an increase in the electron density on the oxygen of cyclic szw catalyst. It was found that the steric hindrance around the catalytic center had relatively little effect on the rate constant k_{cat}. This is in contrast to the urethane reaction, where the steric hindrance around the catalytic center sharply decreased the rate constant k_{cat}.[541] Similar results were obtained in the cyclotrimerization of phenylisocyanate using substituted ammonium car-boxylates as catalysts.[546] These results indicate that initiation of the trimerization reaction proceeds by the nucleophilic attack on the electrophilic carbon atom of the isocyanate by the anionic catalytic center ($-O^-$, $-COO^-$). Cyclotrimerization-oligomerization of toluene diisocyanate in the presence of phenols or amines gave a mixture of isocyanurate oligomers with $\overline{DP}$ = 1 to 3 (see Equations 284 to 286) which contained about 6% free $-NCO$ groups.[547]

N-Cyanourea-terminated oligomers have been found to be very useful for converting into fast-curing thermosets and curing epoxy resins.[548,549] N-Cyanourea compounds can be obtained directly from the reaction of isocyanate and cyanamide. Using a diisocyanate as starting material, an N-cyanourea-terminated resin was synthesized. Furthermore, by reacting a functional diol first with two molecules of diisocyanate and then with two molecules of cyanamide, a structural N-cyanourea oligomer was prepared.[549]

$$
\text{HO–R–OH} + 2\text{OCN–R}'\text{–NCO} \rightarrow \text{OCN–R}'\text{–NH–}\overset{\overset{\displaystyle O}{\|}}{C}\text{O–R–}\overset{\overset{\displaystyle O}{\|}}{C}\text{NH–R}'\text{–NCO}
$$

$$
\xrightarrow{\ 2\text{H}_2\text{N–CN}\ } \text{NC–NH–}\overset{\overset{\displaystyle O}{\|}}{C}\text{–NH–R}'\text{–HN}\overset{\overset{\displaystyle O}{\|}}{C}\text{ORO}\overset{\overset{\displaystyle O}{\|}}{C}\text{NH–R}'\text{–NH}\overset{\overset{\displaystyle O}{\|}}{C}\text{NH–CN} \qquad \text{(295a)}
$$

Under basic conditions, cyanamide undergoes oligomerization to dicyanamide and tri-merizes to melamine upon heating. Reactive oligomers were therefore formed from N-cyanourea-terminated resins containing a similar $-NHCN$ group. The stepwise addition of *N*-cyanourea groups generates a linear polymer with a repeating segment, *N,N'*-biscarbamyl-*N*-cyanoguanidine, which produces a cyclic structure as shown in Equation 295b.

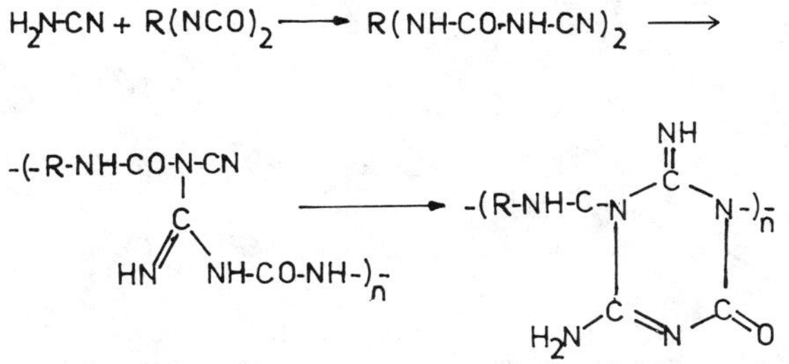
H2N-CN + R(NCO)2
R(NH-CO-NH-CN)2
-(-R-NH-CO-N-CN
C
HN
NH-CO-NH-)n
NH
C
-(R-NH-C-N
N-)n
H2N
C
N
C
O

(295b)

The *N*-cyanourea-terminated oligomers synthesized from a polycaprolactone diol, a diisocyanate, and cyanamide were very useful as adhesive and corrosion-protection coatings for steel substrate.[549]

Mono- and difunctional aryl cyanates undergo trimerization to give thermally stable monomeric and polymeric aryloxy-*s*-triazines, respectively.[550-554]

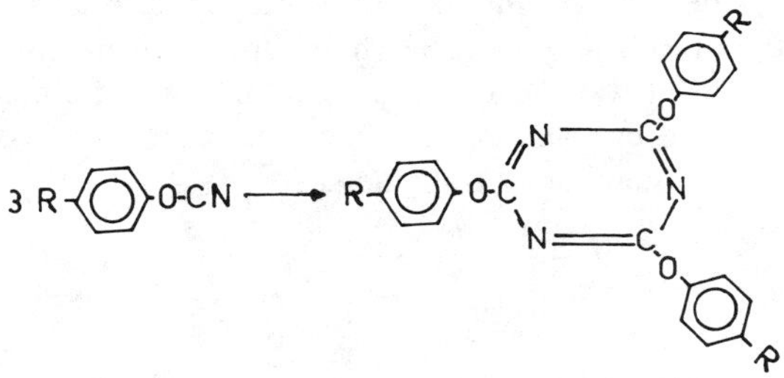
3 R
O-CN
R
O-C
N
C
O
R
N
N
C
O
R

(296)

The first step is a reversible dimerization of two molecules of the cyanate. However, the reaction rate and temperature dependence of the reaction cannot be predictably manipulated by altering the electronic properties of the group *para* to the cyanate. Sodium acetate, triethylamine hydrobromide, and polyphenol catalyze the cyclization oligomerization of cyanates, but the reaction is dramatically catalyzed by zinc stearate.

Aryl cyanates can be used to cross-link oligomers longer in length than bisphenol A, but there is a limit to the length which can be used (MW 4000).

Models for thermotropic liquid crystal dicyanates were prepared in order to offset the shrinkage which accompanies the thermosetting reaction by the order lost in going from the liquid crystalline state to the less-ordered state of the cross-linked polymers. As models for expected polymers, monomeric 1,3,5-*tris*(4-benzylidene oxy-4'-alkylaniline)triazines were synthesized and found to be discotic liquid crystals (R in Equation 296 is $-CH=N-C_6H_4-R'$, where R' is $-C_6H_{13}$ or $-C_{10}H_{21}$).

Trimerization of cyanuryl chloride gives the trichloro-*s*-triazine, which is very reactive.

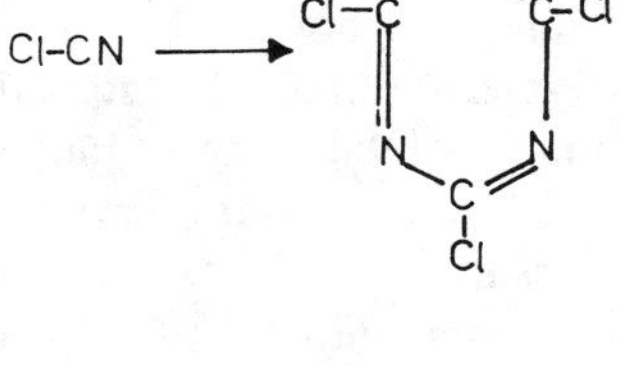
Cl-CN
Cl-C
N
C-Cl
N
N
C
Cl

(297)

This cyclic trimer is a versatile intermediate for obtaining various oligomeric and polymeric materials.[555] For example, trichloro-s-triazine is converted to triallyl cyanurate by reacting with allyl alcohol:[556]

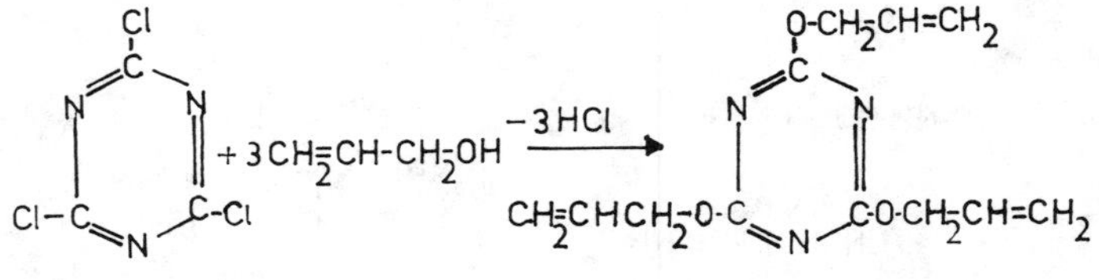

$$\text{(298)}$$

Highly branched, but not yet cross-linked prepolymers are obtained radically (yield, 25%), usually as a solution in monomer, and are cured in the actual processing to finished product. Contraction of the prepolymer during polymerization is very slight, only 1% vs. 12% for polymerizing monomer. Optical articles are molded from these resins since the transparency roughly corresponds to that of PMMA, but the resistance to scratching and abrasion is about 30 to 40 times better.[556]

Addition polymerization of the carbon-nitrogen double bond in acetaldazine, the azomethine analogue of 2,4-hexadiene, initiated by n-BuLi in aprotic solvents, gave only low molecular weight (800 to 1200) polymers.[557] The oligomers were composed of 1,2 as well as 1,4 addition structures (Equation 299). An easy occurrence of a proton transfer from the acetaldazine monomer to the propagating anion was confirmed by comparing the number of charged initiator with the number of oligomers.

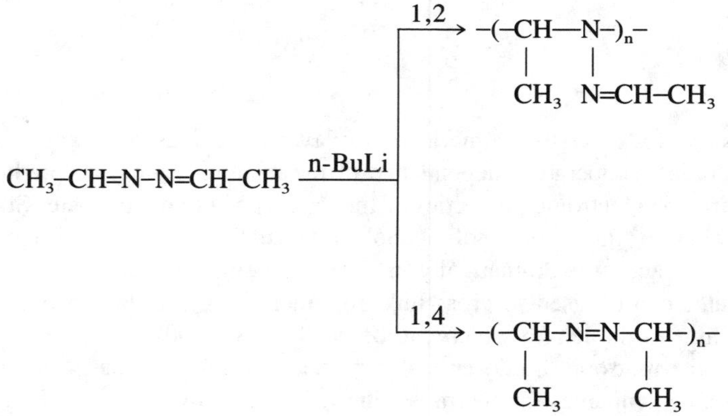

$$\text{(299)}$$

2. Carbon-Oxygen Multiple Bonds

Carbon-oxygen multiple-bond anionic monomers refer chiefly to carbonylic derivatives. Aldehyde polymers are probably the oldest synthetic polymers.[558] Polyoxymethylene, the polymer of formaldehyde, was first described by Butlerov in 1859 and the polymer of chloral was first prepared in 1832 by Liebig. This was about a century before the concept of linear macromolecules was developed. Polyformaldehyde and polychloral were isolated because they were stable at room temperature and above.

True understanding of the polymerization of the aliphatic higher aldehydes began with a clear understanding of the importance of the ceiling temperature for these polymerizations. The polymers of aliphatic aldehydes have ceiling temperatures substantially below room temperature under atmospheric pressure.

Aldehydes are polymerized with ionic initiators to polyacetals, but with a much a lower ΔH_{polym} and, consequently, a lower T_c than when α-olefins of comparable structure are polymerized to polyolefins.

By far the most important polyaldehyde is polyoxymethylene (polyformaldehyde). It is commercially produced by anionic polymerization of purified $CH_2=O$ to high molecular weight products.[559]

Higher aliphatic aldehydes may be polymerized with anionic initiators, but require strong nucleophiles such as alkoxides, while relatively weak nucleophiles such as pyridine are effective for the polymerization of haloaldehydes.[560] It should also be pointed out that in essentially all these polymerizations, the polymer precipitates during polymerization as a crystalline polymer, which adds an additional driving force for enchainment. Higher aliphatic aldehydes conveniently polymerize with anionic initiators at low temperatures, frequently to isotactic polyaldehydes. The range of known polyaldehydes has been extended to poly(n-dodecaldehyde). Polymers with side chains longer than C_3 possess the characteristics of the crystallization of the aliphatic side chain, separate and in addition to the crystallization of the main chain.[561] All isotactic polyaldehydes have a crystal structure with a tetragonal unit cell containing 16 monomer units, which indicates a fourfold helix with four polymer chains in the unit cell (4_1), irrespective of the length of the aliphatic side chain.[562,563] Low molecular weight ($\overline{M}_n$ on the order of 10^3) anionic polymers of higher aldehydes ($> C_4$) are isotactic and highly crystalline materials.[564]

Recent polymerizations with chiral initiating species have resulted in the preparation of isotactic, helical polymers of chloral (trichloroacetaldehyde) that were characterized by macromolecular asymmetry with exceptionally large optical rotations in the solid state.[565,566]

Crystallographic data indicated that polychloral is both isotactic and characterized by a stable 4_1 helical order in the solid state.[567]

Chloral monomer was oligomerized with *achiral* lithium *t*-butoxide under conditions that mimicked the polymerization process.[566] X-ray crystallographic analyses confirmed that the linear dimer-pentamer series exist structurally as low molecular weight homologues of isotactic polychloral, with an approximate repeat arrangement along the acetal backbone (from the *t*-butoxy group to the acetyl terminus) of gauche($+$)-skew($-$) for the *s*-configuration. The conformation of the major tetrameric adduct was found to be very similar to a unit sequence of the 4-helical structure of isotactic polychloral.

The axis of the embryonic helix was found to parallel the c-axis of the crystal, showing that the helix development occurs very early in the growth of the polychloral chain.

When both multiple carbon-oxygen and carbon-nitrogen bonds are present in the same molecule, the former opens first in the presence of basic initiators. For example, anionic polymerization of β-cyanopropioaldehyde initiated by benzophenone-alkali metal complexes resulted in low molecular weight poly(cyanoethyl)oxymethylene

$$NC-CH_2-CH_2-HC=O \rightarrow -(-CH-O-)_n- \qquad (300)$$
$$| $$
$$C_2H_4CN$$

of high stereoregulatory.[568]

The marked influence of initiator concentration on polymer yield and stereoregularity was explained on the basis of the difference in the degree of association of the alcoholate ion pair, i.e., the associated ion pair may form stereoregular polymer and the non- or less-associated ion pair may form atactic polymers. Chain transfer with active hydrogen of β-cyanopropionaldehyde may have been the reason for limiting the molecular weight of polymers. Polymers obtained with cationic or coordinative initiators had the same polyacetal structure, but the molecular weight was higher, while the stereoregularity of chains was lower.

IV. COORDINATION AND MISCELLANEOUS OLIGOMERIZATIONS

A. COORDINATION OLIGOMERIZATIONS

This method is chiefly used to obtain oligo-α-olefins. Typical reactions of chain propagating termination in this case are[569]

1. Chain transfer to monomer or low oligomer olefins
2. Chain transfer to metal-aklyls included in the catalyst system
3. Chain transfer to special agents (e.g., hydrogen, HCl, etc.)
4. Chain termination by hydrid ion transfer from the end of the propagating chain to cation

The reactions of these processes determine the nature of the terminal groups forming oligomers.

What started out in 1952 as an anomalous effect of colloidal nickel on the "Aufbau reaction", i.e., polymerization of ethylene to low molecular weight polyethylene waxes by AlEt$_3$, was transformed by Ziegler et al. into one of the most important chemical discoveries of recent times—the obtaining of stereoregular polymers.[570] Interest in coordination oligomerization as a method of α-olefin synthesis has increased during the last few years.[571-576] Essential to this type of polymerization is insertion of the monomer between the metal-carbon bond at the active site of the catalyst.

Propylene oligomers prepared in the presence of hydrogen and a $(-)$-R-ethylene-bis(tetrahydroindenyl)-zirconium dichloride-aluminoxane catalyst were optically active and highly isotactic.[577] All chains contained a *n*-Pr end group, the other end being either iso-C$_4$H$_9$ or *n*-C$_4$H$_9$, depending on the catalyst used.

As shown in the preceding sections, low molecular weight polymers of epoxides are readily prepared with ordinary acid and base catalysts and have been known for many years. However, only in the last decades has it been possible to make polymers of a wide variety of epoxides, and this progress has resulted from the discovery of some unique and widely effective organometallic catalysts. Some of these catalyst systems are the reaction products of aluminum, zinc, and magnesium organometallics with water, alcohols, or other proton-donating derivatives.[578-582] The ratio of these ingredients can be varied over a wide range, and the best composition will depend on the conditions, the epoxide, and the type of product desired, but, in general, the performed catalyst retains organometal bonds.

Organozinc compounds exhibit a very efficient stereoselective behavior as catalyst in the polymerization of 1-substituted-1,2-epoxide such as propylene oxide.[581]

Chloromethyl oxirane underwent an asymmetric selective (or stereoselective) oligomerization with a binary catalyst systems comprised of Al(C$_2$H$_5$)$_3$ and *N,N'*-disalicylidene (1*R*,2*R*)1,2-cyclohexanediol-diaminatocobalt, and two kinds of linear oligomers were produced with respect to the end groups, namely, epoxy- and halohydrin-ended oligomers.[583] There exists a close mechanistic similarity between α-olefin polymerization to isotactic polymers and ring-opening polymerization of 1,2 epoxides with organometallic initiators. Both of these enchainments have been shown to proceed under the same mechanism of steric control, that is, the enantiomorphic catalyst site control mechanism.[581]

During the reaction of 1-phenoxy-2,3-epoxy propane with catalytic amounts of magnesium perchlorate, a variety of reaction products were formed which did not correspond to a normally distributed oligomer range.[584] Oligo(hydroxyether)s of similar structure and cyclic isomeric dimers of very low reactivity were the main low molecular weight products.

Polymerization of propylene carbonate with diethylzinc — a catalyst specific to the stereoselective polymerization of oxyranes — gave a low molecular weight product (1000 to 4000) with the following structure:[585]

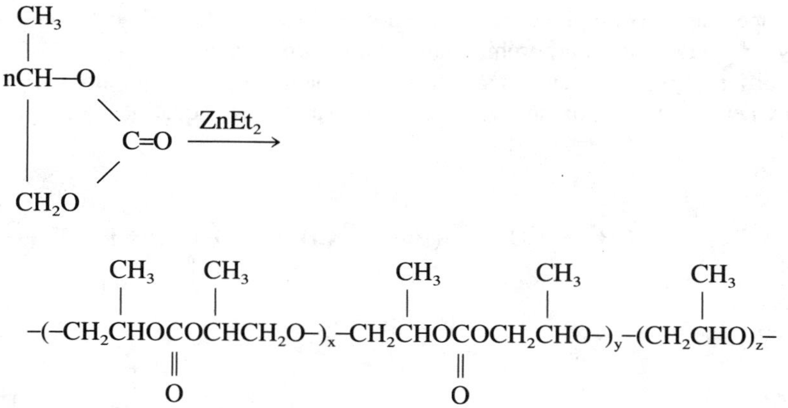

$$\begin{array}{c} CH_3 \\ | \\ nCH-O \\ | \quad\backslash \\ \qquad C{=}O \xrightarrow{\;ZnEt_2\;} \\ | \quad / \\ CH_2O \end{array}$$

$$\begin{array}{ccccc} CH_3 & CH_3 & & CH_3 & CH_3 & CH_3 \\ | & | & & | & | & | \\ -(-CH_2CHOCOCHCH_2O-)_x-CH_2CHOCOCH_2CHO-)_y-(CH_2CHO)_z- & & & (301) \\ \qquad\quad \| & & & \| \\ \qquad\quad O & & & O \end{array}$$

where $x \simeq 0.5$, $y \simeq 0.25$, and $z \simeq 0.25$.

This strongly suggested that the polymerization of propylene carbonate proceeded via 2,7-dimethyl-1,4,6,9-tetraoxa-spiro 4H nonane, DTN, as an intermediate compound. Hence, DTN was synthesized and polymerized with the use of diethylzinc catalyst, and the polymer has the same structure as that shown in Equation 301.

Aluminum porphyrins are very active initiators for the ring-opening polymerization of various heterocyclic monomers, giving the corresponding polymers of uniform controlled molecular weight.[578-584] This novel catalytic system is particularly effective for the ring-opening polymerization of epoxides[578-580] and lactones[581] and the copolymerization of epoxides with carbon dioxide and cyclic anhydrides.[582,583] ^{1}H NMR has demonstrated that the growing species of this initiating system is that shown below

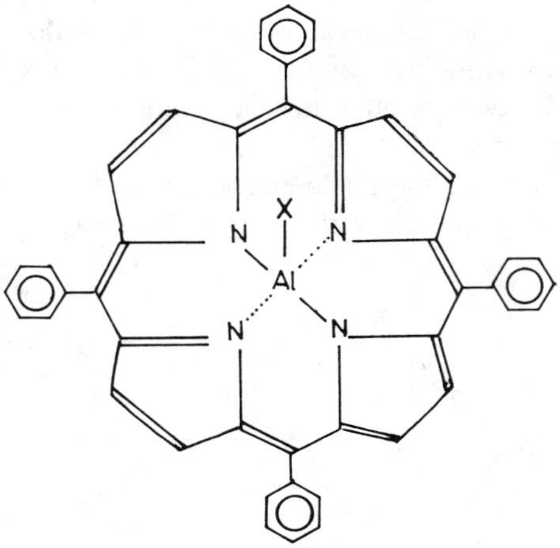

where $X = Cl$, OAr, R, OR, or OCOR and R = alkyl or macromolecular substituent.

The molecular weight of the polymer may be controlled by changing the molar ratio $[M]_o/[I]_o$. Oligoesters and oligoethers having reactive end groups derived from the axial ligand of the catalyst have been prepared by choosing an appropriate ligand:[581]

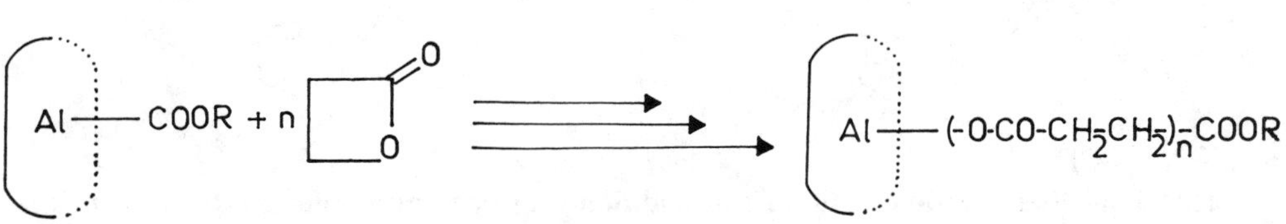

$$(302)$$

The ring opening of β-propiolactone (Equation 302) takes place almost exclusively at the alkyl-oxygen bond to form a porphiratoaluminum carboxylate.[579,582]

Two different types of reactions occurring at the same aluminum atom are involved in the copolymerization of epoxides with CO_2 and anhydrides (Equations 303 and 304),

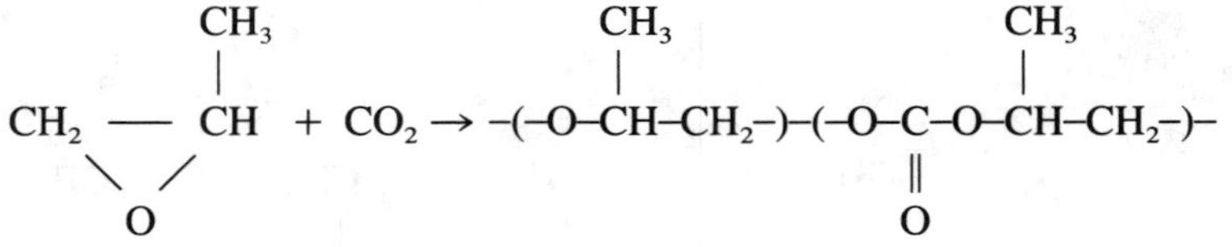

$$CH_2 \underset{O}{\overset{CH_3}{\diagup\!\!\!\diagdown}} CH + CO_2 \rightarrow -(-O-\overset{CH_3}{\underset{|}{CH}}-CH_2-)-(-O-\overset{CH_3}{\underset{\|}{\underset{O}{C}}}-O-\overset{|}{CH}-CH_2-)- \qquad (303)$$

$$CH_2\underset{O}{\overset{CH_3}{\diagup\!\!\!\diagdown}}CH + O=C \underset{O}{\diagdown} C=O \longrightarrow -(-O-\overset{CH_3}{\underset{|}{CH}}-CH_2)-(-O-OC \diagup C O O-\overset{CH_3}{\underset{|}{CH}}\cdot CH_2)- \qquad (304)$$

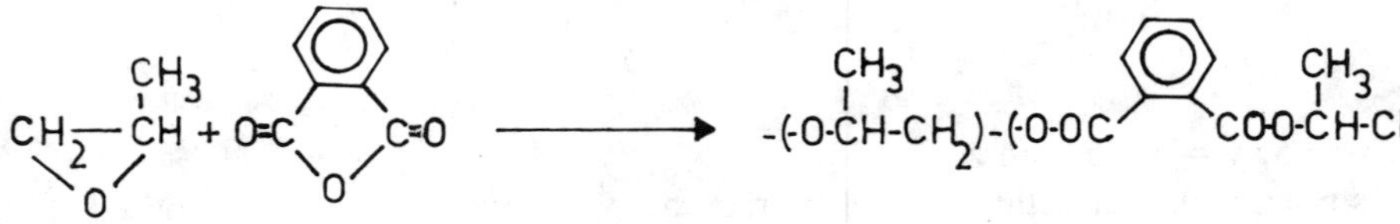

i.e., the reaction of aluminum alkoxide with CO_2 or cyclic acid anhydride and the reaction of aluminum carboxylate with epoxide.[582]

Alkylaluminum porphyrins gave rise to the living polymerization of methacrylic esters which formed the corresponding polymers with narrow MWD and controlled molecular weight. The irradiation of visible light (xenon light, λ = 420 nm) was essential to the polymerization.[584] For example, the polymerization of MMA with methylporphyrin gave, at 30°C, oligomers with molecular weights of several thousands and a $\overline{M}_w/\overline{M}_n$ ratio close to unity (conversion to PMMA, 100%). Elevating the polymerization temperature up to 50°C resulted in a remarkable acceleration of the process without broadening the MWD of the produced oligomer. The polymerization barely took place in the dark under similar conditions. The polymerization was initiated by the addition of the alkyaluminum moiety of the catalyst to the monomer under irradiation to give a (porphyrinato)aluminum enolate as the growing species. The effect of visible light irradiation seemed to appear in both the initiation and propagation steps, indicating the photoenhancement of the nucleophilic reactivities of the initiating and propagating species carrying light-absorbing porphyrin ligand.[584]

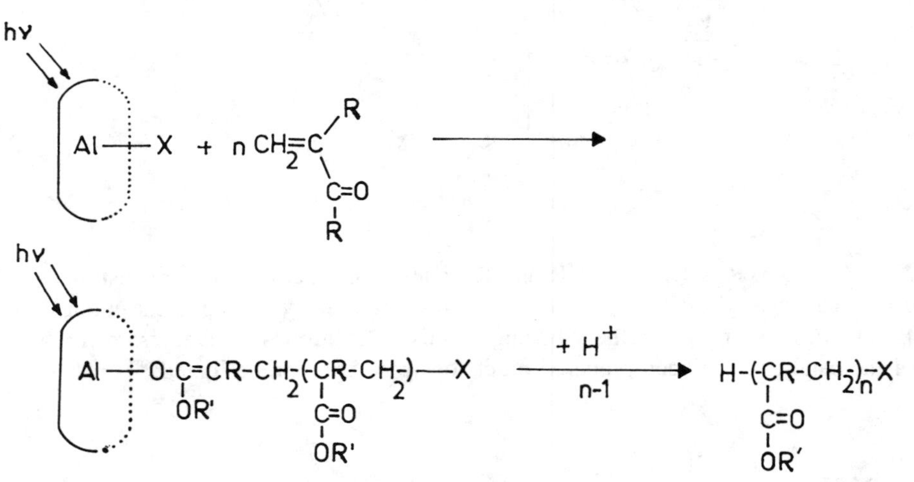

where X = CH_3.

The sequential addition-addition and addition-ring opening polymerizations of methacrylic esters and methacrylic-acrylic esters, epoxides, and lactones, respectively, with these porphyrinic initiators may give novel block copolymers of controlled, uniform blocks of oligomeric lengths.[586-592]

Masked functionality initiators were used to make oligomers and telechelic polymers.[593,594] Thus, the acetal obtained from 3-chloropropanol and ethyl vinyl ether was reacted with lithium and then with 1,1-diphenylethylene to prepare an initiator with a masked hydroxyl group (Scheme 5).[594] Polymerization of methyl methacrylate with this initiator gave, after hydrolysis, monohydroxy oligomer.

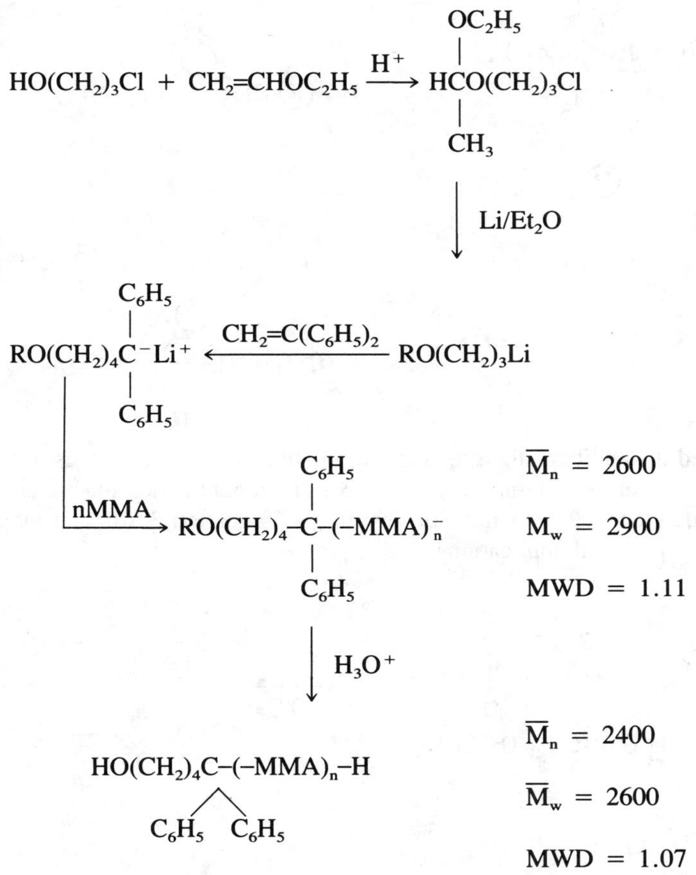

$$HO(CH_2)_3Cl + CH_2=CHOC_2H_5 \xrightarrow{H^+} \underset{\underset{CH_3}{|}}{\overset{\overset{OC_2H_5}{|}}{HCO}}(CH_2)_3Cl$$

$$\Big\downarrow Li/Et_2O$$

$$\underset{\underset{C_6H_5}{|}}{\overset{\overset{C_6H_5}{|}}{RO(CH_2)_4C^-Li^+}} \xleftarrow{CH_2=C(C_6H_5)_2} RO(CH_2)_3Li$$

$$\Big\downarrow^{nMMA}$$

$$\underset{\underset{C_6H_5}{|}}{\overset{\overset{C_6H_5}{|}}{RO(CH_2)_4{-}C{-}(-MMA)_n^-}} \qquad \overline{M}_n = 2600$$

$$M_w = 2900$$

$$MWD = 1.11$$

$$\Big\downarrow H_3O^+$$

$$\underset{\underset{C_6H_5 \quad C_6H_5}{\wedge}}{HO(CH_2)_4C{-}(-MMA)_n{-}H} \qquad \overline{M}_n = 2400$$

$$\overline{M}_w = 2600$$

$$MWD = 1.07$$

SCHEME 5. Functional initiators.

Coupling the living polymer with, e.g., α,ω-dibromo *p*-xylene led to the corresponding dihydroxy polymers (Equation 306); polyfunctional coupling agents gave star polymers (Equations 307 and 308).

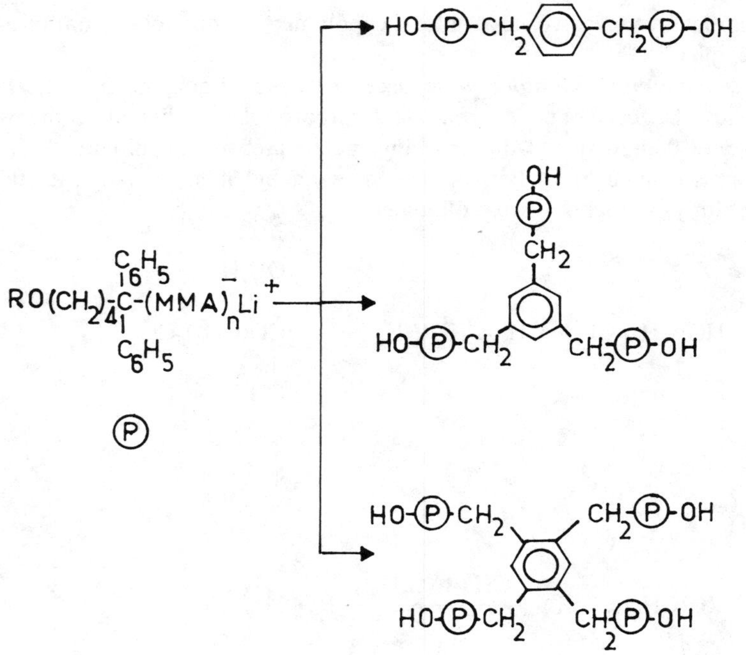

$$HO{-}\boxed{P}{-}CH_2{\bigcirc}{-}CH_2{-}\boxed{P}{-}OH \qquad (306)$$

$$RO(CH_2)_4\overset{\overset{\displaystyle C_6H_5}{|}}{\underset{\underset{\displaystyle C_6H_5}{|}}{C}}{-}(MMA)_n^{-}Li^{+} \longrightarrow \qquad (307)$$

$$\boxed{P}$$

$$HO{-}\boxed{P}{-}CH_2 \qquad CH_2{-}\boxed{P}{-}OH \qquad (308)$$

The marked instability of the growing end in anionic polymerizations of methacrylates, requiring work at very low temperatures in order to suppress side reactions (e.g., thermal cyclization, Equation 309, and initiator destruction, Equation 310), is a major deficiency which limits the practical applications of this process.

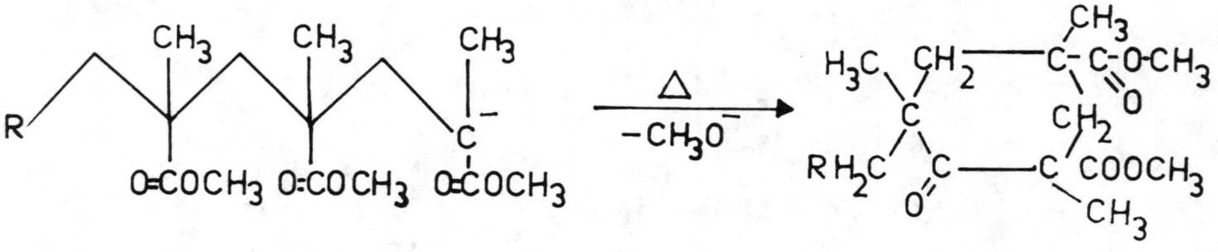

$$(309)$$

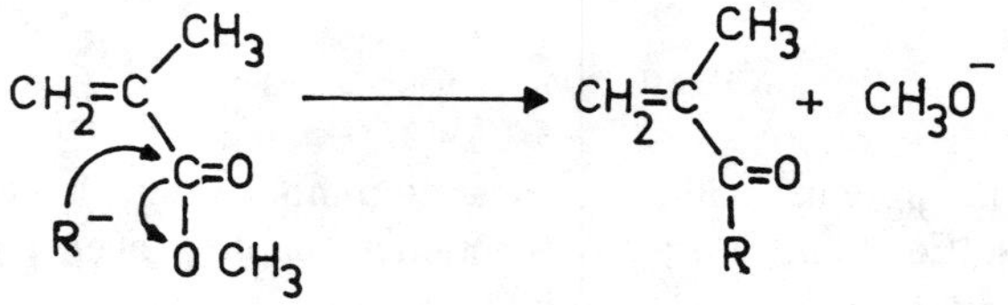

$$(310)$$

Group-transfer polymerization, an elegant demonstration by Webster et al.[595-600] that the nucleophile-assisted "group transfer" reaction of silylketene acetals could be turned into a well-controlled "living" polymerization process of carbonyl-conjugated monomers, gives living polymers of high stability at room temperature or above. This reaction, illustrated in Equations 311 and 312 for MMA, involves the repeated transfer of a trialkylsilyl group from the growing chain end (or, at first, from the initiator) to the incoming monomer:

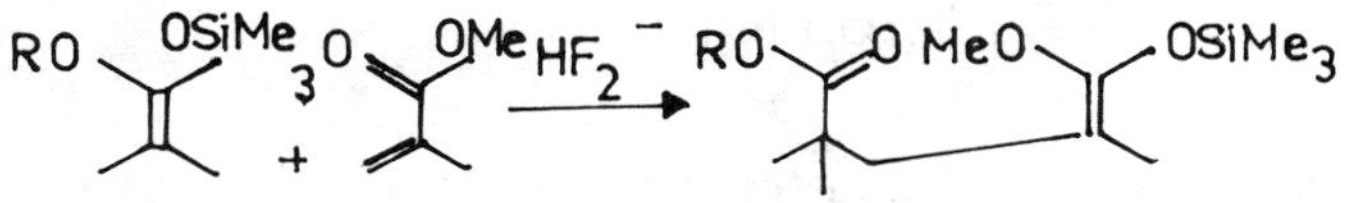

$$\text{(311)}$$

Polymerization is initiated by silylketene acetals, but a catalyst, e.g., a soluble bifluoride, is also required for the reaction to proceed.

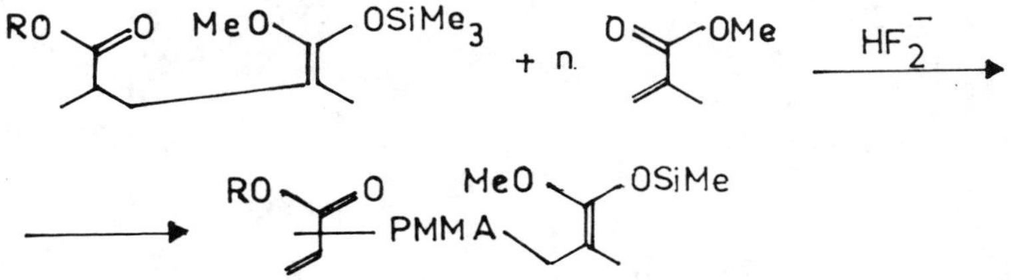

$$\text{(312)}$$

The most reasonable mechanism involves associative, intramolecular silyl transfer facilitated by coordination of the nucleophilic catalyst with the silicon:

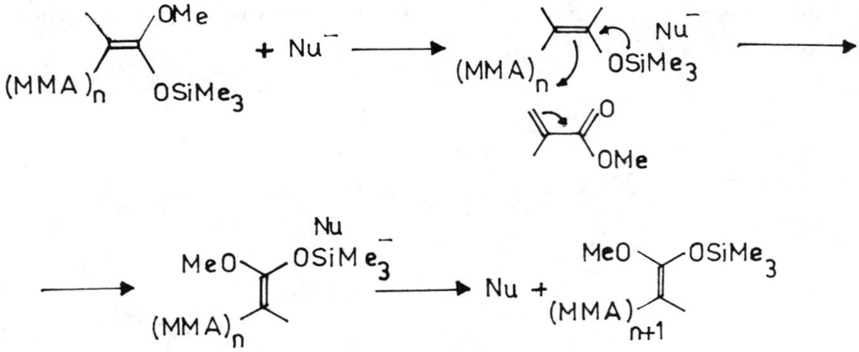

$$\text{(313)}$$

Group-transfer polymerization is a convenient route for the preparation of monofunctional and telechelic methycrylate polymers, as shown in Equations 314 to 317 for hydroxy-ended polymers.[601] The capped hydroxy-functional initiator only reacts at the enol ether silicon (Equation 314).

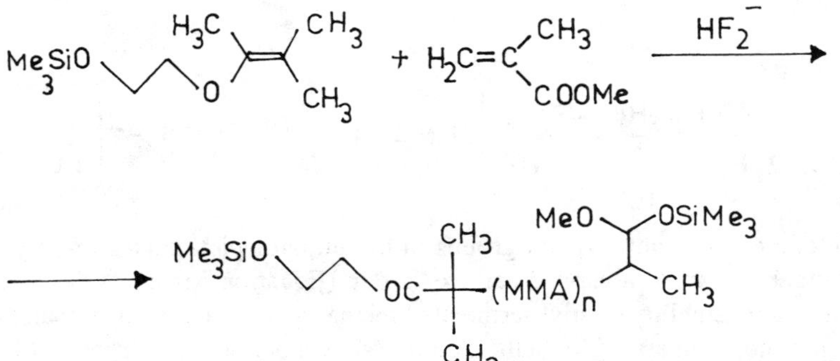

$$\text{(314)}$$

$$A \xrightarrow{\text{MeOH(F}^-)} HO \rightarrow PMMA \qquad (315)$$

Many electrophiles can be used as coupling agents for the preparation of telechelic polymers. Direct coupling of two silyl ketene acetal ends by Br_2-$TiCl_4$ gives polymers with a central head-to-head linkage (Equation 317).

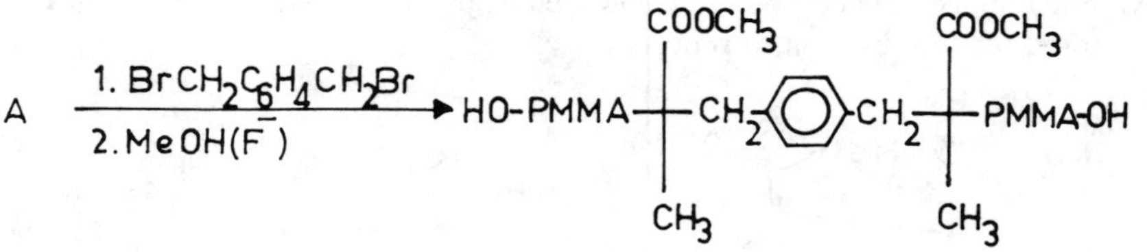

$$(316)$$

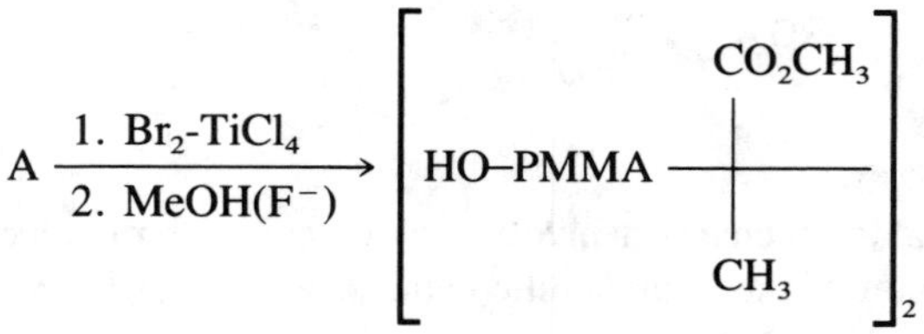

$$(317)$$

Using disilyl ketene acetals, carboxy-terminated polymers were prepared in an analogous manner (Equations 318 and 319).

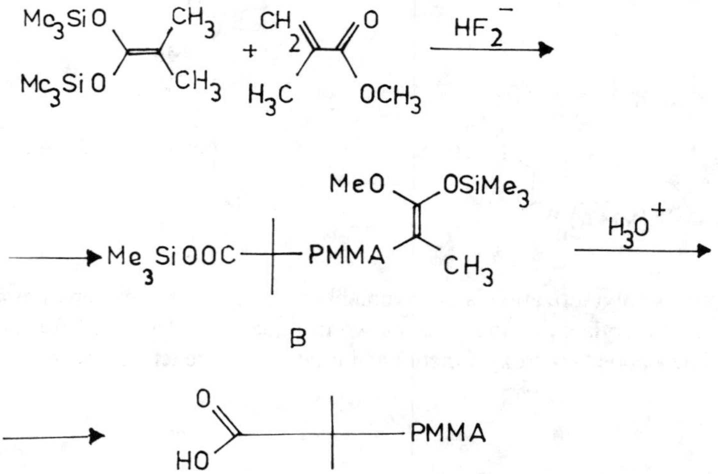

$$(318)$$

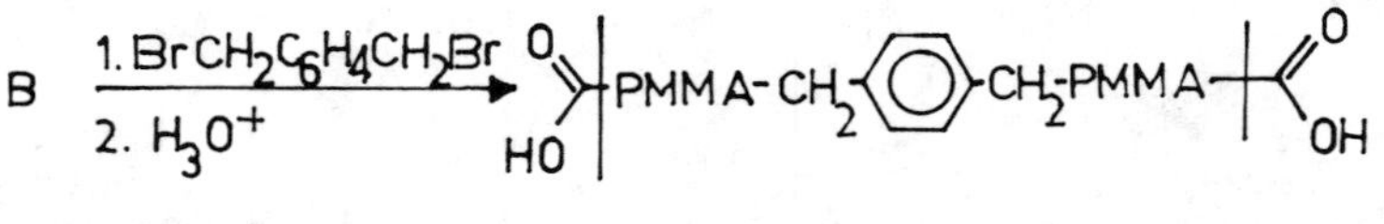

$$(319)$$

Again, only one of the trimethylsilyl groups in the initiator participates in polymerization, leaving the other as part of a capped carboxy group (Equation 318).

The outstanding stability of silyl-terminated methacrylates facilitates preparation of multiblock and graft copolymers.[594,602] In the example shown below,[594] addition of MMA (Equation 320) is followed, after complete polymerization, by addition of *n*-butyl methacrylate

(BMA, Equation 321) which, in turn, is followed by allyl methacrylate (AMA, Equation 322).

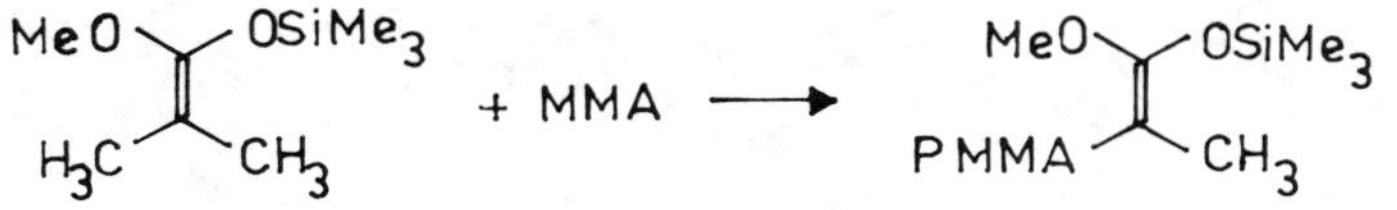

(320)

C

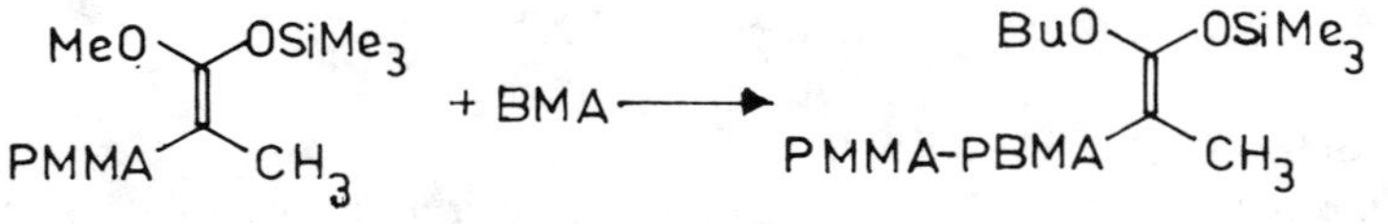

(321)

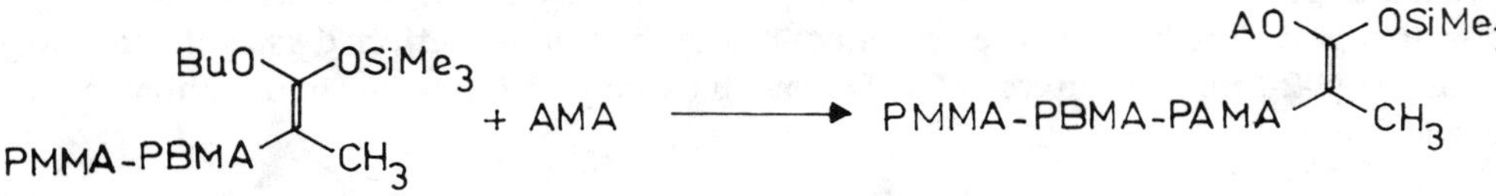

(322)

where $\overline{M}_n = 3800$, $\overline{M}_w = 4060$ (in theory 4060), and $T_g = -19$, $+38°$, and $+108°C$.

Excellent molecular weight control is maintained up to the end of the polymerization, opening the way for the design of carefully tailored polymers and the subsequent determination of structure-property correlations. This example also shows that group-transfer polymerization is not affected by unsaturation in the side chain: the soluble block copolymer formed in Equation 322 contained 11 mol% allyl methacrylate, enough to lead to severe gelation in a free-radical polymerization. Similar to anionic oligomerization, many silicon-masked functional groups can also be present, either as end groups or in the pendant groups.

Recently, PMMA macromonomers were synthesized by two more direct techniques.[603] In the first technique (Equation 323), living PMMA was simply reacted with p-vinylbenzylbromamide in the presence of a catalyst.

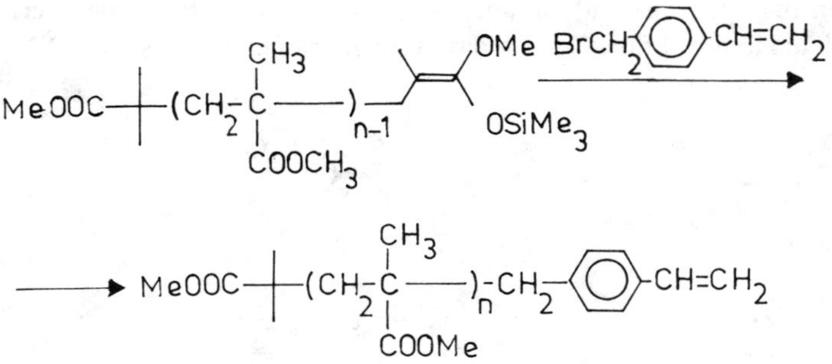

(323)

However, this method resulted in no more than 83% functionalization, whereas 100% functionalization was achieved using vinylphenylketene methylsilyl acetal as the initiator (Equation 324):

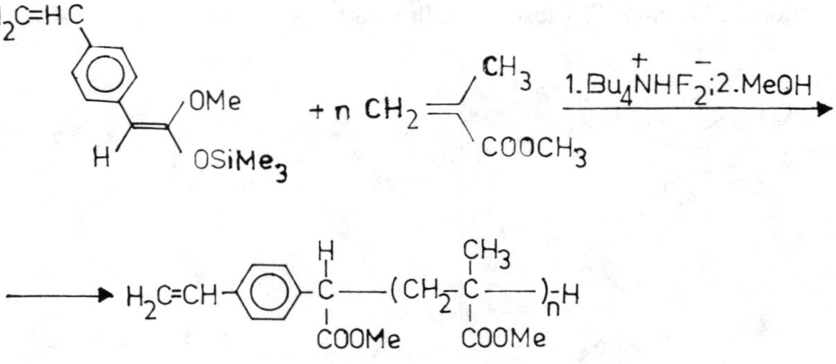

$$(324)$$

where $\overline{M}_n$ = 3600 and MWD = 1.09.

Besides the polymerization of polar monomers such as methacrylates, acrylates, and acrylamides, ring-opening polymerization of propylene sulfide using the group-transfer polymerization procedure has also been reported.[604] The reaction was effected using the initiator trimethylsilyl-2-trimethylsiloxyethyl sulfide and the catalyst trisdimethylaminosulfonium bifluoride:

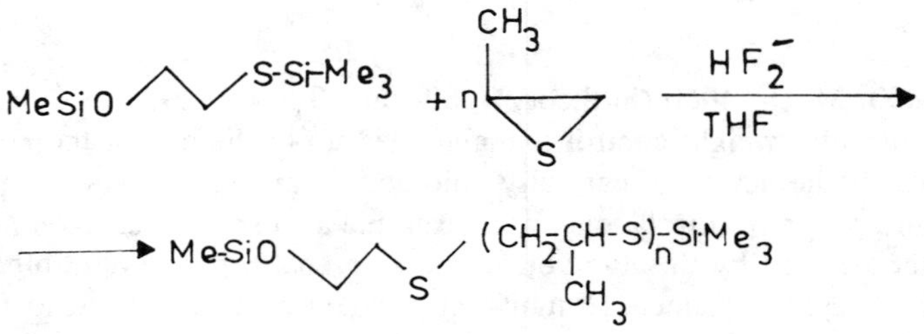

$$(325)$$

where $\overline{M}_n$ = 530 and MWD = 1.12.

B. MISCELLANEOUS METHODS

Miscellaneous methods of obtaining oligomers and monomers have been reviewed recently.[605] From the multitude of such techniques for preparing synthetic oligomers, one should mention the synthesis and cationic ring-opening of carbohydrate-derived dialkoxy-oxolanes which led to novel functionalized poly(oxytetramethylene)s with $\overline{DP}_n$ up to 35:[606,607]

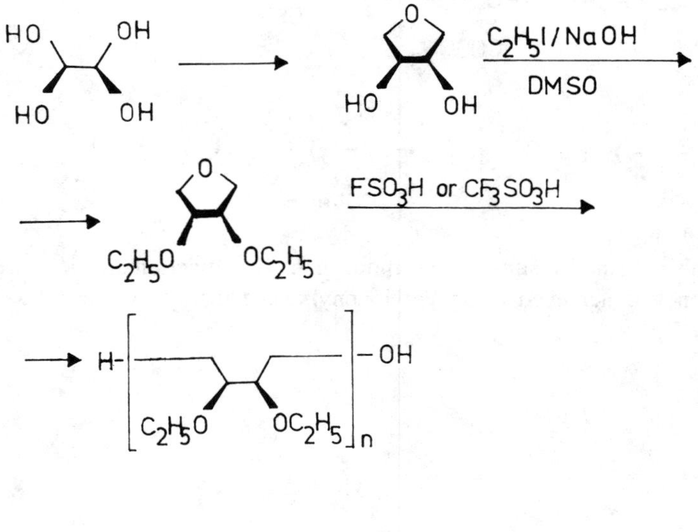

$$(326)$$

$\overline{M}_n$ = 2170.

These oligomers may be regarded as alternatives to poly(oxytetramethylene) prepared by cationic oligomerization of THF, i.e., an alternative to fossil raw materials (gas and petroleum, carbon-limited in the long run) currently used for the production of polymers.

REFERENCES

1. **Elias, H. G.**, *Macromolecules. Synthesis and Materials*, Vol. 2, Plenum Press, New York, 1977, chap. 18.
2. **Ledwith, A. and Sherrington, D. C.**, Kinetics of homogeneous cationic polymerization, in *Comprehensive Chemical Kinetics*, Vol. 15, Bamford, C. H. and Tipper, C. F. H., Eds., Elsevier, Amsterdam, 1976, chap. 2.
3. **Hiemenez, P. C.**, *Polymer Chemistry. The Basic Concepts*, Marcel Dekker, New York, 1984, 328.
4. **Collomb, J., Arland, P., Gandini, A., and Cheradame, H.**, Cationic polymerization and electrophilic reactions promoted by metal salts of strong acids, in *Cationic Polymerizations and Related Processes*, Goethals, E. J., Ed., Academic Press, London, 1984, 49.
5. **Collomb, J., Gandini, A., and Cheradame, H.**, Cationic polymerization induced by metal salts of strong acids: kinetics and mechanism, Prepr. 28th IUPAC Symp. Macromolecules, Amherst, July 12—16, 1982, 136.
6. **Crivello, J. V. and Lam, J. W. H.**, Diaryliodonium salts. A new class of photoinitiators for cationic polymerizations, *Macromolecules*, 10, 1307, 1977.
7. **Crivello, J. V.**, *Recent Progress in the Design of New Photoinitiators for Cationic Polymerization and Related Processes*, Goethals, E. J., Ed., Academic Press, London, 1986, 289.
8. **Tabata, Y.**, Vinyl polymerization, in *Kinetics and Mechanism of Polymerization*, Vol. 1, Part II, Ham, G. E., Ed., Marcel Dekker, New York, 1969, 305.
9. **Hayashi, K.**, Cation radicals via photoillumination and radiation processes, *J. Polym. Sci. Polym. Symp.*, 56, 490, 1976.
10. **Kennedy, J. P.**, *Cationic Polymerization of Olefins: A Critical Inventory*, John Wiley & Sons, New York, 1975, chap. 3, 4.
11. **Edwards, W. R. and Chanberlain, N. F.**, Carbonium ion of branched rearrangement in the cationic polymerization of α-olefins, *J. Polym. Sci. Part A*, 1, 2299, 1963.
12. **Kennedy, J. P., Minckler, L. S., Jr., Wanless, G. G., and Thomas, R. M.**, Intramolecular hydride shift polymerization by cationic mechanism. Introduction and structure analysis of poly(3-methylbutene-1), *J. Polym. Sci. Part A*, 2, 1441, 1964.
13. **Otto, M. and Müller-Cunradi, M.**, German Patent 641284, 1931.
14. **Otto, M. and Müller-Cunradi, M.**, U.S. Patent 2,203,873, 1937.
15. **Plesch, P. H.**, The mechanism of cationic polymerization, *Z. Electrochem.*, 30, 325, 1956.
16. **Marek, M., Pecka, J., and Halaska, V.**, New facts about the polymerization of isobutylene with Lewis acids, in *Cationic Polymerization and Related Processes*, Goethals, E. J., Ed., Academic Press, London, 1984, 17.
17. **Kennedy, J. P. and Thomas, R. M.**, Cationic polymerization at ultralow temperatures, *Adv. Chem. Ser.*, 34, 11, 1962.
18. **Kennedy, J. P. and Maréchal, E.**, *Carbocationic Polymerization*, Interscience, New York, 1982.
19. **Kennedy, J. P.**, New polymers and polymer derivatives by cationic techniques, 29th IUPAC Int. Symp. Macromolecules, Part 1, Plenary and Invited Lectures, Bucharest, 1983, 194 (and literature cited therein).
20. **Kennedy, J. P.**, Novel sequential copolymers by elucidating the mechanism of initiation and termination of carbocationic polymerization, *J. Polym. Sci. Polym. Symp.*, 56, 1, 1976.
21. **Kennedy, J. P.**, New telechelic prepolymers and networks therefrom, in *Cationic Polymerization and Related Processes*, Goethals, E. J., Ed., Academic Press, London, 1984, 335.
22. **Dontsov, A. A., Kanonzova, A. A., and Litvinova, T. V.**, *Rubber-Oligomeric Compositions for Manufacture of Rubber Products*, Khimia, Moscow, 1986 (in Russian).
23. **Minsker, K. S. and Sangalov, Yu. A.**, *Isobutylene and Its Polymers*, Khimia, Moscow, 1986 (in Russian).
24. **Ivan, B. and Kennedy, J. P.**, Quantitative aspects of chain extension of telechelics, *Polym. Bull.*, 2, 351, 1980.
25. **Kennedy, J. P.**, Macromolecular engineering by carbocationic polymerization, *Proc. Robert A. Welch Found. Chem. Res.*, 1983, 26, p. 71.
26. **Kennedy, J. P. and Smith, R. A.**, New telechelic polymers and sequential copolymers by polyfunctional initiator-transfer agents (INIFERS). I. Synthesis and characterization of, α,ω-di(t-chloro)polyisobutylene, *Polym. Prepr. Am. Chem. Soc. Div. Polym. Chem.*, 20, 316, 1979.

27. **Kennedy, J. P. and Smith, R. A.,** New telechelic polymers and sequential copolymers by polyfunctional initiator-transfer agents (INIFERS). III. Synthesis and characterization of poly(α-methylstyrene-b-isobutylene-b-α-methylstyrene), *J. Polym. Sci. Polym. Chem. Ed.,* 18, 1539, 1980.

28. **Kennedy, J. P. and Hiza, M.,** New telechelic polymers and sequential copolymers by polyfunctional initiator-transfer agents (INIFERS). XXX. Synthesis and quantitative terminal functionalization of α,ω-diarylpolyisobutylene, *J. Polym. Sci. Polym. Chem. Ed.,* 21, 3573, 1983.

29. **Kennedy, J. P., Ross, L. R., Lackly, J. L., and Nuyken, O.,** New telechelic polymers and sequential copolymers by polyfunctional initiator-transfer agents (INIFERS). X. Three-arm star telechelic polyisobutylene carrying chlorine, olefin or primary alcohol endgroups, *Polym. Bull.,* 4, 67, 1981.

30. **Ver Strate, G. and Baldwin, F. P.,** Terminally functional saturated hydrocarbon polymers produced directly by cationic polymerization, *Polym. Prepr. Am. Chem. Soc. Polym. Div.,* 17, 808, 1976.

31. **Ver Strate, G. and Baldwin, F. P.,** U.S. Patent 4,278,822, 1981.

32. **Kennedy, J. P., Guhaniyogi, S., and Ross, L. R.,** Three-arm star chlorine-telechelic polyisobutylene and star poly(isobutylene-b-α-methylstyrene), *Org. Coat. Appl. Polym. Sci.,* 46, 178, 1982.

33. **Kennedy, J. P., Guhaniyogi, S., and Percec, V.,** New telechelic polymers and sequential copolymers by polyfunctional initiator-transfer agents (INIFERS). XXVI. Bisphenol- and trisphenol-polyisobutylenes, *Polym. Bull.,* 8, 563, 1982.

34. **Kennedy, J. P., Guhaniyogi, S. C., and Percek, V.,** New telechelic polymers and sequential copolymers by polyfunctional initiator-transfer agents (INIFERS). XXVIII. Glycidyl ethers of bisphenol- and trisphenol-Pir₃, and their curing to epoxy resins, *Polym. Bull.,* 8, 571, 1982.

35. **Kennedy, J. P., Chang, V. S. C., Smith, R. A., and Ivan, B.,** New telechelic polymers and sequential copolymers by polyfunctional initiator-transfer agents (INIFERS). V. Synthesis of α-tert-butyl-ω-isopropenyl polyisobutylene and α,ω-di(isopropenyl)-polyisobutylene, *Polym. Bull.,* 1, 575, 1979.

36. **Kennedy, J. P., Chang, V. S. C., and Francik, W. P.,** New telechelic polymers and sequential copolymers by polyfunctional initiator-transfer-agents (INIFERS). XVIII. Epoxy and aldehyde telechelic polyisobutylenes, *J. Polym. Sci. Polym. Chem. Ed.,* 20, 2809, 1982.

37. **Mohajer, Y., Tyagi, D., Wilkes, G. L., Storey, R. F., and Kennedy, J. P.,** New polyisobutylene-based model isomers. III. Further mechanical and structural studies, *Polym. Bull.,* 8, 47, 1982.

38. **Kennedy, J. P. and Storey, R. F.,** New polyisobutylene-based ionomers: synthesis and model experiments, *Org. Coat. Appl. Polym. Sci.,* 46, 182, 1982.

39. **Ivan, B., Kennedy, J. P., and Chang, V. S. C.,** New telechelic polymers and sequential copolymers by polyfunctional initiator-transfer agents (INIFERS). VII. Synthesis and characterization of α,ω-(dihydroxy)polyisobutylene, *J. Polym. Sci. Polym. Chem. Ed.,* 18, 3177, 1980.

40. **Wondraczek, R. H. and Kennedy, J. P.,** New telechelic polymers and sequential copolymers by polyfunctional initiator-transfer agents (INIFERS). XI. Synthesis extension and crosslinking of oxycarbonyl isocyanate telechelic polyisobutylenes, *Polym. Bull.,* 4, 445, 1981.

41. **Percec, V., Guhaniyogi, S. C., and Kennedy, J. P.,** New telechelic polymers and sequential copolymers by polyfunctional initiator-transfer agents (INIFERS). XXXVI. Synthesis and characterization of various nitrile-telechelic polyisobutylenes, *Polym. Bull.,* 10, 31, 1983.

42. **Chang, V. S. C. and Kennedy, J. P.,** Optimizing urethane synthesis by studying the kinetics of reactions between a model alcohol and isocyanates in various solvents, *Polym. Bull.,* 9, 479, 1983.

43. **Chang, V. S. C. and Kennedy J. P.,** Gas permeability, water absorption, hydrolytic stability and air-oven ageing of polyisobutylene-based polyurethane networks, *Polym. Bull.,* 8, 69, 1982.

44. **Percec, V., Guhaniyogi, S. C., and Kennedy, J. P.,** New telechelic polymers and sequential copolymers by polyfunctional initiator transfer agents (INIFERS). XXIX. Synthesis of α,ω-di(amino-polyisobutylenes, *Polym. Bull.,* 9, 27, 1983.

45. **Nagy, A., Faust, R., and Kennedy, J. P.,** New telechelic polymer and sequential copolymers by polyfunctional initiator-transfer agents (INIFERS). XXXXIV. End-reactive three-arm star polyisobutylenes by continuous polymerization, *Polym. Bull.,* 13, 97, 1985.

46. **Kennedy, J. P. and Smith, A, R.,** New telechelic polymers and sequential copolymers by polyfunctional initiator-transfer agents (INIFERS). II. Synthesis and characterization of α,ω-di(tert-chloro) polyisobutylenes, *J. Polym. Sci. Polym. Chem. Ed.,* 18, 1523, 1980.

47. **Nuyken, O., Pask, S. D., Vischer, A., and Walter, M.,** Some detailed observations on the inifer technique, in *Cationic Polymerization and Related Processes,* Goethals, E. J., Ed., Academic Press, London, 1984, 35.

48. **Pask, S. D., Nuyken, O., Vischer, A., and Walter, M.,** Equilibria and cationic polymerization, in *Cationic Polymerization and Related Processes,* Goethals, E. J., Ed., Academic Press, London, 1984, 25.

49. **Nuyken, O., Pask, S. D., and Walter, M.,** Polymerization of isobutylene by inifer technique. II. Products using cumyl chloride + BCl₃, *Polym. Bull.,* 8, 451, 1982.

50. **Nuyken, O., Pask, S. D., Vischer, A., and Walter, M.,** α,ω-dicholorpoly(e-methylpropene) as an inifer: a modification to the Kennedy-Smith mechanism, *Makromol. Chem..,* 186, 173, 1985.

51. **Nuyken, O. and Pask, S. D.,** Polymerization of isobutene by the inifer technique. III. The effect of 2,6-di-*t*-butyl pyridine on polymerization using TCC + BCl₃, *Polym. Bull.,* 8, 475, 1982.

52. **Nuyken, O., Pask, S. D., and Vischer, A.,** Polymerization of isobutene by the inifer technique. I. Low molecular weight products using 1,4-bis(1-chloro-1-methylethyl)benzene/BCl₃ as initiator, *Makromol. Chem.,* 184, 553, 1983.

53. **Freyer, C . V., Muehlbaker, H. P., and Nuyken, O.,** Telechele des Methylpropene durch kationische Polymerization, *Agnew. Makromol. Chem.,* 145/146, 69, 1986.

54. **Nuyken, O., Pask, S. D., Vischer, A., and Walter, M.,** Recent progress towards an understanding of the inifer mechanism, *Polym. Prepr. Am. Chem. Soc. Div. Polym. Chem.,* 26(1), 44, 1985.

55. **Pask, S. D. and Plesch, P. H.,** Developments in the theory of cationic polymerization. VII. Theoretical attempts at improving initiators for cationic polymerization of alkenes, *Eur. Polym, J.,* 18, 839, 1982.

56. **Kennedy, J. P., Ross, L., Lackey, J. E., and Nuyken, O.,** New telechelic polymers and sequential copolymers by polyfunctional initiator-transfer agents (INIFERS). X. Three-arm star telechelic polyisobutylene carrying chlorine olefin or primary alcohol end groups, *Polym. Bull.,* 4, 67, 1981.

57. **Dittmer, T., Gruber, F., and Nuyken, O.,** Cationic polymerization of bis(1-alkylvinyl)benzenes and related monomers. I. Structural elucidation of 1,1,3-trimethyl substituted polyindane, *Makromol. Chem.,* 190, 1755, 1989.

58. **Dittmer, T., Gruber, F., and Nuyken, O.,** Cationic polymerization of bis(1-alkylvinyl)benzenes and related monomers. II. Controlled synthesis of 1,1,3-trimethyl substituted polyindanes, *Makromol. Chem.,* 190, 1771, 1989.

59. **Tessier, M., Hung, N. A., and Maréchal, E.,** Synthesis of mono- and difunctional oligoisobutylenes. I. Synthesis of mono- and dichloroisobutylenes, *Polym. Bull.,* 4, 111, 1981.

60. **Kennedy, J. P., Feinberg, S. C., and Huang, S. T.,** BCl₃ rediscovered: an efficient coinitiator for cationic olefin polymerization and for synthesis of isobutylene-styrene block copolymers, *Polym. Prepr. Am. Chem. Soc. Div. Polym. Chem.,* 17(1), 194, 1976.

61. **Feinberg, S. C. and Kennedy, J. P.,** The cationic polymerization of styrene using boron halides and the synthesis of styrene-isobutylene block copolymers using BCl₃, *Polym. Prepr. Am. Chem. Soc. Div. Polym. Chem.,* 17(2), 797, 1976.

62. **Tessier, M. and Maréchal, E.,** Synthesis of macro- and di-functional oligoisobutylenes. II. Structure of α,ω-dichloro-oligoisobutylenes obtained by inifer method, *Polym. Bull.,* 10, 152, 1983.

63. **Taha, M., Rigal, G., Pietrasanta, T., Platzer, N., Sudres, P., and Raynal, S.,** Télomérisation de l'isobutylene, *Makromol. Chem.,* 182, 2545, 1981.

64. **Clonet, F., Zurick, V., Franta, E., and Brossas, J.,** Cationic telomerisation of isobutylene, *Polym. Bull.,* 7, 449, 1982.

65. **Onopechenko, A., Cupples, B. L., and Kresge, A. N.,** Boron-fluoride oligomerization of alkenes: structures, mechanisms and properties, *Ind. Eng. Chem. Prod. Res. Div.,* 22, 182, 1983.

66. **Onopechenko, A., Cupples, B. L., and Kresge, A. N.,** The boron fluoride-catalyzed oligomerization of alkenes-structures, mechanisms and properties, *Prepr. Am. Chem. Soc. Div. Pet. Chem.,* 27(2), 331, 1982.

67. **Priola, A., Fattore, V., Arighetti, S., and Mancini, G.,** European Patent Appl. 68,554, 1981; *Chem. Abstr.,* 98, 163757q, 1983.

68. **Dressler, F. H. and Vermaine, S.,** The cationic oligomerization of C-10 Fischer-Tropsch olefins, *Makromol. Chem. Makromol. Symp.,* 13/14, 271, 1988.

69. **Langer, A. W., Steger, J., and Burkhardt, T. J.,** Linear olefin products, U.S. Patent 4,361,714, 1981; *Chem. Abstr.,* 98, 35150z, 1983.

70. **Ionescu, E., Vasilescu, S. D., Matache, S., Toc, V., Cobianu, N., Kercinschi, V., Avram, T., Măciucă, E., Oprea, S., and Mărculescu, B.,** Romanian Patent 86,030, 1985; *Chem. Abstr.,* 104, 235459x, 1986.

71. **Mel'nikov, V. N., Sycheva, O. A., Matkovskii, P. E., and Moreva, T. A.,** Effect of the catalyst composition on the rate and selectivity of low temperature ethylene oligomerization, *Khim. Prom. (Moscow),* 6, 323, 1986 (in Russian).

72. **Marek, M., Roman, M., Doskocilova, D., and Pokorny, S.,** The cationic polymerization of 3-chloro-2-methyl-1-propene *Makromol. Chem.,* 187, 2337, 1986.

73. **Andisio, G., Priola, A., and Rossini, A.,** Structure of dimers and trimers obtained in the cationic oligomerization of propylene, *Makromol. Chem.,* 189, 111, 1988.

74. **Pape, W., Hartung, H., Henlein, G., Stadermann, D., Sangalov, Yu. A., Yasman, Yu. B., Nelkenbaum, Yu. Ya., and Minsker, K. A.,** On the separation of C₄-hydrocarbon mixtures by oligomerization and polymerization. I. Proton acid complexes as initiators for isobutene oligomerization, *Acta Polym.,* 35, 677, 1984.

75. Compagnie Francaise de Raffinage, French Patent 1,337,232, 1963.

76. Farbenfabriken Bayer AG, German Patent 1,720,779, 1968.

77. **Watson, J. M.,** U.S. Patent 3,985,822, 1976.

78. **Sangalov, Yu. A., Yasman, Yu. B., Valeev, F. A., and Minsker, K. S.,** Arenonium ions in Gustavson complexes as initiators of electrophylic polymerization of α-olefins, *Vysokomol., Soedin. Ser. A,* 20, 1339, 1978.

79. **Sangalov, Yu. A., Yasman, Yu. B., Nelkenbaum, E. M., Badretdinova, V. Ya., and Minsker, K. S.,** On the activity of ethyl aluminium chlorides in the presence of HCl in the polymerization of isobutylene, *Vysokomol. Soedin. Ser. A,* 22, 1588, 1980.

80. **Sangalov, Yu. A., Ponomaryov, O. A., Nelkenbaum, Yu. Ya., Romanko, V. G., Petrova, V. D., and Minsker, K. S.,** On the nature of the activity of the systems $C_2H_5AlCl_2$-alcohol in cationic polymerization of olefins, *Vysokomol. Soedin. Ser. A,* 22, 1331, 1978.

81. **Razzank, H., Bouridah, K., Gandini, A., and Cheradame, H.,** Functionalized oligo- and polyisobutylenes, in *Cationic Polymerization and Related Processes,* Goethals, E. J., Academic Press, London, 1984, 355.

82. **Faust, R., Nagy, A., and Kennedy, J. P.,** Living carbocationic polymerization. V. Linear telechelic polyisobutylenes by bifunctional initiators, *J. Macromol. Sci. Chem.,* A24, 595, 1987.

83. **Storey, R. F. and Lee, Y.,** Living carbocationic polymerization of three-arm star polyisobutylene, *Polym. Prepr. Am. Chem. Soc. Div. Polym. Chem.,* 30(2), 162, 1989.

84. **Mishra, N. K. and Kennedy, J. P.,** Living carbocationic polymerization of isobutylene by tertiary alkyl (or aryl) methyl ether/boron trichloride complexes, *J. Macromol. Sci. Chem.,* A24, 933, 1987.

85. **Boettcher, F. P.,** Recent progress in the preparation of functional methacrylate polymers, *J. Macromol. Sci. Chem.,* A22, 665, 1985.

86. **Faust, R. and Kennedy, J. P.,** Living carbocationic polymerization. III. Demonstration of the living polymerization of isobutylene, *Polym. Bull.,* 15, 317, 1986.

87. **Nagy, A., Faust, R., and Kennedy, J. P.,** New telechelic polymers and sequential copolymers by polyfunctional initiator transfer agents (INIFERS). XXXXIV. End-reactive three-arm star polyisobutylenes by continuous polymerization, *Polym. Bull.,* 13, 97, 1985.

88. **Faust, R. and Kennedy, J. P.,** Living carbocationic polymerization. XXI. Kinetic and mechanistic studies of isobutylene polymerization initiated by trimethylpentyl esters of different acids, *Polym. Prepr. Am. Chem. Soc. Div. Polym. Chem.,* 29(2), 69, 1988.

89. **Findley, J.,** Petroleum resins, in *Encyclopedia of Polymer Science and Technology,* Vol. 9, Mark, H. F., Ed., John Wiley & Sons, New York, 1968, 853.

90. **Elias, H. G.,** *Macromolecules. Synthesis and Materials,* Plenum Press, New York, 1977, 877.

91. **Kuntz, I., Powers, K. W., Hsu, C. S., and Rose, K. D.,** Cyclic oligomer formation in the copolymerization of isoprene with isobutylene, *Makromol. Chem. Makromol. Symp.,* 13/14, 337, 1988.

92. **Atamanenko, O. P., Nelkenbaum, E. M., Yasman, Yu. B., Sangalov, Yu. A., Sokolova, N. P., Taits, S. Z., Dumsku, Yu. V., and Vasserberg, V. G.,** Oligomerization of piperylene and piperylene fraction in the presence of aluminium chloride based complex catalysts, *Neftechimyia,* 27, 87, 1987.

93. **Pantukh, B. I., Rozentsvet, V. A., Nagimov, V. G., and Moiseev, V. D.,** Russian Patent 1,229,205, 1986; *Chem. Abstr.,* 105, 116637d, 1986.

94. **Heublein, G., Freitag, W., and Mock, B.,** Copolymerization von Cyclopentadiene mit Butadien, *Acta Polym.,* 30, 446, 1979.

95. **Hasegawa, K., Asami, R., and Higashimura, T.,** Cationic polymerization of alkyl-1,3-butadienes, *Macromolecules,* 10, 522, 1977.

96. **Higashimura, T. and Hasegawa, H.,** Monomer isomerization oligomerization of 2-ethyl-1,3-butadiene by acid catalysts, *J. Polym. Sci. Polym. Chem. Ed.,* 17, 2439, 1979.

97. **Sperri, P. E. and Rosen, M. J.,** Studies on the structure and oxidation products of 1-methyl-3-phenylindane, a dimer of styrene, *J. Am. Chem. Soc.,* 72, 4918, 1950.

98. **Roesen, M. J.,** The dimerization of styrene in aqueous sulfuric acid, *J. Org. Chem.,* 18, 1701, 1953.

99. **Corson, B. B., Heintzelman, W. J., Moe, H., and Rousseau, C. R.,** Reactions of styrene dimers, *J. Org. Chem.,* 27, 1636, 1962.

100. **Hamaya, T.,** Oligomerization of styrene. III. Formation of styrene oligomers with a trifluoroacetyl endgroups, *Makromol. Chem. Rapid Commun.,* 3, 953, 1982.

101. **Sawamoto, M., Masuda, T., Nishii, H., and Higashimura, T.,** Selective dimerization of styrene to 1,3-diphenyl-1-butene catalyzed by trifluoromethane sulfonic acid or acetyl perchlorate, *J. Polym. Sci. Polym. Lett. Ed.,* 13, 279, 1975.

102. **Higashimura, T. and Nishii, H.,** Selective dimerization of styrene to 1,3-diphenyl-1-butene by acetyl perchlorate, *J. Polym. Sci. Polym. Chem. Ed.,* 15, 329, 1977.

103. **Nishihara, K.,** Oligomerization of styrene by trichloroacetyl chloride, *Makromol. Chem.,* 136, 201, 1970.

104. **Hamaya, T. and Tamada, S.,** Selective formation of styrene trimers 1,3,5-triphenyl-1-hexene, *Makromol. Chem.,* 180, 2979, 1979.

105. **Nishii, H. and Higashimura, T.,** Dimerization of methylstyrenes by acetyl perchlorate and trifluoromethane sulfonic acid, *J. Polym. Sci. Polym. Chem. Ed.,* 15, 1179, 1977.

106. **Hasegawa, H. and Higashimura, T.,** Synthesis of linear poly(divinylbenzene) through proton-transfer polyaddition by oxo acids, *Macromolecules,* 13, 1350, 1980.

107. **Higashimura, T.,** Cationic polymerization and oligomerization via dissociated and non-dissociated propagating species, *Polym. Prepr. Am. Chem. Soc. Div. Polym. Chem.,* 20, 161, 1979.

108. **Hasegawa, H. and Higashimura, T.,** Cationic oligomerization of styrene by solid acids. I. Catalysis by poly(styrene suflonic acids) resin, *J. Polym. Sci. Polym. Chem. Ed.,* 18, 611, 1980.
109. **Hayes, M. J. and Pepper, D. C.,** The polymerization of styrene by sulfuric acid. II. The kinetics of polymerization in 1,2-dichloroethane, *Proc. R. Soc. London Ser. A,* 263, 63, 1961.
110. **Nasuse, Y., Sakai, S., Suzuki, T., and Kono, Y.,** Japanese Patent 62,42,938, 1987; *Chem. Abstr.,* 107, 134863o, 1987.
111. **Naruse, Y., Sakai, S., Miagawa, T., and Oshisua, S.,** European Patent Appl. 202,965, 1986; *Chem. Abstr.,* 106, 102826y, 1987.
112. **Sugawara, H. and Okawai, T.,** Japanese Patent 61,243,030, 1986; *Chem. Abstr.,* 107, 7802z, 1986.
113. **Kimura, T., Mori, M., and Hamashima, M.,** Cationic oligomerization of α-methylstyrene by chloroacetic acids, *Kobunshi Ronbunshu,* 43, 9, 1986; *Chem. Abstr.,* 104, 149474f, 1986.
114. **Kiji, J., Konishi, H., Okano, T., and Ajiki, E.,** *p*-Chlorostyrene oligomers. Preparation, structure and stereochemistry, *Angew. Makromol. Chem.,* 149, 189, 1987.
115. **Abdel-Koder, M., Padias, A. B., and Hall, H. K., Jr.,** Cationic polymerization of nitrogen-containing electron-rich vinyl monomers by electrophilic olefins and their cyclobutane cycloadducts, *Macromolecules,* 20, 944, 1987.
116. **Trofimov, B. A., Michaleva, A. I., Morozova, L. V., Vasil'ev, A. N., and Sigalov, M. V.,** Dimerization of 1-vinyl-4,5,6,7-tetrahydroindole under the action of HGl, *Khim. Geterotsikl. Soedin.,* 2, 269, 1983.
117. **Trofimov, B. A., Mikhaleva, A. I., and Morozova, L. V.,** Polymerization of N-vinylpyrroles, *Usp. Khim.,* 54, 1034, 1985.
118. **Priola, A., Gatti, G., and Cesca, S.,** Polymerization of 1-vinylindole and its methyl derivatives—structure of the polymers, *Makromol. Chem.,* 180, 1, 1979.
119. **Trofimov, B. A., Morozeva, L. V., Sigalov, M. V., Mikhaleva, A. I., and Markova, M. V.,** An unexpected mode of cationic oligomerization of 1-vinyl-4,5,6,7-tetrahydroindole, *Makromol. Chem.,* 188, 2551, 1987.
120. **Hayashi, K. and Yamamoto, Y.,** Radiation-induced cationic oligomerization of α-methylstyrene in hydrocarbon solutions, *J. Macromol. Sci. Chem.,* A21, 905, 1984.
121. **Crivello, J. V.,** Belgian Patent 826,670, 1974.
122. **Smith, G. H.,** Belgian Patent 828,841, 1975.
123. **Crivello, J. V. and Laru, J. V. H.,** Diaryliodonium salts. A new class of photoinitiators for cationic polymerization, *Macromolecules,* 10, 1307, 1977.
124. **Lapin, S. C.,** Radiation induced cationic curing of vinyl ether functionalized urethane oligomers, *Polym. Mater. Sci. Eng.,* 60, 233, 1989.
125. **Hult A. B. and Sundell, P. E.,** Mechanism of electron-beam induced cationic polymerization of divinylethers, *Polym. Mater. Sci. Eng.,* 60, 453, 1989.
126. **Lapin, S. C. and House, D. W.,** U.S. Patent 4,751,273, 1988.
127. **Sawamoto, M., Enoki, T., and Higashimura, T.,** End-functionalized polymers by living cationic polymerization. II. Vinyl ether macromers with a poly(vinyl ether) backbone, *Polym. Bull.,* 16, 117, 1986.
128. **Sawamoto, M., Takashi, E., and Higashimura, T.,** End-functionalized polymers by living cationic polymerization. I. Mono- and bifunctional poly(vinyl ethers) with terminal malonate or carboxyl groups, *Macromolecules,* 20, 1, 1987.
129. **Higashimura, T., Aoshima, S., and Sawamoto, M.,** Living cationic polymerization of vinyl monomers: new developments, *Polym. Prepr. Am. Chem. Soc. Div. Polym. Chem.,* 29(2), 1, 1988; **Miyamoto, M., Sawamoto, M., and Higashimura, T.,** Living polymerization of isobutylvinyl ether with the hydrogen iodide/iodine initiating system, *Macromolecules,* 17, 265, 1984.
130. **Aoshima, S. and Higashimura, T.,** Living cationic polymerization of vinyl monomers by organoaluminium halides. I. EtAlCl$_2$/ester initiating systems for living polymerization of vinyl ethers, *Polym. Bull.,* 15, 417, 1986.
131. **Nuyken, O. and Kröner, H.,** Advances in living cationic polymeirzation of vinyl ethers, *Polym. Prepr. Am. Chem. Soc. Div. Polym. Chem.,* 29(2), 87, 1988.
132. **Minoda, M., Sawamoto, M., and Higashimura, T.,** Sequence-regulated oligomers and polymers by cationic polymerization. I. Synthesis of sequence-regulated trimers and tetramers of functional vinyl ethers, *Polym. Bull.,* 23, 133, 1990.
133. **Nenițescu, C. D.,** *Organic Chemistry,* Vol. 1, EDT, Bucharest, 1974.
134. **Hall, H. K., Jr. and Ykman, P.,** Addition polymerization of cyclobutene and dicyclobutene monomers, *J. Polym. Sci. Macromol. Rev.,* 11, 1, 1976.
135. **Lawrence, C. D. and Tipper, C. F. H.,** Some reactions of cyclopropane and a comparison with the lower olefins, *J. Chem. Soc.,* 713, 1955.
136. **Pinazzi, C. P., Brosse, J. C., Pleurdeau, A., and Brossas, J.,** Polymerisation des dérivés du cyclopropane, *Polym. Prepr. Am Chem. Soc. Div. Polym. Chem.,* 13(1), 445, 1972.
137. **Kennedy, J. P.,** *Cationic Polymerization of Olefins: A Critical Inventory,* Interscience, 1974, 204.

138. **Tipch, C. F. H. and Walker, D. A.,** Some reactions of cyclopropane and a comparison with lower olefins. IV. Friedel-Crafts polymerization, *J. Chem. Soc.,* p. 1352, 1959.

139. **Naegele, W. and Haubenstock, H.,** Cationic oligomerization of isopropylcyclopropane, *Tetrahedron Lett.,* 48, 4283, 1965.

140. **Ketley, A. D.,** The cationic polymerization of 3-methylbutene-1 and 1,1-dimethylcyclopropane, *Polym. Lett.,* 1, 313, 1963.

141. **Aoky, S., Harita, Y., Otsu, F., and Ymoto, M.,** The attempted cationic polymerization at 1,1-dimethylcyclopropane and phenylcyclopropane, *Bull. Soc. Chem. Jpn.,* 39, 189, 1966.

142. **Pinazzi, C. P., Brosse, J. P., and Pleurdeau, A.,** Polymerization of bicyclo n.1.0 alkanes, *Makromol. Chem.,* 144, 155, 1971.

143. **Pinazzi, C. P., Pleurdeau, A., and Brosse, J. C.,** Polymérisation de dérivés du cyclopropane, *Prepr. IUPAC Int. Symp. Macromolecules,* Budapest, 1969, Preprints, 2, 273.

144. **Goethals, E. J.,** Cyclic oligomers in the cationic polymerization of heterocycles, *Pure Appl. Chem.,* 48, 335, 1976.

145. **Goethals, E. J.,** The formation of cyclic oligomers in the cationic oligomerization of heterocycles, *Adv. Polym. Sci.,* 23, 103, 1977.

146. **Worsfold, D. J. and Eastham, A. M.,** Cationic polymerization of ethylene oxide. II. BF_3, *J. Am. Chem. Soc.,* 79, 900, 1957.

147. **Latremouille, G. A., Merall, G. T., and Eastham, A. M.,** The cationic polymerization of ethylene oxide. III. Depolymerization of polyglycols by oxonium fluoroborates, *J. Am. Chem. Soc.,* 82, 120, 1960.

148. **Penczek, S., Kubisa, P., Matyjaszewski, K., and Szymanski, R.,** Structure and reactivities in the ring-opening and vinyl cationic polymerization, in *Cationic Polymerization and Related Processes,* Goethals, E. J., Ed., Academic Press, London, 1984.

149. **Kobayashi, S., Morikana, K., and Saegusa, T.,** Superacids and their derivatives. IX. Selective cyclodimerization of ethylene oxide to 1,4-dioxane catalyzed by superacids and their derivatives, *Macromolecules,* 8, 952, 1975.

150. **Libiszowski, J., Szymanski, R., and Penczek, S.,** On the cationic polymerization of oxirane with triphenylmethylium salts, *Makromol. Chem.,* 190, 1225, 1989.

151. **Dale, J., Bergen, G., and Daasvatn, K.,** The oligomerization of ethylene oxide to macrocyclic ethers, including 1,4,7-trioxacyclononane, *Acta. Chem. Scand. Ser. B,* 28, 378, 1974.

152. **Dale, J., Daasvatn, K., and Gronneberg, T.,** The mechanism of cationic cyclooligomerization and polymerization of ethylene oxide, *Makromol. Chem.,* 178, 873, 1977.

153. **Kobayashi, S., Kobayashi, T., and Saegusa, T.,** Cationic oligomerization of ethylene oxide, *Polym. J.,* 15, 883, 1983.

154. **Colchough, R. O., Oze, G., Higginson, W. C. E., Jackson, J. B., and Litt, M.,** The polymerization of epoxides by metal halide catalysts, *J. Polym. Sci.,* 34, 171, 1959.

155. **Pasika, W. M.,** Dimer formation in the trityl cationic polymerization of styrene oxide, *J. Polym. Sci. Part A,* 3, 4287, 1965.

156. **Kondo, S. and Blanchard, L. P.,** Cyclic dimer of styrene oxide in cationic copolymerization, *J. Polym. Sci. Polym. Lett.,* 7, 621, 1969.

157. **Stvelichkova, R. and Tsevi, R. A.,** Polymerization of styrene oxide by nitronium tetrafluoroborane, *J. Polym. Sci. Polym. Chem. Ed.,* 24, 3399, 1986.

158. **Katnik, R. J. and Schaefer, J.,** Structural isomer in ring polymers of propylene oxide, *J. Org. Chem.,* 33, 384, 1968.

159. **Kern, R. J.,** Twelve-membered polyether rings. The cyclic tetramers of some olefin oxides, *J. Org. Chem.,* 33, 388, 1968.

160. **Colchough, R. O. and Wilkinson, K.,** Polymerization of propylene oxide catalyzed by trimethyl aluminium, *J. Polym. Sci. Part C,* 4, 311, 1966.

161. **Katnik, R. J. and Schaefer, J.,** Structural isomer distribution in ring polymers of propylene oxide, *J. Org. Chem.,* 33, 384, 1968.

162. **Haubenstock, H., Swanson, D. B., and Lutz, P.,** Cationic oligomerization of isopropyloxirane, *Polymer,* 29, 1335, 1988.

163. **Dreyfuss, M. P.,** U.S. Patent 3,850,856, 1974.

164. **Young, C. L. and Barber, L. L.,** British Patent Appl., 2,021,606, 1979.

165. **Aelany, D.,** U.S. Patent 4,284,826, 1981.

166. **Simon, H. Yu.,** Characterization of hydroxyl-terminated liquid polymers of epichlorohydrin, *Polym. Prepr. Am. Chem. Soc. Div. Polym. Chem.,* 25(1), 117, 1984.

167. **Okamoto, Y.,** Cationic ring-opening polymerization of epichlorohydrin in the presence of ethylene glycol, *Polym. Prepr. Am. Chem. Soc. Div. Polym. Chem.,* 25(1), 264, 1984.

168. **Slomkovski, S.,** Ring-opening polymerization with kinetically controlled enhancement of macrocyclic oligomers on linear macromolecules, *Makromol. Chem.,* 186, 2581, 1985.

285

169. **Bello, A., Perena, J. M., and Perez, E.,** Spanish Patent 527,757, 1985; *Chem. Abstr.,* 107, 59727y, 1987.
170. **Penczek, S.,** Cationic ring-opening polymerization via activated monomer, in *Prepr. IUPAC Macro. 89,* Kyoto, 1988, 56.
171. **Penczek, S.,** Reactive polyethers by cationic activated monomer mechanism, *Polym. Prepr. Am. Chem. Soc. Polym. Chem. Div.,* 29(2), 38, 1988.
172. **Wojtania, M., Kubisa, P., and Penczek, S.,** Polymerization of propylene oxide by activated monomer mechanism. Suppression of macrocyclic formation, *Makromol. Chem. Makromol. Symp.,* 6, 201, 1986.
173. **Kubisa, P.,** Oligomerization of propylene oxide by activated monomer mechanism, *Makromol. Chem. Makromol. Symp.,* 13/14, 203, 1988.
174. **Mueller, H.,** European Patent 307,811, 1989; *Chem. Abstr.,* 111, 97987y, 1989.
175. **Harris, J. M.,** Laboratory synthesis of polyethylene glycol derivatives, *J. Macromol. Sci. Rev. Macromol. Chem.,* C25, 325, 1985.
176. **Matsuyama, A. Ozawa, H. and Hirose, S.,** Japanese Patent 61,195,158, 1986; *Chem. Abstr.,* 107, 199123y, 1987.
177. **Matsuyama, A., Ozawa, H., and Hirose, S.,** Japanese Patent 61,195,158, 1986; *Chem. Abstr.,* 107, 199122x, 1987.
178. **Matsuyama, A., Ozawa, H., and Hirose, S.,** Japanese Patent 61,195,159, 1986; *Chem. Abstr.,* 106, 120404p, 1987.
179. **Bratychak, M. N., Vastres, V. B., and Puchin, V. D.,** Synthesis of peroxide oligomers by cationic telomerization, *Plast. Massy,* No. 12, 21, 1985.
180. **Gonzales, C. C., Bello, A., and Perena, J. M.,** Oligomerization of exetane and synthesis of polyterephthalates derived from 1,3-propanediol and 3,3'-oxidipropanol, *Makromol. Chem.,* 190, 1217, 1989.
181. **Crivello, J. V. and Lee, J. L.,** The UV cure of epoxy-silicone monomers, *Polym. Mater. Sci. Eng.,* 60, 217, 1989.
182. **Eckberg, R. P. and Riding, K. D.,** Ultraviolet cure of epoxysiloxanes and epoxysilicones, *Polym. Mater. Sci. Eng.,* 60, 222, 1989.
183. **Dreyfuss, P. and Dreyfuss, M. P.,** The formation of cyclic oligomers accompanying the polymerization of oxetane, *Polym. J.,* 8, 81, 1976.
184. **Rose, J. B.,** Cationic polymerization of oxacyclobutanes. I., *J. chem. Soc.,* p. 542, 1956.
185. **Arimatsu, Y.,** Cyclic trimer of 3,3-bis(chloromethyl)oxacyclobutane, *J. Polym. Sci. Part A,* 728, 1966.
186. **Kops, J. and Skovby, H. M.,** Synthesis of polymers from functional oligomers prepared by ring-opening reaction, in *Prepr. IUPAC Macro '88,* Kyoto, 1989, 98.
187. **McKenna, J. M., Wu, T. K, and Pruckmayr, G.,** Macrocyclic tetrahydrofuran oligomers, *Macromolecules,* 10, 877, 1977.
188. **Pruckmayr, G. and Wu, T. K.,** Macrocyclic tetrahydrofuran oligomers. II. Formation of macrocyclics in the polymerization of tetrahydrofuran with triflic acid, *Macromolecules,* 11, 265, 1978.
189. **Matsuda, K., Tanaka, Y., and Sakai, T.,** Japanese Patent 13,520 1974.
190. **Tanaka, Y.,** Cationic polymerization of tetrahydrofuran with fuming sulfuric acid initiator. Effects of superacids, salts and aromatic compounds on $\overline{M}_n$ and yield of the polymer, *J. Macromol. Sci. Chem.,* A11, 2189, 1977.
191. **Dreyfuss, P. and Dreyfuss, M. P.,** Polymerization of cyclic ethers and sulphides, in *Comprehensive Chemical Kinetics,* Vol. 15, *Non-Radical Polymerization,* Bamford, C. H. and Tipper, C. F. H., Eds., Elsevier, Amsterdam, 1976, chap. 4.
192. **Penczek, P., Kubisa, P., and Matyjaszewski, K.,** Cationic ring-opening polymerization. II. Synthetic applications, *Adv. Polym. Sci.,* 68/69, 39, 1985.
193. **Tamashita, Y. and Chiba, K.,** Cationic polymerization of cyclic ethers initiated by macromolecular dioxolenium perchlorate, *Polym. J.,* 4, 200, 1973.
194. **Smith, S., Schulz, W. J., and Newsmark, J.,** in *Ring-Opening Polymerization,* ACS Symp. Ser. 59, Saegusa, T. and Goethals, E. J., Eds., American Chemical Society, Washington, D.C., 1977, 13.
195. **Smith, S. and Hubin, A. J.,** U.S. Patent 3,824,198, 1974; *Chem. Abstr.,* 82, 18327, 1975.
196. **Cunliffe, A. V., Richards, D. H., and Robertson, F.,** Reaction of living polyTHF with amines. II. Other tertiary amines, *Polymer,* 22, 108, 1981.
197. **Hartley, D. B., Hayes, M. S., and Richards, D. H.,** Reactions of living polyTHF with amines. III. Secondary amines, *Polymer,* 22, 1981, 1981.
198. **Cohen, P., Abadie, M. J. M., Shul, F., and Richards, D. H.,** Reactions of living PolyTHF with amines. IV. Primary amines, *Polymer,* 23, 1350, 1982.
199. **Goethals, E. J.,** Polymers with functional end groups by cationic ring-opening polymerization, in Prepr. 28th IPAC Symp. Macromolecular, Amherst, July 12—16, 1982, 204.
200. **Goethals, E. J., Van der Velde, M., and Munir, A.,** The synthesis of block-copolymers with polyamine segments using cationic living poly(1-*tert*-butylaziridine), in *Cationic Polymerization and Related Processes,* Goethals, E. J., Ed., Academic Press, London, 1984, 387.

201. **Tezuka, Y. and Goethals, E. J.**, Synthesis and reactions of poly-tetrahydrofuran with azetidinium salt end groups, *Eur. Polym. J.*, 12, 991, 1982.
202. **Goethals, E. J., D'Haese, F., De Clercq, R., and Van Meirvenne**, Block copolymers and polymer networks by living or pseudo-living cationic ring-opening polymerization, *Polym. Prepr. Am. Chem. Soc. Div. Polym. Chem.*, 29(2), 61, 1988.
203. **Tezuka, Y. and Goethals, E. J.**, Ion-exchange and ring-opening reactions of telechelic poly(tetrahydrofuran)s containing terminal cyclic quaternary ammonium salts, *Makromol. Chem.*, 188, 783, 1987.
204. **Tezuka, Y. and Goethals, E. J.**, Synthesis of star and model network polymers from poly(tetrahydrofuran)s with azetidinium end groups and multifunctional carboxylated, *Makromol. Chem.*, 188, 791, 1987.
205. **Kress, H. J., Stix, W., and Heitz, W.**, Telechelics. VII. Polytetrahydrofuran with acetate end groups, *Makromol. Chem.*, 185, 173, 1984.
206. **Kress, H. J. and Heitz, W.**, Polytetrahydrofuran with acrylate and methacrylate end groups, *Makromol. Chem. Rapid. Commun.*, 2, 427, 1981.
207. **Percec, V., Pugh, C., and Pask, S. D.**, Macromonomers, oligomers and telechelic polymers, in *Comprehensive Polymer Science*, Vol. 6, Eastmond, G. C., Ledwith, A., Russo, S., and Sigwalt, P., Eds., Pergamon Press, Oxford, 1988, 281.
208. **Andrews, J. H. and Semlyen, J. A.**, Equilibrium ring concentrations and the statistical conformations of polymer chains. VII. Cyclics in poly(1,3-dioxolane), *Polymer*, 13, 142, 1972.
209. **Miki, T., Higashimura, T., and Okamura, S.**, Trioxane formation in the solid-state polymerization of tetraoxane, *J. Polym. Sci. Polym. Lett.*, 5, 65, 1967.
210. **Keler, T., Schlotterbeck, D., and Jaacks, V.**, Ring formation in the polymerization of 1,3-dioxolane, in *23rd IUPAC Macromolecular Preprint*, Boston, 2, 1971, 645.
211. **Boehlke, K., Weyland, P., and Jaacks, V.**, Uber den Mechanismus der cationischen Polymerisation von 1,3-Dioxolan, in *23rd IUPAC Macromolecular Preprint*, Boston, 2, 1971, 641.
212. **Miki, T., Higashimura, T., and Okamura, S.**, Rates of polymer formation and monomer consumption in the solution polymerization of trioxane catalysed by $BF_3 \cdot O(C_2H_5)_2$, *J. Polym. Sci. Part A*, 5, 95, 1967.
213. **Miki, T., Higashimura, T., and Okamura, S.**, Polymerization of tetraoxane at low concentration catalysed by $BF_3 \cdot O(C_2H_5)_2$, *J. Polym. Sci. Part A*, 5, 2997, 1967.
214. **Higashimura, T., Tanaka, A., Miki, T., and Okamura, S.**, Copolymerization of tetraoxane with styrene catalysed by $BF_3 \cdot O(C_2H_5)_2$, *J. Polym. Sci. Part A*, 5, 1937, 1967.
215. **Plesch, P. H. and Westerman, P. H.**, The polymerization of 1,3-dioxepan. I. Structure of polymer and thermodynamics of its formation and a note on 1,3-dioxan, *Polymer*, 10, 105, 1969.
216. **Okado, M., Kozawa, S., and Tamashita, Y.**, Kinetic studies on the polymerization of 1,3-dioxepane initiated with triethyl oxonium tetrafluoroborate, *Makromol. Chem.*, 127, 271, 1969.
217. **Gong, M. S. and Hall, H. K., Jr.**, Trialkylsilyl triflates, novel initiators for cationic polymerization, *Macromolecules*, 19, 3011, 1986.
218. **Hall, H. K., Padias, A. B., Atsumi, M., and Way, T. F.**, Controlled polymerization of 1,3-dioxepane by trimethylsilyl methanesulfonate, *Polym. Prepr. Am. Chem. Soc. Div. Polym. Chem.*, 29(2), 36, 1988.
219. **Reibel, L. and Franta, E.**, Preparation of block copolymers containing 1,3-dioxolane and 1,3-dioxepane, *Polym. Prepr. Am. Chem. Soc. Div. Polym. Chem.*, 26(1), 55, 1985.
220. **Kawakami, Y. and Yamashita, Y.**, Macrocyclic formals. III. Two-stage polymerization of 1,3-dioxacycloalkanes, *Macromolecules*, 10, 837, 1977.
221. **Yamashita, Y., Mayumi, J., Kawakami, Y., and Ito, K.**, Ring-chain equilibrium of macrocyclic formals, *Macromolecules*, 13, 1075, 1980.
222. **Jacobson, H. and Stockmayer, W. H.**, Molecular weight distribution in high polymers, *J. Chem. Phys.*, 18, 1600, 1950.
223. **Kops, J. and Spanggaard, H.**, Polymerization of *cis*- and *trans*-7,9-dioxabicyclo 4.3.0 nonane, *Makromol. Chem.*, 177, 299, 1975.
224. **Okado, M., Sumitomo, H., Ito, K., Goto, S., and Atsumi, M.**, Cationic ring-opening oligomerization of the two stereoisomers of 4-bromo-6,8-dioxabicyclo 3.2.1 octan-7-one, *Polym. J.*, 20, 55, 1988.
225. **Johns, D. B., Lenz, R. W., and Luecke, A.**, in *Ring-Opening Polymerization*, Ivin, K. J. and Saegusa, T., Eds., Elsevier, New York, 1984, 461.
226. **Ludwig, E. B. and Belen'kaya, B. G.**, Investigation of the mechanism of cationic polymerization of ε-caprolactam, *J. Macromol. Sci.*, 8, 819, 1974.
227. **Belen'kaya, B. G., Lyndwig, E. B., Izummikov, A. L., and Kulvelis, Y. I.**, Features of cationic polymerization of ε-caprolactone in the presence of alcohols, *Vysokomol. Soedin. Ser. A*, 24, 288, 1982.
228. **Lyndwig, E. B., Belen'kaya, B. G., Izyumnikov, A. L., Rogozhkina, E. D., Vel'ts, A. A., and Vauchska, Yu. P.**, Synthesis of bifunctional ε-caprolactone oligomers under the action of *p*-toluenesulfonic acid, *Vysokomol. Soedin. Ser. A*, 28, 1774, 1986.
229. **Estrina, G. A., Karateev, A. M., and Rozenberg, B. A.**, Catalytic reaction of ε-caprolactone with diols, *Vysokomol. Soedin. Ser. A*, 31, 1030, 1989.

230. **Lindwig, E. B., Grigor'eva, A. V., and Boiko, O. T.,** Russian Patent 1,219,597, 1986; *Chem. Abstr.,* 105, 98132d, 1986.

231. **Sigwalt, P. and Spaasky, N.,** in *Ring-Opening Polymerization,* Ivin, K. J. and Saegusa, T., Eds., Elsevier, New York, 1984, 603.

232. **Van Ooteghen, D. R. and Goethals, E. J.,** Cationic polymerization of cyclic sulfides. V. Polymerization of propylene sulfide with tetrafluoroborate, *Makromol. Chem.,* 175, 1513, 1974.

233. **Lambert, J. L., Van Oothegen, D., and Goethals, E. J.,** Isolation and identification of oligomers formed during cationic polymerization of propylene sulfide, *J. Polym. Sci. Part A,* 9, 3055, 1971.

234. **Van Craeynest, A. and Goethals, E. J.,** Cationic polymerization of cyclic sulfides, *Eur. Polym. J.,* 12, 851, 1976.

235. **Noshay, A. and Price, C. C.,** The polymerization of styrene sulfide, *J. Polym. Sci.,* 54, 533, 1961.

236. **Jones, G. D.,** in *The Chemistry of Cationic Polymerization,* Plesch, P. H., Ed., Pergamon Press, Oxford, 1963, 513.

237. **Goethals, E. J.,** Factors influencing the living character of cationic ring-opening polymerizations, *J. Polym. Sci. Polym. Symp.,* 56, 271, 1976.

238. **Goethals, E. J. and Munin, A,** Reactions of cationic living poly(*tert*-butyl aziridine, *J. Polym. Sci. Polym. Chem. Ed.,* 19, 1985, 1981.

239. **Goethals, E. J. and Vlegels, A.,** Synthesis of a polyamine macromer by cationic polymerization, *Polym. Bull.,* 4, 521, 1981.

240. **Goethals, E. J., Van de Velde, M., Eckant, G., and Bouquet, G.,** Polymerization and copolymerization of N-alkylaziridines, in *Ring-Opening Polymerization,* Gorath, J. E. Ed., ACS Symp. Ser. 286, American Chemical Society Washington, D.C., 1985, 219.

241. **Dick, C. R.,** Reaction of aziridines. I. Mechanism of piperazine formation from aziridines, *J. Org. Chem.,* 32, 72, 1967.

242. **Nemetkin, N. S., Grushevenko, I. A., and Perchenko, V. D.,** Transformations of β-(N-ethylenimino)-ethylsilanes at elevated temperatures in the presence of nucleophile and electrophile reactants, *Dokl. Akad. Nauk S.S.S.R. Ser. Khim.,* 162, 347, 1965.

243. **Jones, G. D., MacWilliams, D. C., and Braxtor, N.,** Species in the polymerization of ethylenimine and N-methylethylenimine, *J. Org. Chem.,* 30, 1994, 1965.

244. **Geothals, E. J., Schacht, E. H., and Bruggeman, P.,** Cationic polymerization of aziridines, in Prepr. Int. Symp. Cationic Polymerization, Rouen, 1973, C-30.

245. **Hansen, G. R. and Burg, T. E.,** Unique synthesis of 1,4,7,10-tetraazacyclododecane, *J. Heterocycl. Chem.,* 5, 305, 1968.

246. **Lavagnino, E. R., Chauvette, R. R., Cannon, W. N., and Kornfield, E. C.,** Conidine. Synthesis, polymerization and derivatives, *J. Am. Chem. Soc.,* 82, 2609, 1960.

247. **Toy, M. S. and Price, C. C.,** *d*- and *l*-Polyconidine, *J. Am. Chem. Soc.,* 82, 2613, 1960.

248. **Bossaer, P. K., Goethals, E. J., Hockett, P. J., and Pepper, D. C.,** Cationic block copolymerization initiated by polystyrene perchlorate, *Eur. Polym. J.,* 13, 489, 1977.

249. **Pillai, V. N. R. and Mutter, M.,** New easily removable poly(ethylene glycol) supports for the liquid-phase method of peptide synthesis, *J. Org. Chem.,* 45, 5364, 1980.

250. **Harwood, H. J. and Patel, N. K.,** The structure of polymers obtained from arbuzov-Michaelis. Reaction involving cyclic phophonites, *Macromolecules,* 1, 233, 1968.

251. **Vogt, W. and Ahmad, N. V.,** Polymere Ester von Sauren des Phosphors. V. Ring öffnen de Polymerisation des 2-Phenoxy-1,3,2-dioxaphospholan durch Michaelis-Arbuzov-Reaction, *Makromol. Chem.,* 178, 1711, 1977.

252. **Kobayashi, S., Huang, M. Y., and Saegusa, T.,** Cationic ring-opening-polymerization of 2-phenyl-1,3,6,2-trioxapholpholane, *Polym. Bull.,* 4, 85,1981.

253. **Kobayashi, S., Suzuki, M., and Saegusa, T.,** Cationic ring-opening-polymerization of 2-phenyl-1,2-oxaphospholane (dioxophostone). Synthesis of poly(phosphine oxide) and polyphosphine, *Polym. Bull.,* 4, 315, 1981.

254. **Kobayashi, S., Suzuki, M., and Saegusa, T.,** A new reduction method of poly(phosphine oxide) to polyphosphine. Preparation of poly(*p*-phenyl)trimethylenephosphine, *Polym. Bull.,* 8, 417, 1982.

255. **Kobayashi, S., Suzuki, M., and Saegusa, T.,** New phosphorous-containing polymers including poly-(phosphine oxide)s and polyphosphinies, in Prepr. 28th IUPAC Symp. Macromolecules, Amherst, July 12—16, 1982, 174.

256. **Kobayashi, S.,** Ring-opening polymerization of new cyclic phosphorous monomers, *Polym. Prepr. Am. Chem. Soc. Div. Polym. Chem,,* 25(1), 255, 1984.

257. **Kaluzynski, K.,** A new class of synthetic polyelectrolytes. Acidic polyesters of phosphoric acid poly(hydroxyalkylene phosphates), *Macromolecules,* 9, 365, 1976.

258. **Penczek, S.,** Polymerization of cyclic esters of phosphoric acid, *Pure Appl. Chem.,* 48, 363, 1976.

259. **Lapiensis, G. and Penczek, S.,** Kinetics and thermodynamics of the polymerization of the cyclic phosphate esters. II. Cationic polymerization of 2-methoxy-2-oxo-1,3,2-dioxaphosphorinane (1,3-propylene methyl phosphate), *Macromolecules,* 7, 166, 1974.

260. **Penczek, S.,** Mechanism of ionic polymerization of cyclic esters of phosphoric acid (a new route to models of biopolymers), *J. Polym. Sci. Polym. Symp.,* 67, 149, 1980.

261. **Penczek, S., Baran, J., Pretula, J., and Lopiensis, G.,** New polymers related to biopolymers by ring-opening polymerization, in Prepr. 28th IUPAC Symp. Macromolecules, Amherst, July 12—16, 1982, 203.

262. **Penczek, S., Łapienis, G., and Klosinski, P.,** Models of biopolymers with polyphosphate backbone (nucleic and teichoic analogues), in *Plenary and Invited Lectures, 29th IUPAC Int. Symp. Macromolecules,* Bucharest, 1983, 223.

263. **Kobayashi, S., Suzuki, M., and Saegusa, T.,** Ring-opening polymerization and deoxothiolphostones: synthesis of poly(phosphine sulfide)s, *Macromolecules,* 19, 462, 1986.

264. **Noll, W.,** *Chemistry and Technology of Siloxanes,* Academic Press, New York, 1968.

265. **McGrath, J. E., Riffle, J. I., Banthia, A. K., Yligor, I., and Wilkes, G. L.,** An overview of the polymerization of cyclosiloxanes, in *Initiation of Polymerization,* Bailey, T. E., Jr., Ed., ACS Symp. Ser. 212, 1983, 145.

266. **Kantor, S. W., Grubb, W. T., and Osthoff, R. C.,** The mechanism of the acid- and base catalysed equilibrium of siloxanes, *J. Am. Chem. Soc.,* 76, 5190, 1954.

267. **Yligör, I. and McGrath, J. E.,** Synthesis of organofunctional polysiloxanes by cationic ring-opening polymerization, *Polym. Prepr. Am. Chem. Soc. Div. Polym. Chem.,* 26(1), 57, 1985.

268. **Eaborn, C.,** *Organosilicon Compounds,* Butterworths, London, 1960.

269. **Hurd, D. T.,** On the mechanism of the acid catalyzed rearrangement of siloxane linkages in organo-polysiloxanes, *J. Am. Chem. Soc.,* 77, 2998, 1955.

270. **Kojima, K., Gore, C. R., and Marvel, C. S.,** Preparation of polysiloxanes having terminal carboxyl groups, *J. Polym. Sci. Part A,* 4, 2325, 1966.

271. **Speier, J. L. I., David, M. P., and Eynon, B. A.,** Dehydration of 1,3-bis(hydroxyalkyl)tetramethylsiloxanes, *J. Org. Chem.,* 25, 1637, 1960.

272. **Chojnowski, J. and Wilczek, L.,** Mechanism of the polymerization of hexamethylcyclotrisiloxane (D3) in the presence of a strong protonic acid, *Makromol. Chem.,* 180, 117, 1979.

273. **Lebrun, J. J., Sauvet, G., and Sigwalt, P.,** Polymerization of octamethylcyclotetrasiloxane initiated by a superacid, *Makromol. Chem. Rapid. Commun.,* 3, 757, 1982.

274. **Wilczek, L. and Chojnowski, J.,** Studies of siloxane-acid model system: hexamethyldisiloxane-trifluoroacetic acid, *Makromol. Chem.,* 184, 77, 1983.

275. **Yligör, I., Riffle, J. S., and McGrath, E. J.,** Reactive difunctional siloxane oligomers. Synthesis and characterization, in *Reactive Oligomers,* ACS Symp. Ser. Harris, F. W. and Spinelli, H., Eds., American Chemical Society, Washington, D.C., 1982, 282.

276. **McGrath, J. E., Matzner, M., Noshay, A., and Robeson, L. M.,** Review on block and graft copolymers, in *Encyclopedia of Polymer Science and Technology,* John Wiley & Sons, New York, 1977, Suppl. 2, 129.

277. **Wright, P. T.,** in *Ring-Opening Polymerization,* Ivin, K. and Saegusa, T., Eds., Elsevier, New York, 1984, 1055.

278. **McGrath, J. E.,** Ring-opening polymerization, an introduction, in *Ring-Opening Polymerization: Kinetics and Synthesis,* McGrath, J. E., Ed., ACS Symp. Ser., American Chemical Society, 1985, 286.

279. **Chujr, Y. and McGrath, J. E.,** Synthesis of α,ω-difunctional fluorine containing polysiloxane oligomer *via* hydrosilation reaction, *Polym. Prepr. Am. Chem. Soc. Div. Polym. Chem.,* 24(2), 47, 1983.

280. **Yligör, I., Ward, R. S., and Riffle, J. S.,** Synthesis and characterization of poly(2-ethyloxazoline)-polydimethyl-siloxane-poly(2-ethyloxazoline) triblock-copolymers, *Polym. Prepr. Am. Chem. Soc. Div. Polym. Chem.,* 28(2), 369, 1987.

281. **Steckle, W. P., Yiligör, E., Riffle, J. S., Spinu, M., Yiligör, I., and Ward, R. S.,** Synthesis and characterization of thermally and chemically stable hydroxyl terminated telechelic siloxane oligomers, *Polym. Prepr. Am. Chem. Soc. Div. Polym. Chem.,* 28(1), 254, 1987.

282. **Jewel, B. S., Riffle, J. S., Allison, D., and McGrath, J. E.,** Synthesis and characterization of telechelic polydimethylsiloxanes *via* carboxylic acid functionality, *Polym. Prepr. Am. Chem. Soc. Div. Polym. Chem.,* 30(1), 294, 1989.

283. **DeSimone, J. M., York, G. A., Wilson, G. R., Smith, S. D., Marand, H., Gozdz, A. S., Bowden, M. J., and McGrath, E. J.,** Synthesis and characterization of poly(1-butenesulfone)-g-poly(dimethylsiloxane): a new electron-beam resist for two-layer lithography, *Polym. Prepr. Am. Chem. Soc. Div. Polym. Chem.,* 30(2), 134, 1989.

284. **Boutevin, B. and Robin, J. J.,** Synthèse de polysiloxanes téléchélique. II. Synthèse de diamines aromatiques, *Makromol. Chem.,* 190, 287, 1989.

285. **Andolino-Brandt, P. J., Subramanian, R., Sormani, P. M., and Ward, T. C.,** ^{29}Si NMR functional polysiloxane oligomers, *Polym. Prepr. Am. Chem. Soc. Div. Polym. Chem.,* 26(2), 213, 1985.

286. **Penczek, P.,** Cationic ring-opening polymerisation, *Adv. Polym. Sci.,* 68/69, 216, 1985.

287. **Kogan, E. V., Ivanova, A. G., Rethkhsfeld, V. D., Smirnov, V. O., Smirnov, N. I., and Gruber, V. N.,** Polymerization of octamethylcyclotetrasiloxane in the presence of acid catalysts, *Vysokomol. Soedin.,* 5, 249, 1963.

288. **Kendrick, T. C.,** The acid-catalysed polymerization of cyclosiloxanes. I. The kinetics of the polymerization of octamethylcyclotetrasiloxane catalysed by anhydrous ferric chloride-hydrogen chloride, *J. Chem. Soc.,* 2027, 1965.

289. **Sauvet, G., Lebrun, J. J., and Sigwalt, P.,** Cationic polymerization of cyclosiloxanes: the various processes involved, in *Cationic Polymerization and Related Processes,* Goethals, E. J., Ed., Academic Press, London, 1984, 237.

290. **Sigwalt, P., Nicol, P., and Masure, M.,** Living polymers in cationic polymerization of cyclosiloxanes, *Polym. Prepr. Am. Chem. Soc. Div. Polym. Chem.,* 29(2), 27, 1989.

291. **Chojnowski, J., Sciborek, M., and Kowalski, J.,** Mechanism of the formation of macrocycles during the cationic polymerization of cyclosiloxanes. End to end ring closure versus ring expansion, *Makromol. Chem.,* 178, 1351, 1977.

292. **Morton, M.,** *Anionic Polymerization: Principles and Practice,* Academic Press, New York, 1983, chap. 3.

293. **Percec, V. and Pugh, C.,** Oligomers, in *Encyclopedia of Polymer Science and Engineering,* Vol. 10, 2nd ed. Mark, H. F., Bikales, N. M., Overberger, C. C., and Menges, G., Eds., John Wiley & Sons, New York, 1987, 432.

294. **Hu, H. P., Chen, Z., and Bergbreiter, D. E.,** Entrapment of functionalyzed ethylene oligomers in polyethylene, *Macromolecules,* 17, 2111, 1984.

295. **Bergbreiter, D. E. and Blanton, J. R.,** Diphenylphosphinated ethylene oligomers as polymeric reagents for synthesis of alkyl derivatives from alcohol, *J. Chem. Soc. Chem. Commun.,* 337, 1985.

296. **Bergbreiter, D. E. and Chandran, R.,** Polyethylene-entrapped Nickel(0) diene cyclo-oligomerization catalysts, *J. Chem. Soc. Chem. Commun.,* 1396, 1985.

297. **Bergbreiter, D. E. and Chandran, R.,** Use of functionalized ethylene oligomers to prepare recoverable, recyclable nickel(0) diene cyclo-oligomerization catalysts, *J. Org. Chem.,* 51, 4754, 1986.

298. **Bergbreiter, D. E., Chen, L., and Chandran, R.,** Recyclable polymer-bound lanthanide diene polymerization catalyst, *Macromolecules,* 18, 1055, 1985.

299. **Bergbreiter, D. E. and Chandran, R.,** Concurrent catalytic reductions/stoichiometric oxidation using oligomerically ligated catalyst and polymer bond reagents, *J. Am. Chem. Soc.,* 107, 4792, 1985.

300. **Bergbreiter, D. E. and Blanton, J. R.,** Functionalized ethylene oligomers as phase-transfer catalysts, *J. Org. Chem.,* 50, 5828, 1985.

301. **Poteat, J. L. and Bergbreiter, D. E.,** Chemistry of copper(II) carboxylates at polyethylene solution-interfaces, *Polym. Prepr. Am. Chem. Soc. Div. Polym. Chem.,* 30(1), 224, 1989.

302. **Hay, J. N., McCabe, J. M., and Robb, J. C.,** Kinetics of reaction of metal alkyls with alkenes. VIII. Oligomerization of ethylene by butyllithium-N,N,N',N'-tetramethylethylenediamine, *J. Chem. Soc. Faraday Trans. 1* 68, 1127, 1972.

303. **Magnin, H., Rodriguez, F., Abadie, M., and Schué, F.,** Oligomérization anionique de l'ethylène, I. Etude cinètique, *J. Polym. Sci. Polym. Chem. Ed.,* 15, 875, 1977.

304. **Magnin, H., Rodriguez, F., Abadie, M., and Schué, F.,** Oligomérisation anionique de l'ethylène, II. Etude des paramètres thermodynamiques, *J. Polym. Sci. Polym. Chem. Ed.,* 15, 897, 1977.

305. **Magnin, H., Rodriguez, F., Abadie, M., and Schué, F.,** Oligomérisation anionique de l'ethylène, III. Application à la synthèse d'alcools à longue chain, *J. Polym. Sci. Polym. Chem. Ed.,* 15, 901, 1977.

306. **Rodriguez, F., Abadie, M., and Schué, F.,** Oligomerization of ethylene by tertiary butyllithium-*N,N,N',N'*-tetramethylethylenediamine (TMEDA), *J. Polym. Sci. Polym. Chem. Ed.,* 14, 773, 1976.

307. **Eberhard, G. G. and Butte, W. A.,** A catalytic telomerization reaction of thylene with aromatic hydrocarbons, *J. Org. Chem.,* 29, 2928, 1964.

308. **Eberhard, G. G. and Davis, W. R.,** Polymerization and telomerization reaction of olefins with a tertiary amine-coordinated lithiumalkyl catalyst, *J. Polym. Sci. Part A,* 3, 3753, 1965.

309. **Swarc, M.,** *Carbanions, Living Polymers and Electron Transfer Processes,* Interscience, New York, 1968.

310. **Morton, M. and Fetters, L. J.,** Anionic polymerization of vinyl monomers, *Rubber Chem. Technol.,* 48, 359, 1975.

311. **Flory, P. J.,** Molecular weight distribution in anionic polymerization, *J. Am. Chem. Soc.,* 62, 1561, 1940.

312. **Kume, S., Takahashi, A., Nishikawa, G., Hatano, M., and Kambara, S.,** Anionic telomerization of butadiene with aromatic hydrocarbons. I. *Makromol. Chem.,* 84, 137, 1965.

313. **Kume, S., Tahakashi, A., Nishikawa, G., Hatano, M., and Kambara, S.,** Anionic telomerization of butadiene with aromatic hydrocarbons. II. Solvent effects on rate constants, *Makromol. Chem.,* 84, 147, 1965.

314. **Kume, S., Takahashi, A., Nishikawa, G., Hatano, M., and Kambara, S.,** Anionic telomerization of butadiene with aromatic hydrocarbons. III. Chemical structure of telomer, *Makromol. Chem.,* 98, 109, 1966.

315. **McElory, B. J. and Mekley, J. H.,** U.S. Patent 3,678,121, 1972.

316. **Morrison, R. G. and Kamienski, C. W.,** U.S. Patent 3,725,368, 1973.

317. **Kamienski, C. W. and Eastham, J. F.,** U.S. Patent 3,742,077, 1973.

318. **Kamienski, C. W. and Merkley, J. H.,** U.S. Patent 3,751,501, 1973.

319. **Morton, M.,** *Anionic Polymerization: Principles and Practice,* Academic Press, New York, 1983, chap. 8.

320. **Michel, R. H. and Baker, W. P.,** The structure of anionically prepared poly-9-vinylanthracene, *J. Polym. Sci. Part B,* 2, 163, 1964.

321. **Eisenberg, A. and Rembaum, A.,** The occurrence of chain transfer in the anionic polymerization of 9-vinylanthracene, *J. Polym. Sci. Part B,* 2, 157, 1964.

322. **Gatzke, A.,** Chain transfer in anionic polymerization. Determination of chain transfer constants by using ^{14}C labeled chain transfer agents, *J. Polym. Sci. Polym. Chem. Ed.,* 7, 2281, 1969.

323. **McGarrity, J. F. and Prodolliet, J.,** High-field rapid injection NMR. Observation of unstable primary ozonide intermediates, *J. Org. Chem.,* 49, 4465, 1984.

324. **Bucca, D., Gordon, B., III, and Ogle, C.,** Rapid injection NMR studies of styrene with n-butyllithium, *Polym. Prepr. Am. Chem. Soc. Div. Polym. Chem.,* 30(1), 135, 1989.

325. **Szwark, M., Levy, M., and Milkovich, R.,** Polymerization initiated by electron transfer to monomer— a new method of formation of block polymers, *J. Am. Chem. Soc.,* 78, 2656, 1956.

326. **Leonard, J. and Malhorta, S. L.,** Polymerization of α-methylstyrene at high temperatures in tetrahydrofuran with potassium as initiator. I. Thermodynamic study and GPC analyses of polymers, *J. Macromol. Sci. Chem.,* A10, 1279, 1976.

327. **Schmitt, B. J. and Schulz, G. V.,** Polymerization kinetische und konductometrische Messungen an Polystyrylionenpaaren in lineum grosseren Temperatur Bereich, *Makromol. Chem.,* 175, 3261, 1974.

328. **Leonard, J. and Malhotra, S. L.,** Polymerization of α-methylstyrene in THF with potassium as initiator. II. Gel permeation analysis of polymers, *J. Macromol. Sci. Chem.,* A11, 1867, 1977.

329. **Malhotra, S. L. and Leonard, J.,** Polymerization of α-methylstyrene in p-dioxane with potassium as intitiator. III. GPC analysis of polymers, *J. Macromol. Chem.,* A11, 1907, 1977.

330. **Malhotra, S. L., Leonard, J., and Thomas, M.,** Polymerization of α-methylstyrene with potassium as initiator. III. Thermodynamic studies and GPC analyses of the polymers, *J. Macromol. Chem.,* A11, 2213, 1977.

331. **Richards, D. H. and Williams, R. L.,** Structure of the tetramer of α-methylstyrene, *J. Polym. Sci.,* 11, 89, 1973.

332. **Rupprecht, R., Zilliox, J. G., Franta, E., and Brossas, J.,** Preparation et etude des oligomeres du styrene, *Eur. Polym. J.,* 15, 11, 1979.

333. **Bergeron, J. Y. and Leonard, J.,** Mechanistic aspects of the reversibility and irreversibility of the anionic polymerization of α-methylstyrene, in *Recent Advances in Anionic Polymerization,* Hogen-Esch, T. E. and Smid, J., Eds., Elsevier, New York, 1987, 147.

334. **Quirk, R. P. and Cheng, P. L.,** The polymerization of styrene in the presence of lithium, *Makromol. Chem.,* 183, 2071, 1982.

335. **Quirk, R. P., Chen, W. C., and Cheng, P. L.,** Synthesis of reactive oligomers using anionic polymerization techniques, *Polym. Prepr. Am. Chem. Soc. Div. Polym. Chem.,* 25(1), 144, 1984.

336. **Wyman, D. P., Allen, V. R., and Altares, T.,** Reaction of polystyryllithium with carbon dioxide, *J. Polym. Sci. Part A,* 2, 4545, 1964.

337. **Beak, P. and Kokko, B. J.,** A modification of the Sheverdina-Kocheshkov amination: the use of methoxyamine-methyllithium as a convenient synthetic equivalent for NH_2^+, *J. Org. Chem.,* 47, 2822, 1982.

338. **Quirk, R. P. and Cheng, P. L.,** Functionalization of polymeric organolithium compounds. Amination of poly(styryl)lithium, *Polym. Prepr. Am. Chem. Soc. Div. Polym. Chem.,* 24(2), 461, 1983.

339. **Rupprecht, R. and Brossas, J.,** Synthèse d'alcools terpeniques par oligomérisation anionique en phase heterogene, *J. Polym. Sci. Symp.,* 52, 67, 1975.

340. **Catala, J. M., Pujol, J. M., and Brossas, J.,** New synthesis of polysulfide polymers: kinetic study, *Makromol. Chem.,* 188, 2517, 1987.

341. **Brossas, J. and Clonet, G.,** Synthèse de polymeres mono- et polyphosphonés, *C. R. Acad. Sci. Ser. C,* 280, 1459, 1975.

342. **Reeb, R., Vinchon, Y., Riess, G., Catala, J. M., and Brossas, J.,** Etude de l'oxidation du dimère dicarbanionique du diphenyl-1,1-éthylène. Synthese du dihydropéroxy-1,4-tetraphényl-1,1,4,4-butane, *Bull. Soc. Chim. Fr.,* 216, 2717, 1975.

343. **Catala, J. M., Riess, G., and Brossas, J.,** Synthese de polymères ω-hydroperoxyde, *Makromol. Chem.,* 178, 1249, 1977.

344. **Catala, J. M., Pujol, J. M., and Brossas, J.,** Synthesis and characterization of polysulphide polymers, in *Recent Advances in Anionic Polymerization,* Hogen-Esch, T. E. and Smid, J., Eds., Elsevier, New York, 1987, 431.

345. **Catala, J. M. and Brossas, J.,** Synthesis of controlled polymeric structures in heterogeneous systems through living species, *Polym. Prepr. Am. Chem. Soc. Div. Polym. Chem.,* 29(2), 50,1988.

346. **Tomai, M. and Kakiuchi, H.,** Polymerization in dipolar solvents. I. Polymerization of styrene by Grignard reagents, *Kogyo Kagaku Zasshi,* 73, 2367, 1970.

347. **Yamazaki, N., Nakahama, S., and Hirao, A.,** An attempt at the preparation of macrocyclic polymer, *J. Macromol. Sci. Chem.,* A9, 551, 1975.

348. **Nakahama, S., Hirao, A., and Ohira, Y.,** Dibenzylmagnesium-initiated anionic oligomerization of *trans*-stilbene, 1,1,-diphenylethylene and α-methylstyrene in hexamethylphosphortriamide, *J. Macromol. Sci. Chem.,* A9, 563, 1975.

349. **Yasuda, H., Tatsumi, K., and Nakamura, A.,** New aspects of carbanion chemistry. Structure of pentadienyl anions and butanediyl dianions and their roles in organic and inorganic syntheses, in *Recent Advances in Anionic Polymerization,* Hogen-Esch, T. E. and Smid, J., Eds., Elsevier, New York, 1987, 59.

350. **Angood, A. C., Hurley, S. A., and Tait, P. J.,** Anionic polymerization initiated by diethylamide in organic solvents. I. The use of lithium diethylamide as a polymerization catalysts and the effect of solvent on the polymerization of isoprene and styrene, *J. Polym. Sci. Polym. Chem. Ed.,* 11, 2777, 1973.

351. **Yoshino, N., Yamaki, Y., Nagasaki, Y., and Tsuruta, T.,** Studies of functional diene oligomers with an amino end group. I. Synthesis of functional oligomers of isoprene with one diisopropylamino end group, *Makromol. Chem.,* 184, 737, 1983.

352. **Nagasaki, Y., Yoshino, T., and Tsuruta, T.,** Studies of functional diene oligomers with an amino end group. II. Reaction mechanism of the isoprene oligomerization, *Makromol. Chem.,* 186, 1335, 1985.

353. **Nagasaki, Y., Hyguchi, A., Goan, H., Yoshino, N., and Tsuruta, T.,** Studies of functional oligomers with an amino end group. III. Synthesis of oligoisoprenes containing a secondary amino end group, *Makromol. Chem.,* 190, 53, 1989.

354. **Hamana, H., Hagiwara, T., and Navrita, T.,** Synthesis of isoprene oligomers by lithium N-alkylbenzylamide/N-alkylbenzylamine initiators, in *Prepr. IUPAC Int. Symp. Macromolecules,* Kyoto, 1988, 190; **Jinjie, L. and Bin, L.,** Study on the kinetics of anionic telomerization of butadiene, in *Prepr. IUPAC Int. Symp. Macromolecules,* Kyoto, 1988, 137.

355. **Lubnin, A. V., Osetrova, L. V., and Podkorytov, I. S.,** Reaction of ethyllithium with 1,3-butadiene. Analysis of oligomer terminal groups by the method of carbon-13 NMR multipulse spectroscopy, *Zh. Obshch. Khim.,* 59, 1159, 1989.

356. **Dolinskaya, E. R., Erusalimskii, G. B., Rozinova, O. A., and Kormer, V. A.,** Structural results of complexation of polybutadienyllithium and its analogs with various electron donors, *Vysokomol. Soedin. Ser. A,* 29(12), 2521, 1987.

357. **Hiroshi, H., Tomoyoshi, K., Tokio, H., and Tadoshi, N.,** Isoprene oligomerization with lithiated diethylenetriamine catalyst, *Kobunshi Robunshu,* 46, 277, 1989; *Chem. Abstr.,* 111, 97786g, 1989.

358. **Ivan, G. and Lirngoni, R.,** U.S. Patent 3,966,638, 1976; *Chem. Abstr.,* 85, 94953t, 1976.

359. **Makowski, H. S. and Lynn, M.,** Butyllithium polymerization of butadiene. III. Effect of inactive lithium compounds, *J. Macromol. Sci. Chem.,* A2, 638, 1968.

360. **Lynn, M. and Makowski, H. S.,** French Patent 1,492,154, 1967; *Chem. Abstr.,* 68, 106145v, 1968.

361. **Randall, J. C. and Silas, R. S.,** Model butadiene polymerization. II. Monomer distributions in butadiene oligomers, *Macromolecules,* 3, 491, 1970.

362. **Kaemph, B., Maillard, A., Schue, F., Sledz, J., Sommer, J., and Tanelian, Ch.,** Action of butyllithium isomers on 2-phenyl-1,3-butadiene in hydrocarbon media. Structure and composition of the oligomers obtained, *Bull. Soc. Chim. Fr.,* 1153, 1972.

363. **Tanaka, Y., Sato, H., Mita, K., and Shimizu, M.,** Japanese Patent 61,136,507, 1986; *Chem. Abstr.,* 106, 19229k, 1987.

364. **Jacobi, M. M., Stadler, R., Stibal, E., and Gromski, W.,** Synthesis and characterization of deuterated low-molecular weight polybutadienes, *Makromol. Chem. Rapid Commun.,* 7, 443, 1986.

365. **Skornyakov, A. S. and Kzol, V. A.,** Kinetics of the oligomerization of diene hydrocarbons under the action of metallic sodium in the presence of triisobutylaluminium, *Vysokomol. Soedin. Ser. A,* 28, 1875, 1986.

366. **Noshay, A. and McGrath, E. J.,** *Block Copolymers. Overview and Critical Survey,* Academic Press, New York, 1977.

367. **Morton, M.,** *Anionic Polymerization: Principles and Practice,* Academic Press, New York, 1983, chap. 9-11.

368. **Bywater, S.,** Preparation and properties of star-branched polymers, *Adv. Polym. Sci.,* 30, 89, 1979.

369. **Bamford, C. H., Eastmond, G. C., Woo, J., and Richards, D. H.,** Formation of polystyrene-poly(methyl methacrylate) block copolymers by anion to free radical transformation, *Polymer,* 23, 643, 1982.

370. **Antonov, V. L., Arestyakubovich, A. A., Baseva, R. V., Ermakova, J. L., Zolotarev, V. L., Kristal'nyi, E. V., and Krol, V. A.,** Preparation of SKDSR rubber over an organosodium catalyst, *Proms. Sint. Kauch.,* 6, 15, 1983.

371. **Anton, E., Griehl, V., Stubenranch, D., Schulz, H., and Singer, H.,** East German Patent 155,995, 1982; *Chem. Abstr.,* 98, 107975x, 1983.

372. **Naumova, S. F., Yurina, O. D., Kazec, A. N., and Radkevich, S. E.,** Polymerization of 1,3-cyclohexadiene in the presence of an n-butyllithium-tetrahydrofuran complex, *Vestsi. Akad. Navuk. B.S.S.R. Ser. Khim. Navuk,* 2, 80, 1983.

373. **Rhodes, R. P. and Guthrie, D. A.**, U.S. Patent 3,356,664, 1967; *Chem. Abstr.*, 68, 30890m, 1968.
374. **Geiderikh, M. A., Davydov, B. E., Zaliznaya, N. F., and Oreshkina, G. A.**, Anionic polymerization of phenyl acetylene, *Vysokomol. Soedin. Ser. B*, 11, 870, 1969.
375. **Misin, V. M., Kisilitso, P. P., Bolondaeva, N. S., and Cherkashin, M. I.**, Polymerization of tolan and diphenylbutadiene, *Vysokomol. Soedin. Ser. A*, 18, 1726, 1976.
376. **Wentworth, S. E. and Bergquist, P. R.**, Regiospecifically substituted poly(phenylacetylene) as potential chemical detectors, *J. Polym. Sci. Polym. Chem. Ed.*, 23, 2197, 1985.
377. **Benes, M., Pleska, J., and Wichterle, O.**, Polydicyanoacetylene: preparation and properties, *J. Polym. Sci. Part A*, 1, 325, 1963.
378. **Goode, W. E., Owens, F. H., and Myers, W. L.**, Crystalline studies of acrylic polymers, *J. Polym. Sci.*, 47, 75, 1960.
379. **Owens, F. H., Myers, W. L., and Zimmerman, F. E.**, The reaction of acrylates and methacrylates with organomagnesium compounds, *J. Org. Chem.*, 26, 2288, 1961.
380. **Müller, A. H. E.**, Kinetics and mechanisms in the anionic polymerization of methacrylic esters, in *Recent Advances in Anionic Polymerization*, Hogen-Esch, T. E. and Smid, J., Eds., Elsevier, New York, 1987, 205.
381. **Hatada, K., Ute, K., Tanaka, K., Kitayama, T., and Okamoto, Y.**, Mechanism of polymerization of MMA by Grignard reagents and preparation of highly isotactic PMMA with narrow molecular weight distribution, *Polym. Prepr. Am. Chem. Soc. Div. Polym. Chem.*, 27(1), 151, 1986.
382. **Hatada, K., Ute, K., Tanaka, K., Kitayama, T., and Okamoto, Y.**, Mechanism of polymerization of MMA by Grignard reagents and preparation of highly isotactic PMMA with narrow MWD, in *Recent Advanves in Anionic Polymerization*, Hogen-Esch, T. E. and Smid, J., Eds., Elsevier, New York, 1987, 195.
383. **Hatada, K., Kitayama, T., Okahata, S., and Yuki, H.**, Mechanism of polymerization of MMA by Grignard reagents, *Polym. J.*, 13, 1045, 1981.
384. **Novoselova, A. V., Erussalimsky, B. L., Krasulina, V. N., and Zascherinski, E. V.**, The reactive roles of oligomerization and polymerization in the acrylonitrile-toluene-butyllithium system, *Vysokomol. Soedin. Ser. A*, 13, 87, 1971.
385. **Tsvetanov, Kh. and Docheva, D.**, Ion equilibria in the solution of models of "living" methacrylonitrile oligomers with lithium (+) counterion, *J. Polym. Sci. Polym. Chem. Ed.*, 24, 2253, 1986.
386. **Adler, J. H., Lochman, L., Docheva, D., and Tsvetanov, Kh.**, Interaction of alkoxides. XIII. On the nature of the active centers in the anionic polymerization of methacrylonitrile. Interaction with lithium *tert*-butoxide, *Makromol. Chem.*, 187, 1253, 1986.
387. **Erussalimski, B. L., Krasnoselskaya, J. G., Krasulina, V. N., Novoselova, A. V., and Zashtsherinsky, E. V.**, Mechaisnm of the side reactions in anionic systems with low efficiency of the initiation, *Eur. Polym. J.*, 6, 1391, 1970.
388. **Tsvetanov, C. and Panayotov, I. M.**, On the nature of the active centers in the initial stages of methacrylonitrile anionic polymerization. I. Spectral studies, *Eur. Polym. J.*, 11, 209, 1975.
389. **Tsvetanov, C. and Panayotov, I. M.**, Electronic spectra of "living" oligomers in the anionic polymerization of acrylonitrile and methacrylonitrile, *Polym. Prepr. Am. Chem. Soc. Div. Polym. Chem.*, 27(1), 147, 1986.
390. **Tsvetanov, C., Muller, A. H. E., and Schulz, G. V.**, Dependence of the propagation rate constants on the degree of polymerization in the initial stage of the anionic polymerization of MMA in THF, *Macromolecules*, 18, 863, 1985.
391. **Allen, R. D. and McGrath, J. E.**, Synthesis of tactic poly(alkyl methacrylate) homo and copolymers, *Polym. Prepr. Am. Chem. Soc. Div. Polym. Chem.*, 25(2), 9, 1984.
392. **Allen, R. D., Smith, S. D., Long, T. E., and Grath, J. E.**, Synthesis of alkyl(methacrylate) copolymers by anionic polymers, *Polym. Prepr. Am. Chem. Soc. Div. Polym. Chem.*, 26(1), 247, 1985.
393. **Allen, R. D., Long, T. E., and McGrath, J. E.**, Preparation of high purity anionic polymerization grade alkylmethycrylate monomers, *Polym. Bull.*, 15, 127, 1986.
394. **Ouhadi, T., Forte, R., Jérome, R., Fayt, R., and Teyssié, Ph.**, Luxembourg Patent 85,627, 1984.
395. **Teyssié, Ph., Fayt, R., Jacobs, C., Jérome, R., Leemans, L., and Varshney, S.**, New prospects in living anionic polymerization of methacrylic and acrylic esters, *Polym. Prepr. Am. Chem. Soc. Div. Polym. Chem.*, 29(2), 52, 1988.
396. **Fayt, R., Forte, R., Jacobs, C., Jerome, R., Ouhadi, T., Teyssié, Ph., and Varshney, S. K.**, New initiator system for the "living" polymerization of *tert*-alkyl acrylates, *Macromolecules*, 20(6), 1442, 1987.
397. **Jérome, R., Forte, R., Varshney, S. K., Fayt, R., and Teyssié, Ph.**, in *Recent Advances in Mechanistic and Synthetic Aspects of Polymerizations*, Fontanille, M., Ed., Reidel, Dordrecht, 1987, 101.
398. **Müller, A. H. E., Lochmann, L., and Trekoval, J.**, Equilibria in the anionic polymerization of MMA. I. Chain-length dependence of the rate and equilibrium constants, *Makromol. Chem.*, 187, 1473, 1986.

399. **Lochmann, L., Janata, M., Machova, L., Vlacek, P., Mitera, J., and Müller, A. H. E.,** The living anionic polymerization of methacrylate and acrylate esters. An investigation of model and polymerization systems, *Polym. Prepr. Am. Chem. Soc. Div. Polym. Chem.,* 29(2), 29, 1988.

400. **Lochmann, L., Rodova, M., Trekoval, J.,** Anionic polymerization of methacrylate esters initiated with esters of α-metallocarboxylic acids, *J. Polym. Sci. Polym. Chem. Ed.,* 12, 2091, 1972.

401. **Lochmann, L., Pokorny, S., Trekoval, J., Adler, H. J., and Berger, W.,** Untersuchung des Anfangsstadiums der anionischen Polymerisation von Methylmethacrylat Einfluss von Reacktions-bedingungen auf die Umsetzung von Methyl-2-sodioisobutyrat mit Methylmethacrylat, *Makromol. Chem.,* 184, 2021, 1983.

402. **Lochmann, L., Kolarik, J., Doskocilova, D., Vozka, S., and Trekoval, J.,** Metallo esters. VII. Stabilizing effect of sodium *tert*-butoxide on the growth center in the anionic polymerization of methacrylic esters, *J. Polym. Sci. Polym. Chem. Ed.,* 17, 1727, 1979.

403. **Hatada, K., Kitayama, T., Okamoto, Y., and Ute, K.,** Controlled polymer structures by anionic mechanisms, *Polym. Prepr. Am. Chem. Soc. Div. Polym. Chem.,* 26(1), 249, 1985.

404. **Hatada, K., Ute, K., Tanaka, K., Okamoto, Y., and Kitayama, T.,** Living and highly isotactic polymerization of MMA by *t*-C₄H₉MgBr in toluene, *Polym. J.,* 18, 1037, 1986.

405. **Hatada, K., Ute, K., Imanaki, M., and Fuji, N.,** Two dimensional NMR spectra of isotactic PMMA prepared with *t*-C₄H₉MgBr and detailed examination of tacticity, *Polym. J.,* 19, 425, 1987.

406. **Hatada, K., Ute, K., Tanaka, K., Yamamoto, M., and Mishimura, T.,** Stereoregular polymerization and oligomerization of MMA by *t*-C₄H₉MgBr, in *Prepr. IUPAC Int. Symp. Macromolecules,* Kyoto, 1988, 103.

407. **Hatada, K., Kitayama, T., Ute, K., Masuda, E., Shinozaki, T., and Yamamoto, M.,** Formation of living PMMAs with high stereoregularity and their utilization to block and graft copolymer synthesis, *Polym. Prepr. Am. Chem. Soc. Div. Polym. Chem.,* 29(2), 54, 1988.

408. **Hatada, K., Ute, K., Nishimura, T., and Tanaka, K.,** Stereoregularities and NMR parameters of the polymer and oligomer of MMA prepared with *t*-C₄H₉MgBr, in *Prepr. IUPAC Int. Symp. Macromolecules,* Kyoto, 1988, 747.

409. **Smith, S. D.,** Synthesis of poly(methyl methacrylate) macromonomers via anionic polymerization, *Polym. Prepr. Am. Chem. Soc. Div. Polym. Chem.,* 29(2), 48, 1988.

410. **Hatada, K., Ute, K., Tanaka, K., and Kitayama, T.,** Stereochemistry of the oligomerization of methyl methacrylate by *t*-C₄H₉MgBr in toluene at −78°C, *Polym. J.,* 11, 1325, 1987.

411. **Braun, D., Herner, M., Johnsen, V., and Kern, W.,** Polymerisation von Methylmethacrylat mit alkaliorganischen Verbindungen. Über die stereospezifische Polymerisation mit alkaliorganischen Verbindungen. V, *Makromol. Chem.,* 51, 15, 1962.

412. **Andrews, G. D., Anderson, B. C., Arthur, P., Jacobson, H. W., Melby, L. R., Playtis, A. J., and Sharkey, W. H.,** Anionic polymerization of methacrylates. Novel functional polymers and copolymers, *Macromolecules,* 14, 1599, 1981.

413. **Huinh-Ba, G. and McGrath, J. E.,** Pyridine-mediated homo- and copolymerization of alkylmethacrylate, in *Recent Advances in Anionic Polymerization,* Hogen-Esch, T. E. and Smid, J., Eds., Elsevier, New York, 1987, 173.

414. **Ito, H., Miller, D. C., and Willson, C. G.,** Polymerization of methyl-α-(trifluoromethyl)acrylate and α-(trifluoromethyl)acrylonitrile and copolymerization of these monomers with methylmethacrylate, *Macromolecules,* 15, 915, 1982.

415. **Hatada, K., Sujino, H., Ise, H., Kitayama, T., Okamoto, Y., and Yuki, H.,** Heterocyclic polymers of α-substituted acrylic acid esters, *Polym. J.,* 12, 55, 1980.

416. **Ricca, G. and Severini, F.,** A carbon-13 and proton NMR study of the pyridine-initiated oligomerization of maleic anhydride and of the polymer structure, *Polymer,* 29, 880, 1988.

417. **Cram, D. J. and Sogah, D. Y.,** Chiral complexes polymerize methacrylate esters to give helical polymers that mutarotate by uncoiling, *J. Am. Chem. Soc.,* 107, 8301, 1985.

418. **Okamoto, Y., Suzuki, K., Ohta, K., Hatada, K., and Yuki, H.,** Optically active poly(triphenylmethyl methacrylate) with one handed helical conformation, *J. Am. Chem. Soc.,* 101, 4763, 1979.

419. **Okamoto, Y., Suzuki, K., and Yuki, H.,** Asymmetric polymerization of triphenylmethyl methacrylate by optically active anionic catalysis, *J. Polym. Sci. Polym. Chem. Ed.,* 18, 3043, 1980.

420. **Okamoto, Y., Yoshima, E., Nakano, T., and Hatada, K.,** Mechanism of asymmetric polymerization of triphenylmethyl methacrylate separation and optical resolution of oligomers, *Chem. Lett.,* 759, 1987.

421. **Okamoto, Y.,** Asymmetric polymerization of methacrylates and optical resolution of polymers, in *Prepr. IUPAC Int. Symp. Macromolecules,* Kyoto, 1988, 42.

422. **Okamoto, Y., Nakano, T., Yashima, E., and Hatada, K.,** Asymmetric oligomerization of triphenylmethyl methacrylate, in *Prepr. IUPAC Int. Symp. Macromolecules,* Kyoto, 1988, 692.

423. **Okamoto, Y., Nakano, T., and Hatada, K.,** Optical activity of optical activity of isotactic oligomers of methyl methacrylate, *Polym. J.,* 21, 199, 1989.

424. **Wulff, G., Sezepan, R., and Steigel, A.,** Chirality of polyvinyl compounds. IV. Stereoregulation during the oligomerization of trityl methacrylate, *Tetrahedron Lett.,* 27, 1991, 1986.

425. **Hatada, K., Ute, K., Nishimura, T., Kashiyama, M., Saito, T., and Takeuchi, M.,** Stereoregular oligomers of methyl methacrylate. IV. Supercritical fluid chromatography of the isotactic oligomers of MMA, *Polym. Bull.,* 23, 157, 1990.

426. **Tiess, C. F. and Hogen-Esch, T. E.,** Oligomerization stereochemistry of vinyl monomers. V. Methyl, isopropyl, and *tert*-butyl acrylates, *J. Polym. Sci. Polym. Chem. Ed.,* 17, 281, 1979.

427. **Soum, A. and Hogen-Esch, T. E.,** Stereochemical kinetics of anionic vinyl polymerization. II. ^{13}C NMR analysis of poly(2-vinylpyridine) and poly(4-vinylpyridine) terminated with labeled end groups, *Macromolecules,* 18, 690, 1985.

428. **Khan, I. M. and Hogen-Esch, T. E.,** Role of E- and Z-carbanions in the stereochemistry of anionic vinyl polymerization, in *Recent Advances in Anionic Polymerization,* Hogen-Esch, T. E. and Smid, J., Eds., Elsevier, New York, 1987, 261.

429. **Khan, I. M. and Hogen-Esch, T. E.,** Oligomerization of vinyl monomers. XVI, *Makromol. Chem. Rapid Commun.,* 4, 569, 1983.

430. **Huang, S. S., Mathis, C., and Hogen-Esch, T. E.,** Oligomerization of vinyl monomers. IX. ^{13}C NMR and chromatographic studies of oligomers of 2-vinylpyridine, *Macromolecules,* 14, 180, 1981.

431. **Soum, A. H., Huang, S. S., and Hogen-Esch, T. E.,** Stereochemical analysis of vinyl polymers. ^{13}C NMR of labelled terminal CH_3 groups, *Polym. Prepr. Am. Chem. Soc. Div. Polym. Chem.,* 24(2), 149, 1983.

432. **Meverden, C. and Hogen-Esch, T. E.,** Oligomerization of vinyl monomers. XV. Thermodynamics of intramolecular cation coordination in the anionic polymerization of 2-vinyl pyridine, *Makromol. Chem. Rapid Commun.,* 4, 563, 1983.

433. **Hogen-Esch, T. E. and Jenkins, W. L.,** NMR and conductometric studies of 2-pyridyl-substituted carbanions. II. Effects of cation size and coordination, *J. Am. Chem. Soc.,* 103, 3666, 1981.

434. **Hogen-Esch, T. E. and Dimov, D.,** Stereochemistry of anionic vinyl polymerization. Stereochemical roles of E- and Z-isomers, *Polym. Prepr. Am. Chem. Soc. Div. Polym. Chem.,* 30(2), 429, 1989.

435. **Khan, I. M. and Hogen-Esch, T. E.,** Stereoregular oligomerization and polymerization of 3-methyl-2-vinyl pyridine, *Polym. Prepr. Am. Chem. Soc. Div. Polym. Chem.,* 26(2), 273, 1985.

436. **Yuki, H., Hatada, K., Niinomi, T., and Kikuki, Y.,** Stereospecific polymerization of trityl, diphenyl-methyl, and benzyl methacrylates, *Polym. J.,* 1, 36, 1970.

437. **Toreki, W., Hogen-Esch, T. E., and Butler, G. B.,** Synthesis of macrocyclic poly(2-vinylpyridine), *Polym. Prepr. Am. Chem. Soc. Div. Polym. Chem.,* 28(2), 343, 1987.

438. **Toreki, Wm., Hogen-Esch, T. M., and Butler, G. B.,** Synthesis of macrocyclic poly(2-vinyl pyridine) and macrocyclic(N-alkyl-2-vinylpyridinium) bromide, a cyclic polyelectrolyte, *Polym. Prepr. Am. Chem. Soc. Div. Polym. Chem.,* 29(2), 17, 1988.

439. **Hogen-Esch, T. E. and Toreki, W.,** Synthesis of macrocyclic and star-vinyl polymers, *Polym. Prepr. Am. Chem. Soc. Div. Polym. Chem.,* 30(1), 129, 1989.

440. **Buese, M. A. and Hogen-Esch, T. E.,** The dimerization of vinyl phenyl sulfoxide as a model to elucidate the stereochemistry of poly(vinyl phenyl sulfoxide) stereoselectivity and stereoelectivity, in *Proc. 28th IUPAC Macromolecular Symp.,* Amherst, 1982, 131.

441. **Boor, J., Jr. and Finch, A. M.,** Polymerization of phenyl and methyl vinyl sulfones with anionic type initiators, *J. Polym. Sci. Part A,* 9, 249, 1971.

442. **Mulvaney, J. E. and Ottaviani, R. A.,** Anionic polymerization of vinyl sulfoxides, *J. Polym. Sci. Part A,* 8, 2293, 1970.

443. **Buese, M. A. and Hogen-Esch, T. E.,** The dimerization of vinyl phenyl sulfoxides, *Macromolecules,* 17, 118, 1984.

444. **Buese, M. A. and Hogen-Esch, T. E.,** Oligomerization of vinyl sulfoxide monomer, *J. Am. Chem. Soc.,* 107, 4509, 1985.

445. **Buese, M. A. and Hogen-Esch, T. E.,** Stereochemistry of the anionic polymerization of vinyl phenyl sulfoxide, *Polym. Prepr. Am. Chem. Soc. Div. Polym. Chem.,* 27(1), 155, 1986.

446. **Buese, M. A. and Hogen-Esch, T. E.,** Stereoelective anionic polymerization of chiral vinyl monomers *via* interconverting ion pair epimers, in *Recent Advances in Anionic Polymerization,* Hogen-Esch, T. E. and Smid, J., Eds., Elsevier, New York, 1987, 231.

447. **Lyons, A. R.,** Polymerization of vinyl ketones, *J. Polym. Sci. Macromol. Rev.,* 6, 251, 1972.

448. **Yasuda, Y., Kawabata, N., and Tsuruta, T.,** Elementary reactions of methyl alkyl in anionic polymerization. III. Reaction made of n-butyl magnesium bromide with α,β-unsaturated carbonyl compounds, *J. Macromol. Sci. Chem.,* A-1, 669, 1967; **Lyons, A. R. and Catteral, E.,** The organo-metal initiated polymerization of vinylketones. I. The butyllithium initiated polymerization of methyl isopropenyl ketone, *Eur. Polym. J.,* 7, 839, 1971.

449. **Ito, H. and Renaldo, F.,** Anionic polymerization of isopropenyl *t*-butyl ketone, *Polym. Prepr. Am. Chem. Soc. Div. Polym. Chem.,* 30(2), 164, 1989.
450. **Bell, B. C. and Hoge-Esch, T. E.,** Stereochemistry of the anionic oligomerization of *t*-butyl vinyl ketone, *Polym. Prepr. Am. Chem. Soc. Div. Polym. Chem.,* 26(1), 152, 1985.
451. **Bamford, C. H., Eastmond, G. C., and Imanishi, Y.,** Studies in α-peptide formation by hydrogen migration polymerization. I. Oligomers from maleamide, mesaconic acid α-methylester β-amide and mesaconamide, *Polymer,* 8, 651, 1967.
452. **Camino, G., Costa, L., and Trossarelli, L.,** Formation of a cyclic oligomer by hydrogen transfer polymerization of acrylamide, *J. Polym. Sci. Polym. Chem. Ed.,* 18, 377, 1980.
453. **Völker, Th., Neuman, A., and Bauman, U.,** Anionic oligomerization, *Makromol. Chem.,* 63, 182, 1963.
454. **Shimimura, T., Ono, K., Tsuchida, E., and Shinohara, I.,** Anionic oligomerization of alkyl methacrylate by sodium methoxide, *Kogyo Kagaku Zasshi,* 71, 1070, 1968; *Chem. Abstr.,* 69, 97251t, 1968.
455. **Feit, B. A., Wallach, J., and Zilkha, A.,** Anionic polymerization and oligomerization of methacrylonitrile by alkali metal alkoxides, *J. Polym. Sci. Part A,* 2, 473, 1964.
456. **Feit, B. A., Heller, E., and Zilkha, A.,** Anionic oligomerization of methacrylonitrile, *J. Polym. Sci. Part A,* 4, 1499, 1966.
457. **Teichman, G. and Zilkha, A.,** Anionic oligomerization of acrylic monomers, *J. Polym. Sci.,* 43, 25, 1962.
458. **McClure, J. D.,** French Patent 1,411,003, 1965.
459. **Feldman, J. and Saffer, B. A.,** French Patent 1,388,144, 1965.
460. **Feit, B. A.,** Anionic polymerization of acrylic esters, *Eur. Polym. J.,* 3, 523, 1967.
461. **McClure, J. D.,** U.S. Patent 3,277,745, 1966.
462. **McDonald, R. N. and Chowdhung, A. K.,** Gas-phase anionic oligomerization of methacrylonitrile initiated by F$_3$C$^-$, NCCH$_2$ and allyl anions, *J. Am. Chem. Soc.,* 105, 2194, 1983.
463. **Feit, B. A.,** Base catalysed oligomerization of vinyl monomers. III. A general route to the stepwise synthesis of vinyl oligomers, *Eur. Polym. J.,* 8, 321, 1972.
464. **Berlin, A. A., Korolev, B. G., Kefeli, T. Ya., and Sivergin, Yu, M.,** *Acrylated Oligomers and Materials Based on Them,* Khimia, Moscow, 1983, chap. 1 (in Russian).
465. **Berlin, A. A., Kefeli, T. Ya., and Korolev, A. V.,** *Polyesteracrylates,* Nauka, Moscow, 1967, 54 (in Russian).
466. **Nishikubo, T., Takehara, E., Saita, S., and Matsumura, T.,** Study of photopolymers. XXX. Syntheses of new di(meth)acrylate oligomers by addition reactions of epoxy compounds with active esters, *J. Polym. Sci. Polym. Chem. Ed.,* 25, 3045, 1987.
467. **Berlin, A. A., Kefeli, T. A., Sivergin, Yu. M., Kireyeva, S. M., Marshavina, N. L., and Filippovskaya, Yu. M.,** Correlation between the chemical nature and the characteristics of polymers based on oligocarbonate acrylates, *Plast. Massy,* No. 5, 33, 1974.
468. **Howard, D. D. and Heize, R. E.,** Ultraviolet cure of acrylourethane in air; influence of oligomer structure, *Polym. Mater. Sci. Eng.,* 60, 562, 1989.
469. **Berlin, A. A., Matveeva, N. G., and Pankova, E. S.,** Oligoester acrylates, *Vysokomol. Soedin. Ser. A,* 9, 1326, 1967.
470. **Novitskyi, E. G. and Slugyna, N. D.,** Synthesis of polymerizable organosilicon oligomers ω,ω'-bis(2-methacryloxyethyl) poly(dialkyl (and alkylaryl)siloxanes), *Plast. Massy,* No. 8, 21, 1965.
471. **Suzuki, T. and Okawa, T.,** Polydimethylsiloxane monomers having alkenyl functionalities prepared by ''initiator method'', Prepr. IUPAC Int. Symp. Macromolecules, Kyoto, 1988, 131.
472. **Rempp, P. and Franta, E.,** Synthesis and applications of macromonomers, in *Recent Advances in Anionic Polymerizations,* Hogen-Esch, T. E. and Smid, J., Eds., Elsevier, New York, 1987, 353.
473. **Anderson, B. C. and Andrews, G. D.,** Anionic polymerization of methacrylates. Novel functional polymers and copolymers, *Macromolecules,* 14, 1599, 1981.
474. **Rempp, P., Lutz, P., Masson, P., and Franta, E.,** Oligomerization of alkylmethacrylate monomers, *Makromol. Chem. Suppl.,* 8, 3, 1984.
475. **Lutz, P., Masson, P., Beinert, G., and Rempp, P.,** Synthesis and characterization of polyalkylmethacrylate macromonomers, *Polym. Bull.,* 12, 79, 1984.
476. **Rao, P. R., Masson, P., Lutz, P., Beinert, G., and Rempp, P.,** Synthesis and characterization of polyvinylpyridine macromonomers, *Polym. Bull.,* 11, 115, 1984.
477. **Takaki, M., Asami, R., Tanaka, S., Hayashi, H., and Hogen-Esch, T. E.,** Preparation of (p-vinyl benzyl)poly(2-vinylpyridine) monomers, *Macromolecules,* 19, 2900, 1986.
478. **Severini, S., Pegoraro, M., and Saija, L.,** Copolymerization of 4-vinylpyridine monomers with butyl acrylate, *Angew. Makromol. Chem.,* 133, 111, 1985.
479. **Hoover, J. M., Smith, S. D., Smith, S. M., DeSimone, J. M., Ward, T. C., and McGrath, J. E.,** Gas permeability of well-defined poly(alkyl methacrylate-7-poly(dimethylsiloxane) graft copolymers, *Polym. Prepr. Am. Chem. Soc. Div. Polym. Chem.,* 28(2), 390, 1987.

480. **Penczek, S. and Kazanskii, K. S.**, Ionic polymerization of heterocycles, *Vysokomol. Soedin. Ser. A*, 25, 1347, 1983.

481. **Marton, M. and Kammereck, R. F.**, Nucleophilic substitution at bivalent sulfur. Reaction of alkyllithium with cyclic sulfides, *J. Am. Chem. Soc.*, 92, 3217, 1970.

482. **Marton, M., Kammereck, R. F., and Fetters, L. J.**, Synthesis and properties of block polymers. II. Poly(α-methylstyrene)-poly(propylene sulfide)poly(α-methylstyrene), *Macromolecules*, 4, 11, 1971.

483. **Komarov, B. A., Volkov, V. P., Boiko, G. N., Naidovskii, E. S., and Rozenberg, B. A.**, Fluorine-19 NMR study of the direction of epoxide ring opening in the noncatalytic interaction of α-oxides with phenols and alcohols, *Vysokomol. Soedin. Ser. A*, 25, 1431, 1983.

484. **Hofman, A., Slomkowski, S., and Penczek, S.**, Structure of active centers and mechanism of the anionic polymerization of lactones, *Makromol. Chem.*, 185, 91, 1984.

485. **Sosnowski, S., Duda, A., Slomkowski, S., and Penczek, S.**, Determination of the structure of active centers in the anionic polymerization by ^{31}P NMR, introducing a P-containing end group, *Makromol. Chem. Rapid Commun.*, 5, 551, 1984.

486. **Penczek, S. and Slomkowski, S.**, Progress in anionic ring-opening polymerization, *Polym. Prepr. Am. Chem. Soc. Div. Polym. Chem.*, 27(1), 171, 1986.

487. **Sebenda, J. and Hauer, J.**, Living polymerization of lactams and synthesis of monodisperse polyamides, *Polym. Bull.*, 5, 529, 1981.

488. **Frunze, T. M. et al.**, The nature of the counter ion in the anionic polymerization of ϵ-caprolactam, in *Abstr. IUPAC Symp. Macro '83*, Bucharest, 1983, 233.

489. **Sekiguki, H. and Boi, C.**, Kinetical analysis of the ''activated monomer'' type polymerization, in *Abstr. IUPAC Int. Symp. Macromolecules*, Kyoto, 1988, 95.

490. **Amonyal, M., Coutin, B., and Sekiguki, H.**, Polymerization of α-aminoisobutyric acid NCA: interpretation according to the hypothesis of a multiple mechanism, *J. Macromol. Sci. Chem.*, A20, 675, 1983.

491. **Harwood, H. J.**, Mechanism of N-carboxy anhydride polymerization, in *Ring-Opening Polymerization. Kinetics, Mechanisms, and Synthesis*, ACS Symp. Ser. 286, American Chemical Society, Washington, D.C., 1985, chap. 5.

492. **Sosnowski, S., Slomkowski, S., and Penczek, S.**, Kinetics of anionic polymerization of ϵ-caprolactone (ωCL). Propagation of poly-Σ Cl$^-$K$^+$ ion pairs, *J. Macromol. Sci. Chem.*, A20, 979, 1983.

493. **Boileau, S.**, Anionic polymerization of cyclosiloxanes with cryptates as counterions, in *Ring-Opening Polymerization. Kinetics, Mechanisms, and Synthesis*, ACS Symp. Ser. 286, McGrath, E. J., Ed., American Chemical Society, Washington, D.C., 1985, chap. 2.

494. **Arkhipovitch, G. N., Dubravskii, S. A., Kazanskii, K. S., and Shupik, A. N.**, Formation of complexes of Na$^+$ ions with polyethylene-glycol, *Vysokomol. Soedin. Ser. A*, 23, 1653, 1981.

495. **Dimov, D. K., Panayotov, I. M., Lazarov, V. N., and Tsvetanov, C. B.**, Complex formation between poly(ethylene oxide) and alkali picrates, *J. Polym. Sci. Polym. Chem. Ed.*, 20, 1389, 1982.

496. **Tien, C. F. and Hogen-Esch, T. E.**, Oligomerization stereochemistry of vinyl monomers. II. Effect of ion pair structure on the methylation stereochemistry of 1,3-di(2-pyridyl) butane anion, *J. Am. Chem. Soc.*, 98, 7109, 1976.

497. **Müller, A. H.**, Anionic polymerization of β-propiolactone, in *Anionic Polymerization: Kinetics, Mechanisms, and Synthesis*, McGrath, J. E., Ed., ACS Symp. Ser. 166, American Chemical Society, Washington, D.C., 1981, 441.

498. **Slomkovski, S. and Penczek, S.**, Macroions and macroion pairs in the anionic polymerization of β-propiolactone, *Macromolecules*, 13, 229, 1980.

499. **Slomkovski, S.**, Ring-opening polymerization with kinetically controlled enhancement of macrocyclic oligomers or linear macromolecules, *Makromol. Chem.*, 186, 2581, 1985.

500. **Penczek, S., Lapiens, G., and Klosinski, P.**, Synthetic analogues of phosphorous containing biopolymers, *Pure Appl. Chem.*, 56, 1309, 1984.

501. **Pretula, J., Kaluzinski, K., and Penczek, S.**, Living reversible anionic polymerization of N,N-diethylamine-1,3,2-dioxaphosphorinan, *J. Polym. Sci. Polym. Chem. Ed.*, 22, 1251, 1984.

502. **Bailey, F. E. and Koleske, J. V.**, *Polyethylene Oxide*, Academic Press, New York, 1976, chap. 1.

503. **McGrath, J. E.**, Ring-opening polymerization: introduction, in *Ring-Opening Polymerization. Kinetics, Mechanisms, and Synthesis*, ACS Symp. Ser. 286, McGrath, J. E., Ed., American Chemical Society, Washington, D.C., 1985, chap. 1.

504. **Penczek, S.**, Progress in anionic polymerization, in *Anionic Polymerization: Kinetics, Mechanisms, and Synthesis*, ACS Symp. Ser. 166, McGrath, J. E., Ed., American Chemical Society, Washington, D.C., 1981, 271.

505. **Masson, P., Beinert, G., Franta, E., and Rempp, P.**, Synthesis of polyethylene oxide monomers, *Polym. Bull.*, 7,17, 1982.

506. **Kobayashi, S., Kaku, M., Mizutani, T., and Saegusa, T.**, Preparation of ring-opening polymerizable monomers and its copolymerization leading to graft copolymer, *Polym. Bull.*, 9, 169, 1983.

507. **Hamaide, T., Revillon, A., and Gyot, A.,** Reactivité de macromères du polyethyléne en copolymerisation radicalaire, *Eur. Polym. J.,* 20, 855, 1984; **Sigwalt, P.,** Ring-opening polymerization of heterocyclics with organometallic catalysts, *Angew. Makromol. Chem.,* 94, 161, 1101, 1981.

508. **Kazanskii, K. S. and Ptitsyna, N. V.,** New functional poly(ethylene oxide)s: synthesis and application, *Makromol. Chem.,* 190, 225, 1989.

509. **Guhaniyogi, S. G., Kennedy, J. P., and Ferry, W. M.,** Carbocationic polymerization in the presence of sterically hindered bases. III. Polymerization of isobutylene by the cumyl chloride/BCl₃ system, *J. Macromol. Sci. Chem.,* A18, 25, 1982.

510. **Masson, P., Franta, E., and Rempp, P.,** Synthèse et homopolymérisation de macromères de polystyrène, *Makromol. Chem. Rapid Commun.,* 3, 499, 1982.

511. **Berlinova, I. V. and Panayotov, I. M.,** Synthesis of amphiphilic star-shaped and graft copolymers based on polystyrene and poly(ethylene oxide), *Makromol. Chem.,* 190, 1515, 1989.

512. **Chujo, Y., Tatsuda, T., and Yamashita, Y.,** Synthesis of polyamide-poly(methyl methacrylate) graft copolymers by polycondensation reactions of macromonomers, *Polym. Bull.,* 5, 361, 1981.

513. **Kazanskii, K. S., Solovyanov, A. A., and Entelis, S. E.,** Polymerization of ethylene oxide by alkali metal-naphthalene complexes in THF, *Eur. Polym. J.,* 7, 1421, 1971.

514. **Cabasso, I. and Zilkha, A.,** Observations on the anionic polymerization of ethylene oxide by alkali metal naphthalene and anthracene complexes, *J. Macromol. Sci. Chem.,* A8, 1313, 1974.

515. **Ito, K., Hashizuka, Y., and Yamashita, Y.,** Equilibrium cyclic oligomer formation in the anionic polymerization of ε-caprolactone, *Macromolecules,* 10, 821, 1977; **Yamashita, Y.,** Anionic polymerization of lactones for the block copolymer syntheses, *Polym. Prepr. Am. Chem. Soc. Div. Polym. Chem.,* 21(1), 51, 1982.

516. **Isobe, T., Miho, T., and Yamamoto, E.,** Japanese Patent 62,179,525, 1987; *Chem. Abstr.,* 108, 95151k, 1988.

517. **Sebenda, J.,** Recent progress in the polymerization of lactams, *Prog. Polym. Sci.,* 6, 123, 1978 (and references cited therein).

518. **Joyce, R. M. and Ritter, D. M.,** U.S. Patent, 2,251,519, 1941.

519. **Andrews, J. M., Jones, F. R., and Semlyen, J. A.,** Equilibrium ring concentrations and the statistical conformations of polymer chains. XII. Cyclics in molten and solid nylon-6, *Polymer,* 15, 420, 1974.

520. **Bergwerf, W., Wagner, W., and Marius, K.,** British Patent 1,113,293, 1968; *Chem. Abstr.,* 70, 29561, 1969; **Biagini, E., Costa, G., Russo, S., Turturro, A., and Riva, F.,** Cyclic oligomer content in anionic poly(ε-caprolactam), *Makromol. Chem. Rapid Commun.,* 6, 207, 1986.

521. **King, C.,** U.S. Patent, 3,418,393, 1968; *Chem. Abstr.,* 70, 48558, 1969; **Tsuruta, T.,** Stereoelective and asymmetric-selective (or stereoelective) polymerizations, *J. Polym. Sci. Part D,* 6, 179, 1972.

522. **Sigwald, P.,** Stereoselection and stereoelectivity in the ring-opening polymerization of epoxides and episulfides, *Pure Appl. Chem.,* 48, 257, 1976.

523. **Wanigatunga, S. and Wagener, K. B.,** Segmented copolymers with monodisperse segments, *Polym. Prepr. Am. Chem. Soc. Div. Polym. Chem.,* 29(2), 63, 1988; **Sepulchre, M., Momtaz, A., and Spassky, N.,** Chiral anionic coordinated polymerization of thiiranes: enantiomeric resolution and disulfide linkage polymerization, in *Recent Advances in Anionic Polymerization,* Hogen-Esch, T. E. and Smid, J., Eds., Elsevier, New York, 1987, 297.

524. **Boileau, S., Champetier, G., and Sigwald, P.,** Polymérisation anionique du sulfure de propylène, *Makromol. Chem.,* 69, 180, 1963; **Wagener, K. B. and Wanigatunga, S.,** Synthesis and characterisation of poly(oxyethylene-b-poly(pivalolactone) telechelomer, in *Chemical Reactions on Polymers,* ACS Symp. Ser. 364, Behnam, J. L. and Kinstle, J. F., Eds., American Chemical Society, Washington, D.C., 1988, chap. 12.

525. **Lenz, R. and Bigdelli, E.,** Polymerization of α,ω-disubstituted β-propiolactones and lactames. XIV. Substituent, solvent and counterion effects in the abionic polymerization of lactones, *Macromolecules,* 11, 493, 1978; **Cooper, W., Horgan, D. R., and Wragg, R. T.,** Initiators for the polymerisation of porpylene sulfide, *Eur. Polym. J.,* 5, 71, 1969.

526. **Ito, K., Tomida, M., and Yamashita, Y.,** Ring-chain equilibrium in the anionic polymerization of δ-valerolactame, *Polym. Bull.,* 1, 569, 1979; **Cooper, W.,** Poly(alkylene sulfides), *Br. Polym. J.,* 3, 28, 1971.

527. **Lee, C. L., Fryl, C. L., and Johannson, O. K.,** Selective polymerization of reactive cyclosiloxanes to give non-equilibrium molecular weight distributions. Monodisperse siloxane polymers, *Polym. Prepr. Am. Chem. Soc. Div. Polym. Chem.,* 10(2), 1361, 1969.

528. **Boileau, S.,** Anionic polymerization of cyclosiloxanes with cryptates as counterions, in *Ring-Opening Polymerization. Kinetics, Mechanisms, and Synthesis,* ACS Symp. Ser. 286, McGrath, J. E., Ed., American Chemical Society, Washington, D.C., 1985, chap. 2.

529. **Yiligor, I., Tyagi, D., Sha'ban, A., Steckle, W. S., Jr., Wilkes, G. L., and McGrath, J. E.,** Segmental organosiloxane copolymers, *Polymer,* 25, 1800, 1984.

530. **Sormani, P. M., Minton, R. J., and McGrath, J. E.,** Anionic ring-opening polymerization of octamethylcyclotetrasiloxane in the presence of 1,3-bis(aminopropyl)-1,1,3,3-tetramethyldisoloxane, in *Ring-Opening Polymerization. Kinetics, Mechanisms, and Synthesis,* ACS Symp. Ser. 286, McGrath, J. E., Ed., American Chemical Society, Washington, D.C., 1985, chap. 11.

531. **Boutevin, B. and Youssef, B.,** Synthese des polysiloxanes téléchélique. I. Synthèse des diols et diépoxides, *Makromol. Chem.,* 190, 277, 1989.

532. **Stark, F. O., Falender, J. R., and Wright, A. P.,** in *Comprehensive Organometallic Chemistry,* Wilkinson, G., Ed., Pergamon Press, Oxford, 1983.

533. **Kawakami, Y. and Yamashita, Y.,** Synthesis and applications of polysiloxane macromers, in *Ring-Opening Polymerization. Kinetics, Mechanisms, and Synthesis,* ACS Symp. Ser. 286, McGrath, J. E., Ed., American Chemical Society, Washington, D.C., 1985, chap. 19.

534. **De Simone, J. M., York, G. A., Wilson, G. R., Smith, S. D., Marand, H., Gozdz, A. S., Bowder, M. J., and McGrath, J. E.,** Synthesis and characterization of poly(1-butene sulfone)-g-poly(dimethylsiloxane): a new electronbeam resist for two-layer lithography, *Polym. Prepr. Am. Chem. Soc. Div. Polym. Chem.,* 30(2), 134, 1989.

535. **Simionescu, C. I., Negulescu, I. I., Totolin, M. I., and Bloos, G.,** Cold abiotic synthesis of some precellular components, in *Protein Structure and Evolution,* Fox, J. L., Deyl, Z., and Blazej, A., Eds., Marcel Dekker, New York, 1976, chap. 10.

536. **Mattews, C. N.,** Unconventional but universal heteropolypeptides from hydrogen cyanide and water, *Polym. Prepr. Am. Chem. Soc. Div. Polym. Chem.,* 27(1), 103, 1986.

537. **Simionescu, C. I., Denes, F., and Negulescu, I. I.,** Abiotic synthesis and properties of some protobiopolymers, *J. Polym. Sci. Polym. Symp.,* 64, 281, 1978.

538. **Kresta, J. E. Shen, C. S., and Frisch, K. C.,** Cyclooligomerization of isocyanates, *Polym. Prepr. Am. Chem. Soc. Div. Coat. Plast.,* 36(2), 674, 1976.

539. **Kresta, J. E. Shen, C. S., and Frisch, K. C.,** Polymerization and cyclotrimerization of isocyanates catalysed by cyclic sulfonium zwitterions, *Makromol. Chem.,* 178, 2495, 1977.

540. **Kresta, J. E., Hsieh, K. H., Shen, C. S., and Frisch, K. C.,** Control of the cyclooligomerization of isocyanates by novel anionic initiators, *Polym. Prepr. Am. Chem. Soc. Div. Polym. Chem.,* 21(1), 72, 1980.

541. **Kresta, J. E., and Hsieh, K. H.,** Cyclotrimerization of isocyanates—catalysis and mechanism, *Polym. Prepr. Am. Chem. Soc. Div. Polym. Chem.,* 21(2), 126, 1980.

542. **Kresta, J. E. and Shen, C. S.,** U.S. Patent 4,111,914, 1978.

543. **Kresta, J. E. and Shen, C. S.,** Oligomerization of isocyanates by cyclic sulfonium zwitterions, *Polym. Bull.,* 1, 325, 1979.

544. **Hatch, M. J., Yoshimine, M., Schmidt, D. L., and Smith, H. B.,** U.S. Patent, 3,636,052, 1972.

545. **Schmidt, D. L., Smith, H. B., Yoshimine, M., and Hatch, M. J.,** Preparation and properties of polymers from aryl cyclic, sulfonium zwitterions, *J. Polym. Sci. Part A,* 10, 2951, 1972.

546. **Kresta, J. E., Lin, I. S., Shen, C. S., Hsieh, K. H., and Frisch, K. C.,** Polymerization and oligomerization of isocyanates, *Prepr. Am. Chem. Soc. Div. Org. Coat. Plast.,* 39, 540, 1978.

547. **Golev, V. G., Sorokina, T. K., Barkina, E. S., and Bondarenko, A. M.,** Cyclotrimerization of toluene diisocyanate in a butyl acetate solution, *Lakokras. Mater. Ikh Primen.,* 2, 11, 1988; *Chem. Abstr.,* 108, 223138a, 1988.

548. **Lin, S. C.,** U.S. Patent 4,379,728, 1983.

549. **Lin, S. C.,** N-cyanourea terminated resins, *Polym. Prepr. Am. Chem. Div. Polym. Chem.,* 25(1), 112, 1984.

550. **Korshak, V. V., Vinogradova, S. V., Pankratov, V. A., and Puchin, A. G.,** Polycyclotrimerization of arylcyanates and the investigation of polymer properties, *Dokl. Acad. Nauk S.S.S.R.,* 202, 347, 1972.

551. **Vinogradova, S. V., Parikratov, V A., Puchin, A. G., and Korshak, V. V.,** Dicyan esters of bisphenols, *Izv. Akad. Nauk S.S.S.R. Ser. Khim.,* 837, 1971.

552. **Cercena, J. L. and Huang, S. J.,** The use of arylcyanates as reactive terminal groups for thermally polymerizable oligomers, *Polym. Prepr. Am. Chem. Soc. Div. Polym. Chem.,* 25(1), 114, 1984.

553. **Wertz, D. H. and Prevorsek, D. C.,** SIPNS: a new class of high performance plastics, *Plast. Eng.,* 40, 31, 1984.

554. **Huang, S. G., Feldman, J. A., and Cercena, J. L.,** Polymeric and monomeric aryloxy-s-triazines, *Polym. Prepr. Am. Chem. Soc. Div. Polym. Chem.,* 30(1), 348, 1989.

555. **Pogosyan, G. M., Pankratov, V. A., Zaplishnyi, V. N., and Matsoyan, S. G.,** *Polytriazines,* Armean Academy Press, Erevan, 1987, chap. 5.

556. **Elias, H. G.,** *Macromolecules. II. Synthesis and Materials,* Plenum Press, New York, 1977, 928.

557. **Kamaki, M. and Murahashi, S.,** Polymerization of acetaldazine, *Polym. J.,* 4, 651, 1973.

558. **Vogl, O.**, Kinetics of aldehyde polymerization, in *Comprehensive Chemical Kinetics*, Bamford, C. H. and Tipper, C. F. H., Eds., Elsevier, Amsterdam, 1976, chap. 5.

559. **MacDonald, R. N.**, U.S. Patent 2,708,994, 1956.

560. **Kubisa, P., Negulescu, I., Hatada, K., Lipp, D., Starr, J., Yamada, B., and Vogl, O.**, New development in cationic and anionic aldehyde polymerization, *Pure Appl. Chem.*, 38, 275, 1978.

561. **Negulescu, I. and Vogl, O.**, Polymerization of higher aliphatic aldehydes, *J. Polym. Sci. Polym. Lett. Ed.*, 13, 17, 1975.

562. **Natta, G., Corradini, P., and Bassi, I. W.**, Cationic oligomerization of aldehydes, *J. Polym. Sci.*, 51, 505, 1961.

563. **Wood, J. S., Negulescu, I., and Vogl, I.**, Higher aliphatic polyaldehydes. Crystallinity, crystal structure and melting transitions of isotactic poly(n-heptaldehyde), *J. Macromol. Sci. Chem.*, A11, 171, 1977.

564. **Negulescu, I. and Vogl, O.**, Anionic polymerization of higher aldehydes, *J. Polym. Sci. Polym. Chem. Ed.*, 14, 2996, 1976.

565. **Corley, L. S., Jaycox, G. D., and Vogl, O.**, Haloaldehyde polymers. XXX. Macromolecular asymmetry as the basis of optical activity in polymers, *J. Macromol. Sci. Chem.*, A25, 519, 1988.

566. **Jaycox, G. D., Xi, F., Vogl, O., Hatada, K., Ute, K., and Nishimura, T.**, Oligomerization of chloral with lithium *t*-butoxide: stereochemistry of early chain growth steps, *Polym. Prepr. Am. Chem. Soc. Div. Polym. Chem.*, 30(2), 167, 1989.

567. **Kubisa, P., Corley, L. S., Kondo, T., Jacovic, M., and Vogl, O.**, Anionic oligomerization of aldehydes, *Polym. Eng. Sci.*, 21, 829, 1981.

568. **Sumitomo, H. and Hashimoto, K.**, Polymerization of α-cyanopropionaldehyde. V. Anionic polymerization initiated by benzophenone-alkali metal complexes, *J. Polym. Sci. Part A*, 7, 1331, 1969.

569. **Berlin, A. A. and Matveyeva, N. G.**, Progress in the chemistry of polyreactive oligomers and some trends in its development. I. Synthesis and physico-chemical properties, *J. Polym. Sci. Macromol. Rev.*, 12, 1, 1977.

570. **Boor, J., Jr.**, The nature of the active site in the Ziegler-type catalyst, *Macromol. Rev.*, 2, 115, 1967.

571. **Avdeikina, E. G., Tlenkopachev, M. A., and Korshak, Yu. V.**, Oligomerization of 2-methylpentene-1 on tungsten-containing metathesis catalysts, *Vysokomol. Soedin. Ser. A*, 28, 552, 1986.

572. **Minott, R., Fink, G., and Fenze, N.**, Ethylene insertion with soluble Ziegler catalyst. III. The system $Cp_2TiMeCl/AlMe_2Cl/^{13}C_2H_4$ studied by carbon-13 NMR spectroscopy. The time development of chain propagation and oligomer distribution, *Angew. Makromol. Chem.*, 154, 1, 1987.

573. **Mathys, G.**, British Patent Appl. 2,205,853, 1988; *Chem. Abstr.*, 110, 233619p, 1989.

574. **Kissin, Y. V.**, Co-oligomerization of ethylene and higher linear alpha olefins. II. Olefin reactivities in various reaction steps of co-oligomerization with nickel ylide-based system, *J. Polym. Sci. Part A*, 27, 623, 1989.

575. **Knudsen, R. D.**, The nickel-catalysed oligomerization of ethylene to alpha-olefins, *Prepr. Am. Chem. Soc. Div. Pet. Chem.*, 34(3), 572, 1989.

576. **Ascenso, J., Dias, A. R., Gomes, P. T., Romao, C. C., Neibecker, D., Tkatchenko, I., and Revillon, A.**, Cationic η³-allyl metal complexes. XII. Oligomerization of styrene with cationic allyl nickel compound catalysts. Products and the influence of phosphines, *Makromol. Chem.*, 190, 2773, 1989.

577. **Pino, P., Cioni, P., Galimbert, M., Wei, J., and Piccolrovazzi, N.**, Asymmetric hydrooligomerization of propylene, in *Proc. Int. Symp. Transition Matals and Organometallic Catal. Olefin Polymers*, Kaminsky, W. and Sinn, H., Springer, Berlin, 1988, 269.

578. **Vanderberg, E. J.**, Oligomerization of higher olefins, *J. Polym. Sci.*, 47, 486, 1960.

579. **Vanderberg, E. J.**, Co-oligomerization of higher olefins, *J. Polym. Sci. Part A*, 7, 525, 1969.

580. **Vanderberg, E. J. and Robinson, A. E.**, Coordination polymerization of trimethylene oxide, in *Polyethers*, Vanderberg, E. J., Ed., American Chemical Society, Washington, D.C., 1975, 101.

581. **Tsuruta, T.**, Stereoselective and asymmetric-selective (or stereoselective) polymerizations, *J. Polym. Sci. Part D*, 6, 179, 1972; **Tsuruta, T.**, Mechanistic features of polymerization of N-carboxy-α-aminoacid anhydrides—comparison with those of 1,2-epoxides, *Pure Appl. Chem.*, 48, 267, 1976.

582. **Wintzer, J., Mueller, H., and Fredtke, M.**, Oligomerization of phenyl glycidyl ether in the presence of zinc chloride, *Polym. Bull.*, 17, 31, 1987.

583. **Tezuka, Y., Ishimori, M., and Tsuruta, T.**, Asymmetric selective oligomerization of chloromethyloxirane, *Makromol. Chem.*, 184, 895, 1983.

584. **Taenzer, W., Mueller, H., Wintzer, J., and Fedtke, M.**, Oligomerization and cyclization of 1-phenoxy-2,3-epoxypropane induced by magnesium perchlorate, *Makromol. Chem.*, 188, 2857, 1987.

585. **Soga, K., Tazuke, Y., Hasoda, S., and Ikeda, S.**, Polymerization of propylene carbonate, *J. Polym. Sci.*, 15, 219, 1977.

586. **Aida, T. and Inoue, S.**, Living polymerization of epoxide catalysed by the porphyrin-Et_2AlCl system. Structure of the living end, *Macromolecules*, 14, 1166, 1981.

587. **Yasuda, T., Aida, T., and Inoue, S.**, Living oligomerization of epoxide, *Makromol. Chem. Rapid. Commun.*, 3, 585, 1982.

588. **Aida, T. and Inoue, S.,** Living polymerization of epoxides with metalloporphyrin and synthesis of block copolymers with controlled chain lengths, *Macromolecules,* 14, 1162, 1981.

589. **Yasuda, T., Aida, T., and Inoue, S.,** Synthesis of end-reactive oligomers with controlled molecular weight by metalloporphyrin catalyst, *Polym. Prepr. Am. Chem. Soc. Div. Polym. Chem.,* 25(1), 146, 1984.

590. **Inoue, S.,** Mechanistic aspects of the polymerization of epoxide lactones and CO_2-epoxide with metalloporphyrins as novel catalyst, *Polym. Prepr. Am. Chem. Soc. Div. Polym. Chem.,* 25(1), 225, 1984.

591. **Inoue, S. and Aida, T.,** in *Ring-Opening Polymerization. Kinetics, Mechanisms, and Synthesis,* ACS Symp. Ser. 286, McGrath, J. E., Ed., American Chemical Society, Washington, D.C., 1985, 137.

592. **Aida, T., Kuroki, M., and Inoue, S.,** Visible-light mediated "living" polymerization of alkyl methacrylates initiated by aluminium porphyrin, *Abstr. IUPAC Int. Symp. Macromolecules,* Kyoto, 1988, 101.

593. **Anderson, B. C., Andrews, G. D., Arthur, P., Jr., Jacobson, H. W., Melby, L. R., Playlis, A. J., and Sharkey, W. H.,** Living polymerization of alkyl methacrylates, *Macromolecules,* 14, 1599, 1981.

594. **Boetcher, F. P.,** Recent progress in the preparation of functional methacrylate polymers, *J. Macromol. Sci. Chem.,* A22, 665, 1985 (and references cited therein).

595. **Webster, O. W., Hertler, W. R., Sogah, D. Y., Farnham, W. B., and Rajanbabu, T. V.,** Synthesis of functional MMA polymers, *J. Am. Chem. Soc.,* 105, 5704, 1983.

596. **Webster, O. W.,** U.S. Patent, 4,417,034, 1983.

597. **Farnham, W. B. and Sogah D. Y.,** U.S. Patent, 4,414,372, 1983.

598. **Farnham, W. B. and Sogah, D. Y.,** U.S. Patent, 4,524,196, 1985.

599. **Webster, O. W.,** U.S. Patent, 4,508,880, 1985.

600. **Farnham, W. B. and Sogah, D. Y.,** Group transfer polymerization. Mechanistic studies, *Polym. Prepr. Am. Chem. Soc. Div. Polym. Chem.,* 27(1), 167, 1986.

601. **Sogah, D. Y. and Webster, O. W.,** Group transfer polymerization, *J. Polym. Sci. Polym. Lett. Ed.,* 21, 927, 1983.

602. **De Simone, Hallstern, A. M., Siochi, E. J., Ward, T. C., and McGrath, J. E.,** Model branched poly(methyl methacrylate)s of controlled molecular weight and architecture: synthesis, *Polym. Prepr. Am. Chem. Soc. Div. Polym. Chem.,* 30(1), 137, 1989.

603. **Asami, R., Kondo, Y., and Takaki, M.,** Synthesis of poly(methyl methacrylate) monomer by group transfer polymerization and polymerization of the monomer, *Polym. Prepr. Am. Chem. Soc. Div. Polym. Chem.,* 27(1), 186, 1986.

604. **Quirk, R. P. and Bidinger, G. P.,** Ring-opening polymerization of propylene sulfide using group transfer polymerization procedure, *Polym. Prepr. Am. Chem. Soc. Div. Polym. Chem.,* 29(2) 120, 1988.

605. **Percek, V., Pugh, C., and Pask, S. D.,** Macromonomers, oligomers and telechelic polymers, in *Comprehensive Polymer Science,* Vol. 6, Eastmond, G. C. et al., Eds., Pergamon Press, Oxford, 1988, 281.

606. **Thiem, J., Strietholt, W. A., and Höring, T.,** Ring-opening polymerization of carbohydrate-derived dialkoxyoxolanes, *Makromol. Chem.,* 190, 1737, 1989.

607. **Thiem, J. and Höring, T.,** Synthesis of functionalized poly(oxytetramethylene)s, *Makromol. Chem.,* 187, 711, 1987.

PART II

Characterization

INTRODUCTION

Randomly or purposely formed through polymerization reactions, oligomers form multicomponent systems within which the molecular species are distinguished through their molecular mass, shape, or chemical composition. The nature, position, and distribution of the functional groups complete the nonhomogeneity of these systems. Thus, the main aims of oligomer characterization are (1) their global separation from the polymer, (2) determination of the value and distribution of the molecular mass, (3) individual isolation of some species, (4) determination of their chemical structure, and, finally, (5) establishment of certain structure-molecular mass relations.

Frequently, oligomers are casually formed during polycondensation or polymerization reactions. In such situations, they represent an undesirable "traveling companion". Classical methods of fractionation based on solubility did not offer possibilities of evidencing and separating the oligomers formed simultaneously with the macromolecular compounds; that is why their presence has rarely been noticed. Only in the last few years, thanks to the introduction of gel permeation chromatography (GPC) techniques, has it been possible to analyze more accurately polymer mixtures for the presence of oligomers. This method permitted not only the separation of oligomers, but also the individual isolation of certain terms. Along with this, the dependence between a certain property and the value of the polymerization degree (DP) of the oligomers has been achieved. Generalizing, one can assume that, by settling the polymerization degree-property relation, representation of quantitative accumulation through new qualitative characteristics may be performed. Such a relation has also shown that oligomers represent a specific "qualitative state", defined through peculiar properties.

The value and distribution of the molecular mass are the major specific features of oligomers and polymers. The former is quantitative, while the latter represents mainly a qualitative attribute. The origins of the molecular mass distribution (MMD) in the oligomeric mixtures has to be found in the specific features characterizing step- and chain-growth reactions.

The nonhomogeneity of the chemical composition reflects some uncontrolled secondary reactions against the prevailing mechanism of the polymerization reaction.

The presence of the final groups and their number and distribution along the main chain of the species, led to the appearance of a new type of nonhomogeneity known as the *functionality distribution*. This extremely important distribution determines directly both the properties and the fields of oligomer utilization.

The distribution of molecular mass and functionality is characteristic of oligomers and polymers; it is the property that makes them *infinitely variable*. The same property prevents the attachment of the attribute of *chemically pure* to polymer or oligomer mixtures and explains why the value of their molecular mass is represented as an *average value*.

Chapter 2.1

SOLUBILITY OF OLIGOMERS

I. GENERAL CONSIDERATIONS

As the characterization of oligomers is mainly performed in solution, a short review of solubility and the quite different behavior of oligomer solutions, compared to that of solutions of low molecular or macromolecular solutes, is appropriate.

A solution is defined as a solid or liquid homogeneous system consisting of two or more components whose relative amounts may vary within broad limits.[1,2]*

The homogeneity of solutions makes them very similar to chemical compounds. Solutions differ from chemical compounds in that the composition of a solution can change within broad limits. Moreover, the properties of a solution display many properties of its individual components, which is not observed for a chemical compound. The variability of the composition of solutions brings them close to mechanical mixtures, but they differ sharply from the latter in their homogeneity. Solutions are thus intermediate between mechanical mixtures and chemical compounds.

A. SOLUTION OF LOW MOLECULAR MASS COMPOUNDS

A crystalline low molecular weight compound dissolves in a liquid as follows. When the crystal is introduced into a liquid in which it can dissolve, separate molecules break away from its surface and, owing to diffusion, become uniformly distributed throughout the entire volume of the solvent. The separation of the molecules from the surface of the solid is due, *inter alia,* to their intrinsic vibrational motion and to attraction by the solvent molecules. The molecules that have passed into the solution collide with the surface of the still undissolved substance, are again attracted to it, and enter the composition of its crystals. It is quite obvious that the molecules will leave the solution at a faster rate when their concentration in the solution is higher. And since the latter grows as the substance dissolves, a moment finally arrives when the rate of dissolution becomes equal to that of crystallization. We now have dynamic equilibrium, when the same number of molecules that dissolve in unit time crystallize from the solution. A solution in equilibrium with the dissolving substance is called a *saturated solution.*

The process of dissolution is attended by a considerable growth in the entropy of a system because its number of microstates grows sharply as a result of the uniform distribution of the particles of one substance in the other. Consequently, notwithstanding the endothermic nature of the dissolving of most substances, the change in the Gibbs energy of a system in dissolution is negative and the process goes on spontaneously.

A useful representation of equilibrium between liquid phases in a binary system is the conventional *phase diagram* obtained by plotting the compositions (e.g., the mole fraction x_1 of component 1 or $x_2 = 1 - x_1$ for component 2) against temperature, i.e., the projection

---

* *A homogeneous system* is defined as one whose chemical composition and physical properties are the same in all parts of the system, or change continuously from one point to another. *A heterogeneous system* is defined as one consisting of two or more homogeneous *phases.* Then a homogeneous system or a phase of a heterogeneous system that consists of several substances is called a solution or mixture. Gibbs[3] was the first to give, in his famous monograph, "The Equilibrium of Heterogeneous Substances", the definition of a phase: "In considering the different homogeneous bodies which can be formed out of any set of component substances, it is convenient to have a term which shall refer solely to any such body without regard to its size or form. The word *phase* has been chosen for this purpose. Such bodies differing in composition or state are called different phases of the matter considered, all bodies which differ only in size and form being regarded as different examples of the same phase."

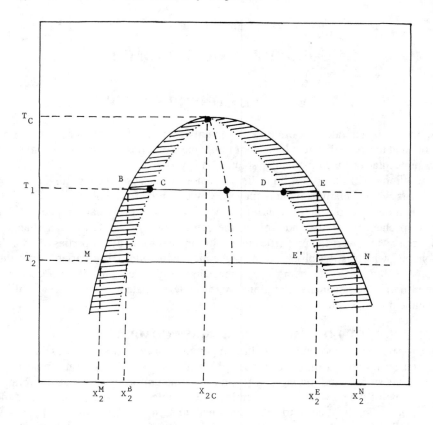

FIGURE 1.   Typical phase diagram in the TX plane for a binary system exhibiting an upper critical solution temperature; binodal (solid curve) and spinodal (dotted). BE and MN are tie lines at temperatures T_1 and T_2. The curve bisecting the tie lines marks the overall compositions giving phases of equal quantity (moles). The critical points are at T_c and X_{2c}. The shaded area corresponds to the region in which the existence of metastable states is possible.

of the three-dimenstional free-energy plot onto the T,X plane. In the schematic representation in Figure 1, the isothermal line BE represents the phases with different compositions that are in equilibrium at given temperatures T_2, T_1, etc. At higher temperatures, the solubility gap BE narrows and, at T_c, it vanishes.

If the one-phase system at $T > T_c$ with the critical composition x_{2c} is cooled, it separates into two phases of only infinitesimally different composition at T_c. For another composition, say $x_2^{(E)}$, the one-phase system can be cooled below T_c to T_1, where an infinitesimal amount of the phase of composition $x_2^{(B)}$ appears; and as the system is cooled further to T_2 along EE', the new phase grows in extent as both phases change composition. Since the coexistence curve maps the terminal points (nodes) of the horizontal *tie lines* connecting the pairs of phases, it is called *binodal* (or *binodial*). The tie lines give the relative amounts of the phases according to the lever rule: e.g., at temperature T_2, the lengths of line segments ME' and E'N are proportional to the number of moles of the phases with compositions $x_2^{(N)}$ and $x_2^{(M)}$, respectively. If the compositions were plotted in mass fractions, the ratio of line segments would give the ratio of the masses of the phases. It is clear that the binodal can also be defined as the boundary of the region of states outside of which the existence of a substance in a two-phase state is impossible.

The dotted curve in Figure 1, called a *spinodal* (or *spinodial*), is the locus of the projection of inflexion points in the ΔG_M vs. x representation onto the T,x coordinate plane.

$$\Delta G_M = \Delta H_M - T\Delta S_M \qquad (1)$$

where ΔG_M, ΔH_M, and ΔS_M are the free energy of mixing, heat of mixing, and entropy of mixing, respectively. Since the entropy of mixing, ΔS_M, is always large and positive, the sign of ΔG_M is determined by the magnitude and sign of ΔH_M. On the other hand, the spinodal can also be identified as the boundary of the region of the states inside of which the existence of a substance in a single-state phase is impossible.

The type of phase plot illustrated in Figure 1, with a single phase splitting into two as the temperature is lowered, is the one most often encountered in ordinary liquid mixtures and also the one commonly studied in polymer-solvent systems. The critical point is then described as an *upper critical solution temperature (UCST)*. However, systems are found (e.g., triethylamine-water) with the phase diagram convex downward, in which case the phase separation occurs on raising the temperature, and T_c is called the *lower critical solution temperature (LCST)*.

B. POLYMER SOLUTIONS

Polymers dissolve very slowly in their solvents, even if ''by eye'' a solution seems to be formed readily. The dissolution process occurs in two stages. First, solvent penetrates into the polymer coils, which for the present remain entangled and/or physically cross-linked (e.g., by hydrogen bridges or by crystallites consisting of subchains of several polymer molecules; we do not mean real chemical cross-links here, where covalent or ionic bonds are involved). The penetration process is time consuming; it is slowest in the most crystalline parts of the material. The first stage is called *swelling*.

The second stage is the decomposition of the swollen polymer into macromolecules; it takes place parallel to the swelling stage and is completed after it. In some cases, during the swelling stage, the polymer decomposes into tiny gel particles, which still consist of millions of tightly cross-linked polymer molecules, which already possess properties similar to those of the final solution; they may have the same refractive index and the same density as the solution itself. A real solution is not formed until all these gel particles are dissolved.

For amorphous polymers, it is sometimes possible to calculate whether a certain liquid is a solvent to a certain polymer. In polymer science, the distinction between good solvents and bad solvents has a special meaning. In a good solvent, there is much less volume expansion on swelling than in a bad one. Whether a polymer readily dissolves in a solvent or not has nothing to do with the solvent quality.

Generally, polymer solutions show large deviations from ideal solution behavior, even at low concentrations (e.g., less than 1%). A solution is said to be ideal when it obeys, for example, Raoult's law: the vapor pressure of the solution is equal to the product of the solute's mole fraction and the vapor pressure of the pure solvent.

Nonideality of polymer solutions is caused, *inter alia,* by the large difference in the molecular size and shape between the molecules of solvent and solute. The expression ''difference in the molecular size'' refers to the qualitative differences determined by the transformation of small molecules (in our case represented by solvent) to the macromolecules. The theories mentioned have not considered the fact that the monomer transformation in the macromolecule occurs stepwise, in a sequence of events from which oligomers may also result. Even taking into account the distribution of the molecular masses, the result obtained cannot represent the real behavior of the polymer solution.

If there is a positive interaction between polymer and solvent, ΔH_M (Equation 1) is negative and solution will occur. When only dispersion forces are involved and ΔH_M is positive, the magnitude of ΔH_M plays an important role in determining whether ΔG_M is negative or not, i.e., whether solution will occur or not.

The heat of mixing is given by

$$\Delta H_M = V_M \phi_1 \phi_2 \left[\left(\frac{\Delta E_1}{V_1} \right)^{1/2} - \left(\frac{\Delta E_2}{V_2} \right)^{1/2} \right]^2 \tag{2}$$

where: V_M = total volume of mixture
ϕ = volume fraction of component 1 or 2 in the mixture
V = molar volume of component 1 or 2
ΔE = energy of vaporization of component 1 or 2

The expression, $(\Delta E/V)^{1/2}$, is called the solubility parameter, δ. Rearranging Equation 2 and introducing the solubility parameter

$$\frac{\Delta H_M}{V_M} = \phi_1 \phi_2 (\delta_1 - \delta_2)^2 \tag{3}$$

For ΔH_M to be small, δ_1 and δ_2 should be similar in magnitude. Thus, in the absence of specific interactions, the most likely solvent for a polymer will be the liquid whose solubility parameter is close to that of the polymer. The solubility parameter of a solvent can be calculated readily from the latent heat of vaporization; that of a polymer, however, cannot be determined directly because most polymers cannot be vaporized. It must, therefore, be estimated indirectly.

An alternate expression for the solubility of polymers is provided by the Flory-Huggins concept.[4,5] According to it, ΔG_M in a solution of linear polydisperse polymer of a single type and a single solvent is given by the following equation:

$$\Delta G_1 = \mu_1 - \mu_1^0 = RT[\ln(1 - \phi) + (1 - x_n^{-1})\phi + \chi^0 \phi^2 + \chi' \phi^3 + \phi'' \phi^3 + ...] \tag{4}$$

where, μ_1 and μ_1^0 are the chemical potential of the solvent, per mole, in the solution and in the pure solvent, respectively; R is the molar gas constant; T is the absolute temperature; x_n is defined as $\overline{V}_{n,x}/\overline{V}_1$, the ratio of the partial molal volume of the polymer to that of the solvent; ϕ is the volume fraction of polymer in the solution; and $\chi^0, \chi',$ and χ'' are "interaction coefficients" and represent a convenient measure of the "solvent power". In order to avoid complicated mathematical calculations, the $\chi' \phi^3$ term and the higher ones have been neglected.

The Flory interaction coefficient, χ, which is a measure of the interaction energy between polymer and solvent, may also be defined by the equation

$$\chi = \frac{\Delta E}{kT} \tag{5}$$

where k is the Boltzman constant, and T the absolute temperature. ΔE represents the difference in the energy of the solvent molecule immersed in pure polymer compared with one surrounded by pure solvent.

A solvent for a given polymer should have $\chi < 0.5$. A liquid having $\chi > 0.5$ belongs to the category of nonsolvents. The critical value χ_c which leads to phase separation depends on the molecular weight of the dissolved polymer molecules. If DP is the degree of polymerization, then

$$\chi_c = \frac{1}{2} + \left(\frac{1}{DP} \right)^{1/2} \tag{6}$$

The interaction parameter χ is also a function of temperature:

$$\chi = A + \frac{B}{T} \tag{7}$$

where A and B are constants, and T is the absolute temperature.

In solution, polymers are known to form multicomponent systems, characterized by partial miscibility. When a homogeneous polymer solution separates into two liquid phases as a result of a lowering in solvent power due to a change in composition or temperature, a small quantity of the *polymer-rich phase* will separate in equilibrium with a relatively large liquid *phase of lower polymer concentration*. In view of the qualitative appreciation of the partial miscibility of such polymeric solutions, several methods have been employed.

Starting from Relation 4, one may calculate the concentration of the polymer in these two phases. The results of these calculations have been subsequently used in the establishment of the principles governing fractionation by solubility methods.

We now consider the Shultz-Flory determination of the *cloud-point curves* (CPC) of a polymer solution in which the macromolecular solute had been prepared by such an efficient fractionation that it could be considered as the unique component.[6,7] Large deviations between theoretical and experimental coexistence curves were found. From these data, the authors concluded that the Flory-Huggins theory is totally inadequate to describe liquid-liquid phase separations of polymer solutions.*

Figure 2 shows a typical phase diagram of polyisobutylene fractions of very different molecular masses in a poor solvent, diisobutylketone.[6] Like typical binary liquid mixtures, this system exhibits an UCST. Apart from the occurrence of the UCST, two characteristics reflecting the polymeric nature of the solute demand attention: (1) the increase in the critical temperature with increasing molecular mass (2) and the concomitant shift of the critical point toward lower concentrations. Since a polymer is never strictly monodisperse, the experimental phase diagrams discussed here are really CPC curves, which are taken to approximate true binodals.

The theoretical curves in Figure 2 were obtained by Flory's method. Even though the experimental critical temperatures are fitted, the theoretical phase diagrams are only qualitatively satisfactory: the predicted solubility gap is too narrow and the critical point is at too low a concentration. This behavior is typical.

A series of studies performed by Köningsveld and Stavermann and co-workers[8-16] shows that an important factor leading to the incorrect shape of the theoretical binodals and spinodals is just this nearly universal assumption that the lattice treatment is basically adequate even though at least one of two coexisting phases is usually extremely dilute in polymer. These authors propose a useful recipe for determining more realistic binodals and spinodals.

On the other hand, one must also consider how far experimental quasibinary CPC curves may be equated with binodals for the molecular size distribution encountered in practice. This is one question that has prompted extensive investigations, notably those by Köningsveld and Stavermann on heterogeneous polymers.[9-16] Their method is basically an iterative computer calculation of coexistence curves using, in addition to the thermodynamic criterion of equality of chemical potentials in phases at equilibrium, the Flory-Huggins free-energy function, an assumed initial molecular mass distribution function for the polymer and the material conservation condition expressed by the following equation

$$f''_x = \frac{V'' \phi''_x}{V' \phi'_x + V'' \phi''_x} \tag{8}$$

---

* The CPC is a quasibinary section of the multidimensional temperature-composition diagram and its shape reflects details of the molecular weight distribution curve.

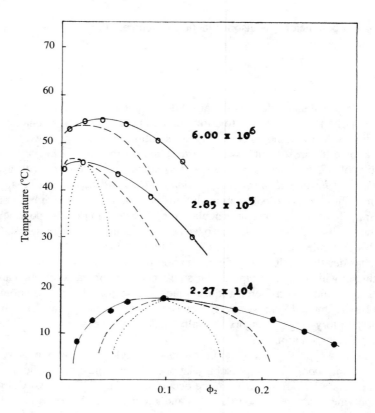

FIGURE 2. Experimental phase diagrams for polyisobutylene fractions (with indicated M_n) in diisobutyl ketone. Also known are theoretical Flory-Huggins binodals (dashed curves) and spinodals (dotted curves). (Adapted from Schultz, A. R. and Flory, P. J., *J. Am. Chem. Soc.*, 74, 4760, 1952.)

where f''_x represents the fraction of x-mer in the concentrated phase, V is the volume of the phase, and ϕ is the volume fraction of the polymer; the prime denotes the more diluted phase and the double prime, the more concentrated phase (the "precipitate").

Using this method, Köningsveld and Stavermann have compiled binodal, spinodal, and cloud-point information for a variety of assumed continuous distributions. From the distributions they examined, Köningsveld and Stavermann arrived at a certain generalization. The initial liquid-liquid critical point was found to be located on the right-hand side of the CPC, which often shows a depression in this place. At constant M_w/M_n, the depression deepens as the M_z/M_w ratio increases. Taking into account that the CPC is a quasibinary section of the multidimensional temperature-composition diagram, and its shape reflects details of the MMD curve, the presence of *low molecular weight fractions* or unusual molecular weight distributions of the polymeric sample can drastically affect the shape of the coexistence curve. This explains the empirical relationship found by McIntyre et al.[17] between the shape of the top of the CPC and the width of the MMD.

The coexistence curve of CPC is of particular interest because it also appears in low molecular weight mixtures. The phase compositions represented by it cannot be examined by direct analysis, *inter alia,* because they are on the verge of appearing or vanishing. Direct measurement by the procedure used for the CPC itself is not feasible either, because the relevant polymer compositions are *a priori* unknown. For similar reasons, this *locus* of the phase compositions coexisting with the cloud points cannot always be found by direct calculation. Therefore, this elusive curve will in the following discussion be called the "shadow curve". Only by extrapolations similar to those needed for the calculation of CPC is it possible to estimate its location. The shadow curve also represent the composition of the first droplets formed for the new phase.

For broad and very skewed distributions, the shadow curves and associated coexistence curves can assume a variety of tortuous forms, although the qualitative relationships still hold. Some examples are given by Köningsveld and Staverman[12] and by Powers.[18] The extreme sensitivity of phase behavior to MMD suggests that CPC may yield important information on the character of a distribution.

The theoretical principles on which the study of polymer miscibility was based have employed as a working method the artificial division of the MMD into a series of "pseudo-components", thereby achieving a simplification of the mathematical calculations. Besides this, the working hypothesis considered the oligomers only as simple low molecular mass homologues of polymers.

Recently, Rätzch et al.[19-22] have applied "continuous thermodynamics" principles in studying polymeric solutions, as well as situations in which the multicomponent systems studied have been composed only of oligomers.

In continuous thermodynamics, dividing the distribution curves into pseudocomponents is avoided, the composition of the multicomponent system being directly represented by the function of the molecular mass distribution.

Application of continuous thermodynamics to the oligomer solution implies several specific modifications of the classical theories. The species of polydisperse oligomer or polymer are identified by a continuous variable, i.e., the molecular mass M instead of the discrete index i.

The above-mentioned continuous thermodynamics allows one to emphasize some specific characteristics which appear when shifting from macromolecules to oligomeric mixtures. This transformation has been achieved through the adequate reduction of $\overline{M}_n$ values by an F factor.

Initially (F = 1), the system represents a macromolecular multicomponent system in which the species with low molecular mass are represented only by solvent. In this case, the CPC and shadow curve show the well-known shape from polymer mixtures (Figure 3a). Reducing F induces the appearance of peculiarities characteristic of oligomeric systems — the occurrence of a "heterogeneous double plait point" at a certain composition of phases. This heterogeneous double plait point corresponds to the coincidence of a metastable and an unstable critical point. The two maxima of the CPC are more readily apparent as the reduction of F values continues. Figure 3b shows the CPC curve and spinodal shortly after the occurrence of the heterogeneous double plait point: the new critical points are still close together. This figure corresponds to F = 0.885. The distance between the two critical points on the spinodal increases and, furthermore, the metastable critical point approaches the CPC. Figure 3c corresponds to F = 0.666; the critical point is already situated beyond this curve, i.e., the metastable critical point has become a stable one. The attached cloud-point curve is also partially situated in the stable range. It intersects the original CPC at the triple point describing the coexistence of three different phases that divides the original and the new CPC into stable and metastable parts.

Application of continuous thermodynamic principles to the study of oligomer multicomponent systems has the advantage of revealing the peculiarities of oligomer behavior in solution. Nevertheless, in the domain of very low degrees of polymerization, these systems have significant peculiarities. Dimers, trimers, and tetramers represent molecular species that can be easily separated. In this domain of the molecular mass, the MMD curve actually represents a sequence of distinctly differentiated peaks. Consequently, estimation of these peaks through a continuous function represents an artificial transformation. It is thus evident that the behavior of oligomers in solution still represents an insufficiently known and debated topic.[23-29]

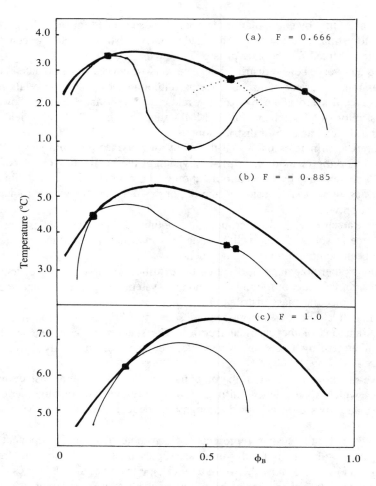

FIGURE 3.   Results of model calculations. Cloud-point curves: stable (heavy lines), metastable (dashed lines), and spinodal (light lines). Critical point (■); triple point (●). (Adapted from Ratzsch, M. T., Kehlen, H., and Thieme, D., *J. Macromol. Sci. Chem.*, A24, 991, 1987.)

REFERENCES

1. **Glinka, M. L.,** *General Chemistry,* Vol. 1, MIR, Moscow, 1981, 220..
2. **Kirilin, V. A., Sychev, V. V., and Sheindlin, A. E.,** *Engineering Thermodynamics,* MIR, Moscow, 1987, chap. 5.
3. *The Collected Works of J. Willard Gibbs,* Vol. 1, Longmans, New York, 1928, 358.
4. **Flory, P. J.,** Thermodynamics of polymer solutions, *J. Chem. Phys.,* 10, 660, 1941.
5. **Huggins, M. L.,** Theory of solutions of high polymers, *J. Am. Chem. Soc.,* 64, 1712, 1942.
6. **Shultz, A. R. and Flory, P. J.,** Phase equilibrium in polymer-solvent systems, *J. Am. Chem. Soc.,* 74, 4760, 1952.
7. **Shultz, A. R. and Flory, P. J.,** Thermodynamic interaction parameters from critical miscibility data, *J. Am. Chem. Soc.,* 75, 3888, 5631, 1953.
8. **Köningsveld, R., Stockmayer, W. H., Kennedy, J. W., and Kleintjens, L. A.,** Liquid-liquid phase separation in multicomponent polymer solutions, *Macromolecules,* 7, 73, 1974.
9. **Köningsveld, R. and Staverman, A. J.,** Liquid-liquid phase separation in multicomponent polymer solutions. I. Statement of the problem and description of methods of calculation, *J. Polym. Sci. Part A,* 6, 305, 1968.

10. **Köningsveld, R. and Staverman, A. J.,** Liquid-liquid phase separation in multicomponent polymer solutions. II. Theoretical state, *J. Polym. Sci. Part A,* 6, 325, 1968.
11. **Köningsveld, R. and Staverman, A. J.,** Liquid-liquid phase separation in multicomponent polymer solutions. III. Cloud-point curve, *J. Polym. Sci. Part A,* 6, 349, 1968.
12. **Köningsveld, R. and Staverman, A. J.,** Liquid-liquid phase separation in multicomponent polymer solutions. IV. Coexistence curve, *Kolloid Z. Z. Polym.,* 218, 114, 1967.
13. **Köningsveld, R. and Staverman, A. J.,** Liquid-liquid phase separation in multicomponent polymer solutions. V. Separation into three liquid phases, *Kolloid Z. Z. Polym.,* 220, 31, 1967.
14. **Köningsveld, R.,** Preparative and analytical aspects of polymer fractionation, *Adv. Polym. Sci.,* 7, 1, 1970.
15. **Köningsveld, R. and Staverman, A. J.,** Polymer fractionation. I. The preparative problem, *J. Polym. Sci. Part A,* 6, 367, 1968.
16. **Köningsveld, R. and Staverman, A. J.,** Polymer fractionation. II. The analytical problem, *J. Polym. Sci. Part A,* 6, 383, 1968.
17. **McIntyre, D., Rounds, N., and Campo-Lopez, E.,** Thermodynamic and structural parameters in tri-block copolymers, *Polym. Prepr.,* 10(2), 531, 1969.
18. **Powers, P. O.,** The solubility of α-methylstyrene copolymers in polybutenes, *Polym. Prepr.,* 15(2), 528, 1974.
19. **Rätzsch, M. T. and Kehlen, H.,** Continuous thermodynamics of polymer solutions: the effect of polydispersity on the liquid-liquid equilibrium, *J. Macromol. Sci. Chem.,* A22, 323, 1985.
20. **Rätzsch, M. T., Kehlen, H., and Thieme, D.,** Polymer compatibility by continuous thermodynamics, *J. Macromol. Sci. Chem.,* A23, 811, 1986.
21. **Kehlen, H., Rätzsch, M. T., and Bergmann, J.,** Application of continuous thermodynamics to the stability of polymer systems, *J. Macromol. Sci. Chem.,* A24, 1, 1987.
22. **Rätzsch, M. T., Kehlen, H., and Thieme, D.,** Oligomer compatibility by continuous thermodynamics, *J. Macromol. Sci. Chem.,* A24, 991, 1987.
23. **Köningsveld, R. and Kleintjens, L. A.,** Liquid-liquid phase separation in multicomponent polymer systems. XV. Thermodynamic aspects of polymer compatibility, *Br. Polym. J.,* 9, 212, 1977.
24. **Olabisi, O., Robeson, L. M., and Shaw, M. T.,** *Polymer-Polymer Miscibility,* Academic Press, New York, 1977, 91.
25. **Köningsveld, R., Onelin, M. H., and Kleintjens, L. A.,** *Polymer Compatibility and Incompatibility,* Solĉ, K., Ed., Harwood Academic, London, 1982.
26. **Kehlen, H. and Ratzsch, M. T.,** Oligomer compatibility by continuous thermodynamics, *Z. Phys. Chem. (Leipzig),* 264, 1153, 1983.
27. **Wohlfarth, C. and Regener, E.,** Calculation of high pressure phase equilibrium in model mixtures of the ethylene-polyethylene system, *Plaste Kautsch.,* 35, 252, 1988.
28. **Akihira, A. and Tasaki, K.,** End-to-end distance in α,ω-diiodo-derivatives of ethylene oxide oligomers: a rotational isomeric state analysis of the X-ray scattering data of Li, Post and Morawetz, *Macromolecules,* 19, 2647, 1986.
29. **Atovmayan, E. G., Baturin, S. M., and Fedotova, T. N.,** Thermodynamic interactions in mixtures of oligomers containing urethane, hydroxyl and ether groups, *Vysokomol. Soedin. Ser. A,* 28, 357, 1986 (in Russian).

Chapter 2.2

MOLECULAR WEIGHT AND STRUCTURE-MOLECULAR WEIGHT RELATIONSHIP

I. INTRODUCTION

Molecular species forming oligomeric mixtures differ through their molecular weight, chemical composition, and functionality. On considering these characteristics as components of the chemical structure, it becomes obvious that the establishment of a structure-property relationship, in the case of a certain oligomeric species, requires a precise method of determination of each of these three characteristics.

II. MOLECULAR WEIGHT

The value of the molecular weight of a mixture formed of oligomeric species will be an *average value*. The frequency variation of the different species, which depends on their molecular weight, may be described by a distribution function characterized by a series of moments, μ, whose values coincide with the number average, weight average, and z average of the molecular weight. These average values may be defined by the following relations:

$$\overline{M}_n = \frac{\mu_1}{\mu_0} \qquad \text{number average} \qquad (1)$$

$$\overline{M}_w = \frac{\mu_2}{\mu_1} \qquad \text{weight average} \qquad (2)$$

$$\overline{M}_z = \frac{\mu_3}{\mu_2} \qquad \text{z average} \qquad (3)$$

Experimentally, these average values may be determined by applying specific methods. Thus, measurement of the coligative properties of oligomeric solutions gives the number-average value of the solute molecular weight. Light scattering and ultracentrifugation allow determination of the weight and z average, respectively.

Another important average of the molecular weight is the viscometric one, denoted as $\overline{M}_v$ and defined by the following relation:

$$\overline{M}_v = \left[\frac{\Sigma n_i M_i^{1+a}}{\Sigma n_i M_i} \right]^{1/a} = \left[\frac{\Sigma w_i M_i^a}{\Sigma w_i} \right]^{1/a} \qquad (4)$$

in which a represent the exponent from the $[\eta] = KM^a$ relation, and may take values from 0.5 to 0.1. If $a = 1$, then $\overline{M}_v = \overline{M}_w$. Starting from Relations 1 to 4, the following relation may be written:

$$\overline{M}_n \le \overline{M}_v \le \overline{M}_w \le \overline{M}_z \qquad (5)$$

There are several methods of determining molecular mass averages, such as titration of the end groups, cryoscopy, ebulioscopy, osmometry, light scattering, gel permeation chromatography (GPC), ultracentrifugation, and others. All these methods have been settled in the chemistry of polymers, being used mainly in the determination of their molecular weight. The preciseness of such methods also depends on the value of the molecular mass. In the field of oligomers, the methods of number-average determination (titration of the end groups, osmometry, ebullioscopy, and gel permeation chromatography) are most efficient. Nevertheless, when applied to oligomers, these methods have been significantly modified, i.e., taking into account the new characteristics of the solute.[1,2] The equipment, working conditions, advantages, and difficulties of these methods have been presented in the literature.[3-7]

This chapter is mainly dedicated to the viscometry and refractometry of oligomeric solutions as simple and efficient means for the determination of the molecular weight of the oligomeric mixtures.

A. VISCOMETRIC MOLECULAR AVERAGE

The dependence between viscosity and concentration of a solution, in which the size of the solute particle is much different from that of the solvent, may be represented by the following relation:

$$\frac{c}{\ln(\eta/\eta_s)} = \frac{1}{[\eta]} + kc \qquad (6)$$

where c is the concentration of the solute in grams per cubic centimeter, η is the viscosity in cP of the solution, η_s is the viscosity, in cP, of the solvent, and $[\eta]$ represents the intrinsic viscosity, which is given by the relation:[8,9]

$$[\eta] = \lim_{c \to 0} \frac{\eta_{sp}}{c} \qquad (7)$$

where

$$\eta_{sp} = \eta_{rel} - 1 = \frac{\eta}{\eta_s} - 1 \qquad (8)$$

The η_{sp} and η_{rel} represent the specific and relative viscosity, respectively.

Application of Relation 7 is conditioned by the absence of the concentration influence upon the value of the solution density; otherwise, the value of intrinsic viscosity, obtained through extrapolating the η_{sp}/c ratio at $c = 0$, becomes uncertain. Such a problem may appear especially in the viscometric analysis of oligomer solutions because, in such situations, more concentrated solutions — compared to polymer solutions — are being used (5 to 10%).[6-9] Such errors may be avoided if the solute concentration is expressed through a weight fraction (grams of polymer or oligomer per gram of solution). If such is the case, the concentration is expressed by the following relation:

$$c = \frac{c}{\dfrac{w}{d_o} + \dfrac{1-w}{d_s}} \qquad (9)$$

where w represents the solute weight, while d_o and d_s represent the solute (oligomer) and solvent density, respectively.

By introducing Relation 9 into Relation 6 and multiplying with

$$\frac{w}{d_o} + \frac{(1 - w)}{d_s}$$

one will get

$$\frac{w}{\ln(\eta/\eta_s)} = \frac{1}{[\eta]_w} + k_w w \tag{10}$$

where

$$k_w = \frac{1}{[\eta]_{d_o}} - \frac{1}{[\eta]_{d_s}} + k \tag{11}$$

and

$$[\eta]_w = d_s[\eta] \tag{12}$$

The above relations contain no element specific to oligomers; they have been determined empirically, starting from experimental data obtained by means of some polymeric solutions. Some linearity deviations, or the appearance of negative value of the slope, within the low concentration domain, were determined by either configuration changes or solvite absorption on the walls of the viscometric capillary.[10-13]

The value of the intrinsic viscosity allows determination of the molecular weight of the solute. In such cases, the viscometric characterization of some polymers or oligomers with a known molecular weight and narrower molecular weight distribution is necessary. In the case of high molecular weight alkanes and poly(oxyethylene glycol), the dependence between intrinsic viscosity and the molecular weight of the solute may be expressed by the following relation:[14,15]

$$[\eta] = A + \frac{B}{M} + KM^a \tag{13}$$

where A, B, K, and a are constants whose values depend on the nature of the solvent-oligomer system. Relation 13 may be written in the following general form:

$$[\eta] = Af(M) + KM^a \tag{14}$$

Constants b and c represent the influence of the chemical composition and of the solute configuration, respectively, upon the value of the function f(M).

In the case of polymers, the dependence between M and $[\eta]$ is expressed through the well-known relation of Mark-Houwink (Equation 15).

$$[\eta] = KM^a \tag{15}$$

Application of this relation in the viscometric characterization of oligomeric solutions requires special attention, due to the fact that constants K and a have been subjected to significant modifications, especially when the value of M ranges between 10^3 and 10^4.[14-22] This means that the determination of the viscometric average of the oligomers molecular weight requires a calibration specific to the analyzed system.

Viscosity of oligomeric solutions is also influenced by the nature of the end groups and by the shape (linear or branched) of the main chain, as well as by the possibility of forming hydrogen bonds.[23-28]

Viscometric studies performed upon oligomeric solutions have taken over, to a great extent, some theoretical aspects characterizing polymer solutions. Thus, the effect of temperature upon the viscosity of dilute solutions of polymers is determined by the increase of chain flexibility and by the improvement of the solvent quality, as well. Quantitatively, a temperature increase induces a lowering of the $[\eta]$ value.[29-31] In the case of oligomers, it has been proved that among bulk viscosity, molecular weight, and temperature, the following relation may be written:

$$V = M^{(\gamma R + \delta)} \exp(\alpha R + \beta) \qquad (16)$$

where V is the bulk viscosity, α, β, γ, and δ are constants, and R = 1/T Ko.

B. THE DETERMINATION OF M_n OF OLIGOMERS BY REFRACTIVE INDEX

Solutions of certain oligomers (isobutylene telomers, oligodienes, and oligoalkenes) show a direct dependence between the refractive index of solution and the 1/M ratio; this dependence is characteristic of the f(n,d) function, being expressed by the following relation:[32]

$$f(n,d) = \frac{a}{M_n} + b \qquad (17)$$

where n is the refractive index, d is the solution density, and a and b are constants.

In certain solutions, Relation 17 can be simplified, as the value of constants a and b are independent of the solution density. For example, in the case of oligoisobutenes, the value of M_n may be obtained by applying the following relation:

$$n^{20} = 1.5034 - \frac{11.391}{M_n} \qquad (18)$$

A number of other procedures were described for determination of the molecular weight of the oligomeric mixtures. The relation between the molecular weight and the absorption of ultrasound at <50 MHz was used for determination of the degree of polymerization of oligo(propylene glycol),[33] and an exact algebraic analysis of proton magnetic resonance spectra provided formulas from which the degree of polymerization of oligo(butylene terephthalate)s were calculated.[34]

III. FUNCTIONALITY

One of the main characteristics of oligomeric mixtures is functionality, denoted as $\langle F \rangle$. Let us define $\langle F \rangle$ by

$$\langle F \rangle = (N/2)/(1/M_n) = NM_n/2 \qquad (19)$$

where N is the total number of functional groups per gram of sample.[35]

Therefore, it is important for oligomer characterization to separate an oligomer mixture into components which have no functionality, or have a mono-, di-, or trifunctionality, etc. Column fractionation, GPC, liquid chromatography, and thin-layer chromatography (TLC) have been used for the determination of the functionality distribution.[36-44] These methods are somewhat tedious in that they are time consumming and require chemical modification of the functional groups in advance, and interpretation of the results is difficult. Nevertheless, determination of $\langle F \rangle$ by TLC has the advantage of greater simplicity. Thus, if we express the fraction of the functional components with w_2 (a value that has been determined by TLC), the following relation may be written:

$$w_2 = x + y \tag{20}$$

where x and y are the weight fractions of mono- and difunctional species, respectively. Then

$$N = (x/M_n) + (y/M_n) \tag{21}$$

By inserting Relation 21 into Relation 19, the following relations can be written:

$$x = 2[1 - (p + 1)\langle F \rangle]w_2 \tag{22}$$

$$y = 2[(1 + p)\langle F \rangle - 1]w_2 \tag{23}$$

Then the functionality distribution can be calculated if one knows w_2, $p = w_1/w_2$, and $w_1 = 1 - w_2$.

IV. STRUCTURE-FUNCTIONALITY-MOLECULAR WEIGHT RELATIONSHIP

If we represent a linear oligomer by the general formula $X-(U)_n-Y$ (where X and Y are different or identical functional groups, and U and n represent the structural unity and polymerization degree, respectively), then, between the value of the oligomer molecular mass and one of its properties, the following relation may be written:[45-47]

$$P = \frac{a}{M} + b \tag{24}$$

where P is the property considered (e.g., refractive index, specific refractivity, specific parachor, specific volume, etc.), M is the molecular weight, and a and b are constants specific to functional group X or Y and the structural unity U, respectively.

Relation 24 may also be applicable to cyclic or branched oligomers; starting from it, one may reach the conclusion that any property of oligomers is determined by the shape, size, and chemical structure of the species forming the mixture under study.

REFERENCES

1. **Cantow, M. J. R. and Johnson, J. F.,** Molecular weight determination, in *Encyclopedia of Polymer Science and Technology,* Vol. 9, John Wiley & Sons, New York, 1968, 182.
2. **Mitchell, J., Jr. and Chin, J.,** Analysis of high polymers, *Anal. Chem.,* 47, 289R, 1975.
3. **Conix, A.,** Molecular weight determination of poly(ethylene terephthalate), *Makromol. Chem.,* 26, 226, 1958.
4. **Uglea, C., Heinisch, P., and Mihăescu, A.,** Characterization of poly(ethylene terephthalate). I. Molecular weight determination, *Mater. Plast. (Bucharest),* 18, 238, 1981 (in Romanian).

5. **Glover, A. C.,** Determination of number average molecular weights by ebulioscopy, in *Adv. Chem. Ser.,* 125, 1, 1973.

6. **Lehrle, R. S.,** Ebulliometry applied to polymer solutions, in *Progress in High Polymers,* Vol. 1, Heywood, London, 1961, 37.

7. **Kretschaner, K. and Helbig, W.,** Microebuliometric molecular weight determination of α-methylstyrene-acrylonitrile co-oligomers using a digital display difference thermometer, *Z. Phys. Chem. (Leipzig),* 268, 711, 1987.

8. **Erikson, J. R.,** Viscosity of oligomer solutions for high solids and UV curable coatings, *J. Coat. Technol.,* 48(620), 1976.

9. **Pietrasanta, Y., Rigal, G., and Schaeffner, P.,** Viscosimetric behaviour of ethyl acrylate/carbon tetrachloride telomers, *Makromol. Chem.,* 182, 1371, 1981.

10. **Boyer, R. F. and Spender, R. S.,** Critical concentration effects in dilute high polymer solutions, *J. Polym. Sci.,* 5, 375, 1950.

11. **Streeter, D. J. and Boyer, R. F.,** Viscosities of extremely dilute polystyrene solutions, *J. Polym. Sci.,* 14, 5, 1954.

12. **Ohrn, O. E.,** Further comments on viscosity measurements, *J. Polym. Sci.,* 19, 199, 1955.

13. **Pepper, D. C. and Rutherford, P. P.,** The viscosity anomaly at low concentrations with polystyrenes of low molecular weight, *J. Polym. Sci.,* 35, 299, 1959.

14. **Rempp, P.,** Viscosity of low molecular weight polymers, *J. Polym. Sci.,* 23, 83, 1957.

15. **Sadron, C. and Rempp, P.,** Intrinsic viscosity of low molecular weight polymers, *J. Polym. Sci.,* 29, 127, 1958.

16. **Seidel, B.,** Viscosity measurements on oligomers from terephthalic acid and glycol, *J. Polym. Sci.,* 55, 411, 1961.

17. **Bianchi, U., Dalpiaz, M., and Patrone, E.,** Viscosity-molecular weight relationship for low molecular weight polymers. I. Polydimethylsiloxane and polyisobutylene, *Makromol. Chem.,* 80, 112, 1964.

18. **Patrone, E. and Bianchi, U.,** Viscosity-molecular weight relationships for low molecular weight polymers. II. Poly(vinylacetate) and poly(methyl methacrylate), *Makromol. Chem.,* 94, 52, 1966.

19. **Fox, T. G., Kisinger, J. B., Mason, H. F., and Schuele, E. M.,** Properties of dilute polymer solutions. I. Osmotic and viscometric properties of solutions of conventional poly(methyl methacrylate), *Polymer,* 3, 71, 1962.

20. **Pietrasanta, Y., Rigal, G., and Schaeffner, P.,** Viscosimetric analysis of ethyl acrylate and vinyl chloride telomers with carbon tetrachloride or methyltrichloroacetate, *Eur. Polym. J.,* 17, 373, 1981.

21. **Aharoni, S. M.,** Monodisperse rodlike oligomers and their mesomorphic higher molecular weight homologues, *Macromolecules,* 20, 2010, 1987.

22. **Ryszkowska, J. and Gruin, J.,** Viscosity and molecular weight of linear oligodiols, *Polimery,* 32, 4679, 1987.

23. **Budtov, V. P. and Romanowskii, G. K.,** Influence of end polar groups onto oligourethane viscosity, *Visokomol. Soedin. Ser. A,* 22, 700, 1980.

24. **Gamburow, G. M., Vainstein, E. F., and Entelis, S. G.,** Model of oligomer molecules in solution, *Dokl. Akad. Nauk S.S.S.R.,* 218, 375, 1974.

25. **Afanasiew, N. V., Iarlicov, B. V., Akutin, M. S., and Sakrin, I. K.,** Viscosity of reactive oligomers, *Tr. Mosk. Him. Tekhnol. Inst. Im. D. I. Mendeleff,* 86, 174, 1975.

26. **Kagemoto, A., Itoi, V., Baba, Y., and Fujishiro, R.,** The heats of dilution of the oligomeric ethylene oxide/alcohol solutions, *Makromol. Chem.,* 150, 255, 1971.

27. **Wolf, B. A. and Blaum, G.,** Thermodynamic properties of liquid oligomer 1/oligomer 2 systems, *Ber. Bunsenges. Phys. Chem.,* 81, 991, 1977.

28. **Salaewa, L. P.,** Viscosity-molecular weight relationship for polymer-oligomer systems, *Strahlentherapie Svoistva,* p. 95, 1979; see *Ref. Zhur.,* 19C, 17C, 23, 1979.

29. **Uglea, C. V., Feldman, D., and Agherghinei, I.,** Characterization of copolymers. IV. Viscosity-temperature relationship of acrylonitrile-vinyl acetate-α-methystyrene copolymers, *Rev. Roum. Chim.,* 16, 1387, 1971.

30. **Uglea, C. V.,** Hemicellulose fractionation, *Makromol. Chem.,* 175, 1535, 1974.

31. **Drexler, L. H.,** A rheological study of the low molecular weight butyl polymers, *J. App. Polym. Sci.,* 14, 1857, 1970.

32. **Ingham, J. D. and Lawson, D. D.,** Refractive index and molecular weight relationship for poly(ethylene oxide), *J. Polym. Sci. Part A,* 3, 2707, 1965.

33. **Hauptmann, P. and Ingo, A.,** Ultrasonic method for determination of molecular weight of oligomers and polymer melts, East German Patent 240,436, 1986.

34. **Nethsinghe, L. P., Plesch, P. H., and Hodge, D. J.,** Finding the DP_n of oligomers from PMR signals, *Eur. Polym. J.,* 22, 643, 1986.

35. **Min, T. I., Miyamoto, T., and Inagaki, H.,** Determination of functionality distributions in telechelic prepolymers by thin-layer chromatography, *Rubber Chem. Technool.,* 50, 63, 1976.

36. **Slonim, I. Ya., Bulai, A. Kh., Urman, Ya. G., Beloded, L. N., Korovin, L. P., Kiselev, V. Ya., and Lazaris, A. Ya.**, NMR and GPC study of molecular weight and functional-group distributions of an oligoesteracrylate based on pentaerythriol, adipic acid and methacrylic acids, *Vysokomol. Soedin. Ser. A,* 27, 2210, 1985.

37. **Vernich, S. S., Gur'eva, N. M., Gorshkov, A. V., Charelishivich, B. I., Zapadinskii, B. I., Evreinov, V. V., and Entelis, S. G.**, Use of exclusion chromatography with combined detection to analyze oligo(epoxy propiolate) heterogeneity, *Vysokomol. Soedin. Ser. B,* 29, 741, 1987.

38. **Gur'yanova, V. V. and Pavlov, A. V.**, Step-by-step analysis of polycondensation oligomers and polymers by liquid chromatography, *J. Chromatogr.,* 365, 197, 1986.

39. **Gorshkov, V. V., Prudskova, T. N., Guryanova, V. V., and Evreinov, V. V.**, Functional type separation of oligocarbonates by liquid chromatography under critical conditions. Effects of pore size, *Polym. Bull.,* 15, 465, 1986.

40. **Maleshovich, A. P., Doroshenko, V. N., and Kozlov, A. A.**, Analysis of functional groups of fractions of the oligomer products of the radiochemical polymerization of epichlorohydrin in the presence of diphenyliodonium fluoroborate, *Vysokomol. Soedin. Ser. B,* 30, 453, 1988.

41. **Goshkov, A. V., Overcem, T., Evreinov, V. V., and Alten, H. A. A.**, Determination of functional type distribution of oligocaprolactone diols by liquid chromatography under critical conditions, *Polym. Bull.,* 18, 513, 1987.

42. **Tsvetkovskii, J. B. and Shlyakhter, R. A.**, Fractionation of polymerizing oligomers according to their functionality, *Zh. Anal. Khim.,* 38, 509, 1983.

43. **Vakhtina, I. A., Shirokova, G. V., and Okuneva, A. E.**, Determination of the amount of monofunctional components in complex oligoesters by thin-layer chromatography, *Vysokomol. Soedin. Ser. A,* 24, 2002, 1982.

44. **Akhmedov, Kh., Chauser, M. G., and Cherkashin, M. I.**, Thin-layer chromatography of polymerization products of epoxypropyl carbazoles, *Izv. Akad. Nauk Tadzh. S.S.R. Otd. Fiz. Mat. Geol. Khim. Nauk,* p. 116, 1982.

45. **Geczy, I.**, Synthetic linear polymers. XXXIV. Change of specific properties of dimethylsiloxane co-oligomers as a function of the chemical compositions and size of the molecule, *Acta Chim. Acad. Sci. Hung.,* 101, 327, 1979.

46. **Geczy, I.**, Synthetic linear polymers. XXXIX. Dependence of the specific properties of dimethylsiloxane co-oligomers containing phenyl groups on the chemical compositions and size of the molecule, *Acta Polym.,* 33, 516, 1982.

47. **Geczy, I.**, The effect of molecular weight and phase state on the properties of oligomeric compounds, *Kolor. Ert.,* 29, 188, 1987.

Chapter 2.3

OLIGOMER FRACTIONATION

In a general sense, the fractionation of a nonhomogeneous mixture consists of the separation of that mixture into its components, using a suitable experimental technique, in order to obtain homogeneous fractions. In the case of oligomers, pleinomers, and macromolecules, four distinct differences between the species of such a mixture can be recognized: in molecular weight, molecular configuration, chemical composition, and functional groups distribution.

Unlike high polymers, oligomers may be separated in individual species, especially when the polymerization degree is low.

In the beginning, fractionation by solubility methods was exclusively used for the fractionation of polymers. Following the synthesis of styrene-divinyl benzene gels,[1] subsequently used as supports in column chromatography, the expected explosive growth in the field of polymer fractionation occurred. The new method, known initially as *gel permeation chromatography* (GPC), offers, besides other advantages, the possibility of separating, identifying, and quantitatively resolving oligomers from their mixtures with simple or macromolecular compounds. The same method has made the determination of molecular weight distribution in polymers a routine operation.

I. OLIGOMER FRACTIONATION BY SOLUBILITY METHODS

The process of fractionation by solubility involves a separation of phases and distribution of mixture species between them: smaller molecules have a greater solubility than larger ones in the same liquid. The appearance of the two phases in the initial solution of the sample may be achieved by any of the following three methods: *addition of nonsolvent, elimination of solvent* by evaporation, and *lowering the temperature* of the system.

Smaller molecules are less discriminating in selecting their own environment than the larger ones, and this is the basis for the partitioning of molecular species between both phases, in different proportions.

The separation process may be applied directly to the sample itself. However, there are situations in which fractionation by solubility involves preparing the sample in an appropriate physical state, in order to facilitate the separation of more homogeneous fractions. Many experimental arrangements have been employed. The sample may be dissolved, finely divided, or deposited on a thin metallic foil or on a support (e.g., sand) in a column. The sample may also be transformed by total precipitation in a *coacervate*. Common to all of these methods is the preparation of the sample in a form which assures the efficiency of the separation process.

The distribution of the mixture species between the two phases has a statistical character, so that the molecular species forming the initial sample will be found in both of them. The difference between the compositions of the two phases is a quantitative one; one of the phases is richer in solute, thus accumulating the species of high molecular weight. Usually, this phase is a *precipitate* (or a viscous liquid), and is called precipitate, polymer-rich, concentrated, gel, or coacervate phase. Due to its higher density, the concentrated phase falls out. The other phase (known as the liquid, diluted, supernatant, or sol phase) contains more solvent and accumulates, preferentially, the lower molecular weight species of the solute.

The main aspects of fractionation by solubility are the mechanism of phase separation, the qualitative and quantitative composition of the two phases, and the selection of the fractionation method, so that the repeated separation of the two phases forms a process of separation according to the molecular weight of the species.

We shall briefly discuss here only the important aspects of the separation by solubility methods.

A. THE SEPARATION PRINCIPLE

Unfortunately, oligomer fractionation through solubility-based methods does not have a suitable theoretical base. The results obtained in the fractionation of systems formed of oligomers or of oligomers and high polymers have been interpreted, up to now, by means of the theories characterizing solutions of high polymers.[2-7] The main shortcoming of such an interpretation is that they greatly simplify things, so that oligomers are considered only as inferior homologues of high polymers.

The system in which fractionation occurs is formed of n components. This fact, and the difference between the size of the solute and solvent particle, determine the complexity of polymer or oligomer fractionation.

The separation of species according to the molecular weight was first interpreted by Brønsted[2] by energetic theory, according to which, in a polymer-solvent (P-S) system, the potential energy of each particle depends on its molecular weight. During selective precipitation, performed through the addition of nonsolvent (NS), the distribution of the species between two phases is controlled by the difference of the potential energies which, in turn, is determined by the difference between the molecular weights of the species; the less mobile ones, having a higher molecular weight, will accumulate in the precipitate (the phase with a lower potential energy), while low molecular weight species, having a higher mobility and a lower potential energy, will accumulate in the liquid phase, which has a lower potential energy. Starting from this energetic description, the relations establishing the quantitative compositions of the two phases have been subsequently determined.[3] The poor agreement between this theory and experimental results is caused by neglect of the entropy-mixing effect.[4]

The second theory, based on a more realistic model, shows that, in solution, the polymeric chain may take various configurations, whose number increases with increasing chain length. Thus, the value of the mixing entropy of the system increases. Such observations together with correct modeling have led to the deduction, by Flory-Huggins, of the principle of polymer fractionation by solubility, based on the following relation:[5-7]

$$W''_{(M)} = \frac{W_{(M)}}{1 + r\exp(-\sigma m)} \tag{1}$$

where $W_{(M)}$ and $W''_{(M)}$ represent the weight of the species having the molecular weight M in the initial sample and in the precipitate, respectively; $r = V'/V''$, where V' and V'' represent the volumes of the diluted and of the concentrated phase, respectively; σ is the distribution coefficient; and m is the length of the macromolecular chain having the molecular weight M expressed through the number of sites occupied in the solution lattice.

Relation 1 shows that the solute present in the precipitate phase has a molecular weight distribution similar to that of the initial sample. For low molecular weight species, the value of the denominator is maximum. Their frequency in the diluted phase also depends on the molecular weight distribution of the initial sample.

The r value from Relation 1 may be expressed by the ratio of the volume fraction ϕ''_i and ϕ'_i of the i species present in the two phases, the concentrated one denoted by ('') and the diluted one, by ('). Thus, the following relation may be written:

$$\frac{\phi''_i}{\phi'_i} = \exp(\sigma m_i) \tag{2}$$

where m_i represents the value of m from Relation 1, corresponding to species i. In the experimental conditions applied for fractionation, σ is positive for each value of m_i. The accumulation of species in the precipitate increases with increasing values of m_i.

Fractionation by solubility can be carried out in two main ways, fractional precipitation and selective extraction, and may be performed by various procedures which, nevertheless, do not change the main principles of separation by solubility.

B. THE EVALUATION OF SEPARATION

The quality of the fractionation process is estimated by its efficiency. The definition of a quantity called "fractionation efficiency", for the comparison of different experimental data, is somewhat arbitrary as long as fractionation is not capable of separating the individual species, but, rather, only a definite number of nonhomogeneous fractions.

The lack of a universal criterion for evaluation of the results obtained by fractionation methods is deeply rooted in the evolution of separation science. Man's constant concerns in this field has resulted in the improvement and diversification of the separation methods in terms of obtaining higher and higher performances. Individualization of some specific domains within the field of separation science brought about the training of specialists with great knowledge in a very limited aspect of separation science. This may explain the absence of general interpretations and corresponding classifications, as well as general criteria for evaluation of the separation results, and could also explain the heterogeneous character of the opinions and interpretations of the separation processes.

When considering separation as a process involving spatial discrimination of the compounds forming a mixture, one notes that this covers a large category of processes, such as distillation, extraction, precipitation, diffusion, and chromatography (in all its forms, such as filtration, decantation, etc.). As regards chromatography, a separation domain in which more numerous theoretical studies have been undertaken, attempts have been made to establish some general criteria for estimating the results of separation.

Resolution,[8] efficiency,[9] the separation number[10] and separation degree,[11] the height to base ratio,[12] the impurity index,[13] the separation function,[14] and the separation entropy[15-17] all represent important criteria in the estimation of the separation process. Figure 1 graphically represents some of the most significant parameters characterizing any process of column separation: t_r (retention time) represents the time necessary for a molecular species to cross, through the column, the distance between the injection system and the detector; t_o represent the time in which a completely excluded molecular species should cover the same distance (usually, this parameter represents the column's dead volume, i.e., the space between the support particles); and w is the width of the peak base. Thus, from Figure 1, $t_r > t_o$, while w, measured tangentially, has to be as low as possible — a condition for satisfactory separation. Consequently, t_r, t_o, and w permit the appreciation of the *column possibilities,* as well as certain *properties of the sample* being analyzed. The former, expressed through column efficiency, which actually represents the *column power,* are measured through the number of *theoretical plates,* denoted by N that is given by Relation 3.

$$N = 16\left(\frac{t_r}{w}\right)^2 \tag{3}$$

The sample characteristics influence the separation through the *capacity factor,* denoted by K' and expressed by the following relation:

$$K' = \frac{t_r - t_o}{t_o} \tag{4}$$

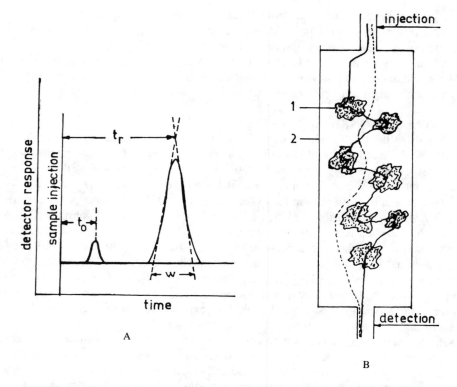

FIGURE 1. Characteristic parameters of column separation. (A) Plotted on the chromatogram; (B) physical significance of these parameters: ···· t_o; —— t_r; 1 — support particle; 2 — column.

which shows how long a species is to be retained in the separation system. The optimum values of K' range between 1 and 8.

When considering a mixture of two components, the results of separation are represented by the ratio between the capacity factors of the two components; this ratio represents the *column selectivity*, α. Figure 2 and Relation 5, whose indices refer to the component number, represent this parameter.

$$\alpha = \frac{K'_2}{K'_1} = \frac{t_2 - t_o}{t_1 - t_o} \tag{5}$$

The capacity factor, efficiency, and selectivity all contribute — with various intensities, as determined by the experimental conditions and the nature of the system — to the quality of separation. The resultant of this influence represents the *resolution* of separation, denoted by R_s and given by Relation 6.

$$R_s = \frac{1}{4} \sqrt{N} \underbrace{\left[\frac{\alpha - 1}{\alpha} \right]}_{\text{selectivity}} \underbrace{\left[\frac{K'}{1 + K'} \right]}_{\text{capacity factor}} \tag{6}$$

$$\underbrace{\phantom{R_s = \frac{1}{4} \sqrt{N}}}_{\text{efficiency}}$$

Special consideration should be given to the entropic criterion for the evaluation of separation quality mainly for the original manner in which it considers the separation. Generally, the separation may be considered as a process of system ordering. The entropy,

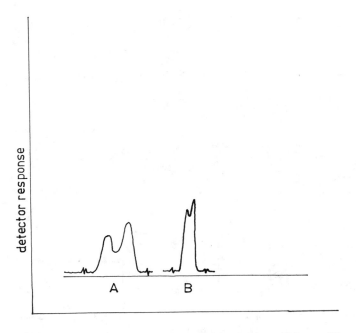

FIGURE 2. (A) Column with high selectivity and low efficiency; (B) column with high efficiency and low selectivity.

which measures the disorder of a system, offers the means for an entropic appreciation of the separation.

The large diversity of methods used in the separation process reduces the general utility of the entropic criterion for the evaluation of separation quality. The example given in Figure 3 illustrates this fact. Diagram A represents a chromatographic separation in which the spatial discrimination of the components also reflects the isolation of the components having different chemical compositions. From a volumetric point of view, the system reveals an expansion (dilution through elution) and, implicitly, an increase of entropy. Nevertheless, at the same time, separation of the chemical species, accompanied by a decrease in entropy, occurs. As a whole, the system being developed according to diagram A undergoes an isoentropic transformation. From a thermodynamic point of view, the chromatography represents mostly a qualitative change of the ordering degree, the system maintaining the amount of its initial disorder.

Diagram B of Figure 3 represents a distillation process. In this case, no volume expansion occurs, while the result of the process is a decrease of the entropy. The two examples given clearly indicate that the entropy cannot be regarded as a general criterion in the evaluation of separation processes. This is why, in the following, we shall only refer to the quality of the fractionation results, i.e., the separation of the more homogeneous fractions that will assure detailed knowledge of the molecular weight distribution of the initial sample. A suitable fractionation depends upon x, the amount of sample separated in each fraction, and n, the number of separated fractions. As optimal working conditions, we shall consider those that will assure the separation of the initial sample into ten fractions, the concentration of the sample initial solution being 1%.

Let us now examine this classical experimental condition of fractionation by solubility methods.

C. EXPERIMENTAL CONDITIONS

The actual relations between x and n (number of fractions) governing fractionation in the upper critical solution (UCST) systems are, however, still in substantial disagreement

SCHEME A   SCHEME B

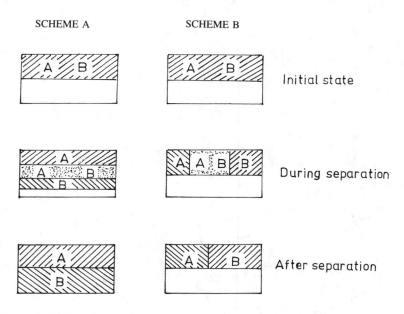

FIGURE 3.   Comparison between two separation processes. Scheme A — chromatographic separation; scheme B — distillation.

with the existing theoretical equations. In some respects, the theory is not yet sufficiently developed to cover all the experimental possibilities. Some results should be mentioned explicitly:

1. The logarithm of the partition coefficient in a particular fractionation experiment depends on x in a nonlinear manner.[18]
2. No dependence of fractionation efficiency on the chemical nature of the solvent has been found so far with common solvents and sufficiently low polymer concentrations, although the phase volume ratio (Equation 2) corresponding to a certain ΔG (Equation 4, Chapter 2.1) varies with the solvent.
3. The partition coefficient extrapolated for the monomer deviates from unity at low concentrations, but the deviation is less pronounced at sufficiently high, overall polymer concentrations.[19]

Extension of the molecular weight distribution into the low molecular weight region by the addition of oligomer[20] answered the question of whether the monomer concentrations in the coexisting phases are practically identical, as expected by classical theories, or larger by a factor of nearly ten in the polymer-rich phase, as extrapolated from the information concerning the polymer. Direct determination[20] yields intermediate values, which demonstrate that the variation of the logarithm of the partition coefficient with the degree of polymerization (DP) cannot be represented by a linear function with sufficient accuracy when a broader range of molecular weights is considered; in this case, it is necessary to replace P by $P^{2/3}$.

Another important parameter conditioning a good fractionation is the solvent-nonsolvent (S-NS) system. In search of the best conditions for determination of the molecular weight distribution, it was still an open question whether the use of S-NS systems showing demixing upon heating had an advantage over the normally applied systems which demix upon cooling. The reason for this uncertainty lay in the fact that the lower critical solution temperature (LCST) can usually be realized with elevated pressures. In this case, the experimental realization of the experimental fractionation becomes difficult. The sensitivity of fractionation at LCST is poor with regard to the separation of oligomers.[21,22] It was, however, unclear whether this observation is inherent for LCST systems or produced by experimental difficulties (e.g., pressure control).[23,24] Due to these experimental difficulties, and also to a reduced efficiency, fractionation methods by solubility could not be used for the qualitative and quantitative separation of oligomers. Nevertheless, some attempts have been made to apply these methods in the determination of the molecular weight distribution in low molecular weight mixtures.[25-28]

The introduction, in the 1960s, of column fractionation procedures and especially of chromatographic precipitation[29] has significantly improved separation quality of solubility methods in the field of low molecular weight species. The simultaneous application of solvent and temperature gradients, along with the modifications proposed by Polacek,[30-34] facilitated the fractionation of some low molecular weight mixtures obtained through the pyrolysis of poly(1,4-*cis*-butadiene) and poly(3,4-isoprene). Under such circumstances, isolation of 9 to 11 fractions, with a molecular weight ranging between 306 and 1260, was achieved.

II. OLIGOMER FRACTIONATION BY GEL PERMEATION CHROMATOGRAPHY

Considering its evolution, the development of polymer fractionation now appears to have been quite slow. Improvement of classical methods or the appearance of some new procedures did not mean the detachment of polymer fractionation from the dependence between solubility and molecular weight.

The qualitative explosion of this domain of separation science began in 1964, when Moore[1] synthesized styrene-divinylbenzene gels, thereby assuring them uniform pore sizes. Starting from these new supports, and backed by some intensive research programs, separation science was enriched with a new method — gel permeation chromatography (GPC).

Extension of GPC to synthetic oligomers was slower. Specialized gel structures had to be developed. There were many advantages to impel such development. Being a column technique, GPC is convenient and versatile. It can be modified in many ways to fit almost all requirements, such as exclusion of oxygen, scaling up and down, and automation. It also is cheap, since the gel can be used over and over again.

The introduction of GPC in separation science was anticipated by some scientists. As early as 1925, Ungerer,[35] Wiegner,[36] and Cernescu[37] demonstrated the gradual exclusion of ammonium ions from natural materials such as zeolite and permutites. Duel[38] was one of the early pioneeres of GPC. He separated natural polymers, such as pectins and algins, from their monomers and recommended that this technique be used for the determination of molecular size for low and high molecular species.

Quite rapid and exact, GPC assures obtaining the experimental data necessary for knowing the molecular weight and molecular weight distribution in polymer or oligomer mixtures.

GPC belongs to the family of liquid chromatographic methods. It separates molecular species according to their size, the separation process occurring in a gel-filled column. The support is a porous, rigidly structured material obtained through cross-linking. In the column, the gel particles are continuously suspended in solvent flow, created by means of a pump. The sample solution is introduced at the top of the column. Size separation occurs inside the gel. Under normal conditions, molecules comparable in size to solvent molecules will distribute through the entire pore volume. Smaller solute molecules diffuse freely into and through the gel pores; they permeate the gel. Thus, these species will travel a longer way through the column and, therefore, the retention time of a certain particle in the column will depend on the molecular size. Smaller solute molecules permeate the gel pores as far as their size permits and move practically without restriction in the solvent contained in the gel. Only very close to the network strands, when the gel segment density is high, does the diffusion rate drop sharply.

Bigger molecules are excluded from the denser parts of the network, but they can diffuse freely through more open passages. The larger a solute molecule is, the fewer apertures suitable for its size it will find.

There may be molecules which are so big that they are completely excluded from the gel. They are flushed through the column first.

Thus, the conclusion to be drawn is that molecular species that may penetrate the gel will leave the column gradually, inversely to the molecular size.

In eluate, the solute is identified by refractometry, UV photometry, flame ionization, or IR absorption.

A species is eluted at a volume exactly equal to the volume available to it in the column. For large, completely excluded molecules, the elution volume V_{EL} is equal to the interstitial volume V_o; for small molecules which can completely penetrate all pores of the gel, it is equal to the total liquid volume of the column, i.e., equal to the sum of V_o and the internal (pore) volume V_I. For molecules of intermediate size, the elution volume is:

$$V_{EL} = V_o + K_d V_I \tag{7}$$

where K_d, the partition constant, is equal to the ratio of the accessible pore volume $V_{I,acc}$, to the total pore volume:

$$K_d = \frac{V_{I,acc}}{V_I} \tag{8}$$

Accumulated experimental data represent the starting point in the attempt of knowing how and why separation occurs in a GPC column. Yet, no suitable answers have been found to these questions, which is probably why this method still does not have an unanimously accepted name. Gel filtration, exclusion chromatography, molecular sieve filtration, and gel permeation chromatography are only some of the names in use. We perfer that of Moore,[1] i.e., gel permeation chromatography.

The theory of GPC is still in a preliminary stage, but it provides a basis for the evaluation of experimental data and for certain predictions. There are still unanswered questions, namely:

1.   Defining the factors influencing the relation between molecular size and elution volume
2.   Establishing the theoretical basis of Relation 7
3.   Explaining the paradox of the coexistence of the tendency of separating species, when a mixture passes through the column, and that of broadening the peak, when the analyzed sample contains only one molecular species

Unfortunately, the experimental performances obtained by applying GPC to the separation of some complex oligomeric mixtures have not been backed up by a suitable theoretical basis. The actual theories on GPC try to explain some peculiarities of gel chromatograms by using the concepts characterizing chromatographic methods. Thus, only partially valuable conclusions are reached, as GPC has been included among chromatographic methods mainly due to its experimental technique and not to its separation mechanism.

At this point, we shall be satisfied with these observations. In view of applying this method to oligomeric mixture separations, it is, nevertheless, necessary to know the mode of transformation of GPC data in valuable information concerning the true composition of

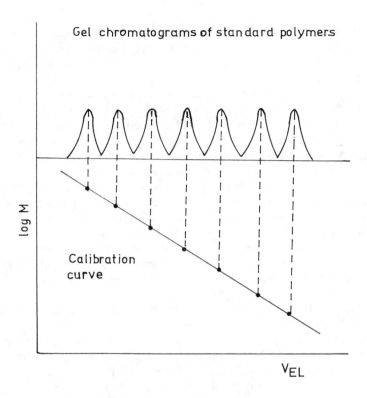

FIGURE 4. Major steps of direct calibration method.

a certain oligomeric mixture. This will enable us to offer a quantitative interpretation of the GPC results. However, one important problem is that GPC is a *relative* method and relies on the validity of *calibration*.

The interpretation of gel chromatograms and especially determination of the molecular weight of the species separated needs a proper calibration curve. Since there are many solute-solvent systems involved in GPC as well as many types of different pore sizes, the best way to obtain an accurate calibration is by experiment rather than by theoretical calculations. Errors in the calibration methods are minimized by obtaining a calibration curve under the same conditions as the sample separation.

The calibration curve for a set of columns can be established experimentally by relating the peak elution volume to molecular weight for a series of narrow molecular weight distribution standards. This is the so-called *direct calibration method*. The accuracy of this method depends on whether the reference compounds and analyzed sample all have a similar conformation and chemical structure. Gel chromatogram determination of each standard compound, plotting of the calibration curve, and their comparison represent the main steps (graphically shown in Figure 4) of the GPC calibration by the direct method.

The main disadvantages of the direct calibration method is the lack of narrow standards. Polystyrene and a few other polymers are the only polymers for which a series of commercial standards of different molecular weight and narrow molecular weight distribution is available. From Figure 4, one can observe that the calibration curve may be represented by the following equation:

$$\log M = A - BV_{EL} \tag{9}$$

where A and B are constants.

Universal calibration procedures represent a tentative calibration associated with the determination of a rigorous way to transform the gel chromatogram into a true molecular weight distribution of the analyzed sample. The *Q factor*[39,40] and *hydrodynamic volume*[41,42] are the major parameters for universal calibration procedures.

The Q factor is the ratio of the molecular weight to the extended chain length calculated from bond lengths and valence angles. Its application assumes a very extended molecule along with the existence of a certain dependence between the "length of the molecular weight" and V_{EL}. This dependence is subsequently considered to be independent of the nature and composition of the S-NS system. Applicable only to rigid polymers, the Q factor led to erroneous results in the field of flexible macromolecules.

The coil conformation adopted by macromolecules in solution suggested the idea of utilizing the hydrodynamic volume as a universal calibration parameter in GPC.[43] The universal calibration procedure utilizes the concept of the hydrodynamic volume of the solute molecule. The hydrodynamic volume can be expressed in terms of the product of the molecular weight M and the intrinsic viscosity of the polymer or oligomer sample. In general, GPC calibration curves for solutes of different types merge into a single plot when the calibration data are plotted as log [η] M vs. elution volume VEL instead of the usual logM vs. elution volume.

The universal calibration method is conceptually sound, but its use is still rather limited. In order to calculate the average molecular weight and distribution of any polymer or oligomer by this method, it is necessary to transform the [η]$_M$ units to molecular weight units for the polymer specified for which Mark-Houwink coefficients in the same solvent at the same temperature are known, or to determine the [η] of each fraction of GPC of the polymer. The universal calibration method should nevertheless be used with caution, since many Mark-Houwink (MH) relations are valuable only over a short molecular weight range.

On the other hand, experimental data obtained for poly(vinyl acetate)-chlorobenzene and poly(methyl methacrylate)-toluene in the $\overline{M}_n$ range of 10^3 to 6.10^3 have values of the exponent a (from the MH relation) close to 0.5.[44] For a polystyrene sample with $\overline{M}_n$ = 11,000 dissolved in benzene, however, a noticeable "unsweep" sets in below 0.2 g/dl concentration. At the lowest concentrations, the errors in the derived values of η_{sp}/c are very great, and the exact course of the upsweep is uncertain.[45] There is no doubt that this "upsweep" is a genuine effect and not simply a consequence of experimental errors in the η_{sp}/c determinations.

In the molecular weight range lower than 10^4 Da, the Gauss statistic is no longer applicable. Actually, it is known that a viscosity theory, based on the fact that the solute has molecular dimensions considerably higher than those of the solvent, cannot represent a valuable tool of viscosity determination for systems of particles having comparable dimensions. This has been observed since 1958, when negative values of intrinsic viscosities were found with certain oligomeric solutions. On the other hand, attention has been drawn[47] to the fact that, in the absence of irreversible adsorption phenomena, universal calibration may be employed with satisfactory results, even in the case of low molecular weights, if the solute does not have aromatic nuclei in its structure.

Apart from interest in the theory of viscosity of polymer and oligomer solutions, these observations have a practical importance in the measurement of intrinsic viscosity for oligomeric solutions. If the molecular weight is very low (10^3 to 10^4), the experimental values of η_{sp}/c may lie on the curved part of the plot, making extrapolation to [η] uncertain. Fortunately, the calibration in the low molecular weight region is aided by the fact that standard materials may be synthesized by known methods.

In the intermediate range of molecular weight (pleinomers and polymers with molecular weights up to about 20,000), the unperturbed dimensions have been shown to be a suitable universal calibration parameter.[48]

Anomalies in the application of the universal calibration method have also been noted[49,50] and attempts are being made to ascertain the reason. Rudin and Hoegy[51] have proposed that deviations from $[\eta]_M$ calibrations arise by not taking into account the concentration of the solutes and that the $[\eta]_M$ correlations are only correct at infinite dilution. This approach appears to be marginally superior to the straightforward $[\eta]_M$ calibration for polymers having different conformations. These authors also state that the deviation of unperturbed dimensions is a correction in the same direction, but is likely to be an overcorrection. A difficulty inherent in this procedure is that the derived relationship is an implicit function and cannot easily be adapted to a routine use.

The observed anomalies with universal calibration are of three types:

1. The unexpectedly large elution volumes determined by irreversible adsorption[52]
2. The deviations specific to systems of oligomer-poor solvents
3. The elution order in the domain of small molecules or oligomers not being governed only by the size of the solute particle

GPC data in the domain of oligomers and small molecules have pointed out the extra mechanisms involved in separation.[53-59] Besides the size exclusion effect, these secondary mechanisms may explain the peculiarities of the shape of actual calibration curves. These mechanisms are the *solvation of solute species, the adsorption* on the pore walls of the packing, and, for swelling packings, the *partition* between the gel and the whole phase.

The theoretical interpretation of the behavior of small molecules or oligomers during separation through GPC are based, partially or totally, on the theories proposed for the GPC of high polymers. Disagreements, observed mainly in the attempt to establish a suitable parameter for GPC calibration, have been caused, in our opinion, by the fact that structural characteristics and solvent-solute interactions dominate over the separation mechanism of exclusion. This observation is supported by experimental data that evidenced the very special behavior of aromatic hydrocarbons.[55,60-68] as well as the influence of functional end groups on the GPC separation of oligomers.[69-75]

The various possibilities of bonding aromatic rings lead to spatial structures with different shapes and dimensions. The aromatic solute has the possibility of offering — when entering inside the gel — various sizes. These sizes depend on the reciprocal position of the aromatic ring in the structure analyzed. Thus, the determination of the dimensional parameter governing the GPC separation of aromatic hydrocarbons acquires a statistical meaning.

Concerning the influence of the solvent, interesting results, listed in Table 1 and presented in Figures 5 to 7, have been obtained.[76]

Dissolved in tetrahydrofuran (THF), the six categories of substances analyzed are distributed on five distinct curves (Figure 5). This behavior has been found for all the calibration parameters. Special mention must be made of the reversal of pericondensed aromatic hydrocarbons. In trichlorobenezene (TCB), the experimental data are distributed on a single, straight line. Again, the organic compounds containing nitrogen have a peculiar behavior. It is possible that, with these compounds, the styrene-divinyl benzene gel manifests a "trap effect" as a consequence of a contraction process under the influence of the =NH group.

The behavior of phenols and their derivatives during separation by GPC has been studied in order to explain some features of the mechanism of phenol-formaldehyde resin synthesis.[77] The oligomers, formed through phenol polycondensation with formic aldehyde, present various structural forms. Calculations revealed that 13,203 linear isomeric structures are possible for a species made up of 10 phenolic units bound together in *ortho*- or *para*-positions by methylenic bridges.[78] On the other hand, resoles have a molecular weight distribution ranging between 18 Da (the molecular weight of the water) and 7000 Da.[79] Very little is known about the conformation and existence of molecular associations formed by phenol and its derivatives with THF.

TABLE 1
Effect of Solvent on Retention Volume of GPC Model Compounds[76]

Compound	Molar vol. (cm³/mol)	Retention in THF/25°C	Volume (cm³/mol) in TCB/145°C
Cata-condensed aromatics			
Benzene	88.9	164.5	166.0
Naphthalene	125.0	162.3	158.4
Anthracene	160.2	158.6	148.8
Peri-condensed aromatics			
Pyrene	171.5	161.3	145.8
Perylene	202.0	162.7	140.3
Coronene	228.0	166.2	136.0
Ovalene	280.0	179.3	129.7
Heterocyclics			
Thiophene	79.0	165.4	170.0
Pyrrole	69.3	149.8	184.4
Indole	102.9	147.0	173.5
Carbazole	144.0	145.0	161.1

Note: THF, tetrahydrofuran; TCB; trichlorobenzene.

Hydroxymethylated derivatives of phenols dissolved in THF differ from mono- or poly-phenols in that they also may form intramolecular hydrogen bonds in polar solvents. IR spectra of the solutions of these compounds in THF have shown that the peak characteristic of the bond between the phenolic or alcoholic group and THF is found at 3280 and 3430 cm^{-1}, respectively. Within the same field (3240 cm^{-1}) there appears the bond characteristic of the intermolecular bond between the phenolic −OH group and the *ortho*-hydroxymethylene group one has an absorption in the same domain (3240 cm^{-1}). Consequently, the IR spectra of the *o*-methylolphenol solution in THF cannot determine the nature of the solute-solvent interactions. One should choose between the two possible structures presented in Figure 8. As the intermolecular hydrogen bonds require the phenolic hydrogen, which has an acid character, conformation 1 (Figure 8) has a reduced probability. Such a hypothesis is also supported by the behavior of this species during separation by GPC.[71]

Further anomalies has been observed in the elution of styrene oligomers.[80] In a polar medium (e.g., dimethylformamide), 1-methyl-3-phenylindane and 1,3-diphenyl-1-butene behave like a monocomponent compound which, depending on the composition of the mixture, leaves GPC columns at different values of V_{EL}. All attempts to explain this behavior have, to date, been unsuccessful.

The conclusion to be drawn from the analysis of these data is that the behavior of aromatic compounds during GPC separations depends first on the solute-solvent interactions being practically independent of the parameter used to represent the solute particle.

Next, we shall analyze the results obtained when utilizing the *molecular volume* of the solute as the calibration parameter.[81] The molecular volume of low molecular compounds may be calculated by the method proposed by Harrison[81] or may be determined by dividing the molecular weight by the density. The calculated value obtained represents the theoretical molecular volume, V_{MT}, while that determined by the molecular weight/density ratio represents the bulk molecular volume, V_{ME}. In terms of solving the same problem, i.e., iden-

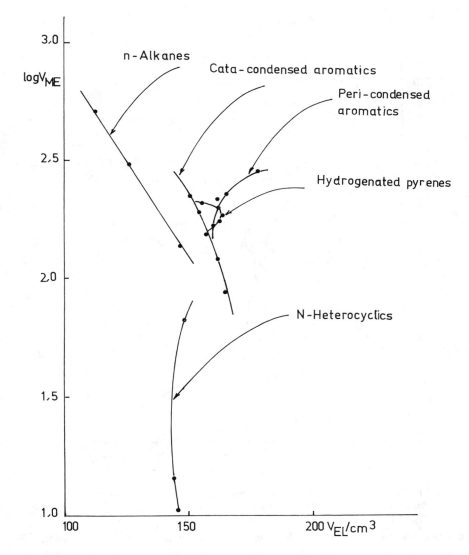

FIGURE 5. GPC data of various organic compounds dissolved in THF plotted in $\log V_{ME}$-V_{EL} coordinates.

tifying a universal calibration parameter, an important number of low molecular weight organic compounds have been analyzed to determine the influence of the support-solvent-solute system on the value of the solute molecular volume. The obtained data revealed the effect of the solute chemical structure — and especially that of the presence of the functional groups — upon behavior during separation of the small organic compounds.[82-89] The interaction between the solute and other components of the system determines to such a great extent the modification of the solute's molecular volume that application of a calibration based on this parameter may lead to errors up to 100%.

Alkanes have been considered as potential standards in the calibration of the GPC separation of oligomers. The absence of solute-solvent interactions and molecular associations support this idea. In the case of *n*-alkanes, a linear calibration curve can be obtained for any of the molecular parameters used.

Also worth mentioning is the fact that chain isomerization determines changes in the behavior of small organic compounds during GPC separation.[91,92] It is known that, during

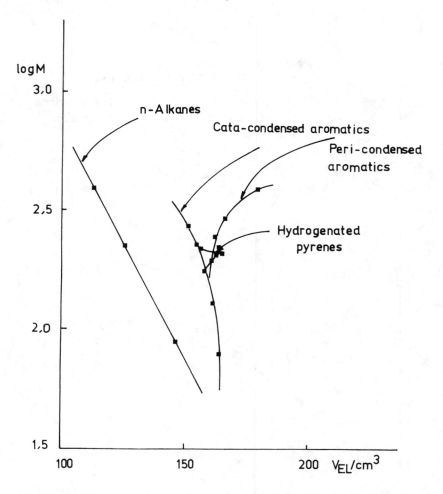

FIGURE 6.   GPC data of various organic compounds dissolved in THF plotted in logM-V_{EL} coordinates.

its free rotation around the simple C–C bond, the molecule adopts some states of both low-energy and higher energy levels. Therefore, rotation around the simple C–C bond is never totally free; the term commonly used is that of the *hindered rotation*. Consequently, there may exist a series of conformers, usually nonisolable, which nevertheless influence the physical properties of the substances.[93] In the case of ethane, for example, a single free rotation around the simple C–C bond may generate an infinite number of conformations. Among these, two may be distinguished by the extreme positions of the hydrogen atoms, i.e., the *eclipsed* and *intercalated* positions (Figure 9). Intercalated conformations are more stable.

If the organic compound is described by the formula $XCH_2–CH_2X$, where X represent a halogen atom or a functional group, two eclipsed and two intercalated conformations are possible (Figure 10).

The presence of intercalated conformations influence the behavior of alkanes or other small organic compounds during separation by GPC; the number of intercalated conformations determines the value of the density, heat of combustion, and refractive index of analyzed compounds.

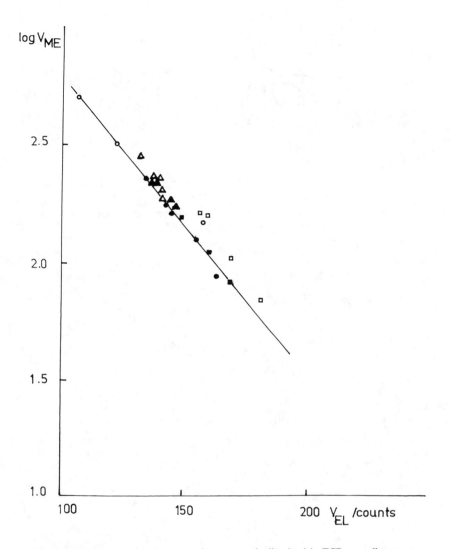

FIGURE 7. GPC data of various organic compounds dissolved in TCB -○-, alkanes; -●-, cata-condensed aromatic hydrocarbons; -△-, peri-condensed aromatic hydrocarbons; -▲-, hydrogenated pyrenes; -□-, N-heterocyclics; -■-, S-heterocyclics.

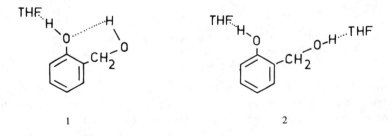

1   2

FIGURE 8. Two possible structures of hydrogen bonds in o-hydroxymethylated phenol — THF system.

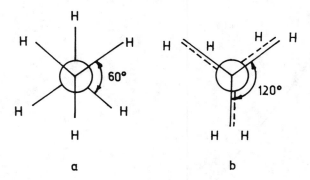

FIGURE 9.   The extreme positions of hydrogen atoms. (a) Staggered positions; (b) eclipsed positions.

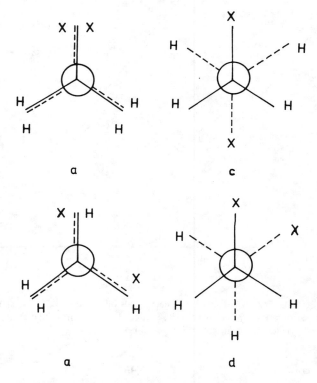

FIGURE 10.   The possible positions of halogen (X) in the structure of a $CH_2X–CH_2X$ type. (a) Eclipsed positions; (c) and (d), staggered positions, *anti* and *gauche*, respectively.

Experimental data[91] reveal that the following empirical relation may be established between the number of intercalated conformations (denoted by Z_g), molecular volume V_{MT}, V_{EL}, and the number of carbon atoms ($\neq C$) from organic compounds:

$$V_{MT} = aZ_g + b(\neq C) + c \qquad (10)$$

where a, b, and c are constants, taking the values -2.22, 16.5, and 35.8, respectively, and

$$V_{EL} = a'\log Z_g + b'\log(\neq C) + c' \qquad (11)$$

where a', b', and c' are constants, taking the values 100, 9.3, and 37.8, respectively.

The above-mentioned data and relations show that molar volume alone does not uniquely describe the elution behavior of branched hydrocarbons, unless the number of carbon atoms is also considered.

Referring further to the universal calibration parameters used for oligomer separation, the relations established between V_{MT}, the effective chain length, and $\neq C$ should be mentioned.[94]

$$\text{Effective chain length, } Å = 2.5 + 1.5(\neq C) \tag{12}$$

$$V_{MT}(\text{ml/mole at 20°C}) = 33.02 + 16.18(\neq C) + 0.0041 \tag{13}$$

Besides the difficulties encountered when utilizing GPC for the separation of small molecules or oligomers, this method also has some intrinsic disadvantages, compared to the other chromatographic methods. Thus, mention must be made, among others, of the low value of the *peak capacity,* n. This term is defined as the maximum number of peaks that can be resolved within a specified range of elution volume. In turn, the V_{EL} of a species is limited to the accessible volume of the pores for the given species. As pointed out by Giddings,[95] GPC is unique in that there is a well-defined limit to peak capacity.

The following relation have been established between n and the number of theoretical plates, N:[95]

$$n = 1 + 0.2 \, N^{0.5} \tag{14}$$

The concept of theoretical plate is used to comparing the separation efficiency of columns. The following equation, originally developed for liquid-liquid chromatography, is still used in GPC.[96]

$$N = 16\left(\frac{V_{EL}}{w}\right)^2 \tag{15}$$

where w represents the peak base width.

As pointed out by Cazes,[97] "the theoretical plate concept is borrowed from that area of chemical engineering involving fractional distillation. A theoretical plate, in the case of distillation, refers to a discrete distillation stage constituting a simple distillation in which complete equilibrium is established between the liquid and vapour phases. In the case of GPC where the two phases are in constant motion, i.e., the solvent in the interstitial volume and the solvent within the gel pores, equilibrium is probably never achieved. The true significance of the theoretical plate is lost. It must be realized, moreover, that the calculated theoretical plate in a chromatographic column represents a smaller separating ability than the theoretical plate in a distillation column by a factor of twenty-five to fifty". Then, the determination of N is GPC is more a measure of how well a column is packed, that is, how much peak spreading it will cause, than how well it will resolve. Moreover, the method of determination of N in the case of GPC is questionable. When solute of low molecular weight is used and its V_{EL} value, introduced in Relation 15, is determined, the value of N thus obtained is meaningless, if referring to a GPC column, due to the fact that the gel is wholly accessible to the reference sample (employed for the determination of N) and partially accessible to the species forming the mixture subjected to analysis. It is obvious that in the case of an oligomeric or macromolecular mixture to be passed through the GPC columns, the support will offer a higher number of theoretical plates for smaller molecular species than for those of higher molecular weight.[98]

In addition, several approximations, although heuristically useful, tend to propagate an ideal view which is no longer valid and which tend to induce overconfidence in the potential of GPC. The assumption of a perfect linearity of the calibration curve, the use of the term "Gaussian" to qualify the shape of the gel chromatogram (this adjective implies more than the simple notion of symmetry), and the poor definition of the baseline due to uncertain limits of integration and the signal-to-noise ratio are examples of excessive idealization or inadequate interpretation of the GPC data.

Although lacking a suitable theoretical base for a precise method of estimating its results, the GPC remains one of the most frequently used methods in the characterization of oligomeric mixtures — especially for a qualitative appreciation of the mixture composition.[99-109] The examples given below demonstrate the potential of classical GPC in the qualitative description of oligomeric mixtures or the presence of oligomers accompanying high polymers.

In polymer dope dyeing, the dyestuff is introduced into the polymerization system during synthesis. For this purpose, either reactive or inert dyes may be used.[110-112] In the first case, the dyestuff becomes part of the main chain by combination with suitable functional groups. In the second case, the dyestuff is dissolved in a polymer molten mass and a physical blend of macromolecular product and dyestuff is obtained. This process is applied in industry for poly(ethylene terephthalate) — PET — dope dyeing. The difference between dope-dyed polyester and colorless polyester obtained under the same experimental conditions is shown by a lower intrinsic viscosity and higher amount of $-COOH$ end groups and diethylene glycol (DEG) units. Efforts have been made to avoid these disadvantages by shortening the contact time between polymer and dyestuff during polycondensation. The results were not successful and polyester fiber and yarn producers now accept these special features of dope dyeing because of the economic characteristics of the dyed polymer obtained.

The gel chromatograms of dope-dyed PET show the systematic influence of dyestuff on the value and distribution of molecular weight in dope-dyed PET.

Polymer was obtained by batch reaction in a laboratory plant. The polycondensation was completed in 270 min. To be able to follow the oligomer transformation during polycondensation, the process was interrupted after 90, 120, 180, and 210 min. The samples obtained in this way will be called "interrupted samples". Thus, we obtained 12 interrupted samples and 3 end products (items 1 and 2 in Table 2 for colorless polymer, items 6 to 10 for blue polymer, and items 10 to 15 for red-colored polymer). The dyestuff (1% by weight for dimethylterephthalate, DMT) was introduced at the beginning of the polycondensation stage. The chemical and molecular characteristics of the resulting products are given in Table 2.

The molecular weight distribution in the synthesized samples was determined by GPC. The elution was carried out at 100°C by using a mixture of nitrobenzene-tetrachloroethane (NB-TCE) (5:95, v/v). The samples were dissolved in phenol:TCE (3:2). Before use, TCE was neutralized and stabilized as follows: freshly distilled TCE was treated twice with 5% aqueous K_2CO_3 solution (10 ml solution K_2CO_3 per 1000 ml TCE) with stirring for 30 min in each treatment. After separation of the aqueous phase, 1 ml of propylene oxide was added per 1000 ml and the TCE was then mixed in the required ratio with NB. The mixture was kept on silica gel until use (10 g silica per 1000 ml), in brown, closed bottles. These experimental conditions were chosen for GPC to avoid the high viscosity of the phenol-TCE mixture and because PET solutions in the TCE-NB mixture (95:5 v/v) are not stable.

Table 2 shows that the colored samples are characterized by a higher content of $-COOH$ end groups than the colorless samples. This finding suggests the occurrence of a degradation reaction during the preparation of dope-dyed PET.

Since the coloring matters used have a 1-aminoanthraquinone structure in which hydrogen atoms of amine groups are partly substituted by different radicals, the higher content of

TABLE 2
Sample Characteristics[110]

No.	Dyestuff	Analytical data				GPC data		
		$\overline{M}_n^a$	COOH[b] (eq/g 10^6)	$[\eta]$ (dl/g)	DEG[c] (%)	$[\eta]$ (dl/g)	$\overline{M}_n$	$\overline{M}_w$
1		1,010	77.0	0.075	3.45	0.080	1,480	2,200
2		4,270	22.0	0.205	1.29	0.140	4,000	7,500
3	CS[d]	6,750	35.0	0.293	1.96	0.240	6,800	10,300
4		16,500	49.6	0.611	1.72	0.690	13,400	20,100
5		21,000	57.0	0.740	1.09	0.760	24,700	58,300
6		1,400	200.0	0.080	6.40	0.110	2,300	3,100
7		1,950	168.0	0.105	1.50	0.120	2,800	3,500
8	ERLS[e]	3,700	98.0	0.180	2.55	0.140	3,700	4,600
9		4,550	96.0	0.225	2.50	0.210	4,270	9,600
10		17,000	78.0	0.620	2.60	0.610	16,500	38,500
11		1,150	137.9	0.082	5.45			
12		1,550	120.0	0.085	1.40			
13	ESGFP[f]	5,200	47.0	0.255	3.00			
14		11,000	58.2	0.365	3.01			
15		17,200	58.7	0.545	3.20			
				Industrial Samples				
I		22,000	28.0	0.760	0.55	0.75	21,500	69,000
II		19,800	68.6	0.725	1.66	0.71	19,500	55,000

[a]   Determined by titration of end groups.
[b]   Determined by potentiometric titration.
[c]   Determined by gas chromatography.
[d]   CS colorless samples.
[e]   Estofil Blue RLS (produced by Sandoz).
[f]   Estofil Red SGFP (produced by Sandoz).

–COOH end groups in colored PET may be explained easily. The basic character of the aromatic amines is decreased by p-π conjugation which involves the unpaired electrons of nitrogen and the π-electrons of the benzene ring. This process gives an extra positive charge at the nitrogen atom, which induces polarization of the ester link. The following reaction may be written:

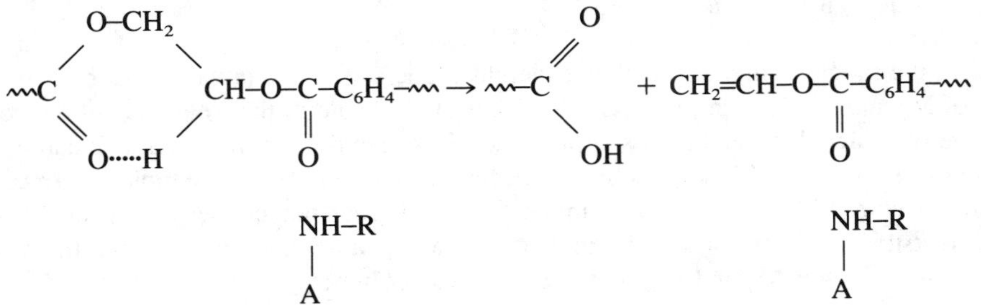

$$\text{~~C} \underset{O\cdots H}{\overset{O-CH_2}{\big\langle}} CH-O-C-C_6H_4\text{~~} \rightarrow \text{~~C} \big\langle_{OH}^{O} + CH_2=CH-O-C-C_6H_4\text{~~} \qquad (16)$$

in which A = anthraquinone rest and R = 1,3,5-trimethylphenyl or –CO–C$_6$H$_5$.

The GPC data confirm the systematic influence of the dyestuff on the molecular weight distribution in dope-dyed PET (Figure 11).[110]

The analysis of product characteristics shows that in the first stage of the polycondensation process (after 90 min) the molecular weight distribution (MWD) of the dyed polymer

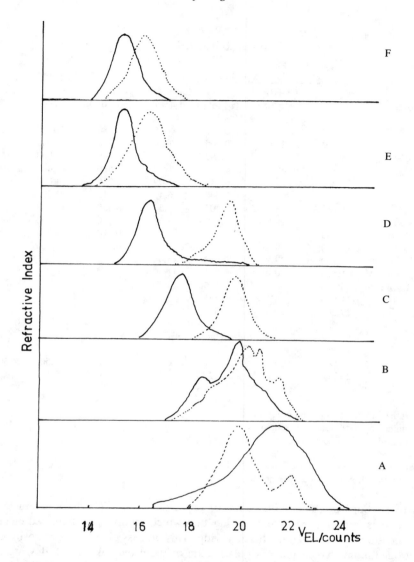

FIGURE 11.   Gel chromatograms of colorless samples (———) and dyed ones (····). (A) After 90 min; (B) after 120 min; (C) after 180 min; (D) after 210 min; (E) after 270 min; (F) industrial samples I and II.

shifted to the range of higher molecular weights. The same results were obtained from the values of M_n and M_w for samples 1 and 6 (Table 2). In Table 3, the frequency of the species per range of molecular weight for samples 1 to 10 is given. From these data, it can be seen that sample 1 contains a higher amount of species with $M_n < 10^3$ than sample 6. Concerning the species with $10^3 < M_n < 10^4$, sample 6 has a higher percentage. As a result of this uneven distribution of species in the molecular weight range in samples 1 and 6, the dyed polymer has a higher M_n and M_w than the colorless polymer.

In the subsequent stages of polycondensation, the presence of dyestuff is accompanied by a systematic shift of the distribution curves of colored PET to the range characteristic of the low molecular weight species. This is shown by the location of the distribution curves in Figure 11 and by the species distribution per range of molecular weights (Table 3). The change in the direction of the value of the molecular weight and MWD of colored PET caused by the dyestuff is the result of the mechanism which dominates the interaction between

TABLE 3
Species Distribution (%) per Range of Molecular Weight in the Analyzed Samples

	$\overline{M}_n$ of colorless samples					$\overline{M}_n$ of colored samples			
No.	10^3	10^4	10^4	$\overline{M}_w/\overline{M}_n$	No.	10^3	10^4	10^4	$\overline{M}_w/\overline{M}_n$
1	20.41	75.68	3.91	1.48	6	6.27	93.73	—	1.34
2	6.49	86.37	7.14	1.87	7	—	98.15	1.85	1.25
3	—	58.97	41.03	1.51	8	—	95.73	4.23	1.24
4	—	21.01	78.99	1.50	9	0.71	77.44	21.85	2.24
5	—	6.50	93.50	2.36	10	—	19.68	80.32	2.33

polymer and dyestuff. In the first stage of polycondensation (after 90 min), the main influence of the dyestuff is to decrease the melt viscosity (in industrial autoclaves, this influence is shown by a decrease in the power required to drive the stirrer). The decrease in melt viscosity is not accompanied by a reduction of polymer molecular weight, but is more a consequence of PET melt plasticization caused by the presence of the dyestuff in the system. This conclusion is shown by the values of the molecular weight of samples 1, 6, and 11 and by the gel chromatograms of samples 1 and 6. As a result of the easier removal of ethylene glycol from the system due to the decrease in melt viscosity, the molecular weight of the colored polymer is higher than that of the colorless one 90 min after the beginning of the polycondensation.

In the later stages of polycondensation, the presence of the dyestuff is accompanied especially by a supplementary degradation effect made evident in the dyed samples by the diminished value of the molecular weight, the higher amount of –COOH end groups and diethylene glycol, and the profiles of the gel chromatograms of samples 7 to 10 (Tables 2 and 3, and Figure 11).

The presence of the dyestuff in PET results in the appearance of some specific peaks in the gel chromatograms of low molecular weight species (V_{EL} = 22.3 counts in curve B_A, and 21.5 counts in curve B, Figure 11). These peaks correspond to M_n values of 1100, 2560, and 1600 Da, respectively.

Gel chromatograms of the colored samples have, as specific characteristics, the appearance at V_{EL} = 23 to 28 counts for some specific peaks (which are not shown in Figure 11) due to the dyestuff. The eluent separated in the fraction collector at V_{EL} = 13 to 23 counts was colorless, while the eluent separated at V_{EL} = 23 to 28 counts was blue. The possibility of separating the dyestuff from the polymer by GPC supports the conclusion that there are no covalent bonds between polymer and dyestuff.

Introduction of some ''colored structures'' in the PET backbone by using the corresponding dyestuff allows the synthesis of some colored products without altering the characteristics of the white polymer.

The insertion of reactive dyestuff N,N-bis(ethoxycarbonylphenyl)-pyromellitic diimide with the chemical structure given in Formula 17 is proved by IR spectroscopy and GPC data.

$$
\begin{array}{ccc}
& O \qquad\quad O & \\
& \| \qquad\quad \| & \\
& C \qquad\quad C & \\
& \diagup \;\diagdown \; \diagup \;\diagdown & \\
H_5C_2OOC{-}C_6H_4{-}N & \quad C_6H_2 \quad & N{-}C_6H_4{-}COOC_2H_5 \qquad (17) \\
& \diagdown \;\diagup \; \diagdown \;\diagup & \\
& C \qquad\quad C & \\
& \| \qquad\quad \| & \\
& O \qquad\quad O &
\end{array}
$$

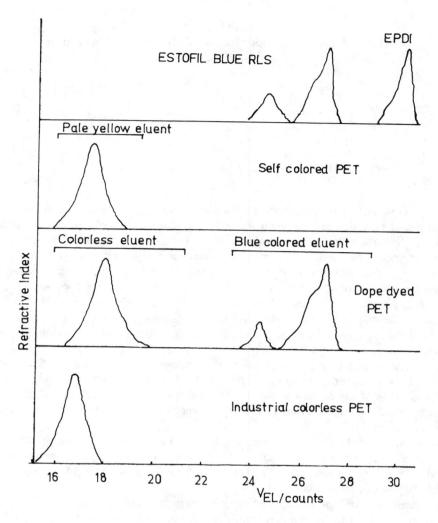

FIGURE 12.   Gel chromatograms of EPDI, *N,N*-bis(ethoxycarbonyl pyromelliticdiimide samples.

In a recent paper, Libert and Maréchal[111] reported that, during the introduction of the functional dye in the main chain of polymers, no change in the structure of the chromophore takes place (its IR spectrum did not change). Comparing the colored polymers with the corresponding monomers, it was observed that the spectra of polymers were slightly broader than that of the monomer. This phenomenon was probably caused by interaction of the chromophores stacked along the backbone. The behavior of dope-dyed PET and self-colored PET (with reactive dyestuff) in the separation by GPC confirms this view. The gel chromatogram of dope-dyed PET shows both a separate, distinct distribution curve for PET (the eluent collected between 16 and 22 counts was colorless) and the characteristic peaks of Estofil Blue RLS®. In this case, the separated eluent between 23 and 28 counts was blue. Gel chromatograms of self-colored PET did not reveal the characteristic peak of EPDI [*N,N*-bis(ethoxycarbonylphenyl)-pyromellitic diimide], while the eluent collected between 16 and 19 counts was yellow (Figure 12).

Mention must also be made of studies applying classical GPC in the determination of the molecular weight distribution within various oligomeric mixtures obtained through the polymerization of vinylic, dienic, or acrylic monomers,[113-132] as well as in oligoquinazolines,[133] oligosulfones,[134,135] various oligoesters,[136-142] oligoethers,[143-146] oligoamides,[147,148] oligourethanes,[149] and resins.[150-160]

The matters discussed up to now have addressed the limitations of GPC. These limitations have tended to improve the performances of this method, mainly in the separation of low molecular weight species. Two major achievements reported to date, the *recycle GPC* and *high-performance GPC (HPGPC)* techniques should also be mentioned.

Recycle GPC is a special technique of column programming for improving resolution by passing the sample through the column, or set of columns, repeatedly. Yet, recycle GPC represents an efficient way of exceeding the intrinsic limits of GPC and also of solving new details of the composition of natural (such as crude oil) or synthetic low molecular weight mixtures. In practice, recycle GPC was introduced by Porath and Bennich[161] and developed by Biesenberger et al.,[162-164] Nakamura et al.,[165] Uglea and Cinru,[166] Bombaugh et al.,[167] and recently by Trolltizsch.[168]

Recycle GPC is performed by two main procedures: direct pumping and alternate pumping. The former shows the disadvantage of creating an over-pressure in the reference cell of the detector, of the approximative measurement of V_{EL}, and of the increase of the axial dispersion. However, with high-efficiency columns, recycle GPC is impractical since the resolution gained by the recycle system would be lost due to band broadening in the pump head. An alternative is to use a low-dead-volume switching valve coupled with two columns and a high-pressure pump.[169]

On the other hand, recycle GPC is also limited to separation of the mixtures of a very few compounds, two to four at the most, with very similar elution behavior. There is the potential of the trailing peak being overtaken (or lapped) by the faster moving peak.

Although recycle GPC has received much attention in the commercial literature, its real utility seems limited to applications requiring the separation of two compounds with almost identical elution behavior.

Conventional GPC does not provide the required resolution in the low molecular weight region for control of the molecular weight distribution in the oligomer-polymer systems. With the advent of high-efficiency columns (particle diameter, <20 μm), the resolution in this molecular weight range of 200 to 10,000 Da has been greatly improved, and conventional GPC becomes HPGPC. Even if the estimation of the performances of chromatographic columns, especially the GPC ones, is not standardized yet, it is generally accepted that if a set of columns has 10^4 to 3.10^4 theoretical plates, then conventional GPC becomes HPGPC. Specific applications of HPGPC include quality control of supplier raw materials, guiding resin synthesis and processing in order to improve end-use properties, and correlating oligomer and polymer molecular weight distributions with end-use properties.

Another field of HPGPC utilization is the characterization of coating systems. Although coating systems and lacquers have been known since cave-dwellers decorated the walls of their abodes, it has only been within the last century that any scientific consideration has been given to them. The organic coating systems commonly called paints, varnishes, and lacquers are usually complex blends of two (or more) phases, a pigment phase and a vehicle phase. It is the vehicle phase that can be fruitfully investigated by HPGPC. This film-forming phase is itself a complex mixture composed of resin, plasticizers, solvents, and various additives.

In recent years, new techniques such as powder, high-solid, water-borne/water-based, and radiation-curable coatings have been developed to meet the challenges of official regulations in the areas of ecology, energy, and consumerism. These constraints call for the design of a carefully tailored molecular weight distribution to minimize the presence of volatile components. They also require the use of water as the major solvent or the use of tailor-made low molecular weight polymers, oligomers, and reactive additives which, when further reacted, produce high molecular weight and cross-linked polymers concomitant with minimization of volatile organic compound emissions. Consequently, the quality control of raw materials and determination of the oligomer/polymer ratio as well as the molecular weight distribution of the binder resins are critically important for the attainment of desired performance properties from these new coating technologies.

Powder coatings are designed to be 100% solids and usually contain a low molecular weight polymer and an oligomeric cross-linking agent to produce a cross-linked polymer. Also, a powder coating contains a small amount of an oligomer flow agent to aid flow and leveling during the baking process. However, the molecular weight distribution of low molecular weight components should be chosen in such a way that (1) the powder particle will not coalesce upon storage; (2) the resin system will melt and flow with appropriate leveling characteristics, optimum for appearance properties, prior to the cross-linking reaction within specific time constraints at a given temperature; and (3) the cross-linking reaction must occur at the appropriate time after the resin has melted, consistent with the development of both good appearance properties and good mechanical properties. Therefore, the molecular weight distribution of polymeric and oligomeric components must be carefully designed and controlled.

HPGPC was used for the characterization of incoming raw materials and also to aid a resin chemist in developing an in-house oligomeric component for a powder-coating system.[170,171]

For *high-solid coatings*, HPGPC is very useful for screening various resins for the optimization of coating viscosity (molecular weight distribution) and cured film properties. High-solid coatings are those which are usually 62.5% or more nonvolatile on a volume basis. These coatings contain oligomers ($M_n \simeq 500$). The key design parameters in high-solid coatings are low viscosity, low volatility, and controlled reactivity.

Generally, the molecular weight distribution of a low molecular weight polyester is a direct result of the molar ratio of diacid and olefin oxide to diol in the reactor. HPGPC results show that, as expected, the molecular weight distribution of the resin increases as the molar ratio of anhydride and propylene oxide to glycol increases. On the other hand, HPGPC can also be used for determination of the level of high molecular weight components in cross-linkers. These compounds would impart less impact resistance to a cured coating, assuming all other parameters are the same.

Radiation-curable coatings (RCCs), via UV radiation, consist of a very low molecular weight multifunctional oligomers diluted with reactive monomers and contain a photosensitizer to promote cross-linking reactions. This type of coating is ideally suited for flat stock such as floor tile and interior wood paneling. The molecular weight distribution of the oligomer must be maximized consistent with acceptable rheological properties. HPGPC is useful for guiding resin synthesis and process development.

Electron beam coatings consist of an unsaturated oligomer mixed with reactive monomers. The molecular weight of the oligomer is very low and HPGPC is again useful for guiding the oligomer synthesis and to monitor the process.

Water-borne coatings (WBC) are prepared in water-miscible organic solvents up to 70 to 80% solid by volume. These coatings usually are in the 5000 to 30,000 molecular weight range and contain polyesters, alkyds, acrylics, and epoxy esters. HPGPC has been used to monitor the molecular weight distribution of raw materials and for periodic characterization of the sample from storage tanks or production tanks.[172]

HPGPC has been also used for the characterization of many epoxy resin formulations[173-175] and in the determination of the amount of oligomers obtained during the thermal degradation of polymeric materials.[176]

The molecular weight distribution of oligomeric mixtures may also be studied by the *semimicro GPC technique*.[177,178] In this procedure, several polystyrene gels of different pore sizes were packed into a 500 × 2.1 mm I.D. column. To enable semimicro GPC to be carried out with a system consisting of a triple piston pump, a microloop injector and a fluorocell with a volume of 1.0 ml were constructed. This improved apparatus was developed because the dead volume of the injector and the cell in the conventional equipment determine a significant loss in terms of column efficiency. The effect of sample amount, injection volume, and mobile flow rate on column efficiency and retention volume were optimized. The sample amount (<500 μg), injection volume (<15 μl), and flow rate range (30 to 70 μl/min) are the optimal operational variables for semimicro GPC. Oligostyrenes, epoxy resins, phenol-formaldehyde resins, and phthalates were analyzed by the semimicro GPC system under given conditions.[177,178] This method showed good accuracy and reproducibility and was preferred to liquid chromatography techniques.

III. OLIGOMER FRACTIONATION BY LIQUID CHROMATOGRAPHIC METHODS

Chromatographic separation is based on the partition between two phases, namely, the stationary and the mobile phase. The chromatographic techniques, require a gas, (gas chromatography) or a liquid (liquid chromatography) as the mobile phase.

Gas-chromatographic methods involve two different procedures, liquid-gas and solid-gas. Liquid chromatographic methods, in turn, are subdivided into ion-exchange, partition, paper, thin-layer, and liquid-solid adsorption methods. The GPC method discussed previously has also been included within liquid chromatography, primarily because of the experimental equipment employed (columns, solid support, etc.) and to a lesser extent, the separation principle.

Both gas chromatographic (GC) and liquid chromatographic (LC) methods have advantages and drawbacks, and, consequently, have specific areas of application. The best resolution is achieved with GC methods, while LC methods enable one to separate thermolabile and high molecular weight compounds not amenable to GC methods. In addition, both methods have their specific detectors. The important problems in GC and LC methods arise from the characteristics of the mobile phases involved. From the point of view of column chromatography requirements related to resolution, selectivity, and methods of sample introduction, the use of gases and liquids as mobile phases means exploiting the two extremes of an ideally mobile phase which would also offer a GC-like resolution for compounds actually separable only by LC methods.

The mobile phase which could cover the gap between the utilization of gases and liquids is that possessing the peculiarities of a gas and a liquid simultaneously, and this is conveniently achieved by a supercritical fluid (SF). Although in the past there were experiments with SFs as chromatographic mobile phases, only the last decade offered the conditions (mainly technical, but psychological, too) for a larger utilization of SFs.[179]

All methods of column chromatography (GC, LC, and SF) have been applied to the separation of oligomeric mixtures as such or in different combinations. Worth mentioning are the results obtained in the identification of N-oligooxyethylenes through GC methods[180] or the quantitative determination of some oligomers from their mixtures with high polymers by combining LC, GPC, and spectral methods.[181-197]

Thin-layer chromatography equipped with a flame ionization detector was applied to the separation and determination of polyethoxylated lauryl alcohol nonionic surfactant[198] and for routine characterization of mixtures of monomeric and oligomeric fatty acids.[199] The same method has been used for determining the functionality distribution in telechelic prepolymers,[200] while, in the determination of oligo(ethylene terephthalate), it was used in combination with HPGPC.[201,202]

In supercritical fluid chromatography, the solubilizing nature of the mobile phase allowed the use of gradient elution as in liquid chromatography. With this gradient technique, oligomer separations were achieved over a wide molecular weight range.[203] For separating oligomers which absorb only at low wavelengths, a CO_2/CH_3CN mobile phase was used which allowed UV detection down to 200 nm.[204,205] As an ideally suited detection assembly for the chromatography of oligomers, a combination of mass-sensitive and evaporative light-scattering detection is suggested.[203]

Other chromatographic procedures, again for the characterization of oligomeric mixtures, may also be mentioned, such as carrier-aided centrifugal partition chromatography,[207] and zone-melting chromatography.[207a]

IV. DETERMINATION OF THE STRUCTURE AND OF THE CHEMICAL COMPOSITION OF OLIGOMERS

The structure and the chemical composition of oligomers are usually determined with the same methods, applying the same principles used for simple substances. Visible or UV absorption spectra have been used by Seidel[208] for determination of the structure of linear and cyclic oligomers separated from poly-(ethylene terephthalate). An IR method is also described[209] for determining the content of −OH groups in oligodiene diols using THF as solvent. The −OH valence vibration appears as a single, broad absorption band at 3470 cm^{-1}. Specific procedures are given for determining the −OH groups in dry samples and in samples containing traces of H_2O. In the latter case, adjustments are made for contributions by H_2O in the absorption region of 3470 cm^{-1}. Use of this method to analyze oligoisoprene diol showed 3.1 to 5.5% deviation between the IR method and chemical analysis for samples containing traces of H_2O and 2.0 to 8.2% deviation for dry samples.

By IR spectroscopy, the mechanism of oligoamide synthesis or reticulation in formaldehyde resins has been determined.[210] IR spectroscopic analysis of linear soluble oligomers formed in the reaction of bismaleimides with diamines (e.g., 4,4′-methylenebisaniline) at a molecular ratio of 2:1, and of cross-linked insoluble polymers formed by continuing the polymerization of oligomers, indicated that the oligomers were formed by addition of the H of primary amines to the double bonds of the maleimides.[211] No addition of the H of secondary amines (formed in the above reaction) to the double bonds of the maleimides and no cleavage of the maleimide rings was detected. Cross-linking of the oligomers occurred apparently by polymerization of the double bonds of the maleimide groups. The kinetics of the copolymerization of bismaleimides and diamines at the molecular ratio of 2:1 and at 150 to 170°C was investigated by following changes in the concentration of double bonds in the system by IR spectroscopy. The obtained data indicated a second-order for the reaction with an activation energy of 122 kJ/mol.[212]

The structural characteristics of oligostyrenes,[213] oligoacrylates,[214] oligoamides, and oligoesters has been determined by IR spectroscopy, too.[215,216]

Low-frequency Raman spectroscopy were reported for monodisperse methylene-oxy-ethylene-methylene triblock oligomers.[217] Low-temperature optical absorption spectroscopy was used for the study of reactive intermediates produced during solid-state polymerization of diacetylene crystals.[218]

Nuclear magnetic resonance (NMR) has been frequently used in the study of oligomers. Thus, ^{13}CNMR spectra of CH_2 carbons of polystyrene exhibited 12 peaks between 42 and 47 ppm which were assigned using the results obtained for symmetrical oligostyrenes with phenyl end groups, which were prepared by oligomerization of styrene in the presence of ethyl Li, followed by fractionation into dimer, trimer, tetramer, and pentamer.[219,220] The presence of diastereoisomers in oligomer mixtures may also be observed by NMR analysis.[221]

Oligo(vinyl chloride) prepared in THF using *tert*-BuMgCl as initiator was analyzed by joint HPLC and ^{13}CNMR spectroscopy.[222] In the case of oligobutadienes[223] and oligobutadienedicarboxylic acids,[224] ^{13}CNMR spectra made it possible to calculate the isomeric compositions of the chain and the number-average molecular weight before and after heating. No rupture of double bonds took place at 150 to 220°C, whereas double bonds of 1,2-units were broken at 260°C as a result of cross-linking.

The chemical shift of internal groups brought about by the presence of end groups has been also determined by ^{13}CNMR for oligosiloxanes,[216,225] oligo(ethylene glycol),[223] hydroxytelechelic polybutadiene,[226] oligomethylenes,[227] oligo(oxy-2,2-dimethylethylenecarbonyl),[228] and oligomethacrylates.[229]

NMR spectroscopy, in combination with chromatographic methods facilitated the study of the microstructure of polyfluorobutadiene,[230] molecular mobility in oligosulfones,[231] and the structure of some oligosulfides,[232,333] oligocarbonates,[234] and vinylic oligomers.[235]

On employing experimental data obtained through GC analysis in combination with mass spectroscopy, or through GPC, HPLC, and mass spectroscopy, several oligomeric additives used in polymer processing[236-240] have been identified, parallel to explaining the mechanism of polyurethane ammonolysis[241] and the structure of oligoesters obtained by the polycondensation of adipic acid with propyleneglycol.[243] The rigid segments in polyurethane elastomers[242] and the monomer reactivity ratio in certain processes of copolymerization have been determined in a similar manner.[244]

NMR spectroscopy helped clarify some aspects of oligourethane reactivity,[245] the delimitation of microphases in styrene-isoprene copolymers,[246] tacticity in acrylonitrile oligomers,[247] and the content of end groups in poly(oxyethylene) telechelics.[248]

NMR data for the polymerization of 1,3-bis(methacryloyloxyethyleneoxycarbonyloxy)-propane showed that both broad-line NMR and pulsed NMR could be used for studying the kinetics of three-dimensional polymerization and the change in molecular mobility during network formation. Broad-line NMR retained its sensitivity to high monomer conversions and could be used for measurements without preliminary calibration. Pulsed NMR was used during the initial stages of polymerization.[249]

Analysis of products of thermal polymerization of phenylacetylene by ^{13}CNMR and mass spectroscopy showed that the end groups in poly(phenyl acetylene) prepared by thermal polymerization were tetrasubstituted cyclohexadiene or benzene rings. Terminal cyclohexadiene rings were formed from chain termination by intramolecular cyclization of the growing chain end. Thus, the low degree of polymerization obtained during thermal polymerization of aryl-acetylenes is caused not by inactivation of the active center from delocalization of an unpaired electron along the conjugation chain, but by termination from intramolecular cyclization of the growing chain.[250]

^{13}CNMR data provided information about the kinetics and mechanism of reaction of bismaleimidodiphenylmethane with diallylbisphenol A.[251] In the case of reaction of polyols with isocyanates,[252] ^{13}CNMR combined with GPC provided information about the formation of urethane, biuret, allophanate, and isocyanurate in the synthesis of polyurethanes.

^{1}H and ^{13}CNMR spectra of linear and branched polysulfide oligomers showed that not all of the 1,2,3-trichloropropane introduced in amounts of 35% into the reaction mixture during polysulfide synthesis in order to stimulate their branching took part in the formation of long-chain branches. The experimentally determined branching degrees of polysulfides did not reach the theoretically predicted values, due to a lower reaction ability of the secondary chloride atoms of trichloropropane, which resulted in the presence of chloride-containing fragments in the polysulfide chains.[233]

Mass spectroscopic data of oligomers enables the determination of end groups, as well as the chemical structure of the products obtained by oligomer-homologue reactions or by degradation of high polymers.[253-258] The same method allows the discrimination of cyclic and linear oligomers into a series of compounds obtained by chain-growth or step-growth polymerization.[259-261]

Oligomers synthesized by specific reactions or obtained by fractionation of low molecular weight polymers are essentially pure compounds and their mass spectra are readily obtainable as such. For such polymer-derived oligomers, additional factors should be considered in evaluating the mass spectral data. At the temperature required to volatilize a polymeric material to give an observable spectrum, the sample may be completely depolymerized to monomer and initiator, and few, if any, species of higher mass than monomer will be seen in the spectrum. If, however, an oligomeric fraction, including monomer, is present in the polymer sample, the oligomer may be volatilized intact and all of the corresponding molecular ions will be observed in the spectra, provided such ions survive the ionizing electron impact. Ion fragments of volatilized species that are formed on electron impact are characteristic of the structure of the repeat unit.[253-258]

A mass spectroscopy study of a variety of oligomers has provided data which can be correlated with the explanations of end-group analyses,[262] and in the determination of the molecular weight distribution.[263-270]

The ^{1}HNMR spectra of methyl methacrylate oligomers prepared through anionic oligomerization were analyzed in relation to the stereospecific conformation of individual molecules,[229] taking into account the structural end-group effects and the characteristics of the geminal methylene proton signals of this series of model compounds. The results obtained for both of the stereoisomers of the dimer molecules were extended for analyzing the more complicated spectra of the trimer molecules. The magnitudes of the coupling constants for the geminal methylene protons of these n-mers were determined as follows: -9.0 Hz for 1-mer, -14.0 Hz for dimers, and -14.5 Hz for trimers.

Cyclic and linear oligomers of poly(ϵ-caprolactam) and poly(butylene isophthalate) with molecular weights below 1450 were identified by fast-atom bombardment mass spectrometry employing collision-activated decomposition.[271]

Spin electronic resonance combined with proton magnetic resonance has been employed in the analysis of some aspects of the mechanism of oligomerization of some vinylic monomers, as well as in the discussion of some structural details of the obtained products.[272-275]

V. DIELECTRIC PROPERTIES

The distribution of the dielectric relaxation times of polymers is known to be higher than that of the small molecules, such behavior probably being induced by the chain length. At present, it is generally accepted that the major relaxation mechanism of polymers (the so-called α-relaxation) depends upon the Brownian micromovement of the segments in the main chain of the polymers. The concept of segment is thought of as referring to the motional unit of the α relaxation, evidencing a common movement. The meaning of such a definition is, nevertheless, doubtful. Yet, one may consider that this sequence is made up of several monomeric units. In the case of simple molecules, the main relaxation is attributed to the rotational relaxation of the whole molecule, while in the case of polymers, it refers solely to a chain sequence. It is obvious that the shifting from small molecules to macromolecules will bring about the appearance of the distribution of dielectric relaxation times. Keeping these considerations in mind, analysis of the dielectric behavior of oligomers becomes necessary for discussing the movement mechanism of sequences within polymeric chains. Total rotation of oligomeric molecules is expected to occur, which would induce a narrow distribution of the dielectric relaxation times. However, data obtained for the oligomers of vinyl acetate[276,277] indicate the presence of a broad distribution of the dielectric relaxation times, which could be interpreted as a consequence of the final group effect.[278]

An attempt to determine the critical length of the chain marking the shift from the relaxation mechanism by rotation of the molecule as a whole to that of sequential rotation led to the obtainment of experimental data showing the complex character of such a process. Thus, the fact that several oligomers (ethylene diamine, ethanol amine, vinyl acetate, and propylene glycol oligomers)[279-281] show a broad distribution of relaxation time, higher even than that of the corresponding polymers, indicates that the mechanism of the relaxation process is complicated by the increase of the chain length, but does not depend on a certain value of the molecular weight. If the oligomer molecules contain functional groups capable of forming hydrogen bonds, they may also influence the value and distribution of the relaxation times.[282]

VI. THERMAL PROPERTIES OF OLIGOMERS

Within a homologous series of linear oligomers, the melting points of the individual members increase more or less rapidly, depending on the type of end groups present, and approach asymptotically the actual melting points of the corresponding high polymers. Nevertheless, several exceptions to this rule may be discussed here, such as the one offerred by linear oligoesters of terephthalic acid and ethylene glycol with the structural formula HO–T–[G–T]$_n$–OH, where G = –O–CH$_2$–CH$_2$–O– and T = –CO–C$_6$H$_4$–CO–. In this case, individual oligomers with n = 2, 3, and 6 show melting point values higher than the one corresponding to poly(ethylene terephthalate).[283]

A regression equation based upon a multiple regression analysis of the experimental data allowed the calculation of the oligomer melting points in the case of oligo(ethylene terephthalate)s removed from the reactor during preparation of poly(ethylene terephthalate).[284] This equation, established for mixed oligomers, was subsequently successfully extended to pure linear oligoesters. The equation was applicable to the intermediate products from almost all stages of the polymer manufacture.[285]

When the melting point of the polymer is higher than the temperature of its decomposition, as in the case polycarboxypiperazine, the melting points of the polymer may be determined by extrapolation from the melting points of the oligomers.[286]

In the case of linear oligoamides with $\overline{M}_n$ = 1500 to 8500 obtained by the polymerization of ϵ-caprolactam or ω-dodecalactam in the presence of hexamethylenediamine, dodecamethylenediamine, or adipic acid as difunctional molecular weight regulators and end-group blockers, the melting points increase with increasing molecular weight, reaching a constant value at $\overline{M}_n$ = 3500 to 4000.[387]

The dependence between the thermal properties and the polymerization degree of oligomers has been studied to gain a better understanding of the mechanism of polymer degradation and of the possibilities of utilizing oligomers as curing agents.[288,289]

Originally, in the case of homologous paraffins, the following relation was postulated for the dependence between melting points and molecular weight:[290]

$$\frac{1}{T_m} = a + \frac{b}{z} \tag{18}$$

where T_m = melting point, a and b are specific constants, and z represents the number of CH$_2$ groups in the chain.

The melting characteristics of oligourethanes were determined by Nagura et al.[291] and Hay[292] in an attempt to document the effect of the final groups on the thermal properties of oligomers. It has been shown that the T_m value is directly proportional to the number of segments comprising the chain.

The melting point variations of homologous series of cyclic oligomers are usually discontinuous and often show an anomalous relationship to ring size. The melting points of the first members are often very much higher than those of the corresponding polymers. The special behavior of cyclic oligomers has been explained by assuming that the melting point of a ring compound depends on the stability of its conformation.[293]

The high crystallinity degree of cyclic oligomers determines, to a great extent, the high value of T_m characterizing cyclic oligomers.[294-296] The obtained experimental data with regard to crystallinity of oligo(ethylene oxide) and oligoester have evidenced the influence of the ordering degree on the thermal properties of oligomers.[297-301] At the same time, the influence of oligomer content on the morphology and crystallinity of polymers has been stressed.[302-304]

REFERENCES

1. **Moore, J. C.,** Gel permeation chromatography. A new method for the determination of molecular weight distribution in high polymers, *J. Polym. Sci. Part A,* 2, 835, 1964.
2. **Brønsted, H. Z.,** Molekülgrosse und Phasenverteilung. I., *Z. Phys. Chem. Bodenstein Festband,* 257, 1931.
3. **Schulz, G. V.,** Über die Löslichkeit und Fällbarkeit hochmolekularer Stoffe. 163 Mitteilung über hochpolymere Verbindungen, *Z. Phys. Chem. Abt. A,* 179, 321, 1937.
4. **Schulz, G. V. and Jirgensons, B.,** Die trennung Polymolekularer gemische durch fraktionierte Fälung. Über die Loslichkeit makromolekularer Stoffe. IX, *Z. Phys. Chem. Abt. B,* 46, 137, 1940.
5. **Meyer, K. H.,** Propriétes de polyméres en solution. XVI. Interprétation statistique de propriétés thermodynamiques de systemes binaires liquides, *Helv. Chim. acta,* 23, 1063, 1960.
6. **Flory, P. J.,** *Principles of Polymer Chemistry,* Cornell University Press, Ithaca, 1953.
7. **Huggins, M. L. and Okamoto, H.,** Theoretical considerations, in *Polymer Fractionation,* Cantow, M. J. R., Ed., Academic Press, New York, 1968, 30.
8. **Jones, W. L. and Kieselbach, I.,** Units of measurement in gas chromatography, *Anal. Chem.,* 30, 1590, 1958.
9. **Kaiser, R.,** Development in gas chromatography from the 1961 literature, *Anal. Chem.,* 189, 1, 1962.
10. **Rony, R. P.,** Extent of separations: a universal separation index, *Sep. Sci.,* 3, 239, 1968.
11. **Cristophe, A. B.,** Valley to peak ratio as a measure for the separation of two chromatographic peaks, *Chromatographia,* 4, 445, 1971.
12. **Glueckauf, E.,** Theory of chromatography. IX. Theoretical plate concept in column separations, *Trans. Faraday Soc.,* 51, 34, 1955.
13. **Giddings, J. C.,** Plate high contribution in gas chromatography, *Anal. Chem.,* 32, 1707, 1960.
14. **De Clerke, K. and Cloete, C. E.,** Entropy as a general separation criterion, *Sep. Sci.,* 6, 627, 1971.
15. **Giddings, J. C.,** Basic approaches to separation, analysis and classification of methods according to underlying transport characteristics, *Sep. Sci. Technol.,* 13, 3, 1978.
16. **Stewart, G. H.,** On the measurement and evaluation of separation, *Sep. Sci. Technol.,* 13, 201, 1978.
17. **Stewart, G. H.,** Chromatography and the thermodynamics of separation, *J. Chromatogr. Sci.,* 14, 69, 1976.
18. **Koningsveld, R.,** Liquid-liquid phase separation in multicomponent systems, *Discuss. Faraday Soc.,* 49, 144, 1970.
19. **Scholte, T. G. and Koningsveld, R.,** Liquid-liquid phase separation in multicomponent systems, *Kolloid Z. Z. Polym.,* 218, 58, 1967.
20. **Kleintjens, L. A., Koningsveld, R. J. R., and Stockmayer, W. H.,** Liquid-liquid phase separation in multicomponent systems, *Br. Polym. J.,* 8, 144, 1976.
21. **Myrat, C. D. and Rowlenson, J. S.,** *Polymer,* 6, 645, 1965.
22. **Baker, C. H., Clemson, C. S., and Allen, G.,** Polymer fractionation at a lower critical solution temperature phase boundary, *Polymer,* 7, 525, 1966.
23. **Koningsveld, R. and Kleintjens, L. A.,** Liquid-liquid phase separation in multicomponent polymer systems. XV. Thermodynamic aspects of polymer compatibility, *Br. Polym. J.,* 9, 212, 1977.
24. **Wolf, B. A., Bieringer, H. P., and Breitenbach, J. W.,** The efficiency of polymer fractionation at lower critical solution temperatures, *Br. Polym. J.,* 10, 156, 1978.
25. **Matkowskyi, R. E., Irjak, V. I., and Diakskowski, F. S.,** Molecular weight distribution of ethylene oligomers obtained in $TiCl_4$–$C_2H_5AlCl_2$-benzene system, *Vysokomol. Soedin. Ser. A,* 19, 2073, 1977.
26. **Gurileva, A., Schlahter, R. A., and Teitelbaum, B. Ia.,** The study of molecular weight distribution in oligocaprolactones, *Vysokomol. Soedin. Ser. A,* 16, 1235, 1974.
27. **Skwarski, T., Larzkiewicz, B., and Mikalajczyk, T.,** Determination of the composition of the technical product of the hydrolytic polymerization of caprolactam. I. Determination of caprolactam and oligomers by extraction and sublimation, *Polymery (Poland),* 18, 28, 1973.
28. **Bledzki, A. and Kwasek, A.,** Nephelometric titration of diallyl phthalate prepolymers, *Chem. Anal. (Poland),* 22, 533, 1977.
29. **Baker, C. A. and Williams, R. J. P.,** A new chromatographic procedure and its application to high polymers, *J. Chem. Soc.,* 2352, 1956.

30. **Polacek, J.**, Fraktionirung von Polymethylmethacrylat durch Fällungs-chromatographie, *Coll. Czech. Chem. Commun.*, 28, 1838, 1963; **Polacek, J.**, Fraction of poly(methyl methacrylate) by precipitation chromatography. II. The influence of fractionation conditions on the course of fractionation, *Coll. Czech. Chem. Commun.*, 28, 3011, 1963; **Polacek, J., Danhelka, J., and Pokorna, Z.**, A fractionation efficiency of the modified method of precipitation chromatography. Influence of fractionated polymer amount, flow rate and temperature cycles, *Coll. Czech. Chem. Commun.*, 42, 2634, 1977; **Polacek, J., Boackova, V., Pokorna, Z., and Sinkylova, E.**, Polymer fractionation by column methods, *Coll. Czech. Chem. Commun.*, 41, 2510, 1976.

31. **Polacek, J., Kossler, J., and Vodenhal, J.**, Fractionation of polydienes by means of precipitation chromatography, *J. Polym. Sci. Part A*, 3, 2511, 1965.

32. **Bohnackova, V., Polacek, J., Grubisic, Z., and Benoit, H.**, Etude de quelques polyisoprènes cycliques par fractionnement et par chromatographie de partage en phase-liquide, *J. Chim. Phys.*, 66, 197, 1969; **Bohnackova, V., Polacek, J., Grubisic, Z., and Benoit, H.**, Etude des quelques polyisoprènes par fractionnement et par chromatographie en phase liquide, *J. Chim. Phys.*, 66, 207, 1969.

33. **Bohnackova, V., Fiserova, E., Grubisic, Z., Polacek, J., and Stolka, M.**, Etude de quelques polybutadienes cyclisées par diffusion de la lumiére, viscosimétrie et chromatographie de partage en phase liquide, *J. Chim. Phys.*, 67, 777, 1970.

34. **Danhelka, J., Polacek, J., and Kossler, J.**, A fractionation efficiency study of the modified method of precipitation chromatography, *Coll. Czech. Chem. Commun.*, 34, 283, 1969.

35. **Ungere, K.**, Beitrag zur Entstehnung von Niederschlsgen mit gesichichteten Strukturen, *Kolloid Z.*, 39, 238, 1928.

36. **Wiegner, G.**, Einige physikalisch-chemische Eigenschaften von Tonen. Basenaustauch oder Ionenaustausch, *J. Chem. Soc. Ind.*, 50, 655, 1931.

37. **Cernescu, N.**, Dissertation 661, Eidgenoessische Technische Hochschule, Zürich, 1933.

38. **Duel, H., Solms, J., and Abyas-Weisz, L.**, Über das verhalten löslicher Polyelektrolyte gegenüber Ionenaustauschern, *Helv. Chim. Acta*, 33, 2171, 1950.

39. **More, J. C. and Hendrickson, J. G.**, Gel permeation chromatography. The nature of the separation, *J. Polym. Sci. Part C*, 8, 233, 1968.

40. **Maley, L. E.**, Application of gel permeation chromatography to high and low molecular weight polymers, *J. Polym. Sci. Part C*, 8, 253, 1968.

41. **Benoit, H., Grubisic, Z., Rempp, P., Decker, D., and Zilliox, J. G.**, A hydrodynamic parameter for universal calibration in GPC, *J. Chim. Phys.*, 63, 1507, 1966.

42. **Grubisic, Z., Rempp, P., and Benoit, H.**, Universal calibration for GPC, *J. Polym. Sci. Part B*, 5, 753, 1967.

43. **Einstein, A.**, *Investigation of the Theory of Brownian Motion*, Dover, New York, 1956.

44. **Patrone, E. and Bianchi, U.**, Viscosity-molecular weight relationships for low molecular weight polymers. II. Poly(vinyl acetate) and poly(methyl methacrylate), *Makromol. Chem.*, 94, 52, 1966.

45. **Pepper, D. C. and Rutherford, P. P.**, The viscosity anomaly at low concentrations with polystyrenes of low molecular weight, *J. Polym. Sci.*, 35, 299, 1959.

46. **Sadron, C. and Rempp, P.**, Viscosité intrinséques de solutions de chaines courtes, *J. Polym. Sci.*, 29, 127, 1958.

47. **Aurenge, J., Gallot, Z., De Vries, A. J., and Benoit, H.**, Utilisation de la chromatographie sur gel dans le domain des molecules organiques de forte mass et des macromolecules de faible masse, *J. Polym. Sci. Symp.*, 52, 217, 1975.

48. **Coll, H. and Prusinowski, L. R.**, Calibration in gel permeation chromatography, *J. Polym. Sci. Part B*, 5, 753, 1967.

49. **Dawkins, J. W.**, High performance GPC of polymers, *Pure Appl. Chem.*, 54, 281, 1982.

50. **Funt, B. L. and Hornof, V.**, An approach to a simplified universal calibration for gel permeation chromatography, *J. Appl. Polym. Sci.*, 15, 2439, 1971.

51. **Rudin, A. and Hoegy, H. L. W.**, Universal calibration in GPC, *J. Polym. Sci. Part A*, 10, 217, 1972.

52. **Mori, S.**, Elution behaviour of oligomers on a polyvinyl alcohol gel column with chloroform-methanol and their mixtures, *J. Liq. Chromatogr.*, 11, 1205, 1988.

53. **Meyerhoff, G.**, Universal calibration in GPC, *Makromol. Chem.*, 89, 282, 1965.

54. **Uglea, C. V. and Andreescu, P.**, Comparative study of universal calibration methods in GPC, *Rev. Roum. Chim.*, 17, 2021, 1972.

55. **Edström, T. and Petro, B. A.**, Gel permeation studies of polynuclear aromatic hydrocarbon materials, *J. Polym. Sci. Part C*, 21, 171, 1968.

56. **Pokorny, S.**, Gel permeation chromatography of low molecular weight compounds and oligomers, *Chem. Listy*, 68, 1027, 1974.

57. **Minarik, M. and Komers, R.**, Gel permeation chromatography of low molecular weight compounds, *Chem. Listy*, 68, 696, 1974.

58. **Krishen, A.**, Separation of small molecules by gel permeation chromatography, *J. Chromatogr., Sci.*, 15, 434, 1977.

59. **Walton, H.**, Ion exchange and liquid column chromatography, *Anal. Chem.,* 50, 36R, 1978.
60. **Thomson, R. E., Sweeney, E. G., and Ford, D. C.,** GPC analysis of model compounds. I. Phenyl and benzo-substituted aromatic compounds, *J. Polym. Sci. Part A,* 8, 1165, 1970.
61. **Quinn, E. J., Ostergould, H. W., Heckles, J. S., and Ziegler, D. C.,** Fractionation of phenol-formaldehyde reaction with GPC, *Anal. Chem.,* 40, 574, 1968.
62. **Harrison, E. K.,** Molecular volume and aromaticity of coal. I. Aromaticity equations, *Fuel,* p. 339, 1965.
63. **Harrison, E. K.,** Molecular volume and aromaticity of coal. II. Evaluation of constants, *Fuel,* 44, 344, 1965.
64. **Harrison, E. K.,** Molecular volume and the aromaticity of coal. III. Evaluations of corrections, *Fuel,* 45, 397, 1966.
65. **Harrison, E. K.,** Molecular volume and the aromaticity of coal. IV. Derivation of structural features, *Fuel,* 47, 265, 1968.
66. **Edwards, G. D. and Ng, Q. Y.,** Elution behaviour of model compounds in GPC, *J. Polym. Sci. Part C,* 24, 105, 1968.
67. **Lambert, A.,** Rationalised sizes of small molecules in GPC, *Anal. Chim. Acta,* 53, 63, 1971.
68. **Chang, T. L.,** Elution behaviour of small molecules in GPC, *Anal. Chim. Acta,* 39, 521, 1967.
69. **Anoop, K. and Tucker, R. G.,** Gel permeation chromatography of low molecular weight materials with high efficiency column, *Anal. Chem.,* 49, 848, 1977.
70. **Bombaugh, K. J., Dark, W. A., and Levangie, R. F.,** Application of gel chromatography to small molecules, *Sep. Sci.,* 3, 375, 1968.
71. **Yoshikawa, T., Kimura, K., and Fujimura, S.,** The gel permeation chromatography of phenolic compounds, *J. Appl. Polym. Sci.,* 15, 2513, 1971.
72. **Batzer, H. and Zahir, S. A.,** Studies in the molecular weight distribution of epoxide resins. I, *J. Appl. Polym. Sci.,* 19, 609, 1975; **Batzer, H. and Zahir, S. A.,** Studies in the molecular weight distribution of epoxide resins. II, *J. Appl. Polym. Sci.,* 21, 1843, 1977.
73. **Wagner, E. R. and Greff, R. J.,** Analysis of resoles by GPC, *J. Polym. Sci. Part A,* 9, 2193, 1971.
74. **Duval, M., Bloch, B., and Kohn, S.,** Analysis of phenol-formaldehyde resols by GPC, *J. Polym. Sci.,* 16, 1585, 1972.
75. **Dauvilliers, J.,** Research on the reactions of the oil soluble formaldehyde resins, *Double Liaison,* 284, 133, 1979.
76. **Bergman, J. G., Duffy, L. J., and Stevenson, R. B.,** Solvent effects in GPC, *Anal. Chem.,* 43, 131, 1971.
77. **Yoshikawa, T., Kimura, K., and Fujimura, S.,** The gel permeation chromatography of phenolic compounds, *J. Appl. Polym. Sci.,* 15, 2513, 1971.
78. **Holingdale, S. H. and Megson, M. J. L.,** Phenol-formaldehyde resins, *J. Appl. Chem.,* 5, 616, 1955.
79. **Kern, W., D'Allasta, G., and Kämmerer, H. K.,** The condensation reaction of phenol with formaldehyde, *Makromol. Chem.,* 8, 252, 1952.
80. **Uglea, C. V.,** Gel permeation chromatography of styrene oligomers, *Makromol. Chem.,* 166, 275, 1973.
81. **Harrison, E. K.,** Molecular volume and aromaticity of coal. IV. Derivation of structural features, *Fuel,* 47, 265, 1968.
82. **Perrault, G., Tremblay, M., Lavertu, R., and Tramblay, R.,** Characterization of low molecular weight difunctional polybutadienes, *J. Chromatogr.,* 55, 121, 1971.
83. **Law, R. D.,** Application of preparative gel permeation chromatography to studies of low molecular weight carboxy-polybutadienes, *J. Polym. Sci. Part A,* 7, 2097, 1969.
84. **Belenkii, B. G.,** Universal calibration for oligomers, *Vysokomol. Soedin. Ser. B,* 16, 507, 1974.
85. **Romanov, A. K., Evreinov, V. V., and Entelis, S. G.,** GPC for oligomers, *Vysokomol. Soedin. Ser. A,* 19, 1172, 1977.
86. **Sadron, C. and Rempp, P.,** Intrinsic viscosity of low molecular weight polymers, *J. Polym. Sci.,* 29, 127, 1958.
87. **Skwortsow, A. M. and Gorbunow, A. A.,** Adsorbtion effects in the chromatographic separation of oligomers, *Vysokomol. Soedin. Ser. A,* 22, 2641, 1980.
88. **Ambler, M. R.,** Gel permeation chromatography of low molecular weight polymers, *J. Polym. Sci. Polym. Lett. Ed.,* 14, 683, 1976.
89. **Linden, C. V.,** Elution and molecular weight parameters in exclusion chromatography, *Polymer,* 21, 171, 1980.
90. **Uglea, C. V.,** *Gel Permeation Chromatography,* Academic Editorial House, Bucharest, 1976.
91. **Shulz, W. W.,** Gel permeation chromatographic elution behaviour of branched alkanes, *J. Chromatogr.,* 55, 73, 1971.
92. **Mann, G., Mühlstadt, H., Braband, J., and Dörning, E.,** Structural characteristics of alkanes, *Tetrahedron,* 23, 3393, 1967.
93. **Nenitzescu, C. D.,** *Organic Chemistry,* Vol. 1, EDP, Bucharest, 1980, 18.
94. **Larsen, A.,** Rationalisation of the sizes of small molecules in GPC, *J. Appl. Chem.,* 20, 305, 307, 1970.

95. **Giddings, J. C.,** Theoretical aspects of GPC, *Anal. Chem.,* 39, 1027, 1967.
96. **Martin, A. J. P. and Synge, R. L. M.,** A new form of chromatogram employing two ligand phases, *Biochem. J.,* 35, 1358, 1941.
97. **Cazes, J. J.,** Gel permeation chromatography, Part II, *J. Chem. Ed.,* 43, A625, 1966.
98. **Bly, D. D.,** Resolution and fractionation in GPC, *J. Polym. Sci. Part C,* 21, 13, 1968.
99. **Schollner, R. and Helwig, J.,** Separation of oligomers by GPC, *Fette, Seifen, Austrichm.,* 70, 770, 1968.
100. **Heitz, W.,** Gel chromatography, *Angew. Chem. Int. Ed. Engl.,* 9, 689, 1970.
101. **Barson, C. A. and Robb, J. C.,** Studies of oligomers by GPC, *Br. Polym. J.,* 3, 53, 1971.
102. **Heitz, W.,** Gel chromatography, *Kontakte,* No. 2, 20, 1977.
103. **Gros, I., Facsko, O., Munteanu, D., and Pape, R.,** Polyester for plasticization of poly(vinyl chloride). I. Structure modification of polyesters from adipic acid and 1,2-propylene glycol, *Plaste Kautsch.,* 30, 445, 1983.
104. **Antipin, L. M., Fedoseevskii, V. V., Mochalov, V. N., Frenkel, A. S., Mecjomedov, G. K., and Shostakovskii, M. F.,** Exclusion chromatographic study of siloxane oligomers containing phenylchloro-tricarbonyl fragments, *Izv. Akad. Nauk S.S.S.R. Ser. Khim.,* 642, 1983.
105. **Hironobu, O., Nishida, Y., and Inouc, A.,** Analysis of styrene oligomers, *Kanzei Chua Bunseikisho Ho,* 23, 67, 1983; *Chem. Abstr.,* 99, 23237r, 1983.
106. **Mori, S.,** Semimicro size exclusion chromatography of oligomers. I. Column packing, *J. Liq. Chromatogr.,* 9, 1317, 1986.
107. **Montando, G., Puglisi, C., Scamporrino, E., and Vitalini, D.,** Separation and characterization of cyclic sulfides formed in the polycondensation of dibromoalkanes with aliphatic dithiols, *Macromolecules,* 19, 2689, 1986.
108. **Romanov, A. K., Evreinov, Y. V., and Entelis, S. G.,** Molecular weight distribution of polyurethanes obtained from oligomers. Relation between molecular weight distribution and final products, *Vysokomol. Soedin. Ser. A,* 28, 1240, 1986.
109. **Ciemniak, G., Lesrak, T., and Balinski, A.,** Chemistry and application of organic isocyanates. IX. Conversion of tolylene-2,4-diisocyanate into urethane derivatives by means of heptane 1,4,7-thiol, *Angew. Makromol. Chem.,* 143, 59, 1986.
110. **Moissonier, R. and Seiler, H.,** German Patent 2,646,902, 1976.
111. **Libert, C. and Maréchal, E.,** Synthèse d'oligoamides colorés par structure, *Eur. Polym. J.,* 16, 951, 1980.
112. **Bykov, A. N., Kirollova, T. M., and Lits, W. P.,** Synthesis and investigations of colored high polymers, *Vysokomol. Soedin.,* 5, 425, 1963.
113. **Ambler, M. R. and Stevenson, W. C.,** Synthesis and characterization of styrene oligomers, *Polymer,* 19, 48, 1978.
114. **Wendler, K.,** Synthesis and characterization of styrene cooligomers, *Plaste Kautsch.,* 29, 502, 1982.
115. **Wesslen, B. and Manson, P.,** Synthesis and chromatographic separation of styrene oligomers, *J. Polym. Sci. Polym. Chem. Ed.,* 13, 2545, 1975.
116. **Klesper, E. and Hartmann, W.,** Chromatographic separation of styrene oligomers, *J. Polym. Sci. Polym. Lett. Ed.,* 15, 9, 1977.
117. **Fujishige, S. and Ohguri, N.,** Characterization of styrene oligomers, *Makromol. Chem.,* 176, 233, 1975.
118. **Teramachi, S., Hasegawa, A., Sato, F., and Takemoto, N.,** MWD in copoly(styrene methacrylate) by GPC, *Macromolecules,* 18, 347, 1985.
119. **Claus, J. and Maréchal, E.,** Physico-chemical study of oligostyrenes synthesized by cationic polymerization, *Rev. Roum. Chim.,* 23, 1325, 1978.
120. **Kuzaew, A. I.,** Determination of the molecular weight values for styrene oligomers by GPC, *Vysokomol. Soedin. Ser. A,* 20, 1146, 1980.
121. **Chang, M. S., French, D. M., and Rogers, P. L.,** Functionality distribution in polybutadiene, *J. Macromol. Sci. Chem.,* A7, 1727, 1973.
122. **Ahad, E.,** Characterization of functionally terminated polybutadiene of low molecular weight, *J. Appl. Polym. Sci.,* 17, 365, 1973.
123. **Suzuki, T., Tsugi, Y., Watanabe, Y., and Tagami, Y.,** Characterization of 2-phenyl-1,3-butadiene-Li oligomers, *Polym. J.,* 11, 937, 1979.
124. **Mladenow, I. T., and Tsenkowa, V. I.,** Separation of 1,4-*cis* polybutadiene on SEPHADEX G-50, *Dokl. Bulg. Akad. Nauk,* 30, 1125, 1977.
125. **Dmitreva, T. S., Valuew, V. I., and Slijachter, R. A.,** Determination of the ramification degree in diene oligomers by GPC, *Vysokomol. Soedin. Ser. V,* 21, 671, 1979.
126. **Valuew, V. I.,** Functionality distribution in oligobutadiene, *Vysokomol. Soedin. Ser. B,* 19, 172, 1977.
127. **Tvetkovskii, I. B.,** MWD of oligodienes, *Vysokomol. Soedin. Ser. A,* 17, 2609, 1975.
128. **Zaripov, I. N., Beresenev, V. V., and Kirpikinikow, P. A.,** MWD in oligoisobutylenes, *Vysokomol. Soedin. Ser. A,* 18, 2228, 1976.

129. **Kennedy, J. P. and Carlson, G. M.,** Silylcyclopentadiene-telechelic polyisobutylene, *J. Polym. Sci. Polym. Chem. Ed.,* 21, 2973, 1983.

130. **Kuzmina, V. K.,** MWD of oligodiendiols, *Kautschuk Rez.,* No. 1, 49, 1974.

131. **Kimura, T.,** Separation and structure of methacrylate telomers, *Polym. J.,* 15, 293, 1983.

132. **Mazurek, A., Scibiorek, M., Chopowski, J., Zavici, B. G., and Zhdanov, A. A.,** Oligomer distribution in poly(methyl methacrylate), *Eur. Polym. J.,* 16, 57, 1980.

133. **Chan, H. S. D. and Still, R. H.,** Synthesis and characterization of quinazoline prepolymers, *J. Appl. Polym. Sci.,* 22, 2173, 1978.

134. **Browden, M. J. and Thomson, I. F.,** GPC of polysulfones, *J. Appl. Polym. Sci.,* 19, 905, 1975.

135. **Nash, D., Wand, D. W., and Pepper, D. C.,** Solution properties of poly(propylene sulphur), *Polymer,* 16, 105, 1975.

136. **Birley, A. W., Dawkins, J. V., and Kyriakos, D.,** Unsaturated polyester-characterization of prepolymers, *Polymer,* 19, 1433, 1978.

137. **Boussias, C. M., Peters, R. H., and Still, R. H.,** Synthesis and characterization of tetramethyleneterephthalate-tetramethylene oxide block copolymers, *J. Appl. Polym. Sci.,* 25, 855, 1980.

138. **Steinka, J. and Wald, H.,** Analysis of oligomers in poly(butylene terephthalate), *Melliand Text. Ber. Int.,* 58, 494, 1977.

139. **Yuki, H., Okamoto, Y., and Doi, Y.,** Synthesis and conformation of oligomeric γ-isobutyl-L-aspartate, *J. Polym. Sci. Polym. Chem. Ed.,* 17, 1911, 1979.

140. **Shiono, S.,** Separation and identification of poly(ethylene terephthalate) oligomers by GPC, *J. Polym. Sci. Polym. Chem. Ed.,* 17, 4123, 1979.

141. **Cuzaeva, A. I., Zaitsewa, N. P., Kobelciuc, Iu. M., and Mosciuskaia, N. K.,** GPC of diglycidylic esters, *Vysokomol. Soedin. Ser. A,* 22, 2763, 1980.

142. **Bledzki, A., Kostanskaia, K. L., and Strumek, J.,** Molecular mass distribution of diallylphthalates prepolymers, *Polim. Tworz. Wielkocz. (Poland),* 28, 267, 1983.

143. **Gurlaewa, A. A.,** Molecular mass distribution of poly(ethylene glycol) oligomers, *Vysokomol. Soedin. Ser. A,* 17, 118, 1975.

144. **Blyahman, Ye. M.,** Influence of the structure and molecular mass of oligomer diglycidol ether of diphenylpropane on the rate of their interaction with n-phenyldiamine, *Vysokomol. Soedin. Ser. B,* 23, 348, 1981.

145. **Schulz, W. W., Kalada, J., and Schulz, D. N.,** Compositional and molecular weight analysis of polyether macromonomers by chromatographic techniques, *J. Polym. Sci. Polym. Chem. Ed.,* 22, 3795, 1984.

146. **Naoya, O., Kohei, S., and Shinichiro, K.,** MWD of poly(p-phenileterephthalamide), *J. Polym. Sci. Polym. Chem. Ed.,* 22, 865, 1984.

147. **Manolova, M. E., Gitsov, I., Velichova, R. S., and Raschow, I. B.,** Separation and characterization of ω-caprolactone by GPC, *Polym. Bull.,* 13, 285, 1985.

148. **Reinisch, G. and Dietrich, K.,** Determination of low molecular weight oligomers of polycaprolactone, *Nuova Chim.,* 49, 72, 1973.

149. **Lorenz, O. and Rose, G.,** GPC separation of polyurethane isomers, *Angew. Makromol. Chem.,* 118, 91, 1983.

150. **Rudin, A., Fyfe, C. A., and Vines, S. M.,** GPC of resole phenolic resins, *J. Appl. Polym. Sci.,* 28, 2611, 1983.

151. **Schultz, G.,** GPC of phenol-formaldehyde resins, *Plaste Kautsch.,* 29, 398, 1982.

152. **Braun, D. and Arndt, J.,** Determination of oligomers in phenol-formaldehyde condensates, *Angew. Makromol. Chem.,* 73, 133, 1978.

153. **Braun, D. and Bayersdorf, F.,** GPC of formaldehyde oligomers, *Angew. Makromol. Chem.,* 81, 147, 1979.

154. **Jacovici, M. and Srebic, M.,** Determination of molecular weight of epoxy adhesives, *C. R. Acad. Sci. C,* 286, 647, 1978.

155. **Braticiak, M. M., Dociak, V. A., and Jivanov, V. V.,** *Vestnik Lvov. Politekh. Inst.,* 131, 26, 163, 1979.

156. **Ravindranath, K. and Grandhi, K. S.,** MWD in epoxy resins, *J. Appl. Polym. Sci.,* 24, 1115, 1980.

157. **Dau, E. and Roller, M. B.,** GPC calibration for epoxy resins, *J. Polym. Sci. Polym. Lett. Ed.,* 21, 875, 1983.

158. **Stacy, C. J.,** MWD in poly(phenylene sulfide) by high temperature GPC, *ACS Polym. Prepr.,* 26(1), 180, 1985.

159. **Haider, B., Vidal, A., Balard, H., and Donnet, J. B.,** Structural differences exhibited by network prepared by chemical photochemical reactions. III. Characterization by inverse GPC, *J. Appl. Polym. Sci.,* 29, 4309, 1984.

160. **Vidas, G. J., Marcizas, D., and Rimtantas, K.,** Application of GPC for investigation of 9-(2,3-epoxypropylcarbazole) oligomers, *J. Liq. Chromatogr.,* 7, 1823, 1984.

161. **Porath, J. and Bennich, H.,** *Arch. Biochem. Biophys. Suppl.,* 1, 152, 1962.

162. **Biesenberger, J. A., Tau, M., Duvedani, I., and Maurer, T.,** Recycle gel permeation chromatography. I. Direct pumping, *J. Polym. Sci. Part B,* 9, 353, 1971.

163. **Biesenberger, J. A., Tau, M., and Duvedani, I.,** Recycle gel permeation chromatography. II. Alternate pumping, *J. Polym. Sci. Part B,* 9, 429, 1971.

164. **Biesenberger, J. A., Tau, M., and Duvedani, I.,** Recycle gel permeation chromatography. III. Separation variables, *J. Appl. Polym. Sci.,* 15, 1549, 1971.

165. **Nakamura, S., Ishiguro, S., Yamada, T., and Morizumi, S.,** Recycle gel permeation chromatographic analysis oligomers and polymer additives, *J. Chromatogr.,* 83, 279, 1973.

166. **Uglea, C. V. and Cincu, C.,** Analysis of isobutyraldehyde oligomers by gel permeation chromatography, *Rev. Chim. (Bucharest),* 30, 369, 1979.

167. **Bombaugh, K. J., Levangie, R. F., and Dark, W. A.,** Application of GPC to small molecules, *J. Chromatogr. Sci.,* 7, 42, 1969.

168. **Trolltizsch, C.,** Preparative GPC isolation of lower acyclic oligomers of isoprene by recycling, *J. Prakt. Chem.,* 328, 454, 1986.

169. **Carr, C. D.,** *Liquid Chromatography at Work,* Varian Instrument, Publ. Rep. 610-22, California, 1974.

170. **Cheng-Yih, K. and Provder, T.,** High performance gel permeation chromatography characterization of oligomers used in coatings systems, *Size Exclusion Chromatography,* ACS Symp. Ser. 138, 1980, 207.

171. **Kuo, C., Prowder, T., and Kah, A. F.,** Application of HPGPC to characterize oligomers and small molecules used in environmentally acceptable coating systems, *Paint Resin,* 53, 26, 1983.

172. **Gordon, J. L.,** Gaps in the current physical chemical chemistry. State-of-the art related to new coating systems, in *Proc. Science of Organic Coatings Workshop,* Myers, R. R., Gordon, J. L., and Lauren, S., Eds., Kent State University, 1978, 65.

173. **Takeuki, T., Ishii, D., and Mori, S.,** Separation of oligomers by HPGPC, *J. Chromatogr.,* 257, 327, 1983.

174. **Nakamura, S., Taguchi, M., Ishigura, S., and Yamada, S.,** High-speed and high-resolution gel permeation chromatography of polymers, oligomers and small molecules, *Anal. Sci.,* 1, 199, 1985.

175. **Russel, D. J.,** Calibration of HPGPC for small epoxy molecules, *J. Liq. Chromatogr.,* 11, 383, 1988.

176. **Chiantori, O. and Guaita, M.,** Applications of HPGPC to the analysis of oligomers, *J. Liq. Chromatogr.,* 9, 1341, 1986.

177. **Chen Weizhuang, and Yai Tongyin,** Study of the molecular weight distribution of three oligomeric mixtures by HPGPC, *Sepu,* 4, 343, 1986 (in Chinese); *Chem. Abstr.,* 106, 67952n, 1987.

178. **Hibi, K., Wada, A., and Mori, S.,** Semimicro size exclusion chromatography: molecular weight measurement of poly(ethylene terephthalate) and separation of oligomers and prepolymers, *Chromatographia,* 21, 635, 1986.

179. **Smith, R. M.,** *Supercritical Fluid Chromatography,* Royal Society of Chemistry Monographs, London, 1988, chap. 1.

180. **Szymonovski, J., Szewczyk, H., Hetper, J., and Beger, J.,** Analysis and identification of N-oligoxyethylene mono- and dialkylamines, *J. Chromatogr.,* 351, 183, 1986.

181. **Elgert, K. F., Henschel, R., Schorn, H., and Kosfeld, R.,** HPLC and ^{13}CNMR spectroscopy of polymers. I. Oligomers of polystyrenes, *Polym. Bull.,* 4, 105, 1981.

182. **Janca, J., Pokorny, S., and Kalal, J.,** Liquid chromatography of liquid rubber, *Plasty Kautsch.,* 11, 334, 1979.

183. **Wachtina, I. A. and Tarakanov, O. G.,** Study of oligomers by HPLC, *Plaste Kautsch.,* 23, 401, 1976.

184. **Braun, D. and Lee, D. W.,** Quantitative determination of oligomers in epoxy resins, *Angew. Makromol. Chem.,* 57, 111, 1977.

185. **Curtis, M. A. and Welb, J. W.,** Fractionation by liquid chromatography of polystyrene oligomers, *Sep. Sci. Technol.,* 15, 1413, 1980.

186. **Munteanu, D., Manoviciu, V., and Manoviciu, I.,** Separation of oligo(1,2-propylene terephthalate)s by gradient elution high performance liquid chromatography, *Rev. Roum. Chim.,* 30, 23, 1985.

187. **Van der Malden, F. P. B., Biemond, M. E. F., and Janssen, P. C. G. M.,** Oligomer separation by gradient elution high performance liquid chromatography, *J. Chromatogr.,* 149, 539, 1978.

188. **Zaborski, L. M., II,** Determination of polyester prepolymers by high performance liquid chromatography, *Anal. Chem.,* 49, 1166, 1977.

189. **Ludwig, F. J., Bardie, A., and Gibbs, A.,** Reversed-phase liquid chromatographic separation of p-*tert* butylphenol-formaldehyde linear and cyclic oligomers, *Anal. Chem.,* 58, 2069, 1986.

190. **Elgert, K. F. and Kasfeld, P.,** HPLC and ^{13}CNMR spectroscopy of polymers. II. Oligomers of poly(vinyl chloride), *Polym. Bull.,* 10, 244, 1983.

191. **Toshima, T., Kawakami, U., Harada, M., Sakada, T., and Terumiki, T.,** Isolation and identification of new oligomers in aqueous solution of glutaraldehyde, *Chem. Pharm. Bull.,* 35, 4169, 1987.

192. **Maj, L. and Laijoki, T.,** Determination of oligomer and polyamines in cured epoxy resins by extraction and HPLC, *Analyst (London),* 113, 239, 1988.

193. **Lazaris, A. Ya. and Beloded, L. N.,** Study of the structure of α,ω-bis(methacryloyl)oligoethylene glycol phthalates, *Zh. Anal. Khim.,* 40, 2259, 1985 (in Russian).

194. **Lazaris, A. Ya. and Beloded, L. N.,** Reversed-phase liquid chromatography of oligoethylene glycol esters. Behaviour of individual oligomer homologs, *Zh. Anal. Khim.,* 41, 345, 1986 (in Russian).

195. **Borsoch, N. A. and Mal'tseva, N. G.**, Separation and determination of linear oligomers of poly(ethylene phthalates) by liquid chromatography, *Zh. Anal. Khim.*, 41, 916, 1986 (in Russian).

196. **Janders, P. and Rozcosma, J.**, Isocratic and gradient-elution liquid chromatography of styrene oligomers on silicagel, *J. Chromatogr.*, 362, 325, 1986.

197. **Bauer, M., Bauer, J., and Much, H.**, Analysis of prepolymers and polymers of aromatic cyanic acid esters. II. HPLC, *Acta Polym.*, 37, 221, 1986.

198. **Sato, T., Saito, Y., and Anazawa, I.**, Polyoxyethylene oligomer distribution of nonionic surfactant, *J. Am. Oil Chem., Soc.*, 65, 996, 1988.

199. **Zeman, I. and Ranny, M.**, Chromatographic analysis of fatty acid dimers. Comparison of gas-liquid chromatography, HPLC and TLC with flame ionization detection, *J. Chromatogr.*, 354, 283, 1986.

200. **Miu, T. I., Miyamoto, T., and Inagaki, H.**, Determination of functionality distribution in telechelic prepolymers by TLC, *Rubber Chem. Technol.*, 50, 63, 1976.

201. **Hudgins, W. R., Theurer, K., and Mariani, T.**, Separation of poly(ethylene terephthalate) oligomers by HPLC and TLC, *J. Appl. Polym. Sci.*, 34, 155, 1978.

202. **Gankina, E. S. and Belenkii, B. G.**, TLC of polymers, *Usp. Khim.*, 47, 1293, 1978.

203. **Schmitz, F. P., Hilgers, H., and Gemmel, B.**, High performance liquid supercritical fluid chromatography separation of vinyl oligomers by gradient-elution, *J. Chromatogr.*, 357, 135, 1986.

204. **Schmitz, F. P. and Hilgers, H.**, Separation by means of supercritical fluid chromatography of 1-vinyl- and 2-vinyl-naphthalene oligomers prepared through radical and anionic initiation, *Makromol. Chem. Rapid Commun.*, 7, 59, 1986.

205. **Schmitz, F. P., Gemmel, B., Leyendecker, D., and Leyendecker, D. S.**, Separation of 2-vinylnaphthalene oligomers by open tubular capillary supercritical fluid chromatography, *J. High Resolut. Chromatogr. Chromatogr. Commun.*, 11, 339, 1988.

206. **Schmitz, F. P. and Gemmel, B.**, Separation of different oligomers by supercritical fluid chromatography using gradient elution, *Fresenius Z. Anal. Chem.*, 330, 216, 1988.

207. **Araki, T., Kubo, Y., Toda, R., Takata, M., Yamamoto, T., Maruyama, W., and Nunogaki, Y.**, Carrier-aided centrifugal partition chromatography for preparative-scale separations, *Analyst*, 110, 913, 1985; **Ikada, H.**, Properties of oligomers, *Kobunski Kako*, 34, 605, 1985 (in Japanese); *Chem. Abstr.*, 104, 26, 1986.

208. **Seidel, B.**, Spectroscopic study of linear and cyclic oligomers from terephthalic acid and glycol, *Z. Elecktrochem.*, 62, 214, 1958.

209. **Pakuro, N. I., Kozeva, N. V., and Polyakov, D. K.**, Study of oligodiene diols by IR-spectroscopy, *Vysokomol. Soedin. Ser. A*, 27, 2196, 1985.

210. **Kulichikhin, S. G., Mihailov, S. V., Kotov, O. V., Cherkasov, M. V., Agapov, O. A., and Vasil'co, V. V.**, Rheokinetic and IR-spectroscopic study of oligoimide formation, *Vysokomol. Soedin. Ser. A*, 30, 707, 1988.

211. **Kushnir, L. V., Bratychak, M. N., Bychkov, V. A., and Puchin, V. A.**, IR spectroscopic study of the hardening of peroxide melamine-formaldehyde oligomers, *Zh. Prikl. Spektrosk.*, 48, 449, 1988.

212. **Yudino, L. V., Levshanov, V. S., and Dolmatov, S. A.**, IR-spectroscopic analysis of bis(maleimide) and aromatic diamine-based thermosetting polymides, *Plast. Massy*, No. 7, 43, 1983.

213. **Tsvetanov, C.**, IR spectroscopy of styrene oligomers, *Eur. Polym. J.*, 10, 557, 1974.

214. **Balard, H. and Meybeck, J.**, IR and UV study of acrylonitrile oligomers, *Bull. Soc. Chim. Fr.*, NO. 11, 1147, 1977; **Deschagarova, E.**, Spectroscopic analysis of technical oligomers, *Angew. Makromol. Chem.*, 81, 193, 1979.

215. **Zahn, H., Rathgeber, P., Rexroth, E., Krzikalla, R., Lauer, W., Spoor, H., Schmidt, F., Seidel, B., and Hildebrand, D.**, Oligoamides and oligoesters, *Angew. Chem.*, 68, 229, 1956.

216. **Harris, R. K. and Robins, M. L.**, S[29]RMN studies of oligomeric and polymeric oligosiloxanes, *Polymer*, 19, 1123, 1978.

217. **Swales, T. G. E., Teo, H. H., Domazy, R. C., Vivas, K., King, T. A., and Booth, C.**, Low-frequency Raman scattering from methylene-oxyethylene-methylene triblock oligomers, *J. Polym. Sci. Polym. Phys. Ed.*, 21, 1501, 1983.

218. **Sixl, H., Gross, H., and Neumann, W.**, Low-temperature spectroscopy of the photopolymerization reaction in diacetylene crystals, *Stud. Inorg. Chem.*, 3, 493, 1986.

219. **Tanaka, Y.**, HNMR spectra of dimers and trimer of styrene, *Makromol. Chem. Rapid Commun.*, 1, 551, 1980.

220. **Sato, H. and Tanaka, Y.**, End groups effect of styrene oligomers on the ^{13}CNMR chemical shift, *Macromolecules*, 17, 1964, 1984.

221. **Kenishi, T., Yoshizaki, T., and Yamakawa, H.**, Determination of stereochemical composition of oligostyrene by ^{13}CNMR, *Polym. J.*, 20, 175, 1988.

222. **Elgert, K. F. and Kosfeld, R.**, Polymerization of vinyl chloride by *tert*-butylmagnesium chloride. II. Structure of oligomers by ^{13}CNMR, *Polym. Bull.*, 10, 175, 1983.

223. **Okada, T.**, NMR spectra of poly(ethylene glycol), *J. Polym. Sci. Polym. Chem. Ed.*, 17, 155, 1979.
224. **Bulai, A. Kh., Slonim, I. Ya., Shurohodina, E. N., Bardakova, Z. M., and Artsis, E. S.**, ^{13}CNMR study of thermal conversions of oligobutadienedicarboxylic acids, *Vysokomol. Soedin. Ser. B*, 30, 380, 1988.
225. **Harris, R. K., Kimber, B. J., Wood, M. D., and Holt, A.**, Si NMR studies of oligo- and polymeric syloxanes, *J. Organometal. Chem.*, 116, 291, 1976.
226. **El Ghafari, M. and Pham Quang, T.**, Hydroxytelechelic polybutadiene. VII. Study by ^{1}HNMR and ^{13}CNMR, *Polymer*, 19, 1150, 1978.
227. **Deveaux, J. and Böhmer, V.**, HNMR spectra of oligo(hydroxy-5-nitro-1,3-phenylene)methylene compounds, *Makromol. Chem.*, 177, 3285, 1976.
228. **Ogawa, E., Kondo, S., and Tanaka, K.**, High resolution NMR studies on oligo-(oxy-2,2-dimethylethylene carbonyl) compounds, *Makromol. Chem.*, 169, 261, 1973.
229. **Fujishige, S.**, HNMR spectra of pure methylmethacrylate oligomers, *Makromol. Chem.*, 179, 2251, 1978.
230. **Toy, M. S. and Stringham, R. S.**, NMR studies of microstructure of polyperfluorobutadiene, *J. Polym. Sci. Polym. Lett. Ed.*, 14, 717, 1976.
231. **Pesryayew, Ye. M.**, Study of molecular mobility in oligosulfones by NMR spectroscopy, *Vysokomol. Soedin. Ser. A*, 22, 2159, 1980.
232. **Z'cova, V. V. and Minkin, V. V.**, Determination of structure in oligomeric polysulfides by ^{13}CNMR, *Zh. Prikl. Spektrosk.*, 41, 318, 1984.
233. **Bordoloi, B. K. and Pearce, E. M.**, Oligomeric alkenyl polysulfide: synthesis and characterization by NMR analysis, *J. Polym. Sci. Polym. Chem. Ed.*, 16, 3293, 1978.
234. **Urman, Ya. G. and Alekseeva, S. G.**, Study of oligomers and low molecular weight fractions of polycarbonates by NMR, *Vysokomol. Soedin. Ser. A*, 22, 929, 1980.
235. **Huang, S. S.**, Oligomerization of vinyl monomers. IX. ^{13}CNMR and chromatographic studies of oligomers of 2-vinylpyridine, *Macromolecules*, 14, 1802, 1981.
236. **Gilbert, J., Startin, J. R., and Wallwork, M. A.**, Gas chromatographic determination of 1,1,1-trichloroethane in vinyl chloride polymers and in foods, *J. Chromatogr.*, 160, 127, 1978.
237. **Gilbert, J. and Startin, J. R.**, Single ion monitoring of styrene in foods by coupled mass spectrometry-automatic headspace gas chromatography, *J. Chromatogr.*, 205, 434, 1981.
238. **Shepherd, M. J. and Gilbert, J.**, Analysis of additives in plastics by HPGPC, *J. Chromatogr.*, 218, 703, 1981.
239. **Gilbert, J., Shepherd, M. J., Startin, J. R., and Wallwork, M. A.**, Identification by gas chromatography-mass spectrometry of vinyl chloride oligomers and their low-molecular-weight components in poly(vinyl chloride) resins for food packaging applications, *J. Chromatogr.*, 237, 249, 1982.
240. **Munteanu, D., Isfan, A., Isfan, C., and Tincul, I.**, Identification of stabilization systems in polyolefins by HPLC, *Mater. Plast. (Bucharest)*, 22, 173, 1985.
241. **Locsei, V., Facsko, O., and Chirilă, T.**, The aminolysis of carbamates. I., *J. Prakt. Chem.*, 324, 816, 1982; **Locsei, V., Facsko, O, and Chirilă, T.**, *J. Prakt. Chem.*, 325, 49, 1983.
242. **Gross, J., Facsko, O., Munteanu, D., and Pape, R. F.**, The determination of hard segments in polyurethane elastomers by HPLC, *Polym. Bull.*, 9, 81, 1983.
243. **Locsei, V., Facsko, O., and Pape, R. F.**, Oligoesters as plastifiants for poly(vinyl chloride), *Plaste Kautsch.*, 30, 445, 1983.
244. **Revillon, A. and Hamaide, T.**, Macromer copolymerization reactivity ratio determined by GPC analysis, *Polym. Bull.*, 6, 235, 1982.
245. **Martin, C. and Gourdenne, A.**, NMR analysis of oligomeric polyurethanes, *ACS Polym. Prepr.*, 17, 773, 1976.
246. **Morese-Séguéla, B., St.-Jacques, M., Renaud, J. M., and Prud'homme, J.**, Microphase separation in low-molecular weight styrene-isoprene diblock copolymers studied by DSC and ^{13}CNMR, *Macromolecules*, 13, 100, 1980.
247. **Balard, H., Fritz, H., and Meybeck, J.**, Determination of the tacticity of polyacrylonitrile and its oligomers by ^{13}CNMR spectroscopy, *Makromol. Chem.*, 178, 2393, 1977.
248. **Zegart, G. and Plannemüller, B.**, ^{13}CNMR analysis of amino and hydroxy endgroups in telechelic poly(oxyethylene)s, *Polym. Bull.*, 4, 467, 1981.
249. **Usmanov, S. M. and Sivergin, Yu. M.**, NMR study of polymerization of oligoester acrylates, *Khim. Fiz.*, 5, 78, 1986 (in Russian).
250. **Chauser, M. G., Anisimova, O. S., Kol'tsowa, L. S., Zaichenko, N. L., and Cherkashin, M. I.**, Analysis of products of thermal polymerization of phenylacetylene by mass- and ^{13}CNMR spectroscopy, *Izv. Akad. Nauk S.S.S.R. Ser. Khim.*, 1988, 67.
251. **Karduner, K. R. and Chattha, M. S.**, ^{13}CNMR investigation of the oligomerization of bismaleimidodiphenylmethane with diallyl-bisphenol A, *Polym. Mater. Sci. Eng.*, 56, 660, 1987.
252. **Sebenic, A., Osredkar, U., and Vizoviseki, I.**, Identification of oligomers in the reaction of polyols with isocyanates by ^{13}CNMR and GPC, *J. Macromol. Sci. Chem.*, A23, 369, 1986.

253. **Wiley, R. H.**, The mass spectral characterization of oligomers, *J. Polym. Sci. Macromol. Rev.*, 14, 379, 1979.

254. **Wiley, R. H. and Carter Cook, J., Jr.**, Field desorbtion mass spectral data for oligomers up to 2400, *J. Macromol. Sci. Chem.*, A10, 811, 1976.

255. **Lüderwald, J., Montando, G., Przybylski, M. and Ringsdorf, H.**, Study of polymer degradation by mass spectrometer, *Makromol. Chem.*, 175, 2423, 1974.

256. **Höcker, H. and Riebel, K.**, Mass spectrometrical behaviour of hydrocarbons, *Makromol. Chem.*, 179, 1765, 1978.

257. **Foti, S., Maravigna, P., and Montando, C.**, Mass spectrometric detection of cyclic oligomers in polyurethanes and polyureas, *Macromolecules*, 15, 883, 1982.

258. **Gleria, M., Audisio, G., Daolio, S., Traldi, P., and Vecchi, E.**, Mass spectrometric studies on cyclo- and polyphosphazenes, *Macromolecules*, 17, 1230, 1984.

259. **Beckewitz, F. and Heusinger, H.**, The mass spectral characterization of oligostyrenes, *Angew. Makromol. Chem.*, 46, 143, 1975.

260. **Burzin, K. and Frenzel, J. P.**, The mass spectral characterization of poly(butylene terephthalate) and its oligomers, *Angew. Makromol. Chem.*, 71, 61, 1978.

261. **Yamammoto, Y.**, Mass spectral data for poly(α-methyl styrene) with low molecular weight, *Polym. J.*, 8, 307, 1976.

262. **Saito, J., Hara, J., Toda, S., and Tanaka, S.**, Analysis for the oligomerization mechanism by field desorption mass spectrometry, *Bull. Chem. Soc. Jpn.*, 56, 748, 1983.

263. **Rudewicz, P. and Munson, B.**, Analysis of complex mixtures of ethoxylated alcohols by probe distillation/chemical ionization mass spectrometry, *Anal. Chem.*, 58, 674, 1986.

264. **Saito, J., Waki, H., and Tanaka, S.**, Application of field desorbtion mass spectrometry to polymer and oligomer analysis, *Prog. Org. Coat.*, 15, 311, 1988.

265. **Kallos, G. J. and Smith, P. B.**, Separation and characterization of acrylic acid oligomers by NMR spectroscopy and thermospray ion-exchange liquid chromatography-mass spectrometry, *J. Chromatogr.*, 408, 349, 1987.

266. **Puglisi, C., Scamporrino, E., and Vitalini, D.**, Mass-spectrometric analysis of the thermal degradation products of poly(*O*-, *m*-, *p*-phenylene sulfide) and of the oligomers produced in the synthesis of these polymers, *Macromolecules*, 19, 2157, 1986.

267. **Laurent, J. L., Garrigue, F., and Mathieu, J.**, Separation and identification of the molecular species in the urea-formaldehyde adducts, *Analysis*, 15, 404, 1987.

268. **Lattimer, R. P. and Hansen, G. E.**, Determination of MWD of polyglycol oligomers by field desorption mass spectroscopy, *Macromolecules*, 14, 776, 1981.

269. **Lattimer, R. P., Harmon, D. J., and Hansen, G. E.**, Determination of MWD of styrene oligomers by field desorption mass spectroscopy, *Anal. Chem.*, 52, 1808, 1980.

270. **Scamporino, M., Puglisi, C., and Vitolini, C.**, Mass spectrometric detection of cyclic sulfides in the polycondensation of dibromoalkanes with dithiols, *J. Polym. Sci. Polym. Symp.*, 74, 285, 1986.

271. **Ballistresi, A., Garozzo, D., Giuffrido, M., Montando, G., Fillippi, A., Guaiato, C., Manaresi, P., and Pilati, F.**, Fast atom bombardment mass spectrometry of oligomers contained in poly (ϵ-caprolactam) and poly(butylene isophthalate), *Macromolecules*, 20, 1029, 1987.

272. **Ooi, T., Mimura, K., Hama, Y., and Shinohara, K.**, ESR study of linear aliphatic oligoesters under γ-irradiation, *J. Polym. Sci. Polym. Chem. Ed.*, 14, 813, 1976.

273. **Hori, Y. and Kispert, L. D.**, ESR evidence of a biradical dimer initiator in diacetylene polymerization, *J. Am. Chem. Soc.*, 101, 3173, 1979.

274. **Tomolo, C., Buoca, G. M., Schilling, F. C., and Bovey, F. A.**, Proton magnetic resonance of linear sarcosine oligomers, *Macromolecules*, 13, 138, 1980.

275. **Hirai, Y., Nunomura, Y., and Imamura, Y.**, PMR for poly(*p*-chloroprene) with low molecular weight, *Bull. Chem. Soc. Jpn.*, 49, 2200, 1976.

276. **Ikada, E., Sugimura, T., and Watanabe, T.**, Dielectric properties of oligomers, *Polymer*, 15, 101, 1975.

277. **Ikada, E., Sugimura, T., and Watanabe, T.**, Dielectric properties of oligomers, *J. Polym. Sci. Polym. Phys. Ed.*, 16, 907, 1978.

278. **Siwergin, J. M.**, Dielectric relaxation in oligoesteracrylates, *Plaste Kautsch.*, 28, 26, 33, 1981.

279. **Ikada, E.**, Dielectric behaviour of ethylenediamine oligomer, *Bull. Inst. Chem. Res. Kyoto Univ.*, 45, 352, 1967.

280. **Ikada, E., Hida, Y., Okamoto, H., Hagino, J., and Koizumi, N.**, Dielectric behaviour of ethanolamine oligomers, *Bull. Inst. Chem. Res. Kyoto Univ.*, 46, 239, 1968.

281. **Ikada, E. and Watanabe, T.**, Dielectric behaviour of acrylonitrile-butadiene oligomers, *J. Polym. Sci. Part A*, 10, 3457, 1972.

282. **Ikada, E., Fukushima, H., and Watanabe, T.**, Dielectric properties of oligomers, *J. Polym. Sci. Polym. Phys. Ed.*, 17, 1789, 1979.

283. **Zahn, H. and Repin, N.,** Synthese linear homologen reihe von cyclischen Äthylenterephthalaten, *Chem. Ber.,* 103, 3041, 1970.

284. **Yamada, T., Mituno, Y., and Yamamura, Y.,** Measurement of the melting points of oligomers obtained in the manufacture of poly(ethylene terephthalate), *Polym. Eng. Sci.,* 28, 377, 1988.

285. **Yamada, T. and Yamamura, Y.,** Relation between characteristics and melting point of oligomers in poly(ethylene terephthalate) manufacture, *Polym. Eng. Sci.,* 28, 381, 1988.

286. **Schwalm, S. R. and Heitz, W.,** Synthesis of telechelic oligo- and poly(carboxypiperazine)s, *Makromol. Chem.,* 187, 1415, 1986.

287. **Raevskayia, E. H., Artsis, E. S., Siling, M. J., Urman, Ya. G., and Bessonova, N. P.,** Synthesis and properties of lactam based oligoamides, *Plast. Massy,* No. 8, 14, 1988.

288. **Krokhmaleva, L. N., Prilutskaya, N. V. and Smekhov, F. M.,** Properties of epoxy polymers prepared using oligoester curing agents, *Plast. Massy,* No. 9, 22, 1988.

289. **Shiono, S., Isamu, K., and Enomoto, J.,** Study of the thermal decomposition behaviour of epoxy oligomers by pyrolysis gas-chromatography, *Kobunshi Ronbunshu,* 40, 351, 1983; *Chem. Abstr.,* 99, 140928c, 1983.

290. **Meyer, K. H. and Van der Wyck, A.,** Die Oligenschaften von Polymeren in Lösung. VIII. Die Bieldung von Micellen in lösungen faserförmiger Molekule, *Helv. Chim. Acta,* 20, 1321, 1937.

291. **Nagura, M., Tagawa, H., and Wada, E.,** Structure of oligourethanes. IV. Melting phenomena of solution grown crystals, *Makromol. Chem.,* 165, 325, 1973.

292. **Hay, N. J.,** On the melting characteristics of oligourethanes, *Makromol. Chem.,* 178, 1601, 1977.

293. **Dale, J.,** Die Konformationen vielgliedriger Ringe, *Angew. Chem.,* 78, 1070, 1966.

294. **Ito, E. and Okajima, S.,** Studies on cyclic tris(ethylene terephthalate), *J. Polym. Sci. Polym. Lett. Ed.,* 7, 483, 1969.

295. **Giuffria, R.,** Microscopic studies of Mylar film and its low molecular weight extracts, *J. Polym. Sci.,* 49, 427, 196.

296. **Asbach, G. I., Drexhage, K. H., Heidemann, G., Glenz, W., and Kilian, H. G.,** Morphology and melting temperature of oligomers, *Makromol. Chem.,* 139, 115, 1970.

297. **Marshall, A.,** Crystallinity of ethylene oxide oligomers, *Eur. Polym. J.,* 17, 885, 1981.

298. **Fraser, M. J., Cooper, D. R., and Both, C.,** Crystallinity and fusion of low molecular weight poly(ethylene oxide): effect of end groups, *Polymer,* 18, 852, 1977.

299. **Aligulijew, R. M.,** Molecular motion in oligoesters, *Plast. Rubber,* 27, 383, 1980.

300. **Cella, R. J.,** Polyester morphology, *J. Polym. Sci. Polym. Symp.,* 42(2), 727, 1973.

301. **Imaeda, M.,** Crystal structures of cyclic oligoesters, *Polym. J.,* 14, 197, 1982.

302. **Weigel, R., Lengyel, H., Schultz, R., and Szekely, I.,** Structure of polyamides. Effect of low molecular weight components upon the crystallinity, *Acta Polym.,* 30, 439, 1979.

303. **Zeluew, Iu. V. and Sivergin, Ju. M.,** Molecular mobility in oligoesteracrylates, *Acta Polym.,* 32, 75, 1981.

304. **Mladenow, I., Vladkowa, T., and Fakirow, S.,** Morphology and crystallization kinetics of oligoamidophosphates, *J. Macromol. Sci. Phys.,* B17, 25, 1980.

INDEX

U

V

W

X

Z